Quantum Field Theory 2nd Edition

Quantum Field Theory
2nd Edition

FRANZ MANDL
and
GRAHAM SHAW

*School of Physics & Astronomy,
The University of Manchester, Manchester, UK*

A John Wiley and Sons, Ltd., Publication

This edition first published 2010
© 2010 John Wiley & Sons, Ltd

Registered office
John Wiley & Sons Ltd, The Atrium, Southern Gate, Chichester, West Sussex, PO19 8SQ, United Kingdom

For details of our global editorial offices, for customer services and for information about how to apply for permission to reuse the copyright material in this book please see our website at www.wiley.com.

Library of Congress Cataloging-in-Publication Data

Mandl, F. (Franz), 1923–2009
 Quantum field theory / Franz Mandl, Graham Shaw. — 2nd ed.
  p. cm.
 Includes bibliographical references and index.
 ISBN 978-0-471-49683-0 (cloth : alk. paper)
 1. Quantum field theory. I. Shaw, G. (Graham), 1942– II. Title.
 QC174.45.M32 2010
 530.14′3—dc22

                                    2010000255

A catalogue record for this book is available from the British Library.

ISBN 978-0-471-49683-0 (H/B)
    978-0-471-49684-7 (P/B)

Set in 10/12pt Times by Integra Software Services Pvt. Ltd., Pondicherry, India

Contents

Preface

Preface to the Second Edition

The first edition of this book aimed to give an easily accessible introduction to QED and the unified theory of electromagnetic and weak interaction. In this edition, we have added five new chapters, giving an introduction to QCD and the methods used to understand it: in particular, path integrals and the renormalization group. At the same time, the treatment of electroweak interactions has been updated to take account of more recent experiments. When the first edition was published in 1984, only a handful of W and Z boson events had been observed and the experimental investigation of high energy electroweak interactions was in its infancy. Now it is a precise science, despite the fact that crucial questions about Higgs bosons and the nature of neutrinos remain unanswered.

The structure of the book is as follows. The first ten chapters deal with QED in the canonical formalism, and are little changed from the first edition. A brief introduction to gauge theories (Chapter 11) is then followed by two sections, which may be read independently of each other. They cover QCD and related topics (Chapters 12–15) and the unified electroweak theory (Chapters 16–19) respectively.

Sadly, my close friend and collaborator, Franz Mandl, passed away on 4th February 2009 after a long illness. He retained his passion for physics, and his active commitment to this project, almost to the end. It was a privilege to work with him, and he remains an inspiration to all who knew him.

November 2009   GRAHAM SHAW

Preface to the First Edition

Our aim in writing this book has been to produce a short introduction to quantum field theory, suitable for beginning research students in theoretical and experimental physics. The main objectives are: (i) to explain the basic physics and formalism of quantum field theory, (ii) to make the reader fully proficient in perturbation theory calculations using Feynman diagrams, and (iii) to introduce the reader to gauge theories which are playing such a central role in elementary particle physics.

The theory has been applied to two areas. The beginning parts of the book deal with quantum electrodynamics (QED) where quantum field theory had its early triumphs. The last four chapters, on weak interactions, introduce non-Abelian gauge groups, spontaneous symmetry breaking and the Higgs mechanism, culminating in the Weinberg-Salam standard electro-weak theory. For reasons of space, we have limited ourselves to purely leptonic processes, but this theory is equally successful when extended to include hadrons. The recent observations of the $W^{\pm}$ and Z° bosons, with the predicted masses, lend further support to this theory, and there is every hope that it is the fundamental theory of electro-weak interactions.

The introductory nature of this book and the desire to keep it reasonably short have influenced both the level of treatment and the selection of material. We have formulated quantum field theory in terms of non-commuting operators, as this approach should be familiar to the reader from non-relativistic quantum mechanics and it brings out most clearly the physical meaning of the formalism in terms of particle creation and annihilation operators. We have developed the formalism only to the level we require in the applications. These concentrate primarily on calculations in lowest order of perturbation theory. The techniques for obtaining cross-sections, decay rates, and spin and polarization sums have been developed in detail and applied to a variety of processes, many of them of interest in current research on electro-weak interactions. After studying this material, the reader should be able to tackle confidently any process in lowest order.

Our treatment of renormalization and radiative corrections is much less complete. We have explained the general concepts of regularization and renormalization. For QED we have shown in some detail how to calculate the lowest-order radiative corrections, using dimensional regularization as well as the older cut-off techniques. The infra-red divergence and its connection with radiative corrections have similarly been discussed in lowest order only. The scope of this book precludes a serious study of higher order corrections in QED and of the renormalization of the electro-weak theory. For the latter the Feynman path integral formulation of quantum field theory seems almost essential. Regretfully, we were not able to provide a short and simple treatment of this topic.

This book arose out of lectures which both of us have given over many years. We have greatly benefited from discussions with students and colleagues, some of whom have read parts of the manuscript. We would like to thank all of them for their help, and particularly Sandy Donnachie who encouraged us to embark on this collaboration.

January 1984

FRANZ MANDL
GRAHAM SHAW

Notes

Acknowledgements

In preparing this new edition, we have again benefited from discussions with many of our colleagues. We are grateful to them all, and especially Jeff Forshaw, who read all the new chapters and made many helpful suggestions. We are also grateful to Brian Martin: Sections 15.1 and 15.3.3 draw heavily on his previous work with one of us (G.S.) and he also generously helped prepare many of the new figures.

Illustrations

Some illustrations in the text have been adapted from diagrams that have been published elsewhere. In a few cases they have been reproduced exactly as previously published. We acknowledge, with thanks, permission from the relevant copyright holders to use such illustrations and this is confirmed in the captions.

Data

Except when otherwise stated, the data quoted in this book, and their sources, are given in "Review of Particle Physics", *Journal of Physics* **G33** (2006) 1 and on the Particle Data Group website http:pdg.lbl.gov, which is regularly updated.

Webpage

This book has its own website:www.hep.manchester.ac.uk/u/graham/qftbook.html. Any misprints or other necessary corrections brought to our attention will be listed on this page. We would also be grateful for any other comments about this book.

1

Photons and the Electromagnetic Field

1.1 Particles and Fields

The concept of photons as the quanta of the electromagnetic field dates back to the beginning of the twentieth century. In order to explain the spectrum of black-body radiation, Planck, in 1900, postulated that the process of emission and absorption of radiation by atoms occurs discontinuously in quanta. Einstein, by 1905, had arrived at a more drastic interpretation. From a statistical analysis of the Planck radiation law and from the energetics of the photoelectric effect, he concluded that it was not merely the atomic mechanism of emission and absorption of radiation which is quantized, but that electromagnetic radiation itself consists of photons. The Compton effect confirmed this interpretation.

The foundations of a systematic quantum theory of fields were laid by Dirac in 1927 in his famous paper on 'The Quantum Theory of the Emission and Absorption of Radiation'. From the quantization of the electromagnetic field one is naturally led to the quantization of any classical field, the quanta of the field being particles with well-defined properties. The interactions between these particles are brought about by other fields whose quanta are other particles. For example, we can think of the interaction between electrically charged particles, such as electrons and positrons, as being brought about by the electromagnetic field or as due to an exchange of photons. The electrons and positrons themselves can be thought of as the quanta of an electron–positron field. An important reason for quantizing such particle fields is to allow for the possibility that the number of particles changes as, for example, in the creation or annihilation of electron–positron pairs.

These and other processes of course only occur through the interactions of fields. The solution of the equations of the quantized interacting fields is extremely difficult. If the interaction is sufficiently weak, one can employ perturbation theory. This has been outstandingly successful in quantum electrodynamics, where complete agreement exists between theory and experiment to an incredibly high degree of accuracy. Perturbation

Quantum Field Theory, Second Edition   Franz Mandl and Graham Shaw
© 2010 John Wiley & Sons, Ltd

theory has also very successfully been applied to weak interactions, and to strong interactions at short distances, where they become relatively weak.

The most important modern perturbation-theoretic technique employs Feynman diagrams, which are also extremely useful in many areas other than relativistic quantum field theory. We shall later develop the Feynman diagram technique and apply it to electromagnetic, weak and strong interactions. For this a Lorentz-covariant formulation will be essential.

In this introductory chapter we employ a simpler non-covariant approach, which suffices for many applications and brings out many of the ideas of field quantization. We shall consider the important case of electrodynamics for which a complete classical theory – Maxwell's – exists. As quantum electrodynamics will be re-derived later, we shall in this chapter, at times, rely on plausibility arguments rather than fully justify all steps.

1.2   The Electromagnetic Field in the Absence of Charges

1.2.1   The classical field

Classical electromagnetic theory is summed up in Maxwell's equations. In the presence of a charge density $\rho(\mathbf{x}, t)$ and a current density $\mathbf{j}(\mathbf{x}, t)$, the electric and magnetic fields $\mathbf{E}$ and $\mathbf{B}$ satisfy the equations

$$\mathbf{\nabla} \cdot \mathbf{E} = \rho \tag{1.1a}$$

$$\mathbf{\nabla} \wedge \mathbf{B} = \frac{1}{c}\mathbf{j} + \frac{1}{c}\frac{\partial \mathbf{E}}{\partial t} \tag{1.1b}$$

$$\mathbf{\nabla} \cdot \mathbf{B} = 0 \tag{1.1c}$$

$$\mathbf{\nabla} \wedge \mathbf{E} = -\frac{1}{c}\frac{\partial \mathbf{B}}{\partial t} \tag{1.1d}$$

where, as throughout this book, rationalized Gaussian (c.g.s.) units are being used.[1]

From the second pair of Maxwell's equations [Eqs. (1.1c) and (1.1d)] follows the existence of scalar and vector potentials $\phi(\mathbf{x}, t)$ and $\mathbf{A}(\mathbf{x}, t)$, defined by

$$\mathbf{B} = \mathbf{\nabla} \wedge \mathbf{A}, \quad \mathbf{E} = -\mathbf{\nabla}\phi - \frac{1}{c}\frac{\partial \mathbf{A}}{\partial t}. \tag{1.2}$$

Eqs. (1.2) do not determine the potentials uniquely, since for an arbitrary function $f(\mathbf{x}, t)$ the transformation

$$\phi \rightarrow \phi' = \phi + \frac{1}{c}\frac{\partial f}{\partial t}, \quad \mathbf{A} \rightarrow \mathbf{A}' = \mathbf{A} - \mathbf{\nabla}f \tag{1.3}$$

leaves the fields $\mathbf{E}$ and $\mathbf{B}$ unaltered. The transformation (1.3) is known as a gauge transformation of the second kind. Since all observable quantities can be expressed in

---

[1] They are also called rationalized Lorentz–Heaviside units. In these units, the fine structure constant is given by $\alpha = e^2/(4\pi\hbar c) \approx 1/137$, whereas in unrationalized gaussian units $\alpha = e^2_{\text{unrat}}/\hbar c$, i.e. $e = e_{\text{unrat}}\sqrt{(4\pi)}$. Correspondingly for the fields $\mathbf{E} = \mathbf{E}_{\text{unrat}}/\sqrt{(4\pi)}$, etc.

terms of **E** and **B**, it is a fundamental requirement of any theory formulated in terms of potentials that it is gauge-invariant, i.e. that the predictions for observable quantities are invariant under such gauge transformations.

Expressed in terms of the potentials, the second pair of Maxwell's equations [Eqs. (1.1c) and (1.1d)] are satisfied automatically, while the first pair [Eqs. (1.1a) and (1.1b)] become

$$-\nabla^2 \phi - \frac{1}{c} \frac{\partial}{\partial t} (\nabla \cdot \mathbf{A}) = \Box \phi - \frac{1}{c} \frac{\partial}{\partial t} \left(\frac{1}{c} \frac{\partial \phi}{\partial t} + \nabla \cdot \mathbf{A} \right) = \rho \qquad (1.4a)$$

$$\Box \mathbf{A} + \nabla \left(\frac{1}{c} \frac{\partial \phi}{\partial t} + \nabla \cdot \mathbf{A} \right) = \frac{1}{c} \mathbf{j} \qquad (1.4b)$$

where

$$\Box \equiv \frac{1}{c^2} \frac{\partial^2}{\partial t^2} - \nabla^2. \qquad (1.5)$$

We now go on to consider the case of the free field, i.e. the absence of charges and currents: $\rho=0, \mathbf{j}=0$. We can then choose a gauge for the potentials such that

$$\nabla \cdot \mathbf{A} = 0. \qquad (1.6)$$

The condition (1.6) defines the Coulomb or radiation gauge. A vector field with vanishing divergence, i.e. satisfying Eq. (1.6), is called a transverse field, since for a wave

$$\mathbf{A}(\mathbf{x}, t) = \mathbf{A}_0 \, e^{i(\mathbf{k} \cdot \mathbf{x} - \omega t)}$$

Eq. (1.6) gives

$$\mathbf{k} \cdot \mathbf{A} = 0, \qquad (1.7)$$

i.e. **A** is perpendicular to the direction of propagation **k** of the wave. In the Coulomb gauge, the vector potential is a transverse vector. In this chapter we shall be employing the Coulomb gauge.

In the absence of charges, Eq. (1.4a) now becomes $\nabla^2 \phi = 0$ with the solution, which vanishes at infinity, $\phi \equiv 0$. Hence Eq. (1.4b) reduces to the wave equation

$$\Box \mathbf{A} = 0. \qquad (1.8)$$

The corresponding electric and magnetic fields are, from Eqs. (1.2), given by

$$\mathbf{B} = \nabla \wedge \mathbf{A}, \quad \mathbf{E} = -\frac{1}{c} \frac{\partial \mathbf{A}}{\partial t}, \qquad (1.9)$$

and, like **A**, are transverse fields. The solutions of Eq. (1.8) are the transverse electromagnetic waves in free space. These waves are often called the radiation field. Its energy is given by

$$H_{\text{rad}} = \frac{1}{2} \int (\mathbf{E}^2 + \mathbf{B}^2) \, d^3\mathbf{x}. \qquad (1.10)$$

In order to quantize the theory, we shall want to introduce canonically conjugate coordinates (like x and p_x in non-relativistic quantum mechanics) for each degree of freedom and subject these to commutation relations. At a given instant of time t, the vector potential $\mathbf{A}$ must be specified at every point $\mathbf{x}$ in space. Looked at from this viewpoint, the electromagnetic field possesses a continuous infinity of degrees of freedom. The problem can be simplified by considering the radiation inside a large cubic enclosure, of side L and volume $V = L^3$, and imposing periodic boundary conditions on the vector potential $\mathbf{A}$ at the surfaces of the cube. The vector potential can then be represented as a Fourier series, i.e. it is specified by the denumerable set of Fourier expansion coefficients, and we have obtained a description of the field in terms of an infinite, but denumerable, number of degrees of freedom. The Fourier analysis corresponds to finding the normal modes of the radiation field, each mode being described independently of the others by a harmonic oscillator equation. (All this is analogous to the Fourier analysis of a vibrating string.) This will enable us to quantize the radiation field by taking over the quantization of the harmonic oscillator from non-relativistic quantum mechanics.

With the periodic boundary conditions

$$\mathbf{A}(0,\ y,\ z,\ t) = \mathbf{A}(L,\ y,\ z,\ t), \text{etc.,} \tag{1.11}$$

the functions

$$\frac{1}{\sqrt{V}}\,\boldsymbol{\varepsilon}_r(\mathbf{k})\,\mathrm{e}^{\mathrm{i}\mathbf{k}\cdot\mathbf{x}}, \quad r = 1,\ 2, \tag{1.12}$$

form a complete set of transverse orthonormal vector fields. Here the wave vectors $\mathbf{k}$ must be of the form

$$\mathbf{k} = \frac{2\pi}{L}\,(n_1,\ n_2,\ n_3), \quad n_1,\ n_2,\ n_3 = 0,\ \pm 1,\ \ldots, \tag{1.13}$$

so that the fields (1.12) satisfy the periodicity conditions (1.11). $\boldsymbol{\varepsilon}_1(\mathbf{k})$ and $\boldsymbol{\varepsilon}_2(\mathbf{k})$ are two mutually perpendicular real unit vectors which are also orthogonal to $\mathbf{k}$:

$$\boldsymbol{\varepsilon}_r(\mathbf{k}) \cdot \boldsymbol{\varepsilon}_s(\mathbf{k}) = \delta_{rs}, \quad \boldsymbol{\varepsilon}_r(\mathbf{k}) \cdot \mathbf{k} = 0, \quad r,\ s = 1,\ 2. \tag{1.14}$$

The last of these conditions ensures that the fields (1.12) are transverse, satisfying the Coulomb gauge condition (1.6) and (1.7).[2]

We can now expand the vector potential $\mathbf{A}(\mathbf{x}, t)$ as a Fourier series

$$\mathbf{A}(\mathbf{x},\ t) = \sum_{\mathbf{k}} \sum_{r} \left(\frac{\hbar c^2}{2V\omega_{\mathbf{k}}}\right)^{1/2} \boldsymbol{\varepsilon}_r(\mathbf{k})\left[a_r(\mathbf{k},\ t)\mathrm{e}^{\mathrm{i}\mathbf{k}\cdot\mathbf{x}} + a_r^*(\mathbf{k},\ t)\mathrm{e}^{-\mathrm{i}\mathbf{k}\cdot\mathbf{x}}\right], \tag{1.15}$$

where $\omega_{\mathbf{k}} = c|\mathbf{k}|$. The summations with respect to r and $\mathbf{k}$ are over both polarization states $r = 1,\ 2$ (for each $\mathbf{k}$) and over all allowed momenta $\mathbf{k}$. The factor to the left of $\boldsymbol{\varepsilon}_r(\mathbf{k})$ has been introduced for later convenience only. The form of the series (1.15) ensures that the vector potential is real: $\mathbf{A} = \mathbf{A}^*$. Eq. (1.15) is an expansion of $\mathbf{A}(\mathbf{x}, t)$ at each instant of time t. The time-dependence of the Fourier expansion coefficients follows, since $\mathbf{A}$ must satisfy the

---

[2] With this choice of $\boldsymbol{\varepsilon}_r(\mathbf{k})$, Eqs. (1.12) represent linearly polarized fields. By taking appropriate complex linear combinations of $\boldsymbol{\varepsilon}_1$ and $\boldsymbol{\varepsilon}_2$ one obtains circular or, in general, elliptic polarization.

wave equation (1.8). Substituting Eq. (1.15) in (1.8) and projecting out individual amplitudes, one obtains

$$\frac{\partial^2}{\partial t^2} a_r(\mathbf{k}, \ t) = - \ \omega_{\mathbf{k}}^2 a_r(\mathbf{k}, \ t). \tag{1.16}$$

These are the harmonic oscillator equations of the normal modes of the radiation field. It
will prove convenient to take their solutions in the form

$$a_r(\mathbf{k}, \ t) = a_r(\mathbf{k}) \exp{(-i\omega_{\mathbf{k}}t)}, \tag{1.17}$$

where the $a_r(\mathbf{k})$ are initial amplitudes at time $t = 0$.

Eq. (1.15) for the vector potential, with Eq. (1.17) and its complex conjugate substituted
for the amplitudes a_r and a_r^*, represents our final result for the classical theory. We can
express the energy of the radiation field, Eq. (1.10), in terms of the amplitudes by
substituting Eqs. (1.9) and (1.15) in (1.10) and carrying out the integration over the volume
V of the enclosure. In this way one obtains

$$H_{\text{rad}} = \sum_{\mathbf{k}} \sum_{r} \hbar\omega_{\mathbf{k}} a_r^*(\mathbf{k}) a_r(\mathbf{k}). \tag{1.18}$$

Note that this is independent of time, as expected in the absence of charges and currents;
we could equally have written the time-dependent amplitudes (1.17) instead, since the time
dependence of a_r and of a_r^* cancels.

As already stated, we shall quantize the radiation field by quantizing the individual
harmonic oscillator modes. As the interpretation of the quantized field theory in terms of
photons is intimately connected with the quantum treatment of the harmonic oscillator, we
shall summarize the latter.

1.2.2   Harmonic oscillator

The harmonic oscillator Hamiltonian is, in an obvious notation,

$$H_{\text{osc}} = \frac{p^2}{2m} + \frac{1}{2}m\omega^2 q^2,$$

with q and p satisfying the commutation relation $[q, p] = i\hbar$. We introduce the operators

$$\left. \begin{array}{c} a \\ a^\dagger \end{array} \right\} = \frac{1}{(2\hbar m\omega)^{1/2}} \ (m\omega q \pm ip).$$

These satisfy the commutation relation

$$\left[a, \ a^\dagger\right] = 1, \tag{1.19}$$

and the Hamiltonian expressed in terms of a and $a^\dagger$ becomes:

$$H_{\text{osc}} = \frac{1}{2}\hbar\omega(a^\dagger a + aa^\dagger) = \hbar\omega\left(a^\dagger a + \frac{1}{2}\right). \tag{1.20}$$

This is essentially the operator

$$N \equiv a^\dagger a, \tag{1.21}$$

which is positive definite, i.e. for any state $|\Psi\rangle$

$$\langle\Psi|N|\Psi\rangle = \langle\Psi|a^\dagger a|\Psi\rangle = \langle a\Psi\,|a\Psi\rangle \geq 0.$$

Hence, N possesses a lowest non-negative eigenvalue

$$\alpha_0 \geq 0.$$

It follows from the eigenvalue equation

$$N|\alpha\rangle = \alpha|\alpha\rangle$$

and Eq. (1.19) that

$$Na|\alpha\rangle = (\alpha - 1)a|\alpha\rangle, \quad Na^\dagger|\alpha\rangle = (\alpha + 1)a^\dagger|\alpha\rangle, \tag{1.22}$$

i.e. $a|\alpha\rangle$ and $a^\dagger|\alpha\rangle$ are eigenfunctions of N belonging to the eigenvalues $(\alpha - 1)$ and $(\alpha + 1)$, respectively. Since α_0 is the lowest eigenvalue we must have

$$a|\alpha_0\rangle = 0, \tag{1.23}$$

and since

$$a^\dagger a|\alpha_0\rangle = \alpha_0|\alpha_0\rangle,$$

Eq. (1.23) implies $\alpha_0 = 0$. It follows from Eqs. (1.19) and (1.22) that the eigenvalues of N are the integers $n = 0, 1, 2, \ldots$, and that if $\langle n|n\rangle = 1$, then the states $|n\pm 1\rangle$, defined by

$$a|n\rangle = n^{1/2}|n - 1\rangle, \quad a^\dagger|n\rangle = (n + 1)^{1/2}|n + 1\rangle, \tag{1.24}$$

are also normed to unity. If $\langle 0|0\rangle = 1$, the normed eigenfunctions of N are

$$|n\rangle = \frac{(a^\dagger)^n}{\sqrt{n!}}\,|0\rangle, \quad n = 0, 1, 2,\ldots. \tag{1.25}$$

These are also the eigenfunctions of the harmonic oscillator Hamiltonian (1.20) with the energy eigenvalues

$$E_n = \hbar\omega\left(n + \frac{1}{2}\right), \quad n = 0, 1, 2, \ldots. \tag{1.26}$$

The operators a and $a^\dagger$ are called lowering and raising operators because of the properties (1.24). We shall see that in the quantized field theory $|n\rangle$ represents a state with n quanta. The operator a(changing $|n\rangle$ into $|n - 1\rangle$) will annihilate a quantum; similarly, $a^\dagger$, will create a quantum.

So far we have considered one instant of time, say $t = 0$. We now discuss the equations of motion in the Heisenberg picture.[3] In this picture, the operators are functions of time. In particular

$$i\hbar\,\frac{da(t)}{dt} = [a(t),\, H_{\text{osc}}] \tag{1.27}$$

---

[3] See the appendix to this chapter (Section 1.5) for a concise development of the Schrödinger, Heisenberg and interaction pictures.

with the initial condition $a(0) = a$, the lowering operator considered so far. Since H_{osc} is time-independent, and $a(t)$ and $a^{\dagger}(t)$ satisfy the same commutation relation (1.19) as a and $a^{\dagger}$, the Heisenberg equation of motion (1.27) reduces to

$$\frac{\mathrm{d}a(t)}{\mathrm{d}t} = -\mathrm{i}\omega a(t)$$

with the solution

$$a(t) = a\,\mathrm{e}^{-\mathrm{i}\omega t}. \tag{1.28}$$

1.2.3   The quantized radiation field

The harmonic oscillator results we have derived can at once be applied to the radiation field. Its Hamiltonian, Eq. (1.18), is a superposition of independent harmonic oscillator Hamiltonians (1.20), one for each mode of the radiation field. [The order of the factors in (1.18) is not significant and can be changed, since the a_r and a_r^* are classical amplitudes.] We therefore introduce commutation relations analogous to Eq. (1.19)

$$\left.\begin{array}{l} [a_r(\mathbf{k}),\, a_s^{\dagger}(\mathbf{k'})] = \delta_{rs}\,\delta_{\mathbf{kk'}} \\[6pt] [a_r(\mathbf{k}),\, a_s(\mathbf{k'})] = [a_r^{\dagger}(\mathbf{k}),\, a_s^{\dagger}(\mathbf{k'})] = 0 \end{array}\right\} \tag{1.29}$$

and write the Hamiltonian (1.18) as

$$H_{\mathrm{rad}} = \sum_{\mathbf{k}}\sum_{r} \hbar\omega_{\mathbf{k}}\left(a_r^{\dagger}(\mathbf{k})a_r(\mathbf{k}) + \frac{1}{2}\right). \tag{1.30}$$

The operators

$$N_r(\mathbf{k}) = a_r^{\dagger}(\mathbf{k})a_r(\mathbf{k})$$

then have eigenvalues $n_r(\mathbf{k}) = 0, 1, 2, \ldots$, and eigenfunctions of the form (1.25)

$$|n_r(\mathbf{k})\rangle = \frac{[a_r^{\dagger}(\mathbf{k})]^{n_r(\mathbf{k})}}{\sqrt{n_r(\mathbf{k})!}}\,|0\rangle. \tag{1.31}$$

The eigenfunctions of the radiation Hamiltonian (1.30) are products of such states, i.e.

$$|\ldots n_r(\mathbf{k})\ldots\rangle = \prod_{\mathbf{k}_i}\prod_{r_i}|n_{r_i}(\mathbf{k}_i)\rangle, \tag{1.32}$$

with energy

$$\sum_{\mathbf{k}}\sum_{r}\hbar\omega_{\mathbf{k}}\left(n_r(\mathbf{k}) + \frac{1}{2}\right). \tag{1.33}$$

The interpretation of these equations is a straightforward generalization from one harmonic oscillator to a superposition of independent oscillators, one for each radiation mode $(\mathbf{k}, r)$. $a_r(\mathbf{k})$ operating on the state (1.32) will reduce the *occupation number* $n_r(\mathbf{k})$

of the mode $(\mathbf{k}, r)$ by unity, leaving all other occupation numbers unaltered, i.e. from Eq. (1.24):

$$a_r(\mathbf{k})|\ldots n_r(\mathbf{k})\ldots\rangle = [n_r(\mathbf{k})]^{1/2}|\ldots, n_r(\mathbf{k})-1, \ldots\rangle. \qquad (1.34)$$

Correspondingly the energy (1.33) is reduced by $\hbar\omega_{\mathbf{k}} = \hbar c|\mathbf{k}|$. We interpret $a_r(\mathbf{k})$ as an annihilation (or destruction or absorption) operator, which annihilates one photon in the mode $(\mathbf{k}, r)$, i.e. with momentum $\hbar\mathbf{k}$, energy $\hbar\omega_{\mathbf{k}}$ and linear polarization vector $\boldsymbol{\varepsilon}_r(\mathbf{k})$. Similarly, $a_r^\dagger(\mathbf{k})$ is interpreted as a creation operator of such a photon. The assertion that $a_r(\mathbf{k})$ and $a_r^\dagger(\mathbf{k})$ are absorption and creation operators of photons with momentum $\hbar\mathbf{k}$ can be justified by calculating the momentum of the radiation field. We shall see later that the momentum operator of the field is given by

$$\mathbf{P} = \sum_{\mathbf{k}}\sum_r \hbar\mathbf{k}\left(N_r(\mathbf{k}) + \frac{1}{2}\right), \qquad (1.35)$$

which leads to the above interpretation. We shall not consider the more intricate problem of the angular momentum of the photons, but only mention that circular polarization states obtained by forming linear combinations

$$-\frac{1}{\sqrt{2}}\left[\boldsymbol{\varepsilon}_1(\mathbf{k}) + i\boldsymbol{\varepsilon}_2(\mathbf{k})\right], \qquad \frac{1}{\sqrt{2}}\left[\boldsymbol{\varepsilon}_1(\mathbf{k}) - i\boldsymbol{\varepsilon}_2(\mathbf{k})\right], \qquad (1.36)$$

are more appropriate for this. Remembering that $(\boldsymbol{\varepsilon}_1(\mathbf{k}), \boldsymbol{\varepsilon}_2(\mathbf{k}), \mathbf{k})$ form a right-handed Cartesian coordinate system, we see that these two combinations correspond to angular momentum $\pm\hbar$ in the direction $\mathbf{k}$ (analogous to the properties of the spherical harmonics $Y_1^{\pm 1}$), i.e. they represent right- and left-circular polarization: the photon behaves like a particle of spin 1. The third spin component is, of course, missing because of the transverse nature of the photon field.

The state of lowest energy of the radiation field is the vacuum state $|0\rangle$, in which all occupation numbers $n_r(\mathbf{k})$ are zero. According to Eqs. (1.30) or (1.33), this state has the energy $\frac{1}{2}\sum_{\mathbf{k}}\sum_r\hbar\omega_{\mathbf{k}}$. This is an infinite *constant*, which is of no physical significance: we can eliminate it altogether by shifting the zero of the energy scale to coincide with the vacuum state $|0\rangle$. This corresponds to replacing Eq. (1.30) by

$$H_{\text{rad}} = \sum_{\mathbf{k}}\sum_r \hbar\omega_{\mathbf{k}}a_r^\dagger(\mathbf{k})\,a_r(\mathbf{k}). \qquad (1.37)$$

[The 'extra' term in Eq. (1.35) for the momentum will similarly be dropped. It actually vanishes in any case due to symmetry in the $\mathbf{k}$ summation.]

The representation (1.32) in which states are specified by the occupation numbers $n_r(\mathbf{k})$ is called the *number representation*. It is of great practical importance in calculating transitions (possibly via intermediate states) between initial and final states containing definite numbers of photons with well-defined properties. These ideas are, of course, not restricted to photons, but apply generally to the particles of quantized fields. We shall have to modify the formalism in one respect. We have seen that the photon occupation numbers $n_r(\mathbf{k})$ can assume all values $0,1,2,\ldots$. Thus, photons satisfy Bose–Einstein statistics. They are *bosons*. So a modification will be required to describe particles obeying Fermi–Dirac statistics (*fermions*), such as electrons or muons, for which the occupation numbers are restricted to the values 0 and 1.

We have quantized the electromagnetic field by replacing the classical amplitudes a_r and a_r^* in the vector potential (1.15) by operators, so that the vector potential and the electric and magnetic fields become operators. In particular, the vector potential (1.15) becomes, in the Heisenberg picture [cf. Eqs. (1.28) and (1.17)], the time-dependent operator

$$\mathbf{A}(\mathbf{x},\ t) = \mathbf{A}^+(\mathbf{x},\ t) + \mathbf{A}^-(\mathbf{x},\ t), \tag{1.38a}$$

with

$$\mathbf{A}^+(\mathbf{x},\ t) = \sum_{\mathbf{k}} \sum_r \left(\frac{\hbar c^2}{2V\omega_{\mathbf{k}}}\right)^{1/2} \boldsymbol{\varepsilon}_r(\mathbf{k}) a_r(\mathbf{k})\ e^{i(\mathbf{k}\cdot\mathbf{x} - \omega_{\mathbf{k}}t)}, \tag{1.38b}$$

$$\mathbf{A}^-(\mathbf{x},\ t) = \sum_{\mathbf{k}} \sum_r \left(\frac{\hbar c^2}{2V\omega_{\mathbf{k}}}\right)^{1/2} \boldsymbol{\varepsilon}_r(\mathbf{k}) a_r^\dagger(\mathbf{k})\ e^{-i(\mathbf{k}\cdot\mathbf{x} - \omega_{\mathbf{k}}t)}. \tag{1.38c}$$

The operator $\mathbf{A}^+$ contains only absorption operators, $\mathbf{A}^-$ only creation operators. $\mathbf{A}^+$ and $\mathbf{A}^-$ are called the positive and negative frequency parts of $\mathbf{A}$.[4] The operators for $\mathbf{E}(\mathbf{x},\ t)$ and $\mathbf{B}(\mathbf{x},\ t)$ follow from Eqs. (1.9). There is an important difference between a quantized field theory and non-relativistic quantum mechanics. In the former it is the amplitudes (and hence the fields) which are operators, and the position and time coordinates $(\mathbf{x},\ t)$ are ordinary numbers, whereas in the latter the position coordinates (but not the time) are operators.

Finally, we note that a state with a definite number v of photons (i.e. an eigenstate of the total photon number operator $N = \Sigma_{\mathbf{k}}\Sigma_r N_r(\mathbf{k})$) cannot be a classical field, not even for $v \to \infty$. This is a consequence of the fact that $\mathbf{E}$, like $\mathbf{A}$, is linear in the creation and absorption operators. Hence the expectation value of $\mathbf{E}$ in such a state vanishes. It is possible to form so-called coherent states $|c\rangle$ for which $\langle c|\mathbf{E}|c\rangle$ represents a transverse wave and for which the relative fluctuation $\Delta\mathbf{E}/\langle c|\mathbf{E}|c\rangle$ tends to zero as the number of photons in the state, $\langle c|N|c\rangle$, tends to infinity, i.e. in this limit the state $|c\rangle$ goes over into a classical state of a well-defined field.[5]

1.3   The Electric Dipole Interaction

In the last section we quantized the radiation field. Since the occupation number operators $a_r^\dagger(\mathbf{k})a_r(\mathbf{k})$ commute with the radiation Hamiltonian (1.37), the occupation numbers $n_r(\mathbf{k})$ are constants of the motion for the free field. For anything 'to happen' requires interactions with charges and currents so that photons can be absorbed, emitted or scattered.

The complete description of the interaction of a system of charges (for example, an atom or a nucleus) with an electromagnetic field is very complicated. In this section we shall consider the simpler and, in practice, important special case of the interaction occurring via

---

[4] This is like in non-relativistic quantum mechanics where a time-dependence $e^{-i\omega t}$ with $\omega = E/\hbar > 0$ corresponds to a positive energy, i.e. a positive frequency.
[5] For a discussion of coherent states see R. Loudon, *The Quantum Theory of Light*, Clarendon Press, Oxford, 1973, pp. 148–153. See also Problem 1.1.

the electric dipole moment of the system of charges. The more complete (but still non-covariant) treatment of Section 1.4 will justify some of the points asserted in this section.

We shall consider a system of N charges $e_1, e_2, \ldots, e_N$ which can be described non-relativistically, i.e. the position of e_i, $i = 1, \ldots, N$, at time t is classically given by $\mathbf{r}_i = \mathbf{r}_i(t)$. We consider transitions between definite initial and final states of the system (e.g. between two states of an atom). The transitions are brought about by the electric dipole interaction if two approximations are valid.

Firstly it is permissible to neglect the interactions with the magnetic field.

Secondly, one may neglect the spatial variation of the electric radiation field, causing the transitions, across the system of charges (e.g. across the atom). Under these conditions the electric field

$$\mathbf{E}_T(\mathbf{r}, t) = -\frac{1}{c} \frac{\partial \mathbf{A}(\mathbf{r}, t)}{\partial t}, \tag{1.39}$$

resulting from the transverse vector potential (1.38) of the radiation field (we are again using the Coulomb gauge $\nabla \cdot \mathbf{A} = 0$), can be calculated at *one* point somewhere inside the system of charges, instead of at the position of each charge.[6] Taking this point as the origin of coordinates $\mathbf{r} = 0$, we obtain for the interaction causing transitions, the electric dipole interaction H_I given by

$$H_I = -\mathbf{D} \cdot \mathbf{E}_T(0, t) \tag{1.40}$$

where the electric dipole moment is defined by

$$\mathbf{D} = \sum_i e_i \mathbf{r}_i. \tag{1.41}$$

Transitions brought about by the interaction (1.40) in first-order perturbation theory are called *electric dipole transitions*. Since $\mathbf{E}_T$, like $\mathbf{A}$ [Eq. (1.38)], is linear in the photon absorption and creation operators, so is H_I. It follows that, in electric dipole transitions, one photon is emitted or absorbed. In the next section it will be shown that the electric dipole approximation is valid, provided the wavelength $\lambda = 2\pi/k$ of the radiation emitted or absorbed in the transition is very large compared to the linear dimensions R of the system of charges: $\lambda \gg R$. For example, for optical transitions in atoms, R is of the order of 1 Å and λ lies in the range 4000–7500 Å. Similarly, for gamma rays emitted by nuclei, R is of the order of a few fermis (1 f $= 10^{-15}$ m) and since $\lambda/2\pi = [197/(E \text{ in MeV})]$ f for a gamma ray of E MeV, the electric dipole approximation is valid up to quite high gamma-ray energies.

If there are selection rules forbidding a transition in the electric dipole approximation, it might still occur via the magnetic interactions or via parts of the electric interactions which are neglected in the dipole approximation. It may happen that a transition is strictly forbidden, i.e. cannot occur in first-order perturbation theory, even when the exact interaction is used as perturbation instead of H_I [Eq. (1.40)]. In such cases, the transition can still occur in higher orders of perturbation theory or, possibly, by some quite different mechanism.[7]

---

[6] In Eq. (1.39) we have written $\mathbf{E}_T$, since we now also have the Coulomb interaction between the charges, which makes a contribution $-\nabla\phi$ to the electric field. [See Eqs. (1.2) and (1.4a) and Section 1.4.]

[7] For selection rules for radiative transitions in atoms, see H. A. Bethe and R. W. Jackiw, *Intermediate Quantum Mechanics*, 2nd edn, Benjamin, New York, 1968, Chapter 11.

Let us now consider in some detail the emission and absorption of radiation in electric dipole transitions in atoms. The atom will make a transition from an initial state $|A\rangle$ to a final state $|B\rangle$ and the occupation number of one photon state will change from $n_r(\mathbf{k})$ to $n_r(\mathbf{k}) \pm 1$. The initial and final states of the system will be

$$\left.\begin{array}{l} |A, n_r(\mathbf{k})\rangle = |A\rangle |n_r(\mathbf{k})\rangle \\ |B, n_r(\mathbf{k})\pm 1\rangle = |B\rangle |n_r(\mathbf{k})\pm 1\rangle \end{array}\right\}, \tag{1.42}$$

where the occupation numbers of the photon states which are not changed in the transition are not shown. The dipole operator (1.41) now becomes:

$$\mathbf{D} = -e \sum_i \mathbf{r}_i \equiv -e\mathbf{x}, \tag{1.43}$$

where the summation is over the atomic electrons and we have introduced the abbreviation $\mathbf{x}$. The transverse electric field $\mathbf{E}_T(0, t)$ which occurs in the interaction (1.40) is from Eqs. (1.38)

$$\begin{aligned} \mathbf{E}_T(0, t) &= -\frac{1}{c} \frac{\partial \mathbf{A}(0, t)}{\partial t} \\ &= i \sum_{\mathbf{k}} \sum_r \left(\frac{\hbar\omega_{\mathbf{k}}}{2V}\right)^{1/2} \boldsymbol{\varepsilon}_r(\mathbf{k}) \left[a_r(\mathbf{k}) \, \mathrm{e}^{-i\omega_{\mathbf{k}}t} - a_r^\dagger(\mathbf{k}) \mathrm{e}^{i\omega_{\mathbf{k}}t}\right]. \end{aligned}$$

Let us consider radiative emission. The transition matrix element of the interaction (1.40) between the states (1.42) then is given by

$$\langle B, n_r(\mathbf{k}) + 1|H_1|A, \, n_r(\mathbf{k})\rangle$$

$$= i\left(\frac{\hbar\omega_{\mathbf{k}}}{2V}\right)^{1/2} \langle n_r(\mathbf{k}) + 1|a_r^\dagger(\mathbf{k})|n_r(\mathbf{k})\rangle \langle B|\boldsymbol{\varepsilon}_r(\mathbf{k}) \bullet \mathbf{D}|A\rangle \, \mathrm{e}^{i\omega_{\mathbf{k}}t}$$

$$= i\left(\frac{\hbar\omega_{\mathbf{k}}}{2V}\right)^{1/2} [n_r(\mathbf{k}) + 1]^{1/2} \langle B|\boldsymbol{\varepsilon}_r(\mathbf{k}) \bullet \mathbf{D}|A\rangle \, \mathrm{e}^{i\omega_{\mathbf{k}}t}, \tag{1.44}$$

where the last line follows from Eq. (1.24).

The transition probability per unit time between initial and final states (1.42) is given by time-dependent perturbation theory as

$$w = \frac{2\pi}{\hbar} \, |\langle B, n_r(\mathbf{k}) + 1|H_1|A, \, n_r(\mathbf{k})\rangle|^2 \, \delta(E_A - E_B - \hbar\omega_{\mathbf{k}}) \tag{1.45}$$

where E_A and E_B are the energies of the initial and final atomic states $|A\rangle$ and $|B\rangle$.[8] The delta function ensures conservation of energy in the transition, i.e. the emitted photon's energy $\hbar\omega_{\mathbf{k}}$ must satisfy the Bohr frequency condition

$$\omega_{\mathbf{k}} = \omega \equiv (E_A - E_B)/\hbar. \tag{1.46}$$

---

[8] Time-dependent perturbation theory is, for example, developed in A. S. Davydov, *Quantum Mechanics*, 2nd edn, Pergamon, Oxford, 1976, see Section 93 [Eq. (93.7)]; E. Merzbacher, *Quantum Mechanics*, 2nd edn, John Wiley & Sons, Inc., New York, 1970, see Section 18.8; L. I. Schiff, Quantum Mechanics, 3rd edn, McGraw-Hill, New York, 1968, see Section 35.

The delta function is eliminated in the usual way from Eq. (1.45) by integrating over a narrow group of final photon states. The number of photon states in the interval $(\mathbf{k}, \mathbf{k} + d\mathbf{k})$, all in the same polarization state $(\boldsymbol{\varepsilon}_1(\mathbf{k})$ or $\boldsymbol{\varepsilon}_2(\mathbf{k}))$, is[9]

$$\frac{V \, d^3 \mathbf{k}}{(2\pi)^3} = \frac{V k^2 \, dk \, d\Omega}{(2\pi)^3}. \tag{1.47}$$

From Eqs. (1.44)–(1.47) we obtain the probability per unit time for an atomic transition $|A\rangle \rightarrow |B\rangle$ with emission of a photon of wave vector in the range $(\mathbf{k}, \mathbf{k} + d\mathbf{k})$ and with polarization vector $\boldsymbol{\varepsilon}_r(\mathbf{k})$:

$$
\begin{aligned}
w_r \, d\Omega = \int & \frac{V k^2 \, dk \, d\Omega}{(2\pi)^3} \frac{2\pi}{\hbar} \, \delta(E_A - E_B - \hbar\omega_{\mathbf{k}}) \\
& \times \left(\frac{\hbar\omega_{\mathbf{k}}}{2 V} \right) [n_r(\mathbf{k}) + 1] |\langle B|\boldsymbol{\varepsilon}_r(\mathbf{k}) \cdot \mathbf{D}|A\rangle|^2.
\end{aligned}
\tag{1.49}
$$

If we perform the integration with respect to $k\,(=\omega_{\mathbf{k}}/c)$ and substitute (1.43) for $\mathbf{D}$, the last expression reduces to

$$w_r \, d\Omega = \frac{e^2 \omega^3 \, d\Omega}{8\pi^2 \hbar c^3} [n_r(\mathbf{k}) + 1]|\boldsymbol{\varepsilon}_r(\mathbf{k}) \cdot \mathbf{x}_{BA}|^2 \tag{1.50}$$

where $\mathbf{x}_{BA}$ stands for the matrix element

$$\mathbf{x}_{BA} \equiv \langle B|\mathbf{x}|A\rangle = \langle B| \sum_i \mathbf{r}_i |A\rangle. \tag{1.51}$$

The most interesting feature of Eq. (1.50) is the occurrence of the factor $[n_r(\mathbf{k}) + 1]$. $n_r(\mathbf{k})$ is the occupation number of photons in the $(\mathbf{k}, r)$ mode present initially, and thus the part of (1.50) proportional to $n_r(\mathbf{k})$ represents induced (or stimulated) emission, i.e. radiation which results from the radiation incident on the atom; classically, we can think of it as resulting from the forced oscillations of the electrons, and this term can be produced from a semiclassical theory of radiation.[10] However, even with no radiation present initially $(n_r(\mathbf{k}) = 0)$, the transition probability (1.50) is different from zero. This corresponds to the spontaneous emission of radiation from an atom, and this cannot be derived from a semiclassical theory of radiation.

Eqs. (1.50) and (1.51) represent the basic result about emission of radiation in electric dipole transitions, and we only briefly indicate some consequences.

---

[9] Since we are using a finite normalization volume V, we should be summing over a group of allowed wave vectors $\mathbf{k}$ [see Eq. (1.13)]. For large V (strictly $V \rightarrow \infty$)

$$\frac{1}{V} \sum_{\mathbf{k}} \rightarrow \frac{1}{(2\pi)^3} \int d^3\mathbf{k}. \tag{1.48}$$

The normalization volume V must of course drop out of all physically significant quantities such as transition rates etc.
[10] See, for example, L. I. Schiff, *Quantum Mechanics*, 3rd edn, McGraw-Hill, New York, 1968, Chapter 11, or Bethe and Jackiw, referred to earlier in this section, Chapter 10.

To sum over the two polarization states for a given **k**, we note that $\boldsymbol{\varepsilon}_1(\mathbf{k})$, $\boldsymbol{\varepsilon}_2(\mathbf{k})$ and $\hat{\mathbf{k}} = \mathbf{k}/|\mathbf{k}|$ form an orthonormal coordinate system. Hence,

$$\sum_{r=1}^{2} |\boldsymbol{\varepsilon}_r(\mathbf{k}) \bullet \mathbf{x}_{BA}|^2 = \mathbf{x}_{BA} \bullet \mathbf{x}_{BA}^* - (\hat{\mathbf{k}} \bullet \mathbf{x}_{BA})(\hat{\mathbf{k}} \bullet \mathbf{x}_{BA}^*)$$

$$= (\mathbf{x}_{BA} \bullet \mathbf{x}_{BA}^*)(1 - \cos^2 \theta)$$

$$= |\mathbf{x}_{BA}|^2 \sin^2 \theta,$$

where the last line but one defines the angle θ which the complex vector $\mathbf{x}_{BA}$ makes with $\hat{\mathbf{k}}$. Hence, from Eq. (1.50)

$$\sum_{r=1}^{2} w_r \, d\Omega = \frac{e^2 \omega^3}{8\pi^2 \, \hbar c^3} \, d\Omega [n_r(\mathbf{k}) + 1]|\mathbf{x}_{BA}|^2 \sin^2 \theta. \tag{1.52}$$

For spontaneous emission, the total transition probability per unit time is obtained from the last equation, with $n_r(\mathbf{k}) = 0$, by integrating over all directions. Since

$$\int \sin^2 \theta \, d\Omega = \frac{8\pi}{3},$$

we obtain

$$w_{\text{total}}(A \rightarrow B) = \frac{e^2 \omega^3}{3\pi \hbar c^3} |\mathbf{x}_{BA}|^2. \tag{1.53}$$

The lifetime τ of an excited atomic state $|A\rangle$ is defined as the reciprocal of the total transition probability per unit time to *all* possible final states $|B_1\rangle$, $|B_2\rangle$, ..., i.e.

$$\frac{1}{\tau} = \sum_{n} w_{\text{total}}(A \rightarrow B_n). \tag{1.54}$$

In particular, if the state $|A\rangle$ can decay to states with non-zero total angular momentum, Eq. (1.54) must contain a summation over the corresponding magnetic quantum numbers.

The selection rules for electric dipole transitions follow from the matrix element (1.51). For example, since **x** is a vector, the states $|A\rangle$ and $|B\rangle$ must have opposite parity, and the total angular momentum quantum number J of the atom and its z-component M must satisfy the selection rules

$$\Delta J = 0, \pm 1, \quad \text{not } J = 0 \rightarrow J = 0, \quad \Delta M = 0, \pm 1.$$

The second selection rule (not $J = 0 \rightarrow J = 0$) applies strictly to one-photon processes, not only in the electric dipole approximation. It is a consequence of the fact that there are no one-photon states with zero angular momentum. To form such a state from the spin 1 of the photon and a unit of orbital angular momentum requires all three components of the spin angular momentum, but because of the transversality of the radiation field, only two of the spin components are available [compare Eq. (1.36)].

Finally, we note that very similar results hold for the absorption of radiation in electric dipole transitions. The matrix element

$$\langle B, n_r(\mathbf{k}) - 1|H_1|A, n_r(\mathbf{k})\rangle$$

corresponding to Eq. (1.44) now involves the factor $[n_r(\mathbf{k})]^{1/2}$ instead of $[n_r(\mathbf{k}) + 1]^{1/2}$. Our final result for emission, Eq. (1.50), also holds for absorption, with $[n_r(\mathbf{k}) + 1]$ replaced by $[n_r(\mathbf{k})]$, $d\Omega$ being the solid angle defining the incident radiation, and the matrix element $\mathbf{x}_{BA}$, Eq. (1.51), representing a transition from an atomic state $|A\rangle$ with energy E_A to a state $|B\rangle$ with energy $E_B > E_A$. Correspondingly, the frequency ω is defined by $\hbar\omega = E_B - E_A$ instead of Eq. (1.46).

1.4   The Electromagnetic Field in the Presence of Charges

After the special case of the electric dipole interaction, we now want to consider the general interaction of moving charges and an electromagnetic field. As this problem will later be treated in a relativistically covariant way, we shall not give a rigorous complete derivation, but rather stress the physical interpretation. As in the last section, the motion of the charges will again be described non-relativistically. In Section 1.4.1 we shall deal with the Hamiltonian formulation of the classical theory. This will enable us very easily to go over to the quantized theory in Section 1.4.2. In Sections 1.4.3 and 1.4.4 we shall illustrate the application of the theory for radiative transitions and Thomson scattering.

1.4.1   Classical electrodynamics

We would expect the Hamiltonian of a system of moving charges, such as an atom, in an electromagnetic field to consist of three parts: a part referring to matter (i.e. the charges), a part referring to the electromagnetic field and a part describing the interaction between matter and field.

For a system of point masses m_i, $i = 1, \ldots, N$, with charges e_i and position coordinates $\mathbf{r}_i$, the Hamiltonian is

$$H_{\mathrm{m}} = \sum_i \frac{\mathbf{p}_i^2}{2m_i} + H_{\mathrm{C}} \tag{1.55a}$$

where H_{C} is the Coulomb interaction

$$H_{\mathrm{C}} \equiv \frac{1}{2} \sum_{\substack{i,j \\ (i \neq j)}} \frac{e_i e_j}{4\pi |\mathbf{r}_i - \mathbf{r}_j|} \tag{1.55b}$$

and $\mathbf{p}_i = m_i \, d\mathbf{r}_i/dt$ is the kinetic momentum of the ith particle. This is the usual Hamiltonian of atomic physics, for example.

The electromagnetic field in interaction with charges is described by Maxwell's equations [Eqs. (1.1)]. We continue to use the Coulomb gauge, $\nabla \cdot \mathbf{A} = 0$, so that the electric field (1.2) decomposes into transverse and longitudinal fields

$$\mathbf{E} = \mathbf{E_T} + \mathbf{E_L},$$

where

$$\mathbf{E_T} = -\frac{1}{c} \frac{\partial \mathbf{A}}{\partial t}, \quad \mathbf{E_L} = -\nabla \phi.$$

(A longitudinal field is defined by the condition $\nabla \wedge \mathbf{E_L} = 0$.) The magnetic field is given by $\mathbf{B} = \nabla \wedge \mathbf{A}$.

The total energy of the electromagnetic field

$$\frac{1}{2} \int (\mathbf{E}^2 + \mathbf{B}^2)\, \mathrm{d}^3 x$$

can be written

$$\frac{1}{2} \int (\mathbf{E_T^2} + \mathbf{B}^2)\, \mathrm{d}^3 x + \frac{1}{2} \int \mathbf{E_L^2}\, \mathrm{d}^3 x.$$

The last integral can be transformed, using Poisson's equation $\nabla^2 \phi = -\rho$, into

$$\frac{1}{2} \int \mathbf{E_L^2}\, \mathrm{d}^3 x = \frac{1}{2} \int \frac{\rho(\mathbf{x},\, t)\rho(\mathbf{x}',\, t)}{4\pi |\mathbf{x} - \mathbf{x}'|}\, \mathrm{d}^3 x\, \mathrm{d}^3 x'. \tag{1.56}$$

Thus the energy associated with the longitudinal field is the energy of the *instantaneous* electrostatic interaction between the charges. With

$$\rho(\mathbf{x},\, t) = \sum_i e_i\, \delta(\mathbf{x} - \mathbf{r}_i(t))$$

Eq. (1.56) reduces to

$$\frac{1}{2} \int \mathbf{E_L^2}\, \mathrm{d}^3 x = \frac{1}{2} \sum_{i,j} \frac{e_i e_j}{4\pi |\mathbf{r}_i - \mathbf{r}_j|}$$

$$= \frac{1}{2} \sum_{\substack{i,j \\ i \neq j}} \frac{e_i e_j}{4\pi |\mathbf{r}_i - \mathbf{r}_j|} \equiv H_C, \tag{1.57}$$

where, in the last line, we have dropped the infinite self-energy which occurs for point charges. The term H_C has already been included in the Hamiltonian H_m, Eqs. (1.55), so we must take as additional energy of the electromagnetic field that of the transverse radiation field

$$H_{\mathrm{rad}} = \frac{1}{2} \int (\mathbf{E_T^2} + \mathbf{B}^2)\, \mathrm{d}^3 x. \tag{1.58}$$

Eqs. (1.55) allow for the instantaneous Coulomb interaction of charges. To allow for the interaction of moving charges with an electromagnetic field, one must replace the matter-Hamiltonian (1.55a) by

$$H'_{\mathrm{m}} = \sum_i \frac{1}{2m_i} \left(\mathbf{p}_i - \frac{e_i}{c} \mathbf{A}_i \right)^2 + H_C \tag{1.59}$$

where $\mathbf{A}_i = \mathbf{A}(\mathbf{r}_i,\, t)$ denotes the vector potential at the position $\mathbf{r}_i$ of the charge e_i at time t. In Eq. (1.59) $\mathbf{p}_i$ is the momentum coordinate canonically conjugate to the position coordinate $\mathbf{r}_i$, in the sense of Lagrangian mechanics, and it is related to the velocity $\mathbf{v}_i = \mathrm{d}\mathbf{r}_i/\mathrm{d}t$ of the *i*th particle by

$$\mathbf{p}_i = m_i \mathbf{v}_i + \frac{e_i}{c} \mathbf{A}_i.$$

It is only for $\mathbf{A} = 0$ that this conjugate momentum reduces to the kinetic momentum $m_i\mathbf{v}_i$. The justification for the form (1.59) for H'_m is that it gives the correct equations of motion for the charges (see Problem 1.2):

$$m_i \frac{d\mathbf{v}_i}{dt} = e_i \left[\mathbf{E}_i + \frac{\mathbf{v}_i}{c} \wedge \mathbf{B}_i \right], \tag{1.60}$$

where $\mathbf{E}_i$ and $\mathbf{B}_i$ are the electric and magnetic fields at the instantaneous position of the ith charge.[11]

We can regroup the terms in Eq. (1.59) as

$$H'_m = H_m + H_I \tag{1.61}$$

where H_I, the interaction Hamiltonian of matter and field, is given by

$$
\begin{aligned}
H_I &= \sum_i \left\{ -\frac{e_i}{2m_ic} (\mathbf{p}_i \bullet \mathbf{A}_i + \mathbf{A}_i \bullet \mathbf{p}_i) + \frac{e_i^2}{2m_ic^2} \mathbf{A}_i^2 \right\} \\
&= \sum_i \left\{ -\frac{e_i}{m_ic} \mathbf{A}_i \bullet \mathbf{p}_i + \frac{e_i^2}{2m_ic^2} \mathbf{A}_i^2 \right\}.
\end{aligned}
\tag{1.62}
$$

In the quantum theory $\mathbf{p}_i$, the momentum canonically conjugate to $\mathbf{r}_i$, will become the operator $-i\hbar\nabla_i$. Nevertheless, the replacement of $\mathbf{p}_i \bullet \mathbf{A}_i$ by $\mathbf{A}_i \bullet \mathbf{p}_i$ in the second line of Eq. (1.62) is justified by our gauge condition $\nabla_i \bullet \mathbf{A}_i = 0$. Eq. (1.62) represents the general interaction of moving charges in an electromagnetic field (apart from H_C). It does not include the interaction of the magnetic moments, such as that due to the spin of the electron, with magnetic fields.

Combining the above results (1.55), (1.58), (1.59) and (1.62), we obtain for the complete Hamiltonian

$$H = H'_m + H_{rad} = H_m + H_{rad} + H_I. \tag{1.63}$$

Just as this Hamiltonian leads to the correct equations of motion (1.60) for charges, so it also leads to the correct field equations (1.4), with $\nabla \bullet \mathbf{A} = 0$, for the potentials.[12]

1.4.2   Quantum electrodynamics

The quantization of the system described by the Hamiltonian (1.63) is carried out by subjecting the particles' coordinates $\mathbf{r}_i$ and canonically conjugate momenta $\mathbf{p}_i$ to the usual commutation relations (e.g. in the coordinate representation $\mathbf{p}_i \rightarrow -i\hbar\nabla_i$), and quantizing the radiation field, as in Section 1.2.3. The longitudinal electric field $\mathbf{E_L}$ does not provide any additional degrees of freedom, being completely determined via the first Maxwell equation $\nabla \bullet \mathbf{E_L} = \rho$ by the charges.

The interaction H_I in Eq, (1.63) is usually treated as a perturbation which causes transitions between the states of the non-interacting Hamiltonian

$$H_0 = H_m + H_{rad}. \tag{1.64}$$

---

[11] For the Lagrangian and Hamiltonian formulations of mechanics which are here used see, for example, H. Goldstein, *Classical Mechanics*, 2nd edn, Addison-Wesley, Reading, Mass., 1980, in particular pp. 21–23 and 346.
[12] See W. Heitler, *The Quantum Theory of Radiation*, 3rd edn, Clarendon Press, Oxford, 1954, pp. 48–50.

The eigenstates of H_0 are again of the form

$$|A, \ldots n_r(\mathbf{k}) \ldots\rangle = |A\rangle \, |\ldots n_r(\mathbf{k}) \ldots\rangle,$$

with $|A\rangle$ and $|\ldots n_r(\mathbf{k}) \ldots\rangle$ eigenstates of H_m and H_{rad}.

Compared with the electric dipole interaction (1.40), the interaction (1.62) differs in that it contains a term quadratic in the vector potential. This results in two-photon processes in first-order perturbation theory (i.e. emission or absorption of two photons or scattering). In addition, the first term in (1.62) contains magnetic interactions and higher-order effects due to the spatial variation of $\mathbf{A}(\mathbf{x}, t)$, which are absent from the electric dipole interaction (1.40). These aspects are illustrated in the applications to radiative transitions and Thomson scattering which follow.

1.4.3   Radiative transitions in atoms

We consider transitions between two states of an atom with emission or absorption of one photon. This problem was treated in Section 1.3 in the electric dipole approximation, but now we shall use the interaction (1.62).

We shall consider the emission process between the initial and final states (1.42). Using the expansion (1.38) of the vector potential, we obtain the matrix element for this transition [which results from the term linear in $\mathbf{A}$ in Eq. (1.62)]

$$\langle B, n_r(\mathbf{k}) + 1 | H_1 | A, n_r(\mathbf{k}) \rangle$$

$$= -\frac{e}{m} \left(\frac{\hbar}{2V\omega_{\mathbf{k}}} \right)^{1/2} [n_r(\mathbf{k}) + 1]^{1/2} \, \langle B | \boldsymbol{\varepsilon}_r(\mathbf{k}) \bullet \sum_i e^{-i\mathbf{k}\bullet\mathbf{r}_i} \, \mathbf{p}_i | A \rangle \, e^{i\omega_{\mathbf{k}}t}. \tag{1.65}$$

Using this matrix element, one calculates the transition probability per unit time as in Section 1.3. Instead of Eqs. (1.50) and (1.51), one obtains:

$$w_r \, d\Omega = \frac{e^2 \omega \, d\Omega}{8\pi^2 m^2 \hbar c^3} \, [n_r(\mathbf{k}) + 1] \, \left| \boldsymbol{\varepsilon}_r(\mathbf{k}) \bullet \langle B | \sum_i e^{-i\mathbf{k}\bullet\mathbf{r}_i} \mathbf{p}_i \, | A \rangle \right|^2. \tag{1.66}$$

These results go over into the electric dipole approximation if in the matrix elements in Eqs. (1.65) and (1.66) we can approximate the exponential functions by unity:

$$e^{-i\mathbf{k}\bullet\mathbf{r}_i} \approx 1. \tag{1.67}$$

This is justified provided the wavelength $\lambda = 2\pi/k$ of the radiation emitted in the transition is very large compared to the linear dimensions R of the system of charges (in our case, of the atom): $\lambda \gg R$. The atomic wavefunctions $|A\rangle$ and $|B\rangle$ restrict the effective values of $\mathbf{r}_i$ to $r_i \lesssim R$, so that $\mathbf{k} \bullet \mathbf{r}_i \lesssim kR \ll 1$. We saw in Section 1.3 that this inequality is generously satisfied for optical atomic transitions. From the equation of motion $i\hbar\dot{\mathbf{r}}_i = [\mathbf{r}_i, H]$ and Eq. (1.46)

$$\langle B | \mathbf{p}_i | A \rangle = m \langle B | \dot{\mathbf{r}}_i | A \rangle = -im\omega \langle B | \mathbf{r}_i | A \rangle.$$

Hence, in the approximation (1.67), Eqs. (1.65) and (1.66) reduce to the electric dipole form, Eqs. (1.44) and (1.50).

If selection rules forbid the transition $|A\rangle$ to $|B\rangle$ via the electric dipole interaction, it may in general still occur via higher terms in the expansion of the exponentials

$$e^{-i\mathbf{k}\cdot\mathbf{r}_i} = 1 - i\mathbf{k}\cdot\mathbf{r}_i + \cdots.$$

With the second term, the expression within the modulus sign in Eq. (1.66) becomes

$$\boldsymbol{\varepsilon}_r(\mathbf{k})\cdot\langle B|\sum_i(-i\mathbf{k}\cdot\mathbf{r}_i)\mathbf{p}_i|A\rangle = -i\sum_{\alpha=1}^{3}\sum_{\beta=1}^{3}\varepsilon_{r\alpha}(\mathbf{k})k_\beta\langle B|\sum_i r_{i\beta}\,p_{i\alpha}|A\rangle,$$

where α, β ($= 1, 2, 3$) label the Cartesian components of the vectors $\boldsymbol{\varepsilon}_r$, $\mathbf{k}$, $\mathbf{r}_i$ and $\mathbf{p}_i$. The matrix element can be written as the sum of an antisymmetric and a symmetric second-rank tensor

$$\langle B|\sum_i r_{i\beta}p_{i\alpha}|A\rangle = \frac{1}{2}\left\{\langle B|\sum_i(r_{i\beta}p_{i\alpha} - r_{i\alpha}p_{i\beta})|A\rangle + \langle B|\sum_i(r_{i\beta}p_{i\alpha} + r_{i\alpha}p_{i\beta})|A\rangle\right\}.$$

The first term contains the antisymmetric angular momentum operator and corresponds to the magnetic dipole interaction. (In practice this must be augmented by the spin part.) The symmetric term corresponds to the electric quadrupole interaction. The parity and angular momentum selection rules for the transitions brought about by these matrix elements are easily determined from their forms. We obtain in this way an expansion into electric and magnetic multipoles, i.e. photons of definite parity and angular momentum. As usual, a better procedure for such an expansion, except in the simplest cases, is to use spherical rather than Cartesian coordinates.[13]

The result (1.66) can again be adapted to the case of absorption of radiation by replacing the factor $[n_r(\mathbf{k}) + 1]$ by $n_r(\mathbf{k})$ and the appropriate re-interpretation of the matrix element, etc.

1.4.4   Thomson scattering

As a second illustration, we consider Thomson scattering, i.e. the scattering of photons of energy $\hbar\omega$ by atomic electrons, with $\hbar\omega$ large compared to the binding energies of the electrons, so that they can be considered as free electrons, but $\hbar\omega$ very small compared to the electron rest energy mc^2. In this case the energy $\hbar\omega'$ of the scattered photon is not changed: $\hbar\omega' = \hbar\omega$, since for small recoil momenta the recoil energy may be neglected.

The scattering from an initial state with one photon of momentum $\hbar\mathbf{k}$ and polarization $\boldsymbol{\varepsilon}_\alpha(\mathbf{k})$ (with $\alpha = 1$ or 2) to a final state with one photon of momentum $\hbar\mathbf{k}'$ and polarization $\boldsymbol{\varepsilon}_\beta(\mathbf{k}')$ (with $\beta = 1$ or 2) can occur in first-order perturbation theory via the term in $\mathbf{A}^2$ in the interaction (1.62). It can also occur in second-order perturbation theory via the term linear in $\mathbf{A}$ in Eq. (1.62), but one can show that

---

[13] See A. S. Davydov, *Quantum Mechanics*, 2nd edn, Pergamon, Oxford, 1976, Sections 81 and 95.

under our conditions the contribution of the second-order process is negligible.[14] The operator $\mathbf{A}^2(0, t)$ can, from Eq. (1.38), be written

$$\mathbf{A}^2(0, t) = \sum_{\mathbf{k}_1 \mathbf{k}_2} \sum_{r, s} \frac{\hbar c^2}{2V(\omega_1 \omega_2)^{1/2}} \, (\boldsymbol{\varepsilon}_r(\mathbf{k}_1) \cdot \boldsymbol{\varepsilon}_s(\mathbf{k}_2))$$

$$\times \left[a_r(\mathbf{k}_1) e^{-i\omega_1 t} + a_r^\dagger(\mathbf{k}_1) \, e^{+i\omega_1 t} \right] \left[a_s(\mathbf{k}_2) e^{-i\omega_2 t} + a_s^\dagger(\mathbf{k}_2) \, e^{+i\omega_2 t} \right],$$

(1.68)

where $\omega_r \equiv c|\mathbf{k}_r|$, $r = 1, 2$. This operator can bring about the transition from the initial state $|\mathbf{k}, \alpha\rangle$ to the final state $|\mathbf{k}', \beta\rangle$ (we use a somewhat simplified, but unambiguous, notation) in two ways: either of the factors in square parentheses can act to absorb the initial photon, and the other factor then creates the final photon. One then obtains the matrix element for this transition from Eq. (1.62)

$$\langle \mathbf{k}', \beta | \frac{e^2}{2mc^2} \mathbf{A}^2(0, t) | \mathbf{k}, \alpha \rangle = \frac{e^2 \hbar}{2mV(\omega\omega')^{1/2}} \, \boldsymbol{\varepsilon}_\alpha(\mathbf{k}) \cdot \boldsymbol{\varepsilon}_\beta(\mathbf{k}') \, e^{i(\omega' - \omega)t}$$

where $\omega = c|\mathbf{k}|$ and. $\omega' = c|\mathbf{k}'|$. The transition probability per unit time for a photon, initially in the state $|\mathbf{k}, \alpha\rangle$, to be scattered into an element of solid angle $d\Omega$ in the direction $\mathbf{k}'$, and with polarization $\boldsymbol{\varepsilon}_\beta(\mathbf{k}')$, is given by

$$w_{\alpha \to \beta}(\mathbf{k}') \, d\Omega = \frac{2\pi}{\hbar} \int \frac{V k'^2 \, dk' \, d\Omega}{(2\pi)^3} \, \delta(\hbar\omega' - \hbar\omega)$$

$$\times \left(\frac{e^2 \hbar}{2mV} \right)^2 \left(\frac{1}{\omega\omega'} \right) \left[\boldsymbol{\varepsilon}_\alpha(\mathbf{k}) \cdot \boldsymbol{\varepsilon}_\beta(\mathbf{k}') \right]^2$$

$$= \frac{c}{V} \left(\frac{e^2}{4\pi mc^2} \right)^2 \left[\boldsymbol{\varepsilon}_\alpha(\mathbf{k}) \cdot \boldsymbol{\varepsilon}_\beta(\mathbf{k}') \right]^2 d\Omega$$

where $|\mathbf{k}'| = |\mathbf{k}|$. Dividing this transition probability per unit time by the incident photon flux *(c/V)*, one obtains the corresponding differential cross-section

$$\sigma_{\alpha \to \beta}(\mathbf{k}') \, d\Omega = r_0^2 \left[\boldsymbol{\varepsilon}_\alpha(\mathbf{k}) \cdot \boldsymbol{\varepsilon}_\beta(\mathbf{k}') \right]^2 d\Omega,$$

(1.69)

where the classical electron radius has been introduced by

$$r_0 = \frac{e^2}{4\pi mc^2} = 2.818 \quad \text{fm}.$$

(1.70)

For an unpolarized incident photon beam, the unpolarized differential cross-section (i.e. the final polarization state is not observed) is obtained from Eq. (1.69) by summing over final and averaging over initial polarization states. We introduce the abbreviations $\boldsymbol{\varepsilon}_\alpha \equiv \boldsymbol{\varepsilon}_\alpha(\mathbf{k})$ and $\boldsymbol{\varepsilon}'_\beta \equiv \boldsymbol{\varepsilon}_\beta(\mathbf{k}')$. Since $\boldsymbol{\varepsilon}_1$, $\boldsymbol{\varepsilon}_2$ and $\hat{\mathbf{k}} = \mathbf{k}/|\mathbf{k}|$ form an orthonormal coordinate system,

$$\sum_{\alpha = 1}^2 \left(\boldsymbol{\varepsilon}_\alpha \cdot \boldsymbol{\varepsilon}'_\beta \right)^2 = 1 - \left(\hat{\mathbf{k}} \cdot \boldsymbol{\varepsilon}'_\beta \right)^2.$$

---

[14] See J. J. Sakurai, *Advanced Quantum Mechanics*, Addison-Wesley, Reading, Mass., 1967, p. 51.

Similarly

$$\sum_{\beta=1}^{2} \left(\hat{\mathbf{k}} \cdot \mathbf{\epsilon}'_{\beta}\right)^2 = 1 - \left(\hat{\mathbf{k}} \cdot \hat{\mathbf{k}}'\right)^2 = \sin^2 \theta$$

where θ is the angle between the directions $\mathbf{k}$ and $\mathbf{k}'$ of the incident and scattered photons, i.e. the angle of scattering. From the last two equations

$$\frac{1}{2} \sum_{\alpha=1}^{2} \sum_{\beta=1}^{2} \left(\mathbf{\epsilon}_{\alpha} \cdot \mathbf{\epsilon}'_{\beta}\right)^2 = \frac{1}{2} \left(2 - \sin^2 \theta\right) = \frac{1}{2} \left(1 + \cos^2 \theta\right) \qquad (1.71)$$

and hence the unpolarized differential cross-section for scattering through an angle θ is from Eq. (1.69) given as

$$\sigma(\theta)\, d\Omega = \frac{1}{2} r_0^2 \left(1 + \cos^2 \theta\right) d\Omega. \qquad (1.69a)$$

Integrating over angles, we obtain the total cross-section for Thomson scattering

$$\sigma_{\text{total}} = \frac{8\pi}{3} r_0^2 = 6.65 \times 10^{-25} \text{ cm}^2. \qquad (1.72)$$

1.5   Appendix: The Schrödinger, Heisenberg and Interaction Pictures

These three pictures (abbreviated S.P., H.P. and I.P.) are three different ways of describing the time development of a system. In this Appendix, we shall derive the relationships between the three pictures. Quantities in these three pictures will be distinguished by the labels S, H and I.

In the S.P., the time-dependence is carried by the states according to the Schrödinger equation

$$i\hbar \frac{d}{dt} |A,\, t\rangle_{\text{S}} = H |A,\, t\rangle_{\text{S}}, \qquad (1.73)$$

where H is the Hamiltonian of the system in the S.P. This can formally be solved in terms of the state of the system at an arbitrary reference time t_0

$$|A,\, t\rangle_{\text{S}} = U_{\text{S}}(t) |A,\, t_0\rangle_{\text{S}} \qquad (1.74)$$

where $U_{\text{S}}(t)$ is the unitary operator:

$$U_{\text{S}}(t) = e^{-iH(t - t_0)/\hbar}. \qquad (1.75)$$

By means of $U_{\text{S}}(t)$ we can carry out a unitary transformation of states and operators (O) from the S.P. to the H.P., in which we define

$$|A\rangle_{\text{H}} = U_{\text{S}}^{\dagger}(t) |A,\, t\rangle_{\text{S}} = |A,\, t_0\rangle_{\text{S}} \qquad (1.76)$$

and

$$O^{\mathrm{H}}(t) = U_{\mathrm{S}}^{\dagger}(t)\, O^{\mathrm{S}}\, U_{\mathrm{S}}(t). \tag{1.77}$$

At $t = t_0$, states and operators in the two pictures are the same. We see from Eq. (1.76) that in the H.P. state, vectors are constant in time; the time-dependence is carried by the Heisenberg operators. From Eq. (1.77)

$$H^{\mathrm{H}} = H^{\mathrm{S}} \equiv H. \tag{1.78}$$

Since the transformation from the S.P. to the H.P. is unitary, it ensures the invariance of matrix elements and commutation relations:

$$_{\mathrm{S}}\langle B,\, t | O^{\mathrm{S}} | A,\, t \rangle_{\mathrm{S}} = {}_{\mathrm{H}}\langle B, t | O^{\mathrm{H}}(t) | A,\, t \rangle_{\mathrm{H}}, \tag{1.79}$$

and if O and P are two operators for which $[O^{\mathrm{S}}, P^{\mathrm{S}}] = $ const., then $[O^{\mathrm{H}}(t), P^{\mathrm{H}}(t)]$ equals the same constant.

Differentation of Eq. (1.77) gives the Heisenberg equation of motion

$$i\hbar\, \frac{\mathrm{d}}{\mathrm{d}t}\, O^{\mathrm{H}}(t) = \left[O^{\mathrm{H}}(t),\, H\right]. \tag{1.80}$$

For an operator which is time dependent in the S.P. (corresponding to a quantity which classically has an explicit time dependence), Eq. (1.80) is augmented to

$$i\hbar\, \frac{\mathrm{d}}{\mathrm{d}t}\, O^{\mathrm{H}}(t) = i\hbar\, \frac{\partial}{\partial t}\, O^{\mathrm{H}}(t) + \left[O^{\mathrm{H}}(t),\, H\right]. \tag{1.81}$$

We shall not be considering such operators.

The I.P. arises if the Hamiltonian is split into two parts

$$H = H_0 + H_{\mathrm{I}}. \tag{1.82}$$

In quantum field theory, H_{I} will describe the interaction between two fields, themselves described by H_0. [Note that the suffix I on H_{I} stands for 'interaction'. It does not label a picture. Eq. (1.82) holds in any picture.] The I.P. is related to the S.P. by the unitary transformation

$$U_0(t) = e^{-iH_0(t - t_0)/\hbar} \tag{1.83}$$

i.e.

$$|A,\, t\rangle_{\mathrm{I}} = U_0^{\dagger}(t)\, |A,\, t\rangle_{\mathrm{S}} \tag{1.84}$$

and

$$O^{\mathrm{I}}(t) = U_0^{\dagger}(t)\, O^{\mathrm{S}}\, U_0(t). \tag{1.85}$$

Thus the relation between I.P. and S.P. is similar to that between H.P. and S.P., but with the unitary transformation U_0 involving the non-interacting Hamiltonian H_0, instead of U involving the total Hamiltonian H. From Eq. (1.85):

$$H_0^{\mathrm{I}} = H_0^{\mathrm{S}} \equiv H_0. \tag{1.86}$$

Differentiating Eq. (1.85) gives the differential equation of motion of operators in the I.P.:

$$i\hbar \frac{d}{dt} O^{I}(t) = [O^{I}(t), H_0].\tag{1.87}$$

Substituting Eq. (1.84) into the Schrödinger equation (1.73), one obtains the equation of motion of state vectors in the I.P.

$$i\hbar \frac{d}{dt} |A, t\rangle_I = H_1^I(t) |A, t\rangle_I\tag{1.88}$$

where

$$H_1^I(t) = e^{iH_0(t - t_0)/\hbar}\, H_1^S\, e^{-iH_0(t - t_0)/\hbar}.\tag{1.89}$$

Finally, from the above relations, one easily shows that the I.P. and H.P. are related by

$$O^{I}(t) = U(t) O^{H}(t) U^{\dagger}(t)\tag{1.90}$$

$$|A, t\rangle_I = U(t) |A\rangle_H\tag{1.91}$$

where the unitary operator $U(t)$ is defined by

$$U(t) = e^{iH_0(t - t_0)/\hbar}\; e^{-iH(t - t_0)/\hbar}.\tag{1.92}$$

The time development of the I.P. states follows from Eq. (1.91). From this equation

$$|A, t_1\rangle_I = U(t_1)|A\rangle_H = U(t_1)U^{\dagger}(t_2)|A, t_2\rangle_I.$$

Hence

$$|A, t_1\rangle_I = U(t_1, t_2)|A, t_2\rangle_I\tag{1.93}$$

where the unitary operator $U(t_1, t_2)$, defined by

$$U(t_1, t_2) = U(t_1)U^{\dagger}(t_2),\tag{1.94}$$

satisfies the relations

$$U^{\dagger}(t_1, t_2) = U(t_2, t_1)\tag{1.95a}$$

$$U(t_1, t_2)U(t_2, t_3) = U(t_1, t_3).\tag{1.95b}$$

Problems

1.1. The radiation field inside a cubic enclosure, which contains no charges, is specified by the state

$$|c\rangle = \exp\left(-\frac{1}{2}|c|^2\right) \sum_{n=0}^{\infty} \frac{c^n}{\sqrt{n!}} |n\rangle$$

where $c = |c|\, e^{i\delta}$ is any complex number and $|n\rangle$ is the state (1.31) in which there are n photons with wave vector $\mathbf{k}$ and polarization vector $\boldsymbol{\varepsilon}_r(\mathbf{k})$ present, and no others. Derive the following properties of the state $|c\rangle$.

(i) $|c\rangle$ is normalized: $\langle c|c\rangle = 1$.

(ii) $|c\rangle$ is an eigenstate of the destruction operator $a_r(\mathbf{k})$ with the complex eigenvalue c:

$$a_r(\mathbf{k})|c\rangle = c|c\rangle.$$

(iii) The mean number $\overline{N}$ of photons in the enclosure in the state $|c\rangle$ is given by

$$\overline{N} = \langle c|N|c\rangle = |c|^2 \tag{A}$$

where N is the total photon number operator.

(iv) The root-mean-square fluctuation ΔN in the number of photons in the enclosure in the state $|c\rangle$ is given by

$$(\Delta N)^2 = \langle c|N^2|c\rangle - \overline{N}^2 = |c|^2. \tag{B}$$

(v) The expectation value of the electric field $\mathbf{E}$ in the state $|c\rangle$ is given by

$$\langle c|\mathbf{E}|c\rangle = -\,\boldsymbol{\varepsilon}_r(\mathbf{k})2\left(\frac{\hbar\omega_{\mathbf{k}}}{2V}\right)^{1/2}|c|\sin(\mathbf{k}\bullet\mathbf{x} - \omega_{\mathbf{k}}t + \delta) \tag{C}$$

where V is the volume of the enclosure.

(vi) The root-mean-square fluctuation ΔE of the electric field in the state $|c\rangle$ is given by

$$(\Delta E)^2 = \langle c|\mathbf{E}^2|c\rangle - \langle c|\mathbf{E}|c\rangle^2 = \frac{\hbar\omega_{\mathbf{k}}}{2V}. \tag{D}$$

We noted in Section 1.2.3 that the expectation value of $\mathbf{E}$ in a state with a definite number of photons is zero, so that such a state cannot represent a classical field, even for very large photon numbers. In contrast, it follows from Eqs. (A)–(D) that the relative fluctuation in photon numbers

$$\frac{\Delta N}{\overline{N}} = \overline{N}^{-1/2}$$

tends to zero as $\overline{N} \to \infty$, and that the fluctuation ΔE becomes negligible for large field strengths, i.e. $|c\rangle$ goes over into a classical state in which the field is well defined as $\overline{N} \to \infty$. The state $|c\rangle$ is called a coherent state and represents the closest quantum-mechanical approach to a classical electromagnetic field. (For a full discussion, see the book by Loudon, quoted at the end of Section 1.2.)

1.2. The Lagrangian of a particle of mass m and charge q, moving in an electromagnetic field, is given by

$$L(\mathbf{x}, \dot{\mathbf{x}}) = \frac{1}{2}m\dot{\mathbf{x}}^2 + \frac{q}{c}\mathbf{A}\cdot\dot{\mathbf{x}} - q\phi$$

where $\mathbf{A} = \mathbf{A}(\mathbf{x}, t)$ and $\phi = \phi(\mathbf{x}, t)$ are the vector and scalar potentials of the electromagnetic field at the position $\mathbf{x}$ of the particle at time t.

(i) Show that the momentum conjugate to $\mathbf{x}$ is given by

$$\mathbf{p} = m\dot{\mathbf{x}} + \frac{q}{c}\mathbf{A} \tag{A}$$

(i.e. the conjugate momentum $\mathbf{p}$ is not the kinetic momentum $m\dot{\mathbf{x}}$, in general) and that Lagrange's equations reduce to the equations of motion of the particle [compare Eq. (1.60)]

$$m\frac{d}{dt}\dot{\mathbf{x}} = q\left[\mathbf{E} + \frac{1}{c}\dot{\mathbf{x}} \wedge \mathbf{B}\right], \tag{B}$$

where $\mathbf{E}$ and $\mathbf{B}$ are the electric and magnetic fields at the instantaneous position of the charge.

(ii) Derive the corresponding Hamiltonian [compare Eq. (1.59)]

$$H = \frac{1}{2m}\left(\mathbf{p} - \frac{q}{c}\mathbf{A}\right)^2 + q\phi,$$

and show that the resulting Hamilton equations again lead to Eqs. (A) and (B).

1.3. For Thomson scattering of an unpolarized beam of photons, obtain the differential cross-section for scattering through an angle θ, with the scattered radiation being linearly polarized in a given direction. By considering two mutually perpendicular such directions, use your result to re-derive Eq. (1.69a) for the unpolarized differential cross-section.

Show that for $\theta = 90°$, the scattered beam is 100% linearly polarized in the direction of the normal to the plane of scattering.

2

Lagrangian Field Theory

In the last chapter we quantized the electromagnetic field by Fourier analysing the classical field into normal modes and imposing harmonic oscillator commutation relations on the normal coordinates. We shall now take the fields at each point in space as the dynamical variables and quantize these directly. This approach generalizes the classical mechanics of a system of particles, and its quantization, to a continuous system, i.e. to fields.[1] One introduces a Lagrangian (actually, as we shall see, it is a Lagrangian density), from which the field equations follow by means of Hamilton's principle. One introduces momenta conjugate to the fields and imposes canonical commutation relations directly on the fields and the conjugate momenta. This formalism provides a systematic quantization procedure for any classical field theory derivable from a Lagrangian. Since this approach is equivalent to that of the last chapter, one can only obtain bosons in this way and a different formalism will be needed for fermions.

Another difference from Chapter 1 is that the theory will now be developed in a manifestly relativistically covariant form, and in Section 2.1 we shall define our relativistic notation. The classical Lagrangian field theory will be developed in Section 2.2, to be quantized in Section 2.3. An important feature of a Lagrangian field theory is that all its symmetry properties and the consequent conservation laws are contained in the Lagrangian density. We shall consider some of these aspects in Section 2.4.

---

[1] The relevant Lagrangian and Hamiltonian mechanics is, for example, developed in H. Goldstein, *Classical Mechanics*, 2nd edn, Addison-Wesley, Reading, Mass., 1980, Chapters 2 and 8, or in L. D. Landau and E. M. Lifshitz, *Mechanics*, Pergamon, Oxford, 1960, Sections 1–7 and Section 40.

Quantum Field Theory, Second Edition   Franz Mandl and Graham Shaw
© 2010 John Wiley & Sons, Ltd

2.1   Relativistic Notation

We shall write x^μ ($\mu = 0, 1, 2, 3$) for the space–time four-vector with the time component $x^0 = ct$ and the space coordinates x^j ($j = 1, 2, 3$), i.e. $x^\mu = (ct, \mathbf{x})$. The components of four-vectors will be labelled by Greek indices, the components of spatial three-vectors by Latin indices.

By means of the metric tensor $g_{\mu\nu}$, with components

$$\left. \begin{array}{l} g_{00} = -g_{11} = -g_{22} = -g_{33} = +1 \\ g_{\mu\nu} = 0 \ \ \text{if} \ \mu \neq \nu \end{array} \right\}, \tag{2.1}$$

we define the covariant vector x_μ from the contravariant x^μ:

$$x_\mu = \sum_{\nu=0}^{3} g_{\mu\nu} x^\nu \equiv g_{\mu\nu} x^\nu. \tag{2.2}$$

In the last expression we have used the summation convention: repeated Greek indices, one contravariant and one covariant, are summed. From Eqs. (2.1) and (2.2) we have $x_\mu = (ct, -\mathbf{x})$.

We also define the contravariant metric tensor $g^{\lambda\mu}$ by

$$g^{\lambda\mu} g_{\mu\nu} = g^\lambda_\nu = \delta^\lambda_\nu \tag{2.3}$$

where δ^λ_ν is the usual Kronecker delta: $\delta^\lambda_\nu = 1$ if $\lambda = \nu$, and $\delta^\lambda_\nu = 0$ if $\lambda \neq \nu$. From Eqs. (2.1) and (2.3), $g^{\mu\nu} = g_{\mu\nu}$.

A Lorentz transformation

$$x^\mu \rightarrow x'^\mu = \Lambda^\mu{}_\nu x^\nu \tag{2.4}$$

leaves

$$x^\mu x_\mu = \left(x^0\right)^2 - \mathbf{x}^2 \tag{2.5}$$

invariant, i.e. $x'^\mu x'_\mu = x^\mu x_\mu$ is a scalar quantity. Hence

$$\Lambda^{\lambda\mu} \Lambda_{\lambda\nu} = \delta^\mu_\nu. \tag{2.6}$$

(In addition the matrix $\Lambda^{\lambda\mu}$ must be real to ensure the reality of the space–time coordinates.)

A four-component object s^μ (s_μ) transforming like x^μ (x_μ) under Lorentz transformations, and hence with $s^\mu s_\mu$ invariant, is a contravariant (covariant) four-vector. An example is the energy–momentum vector $p^\mu = (E/c, \mathbf{p})$. When no confusion can result, we shall often omit the tensor indices, e.g. we may write x for x^μ or x_μ.

The scalar product of two four-vectors a and b can be written in various ways:

$$ab = a^\mu b_\mu = a_\mu b^\mu = g_{\mu\nu} a^\mu b^\nu = \cdots = a^0 b^0 - \mathbf{a} \cdot \mathbf{b}. \tag{2.7}$$

Like $x^2 = x^\mu x_\mu$, so the scalar product ab is an invariant under Lorentz transformations.

The four-dimensional generalization of the gradient operator ∇ transforms like a four-vector. If $\phi(x)$ is a scalar function, so is

$$\delta\phi = \frac{\partial\phi}{\partial x^\mu}\,\delta x^\mu,$$

and hence

$$\frac{\partial\phi}{\partial x^\mu} \equiv \partial_\mu\phi \equiv \phi_{,\mu} \tag{2.8a}$$

is a covariant four-vector. Similarly

$$\frac{\partial\phi}{\partial x_\mu} \equiv \partial^\mu\phi \equiv \phi_{,}{}^{\mu} \tag{2.8b}$$

is a contravariant four-vector. Note that indices following a comma denote differentiation. Finally, we note that the operator $\square$ is a scalar:

$$\partial^\mu\partial_\mu = \frac{1}{c^2}\frac{\partial^2}{\partial t^2} - \nabla^2 \equiv \square. \tag{2.9}$$

2.2   Classical Lagrangian Field Theory

We consider a system which requires several fields $\phi_r(x), r=1,\ldots,N$, to specify it. The index r may label components of the same field [for example, the components of the vector potential $\mathbf{A}(x)$], or it may refer to different independent fields. We restrict ourselves to theories which can be derived by means of a variational principle from an action integral involving a Lagrangian density

$$\mathscr{L} = \mathscr{L}\left(\phi_r, \phi_{r,\alpha}\right) \tag{2.10}$$

where the derivative $\phi_{r,\alpha}$ is defined by Eq. (2.8a). The Lagrangian density (2.10), depending on the fields and their first derivatives only, is not the most general case possible, but it covers all theories discussed in this book and greatly simplifies the formalism.

We define the action integral $S(\Omega)$ for an arbitrary region Ω of the four-dimensional space-time continuum by

$$S(\Omega) = \int_\Omega \mathrm{d}^4x\,\mathscr{L}\left(\phi_r, \phi_{r,\alpha}\right), \tag{2.11}$$

where d^4x stands for the four-dimensional element $\mathrm{d}x^0\,\mathrm{d}^3\mathbf{x}$.

We now postulate that the equations of motion, i.e. the field equations, are obtained from the following variational principle, which is closely analogous to Hamilton's principle in mechanics. For an arbitrary region Ω, we consider variations of the fields,

$$\phi_r(x) \rightarrow \phi_r(x) + \delta\phi_r(x), \tag{2.12}$$

which vanish on the surface $\Gamma(\Omega)$ bounding the region Ω

$$\delta\phi_r(x) = 0 \text{ on } \Gamma(\Omega). \tag{2.13}$$

The fields ϕ_r may be real or complex. In the case of a complex field $\phi(x)$, the fields $\phi(x)$ and $\phi^*(x)$ are treated as two independent fields. Alternatively, a complex field $\phi(x)$ can be decomposed into a pair of real fields, which are then treated as independent fields. We now demand that for an arbitrary region Ω and the variation (2.12–2.13), the action (2.11) has a stationary value, i.e.

$$\delta S(\Omega) = 0. \tag{2.14}$$

Calculating $\delta S(\Omega)$ from Eq. (2.11), we obtain[2]

$$\delta S(\Omega) = \int_\Omega d^4x \left\{ \frac{\partial \mathcal{L}}{\partial \phi_r} \delta\phi_r + \frac{\partial \mathcal{L}}{\partial \phi_{r,\alpha}} \delta\phi_{r,\alpha} \right\}$$
$$= \int_\Omega d^4x \left\{ \frac{\partial \mathcal{L}}{\partial \phi_r} - \frac{\partial}{\partial x^\alpha} \left(\frac{\partial \mathcal{L}}{\partial \phi_{r,\alpha}} \right) \right\} \delta\phi_r + \int_\Omega d^4x \frac{\partial}{\partial x^\alpha} \left(\frac{\partial \mathcal{L}}{\partial \phi_{r,\alpha}} \delta\phi_r \right), \tag{2.15}$$

where the last line is obtained by partial integration, since

$$\delta\phi_{r,\alpha} = \frac{\partial}{\partial x^\alpha} \delta\phi_r.$$

The last term in Eq. (2.15) can be converted to a surface integral over the surface $\Gamma(\Omega)$ using Gauss's divergence theorem in four dimensions. Since $\delta\phi_r = 0$ on Γ, this surface integral vanishes. If $\delta S(\Omega)$ is to vanish for arbitrary regions Ω and arbitrary variations $\delta\phi_r$, Eq. (2.15) leads to the Euler–Lagrange equations

$$\frac{\partial \mathcal{L}}{\partial \phi_r} - \frac{\partial}{\partial x^\alpha} \left(\frac{\partial \mathcal{L}}{\partial \phi_{r,\alpha}} \right) = 0, \quad r = 1, \ldots, N. \tag{2.16}$$

These are the equations of motion of the fields.

In order to quantize this classical theory by the canonical formalism of non-relativistic quantum mechanics, we must introduce conjugate variables. We are dealing with a system with a continuously infinite number of degrees of freedom, corresponding to the values of the fields ϕ_r, considered as functions of time, at each point of space $\mathbf{x}$. We shall again approximate the system by one having a countable number of degrees of freedom and ultimately go to the continuum limit.

Consider the system at a fixed instant of time t and decompose the three-dimensional space, i.e. the flat-space-like surface $t = \text{const.}$, into small cells of equal volume $\delta\mathbf{x}_i$, labelled by the index $i = 1, 2, \ldots$ We approximate the values of the fields within each cell by their values at, say, the centre of the cell $\mathbf{x} = \mathbf{x}_i$. The system is now described by the discrete set of generalized coordinates:

$$q_{ri}(t) \equiv \phi_r(i, t) \equiv \phi_r(\mathbf{x}_i, t), \quad r = 1, \ldots N, \quad i = 1, 2, \ldots \tag{2.17}$$

---

[2] In Eq. (2.15), and thereafter, summations over repeated indices r and α, occurring in products, is implied.

which are the values of the fields at the discrete lattice sites $\mathbf{x}_i$. If we also replace the spatial derivatives of the fields by their difference coefficients between neighbouring sites, we can write the Lagrangian of the discrete system as

$$L(t) = \sum_i \delta\mathbf{x}_i \mathscr{L}_i\Big(\phi_r(i,t), \dot{\phi}_r(i,t), \phi_r(i',t)\Big) \tag{2.18}$$

where the dot denotes differentiation with respect to time. The Lagrangian density in the ith cell, $\mathscr{L}_i$, depends on the fields at the neighbouring lattice sites i' on account of the approximation of the spatial derivatives. We define momenta conjugate to q_{ri} in the usual way as

$$p_{ri}(t) = \frac{\partial L}{\partial \dot{q}_{ri}} \equiv \frac{\partial L}{\partial \dot{\phi}_r(i,t)} \equiv \pi_r(i,t)\delta\mathbf{x}_i \tag{2.19}$$

where

$$\pi_r(i,t) \equiv \frac{\partial \mathscr{L}_i}{\partial \dot{\phi}_r(i,t)}. \tag{2.20}$$

The Hamiltonian of the discrete system is then given by

$$\begin{aligned} H &= \sum_i p_{ri}\dot{q}_{ri} - L \\ &= \sum_i \delta\mathbf{x}_i\Big\{\pi_r(i,t)\dot{\phi}_r(i,t) - \mathscr{L}_i\Big\}. \end{aligned} \tag{2.21}$$

With a view to going to the limit $\delta\mathbf{x}_i \to 0$, i.e. letting the cell size and the lattice spacing shrink to zero, we define the fields conjugate to $\phi_r(x)$ as

$$\pi_r(x) = \frac{\partial \mathscr{L}}{\partial \dot{\phi}_r}. \tag{2.22}$$

In the limit as $\delta\mathbf{x}_i \to 0, \pi_r(i,t)$ tends to $\pi_r(\mathbf{x}_i, t)$, and the discrete Lagrangian and Hamiltonian functions (2.18) and (2.21) become

$$L(t) = \int d^3\mathbf{x}\,\mathscr{L}(\phi_r, \phi_{r,\alpha}) \tag{2.23}$$

and

$$H = \int d^3\mathbf{x}\,\mathscr{H}(x), \tag{2.24}$$

where the Hamiltonian density $\mathscr{H}(x)$ is defined by

$$\mathscr{H}(x) = \pi_r(x)\dot{\phi}_r(x) - \mathscr{L}(\phi_r, \phi_{r,\alpha}), \tag{2.25}$$

and the integrations in Eqs. (2.23) and (2.24) are over all space, at time t. With our Lagrangian density, which does not depend explicitly on the time, the Hamiltonian H is, of course, constant in time. The conservation of energy will be proved in Section 2.4, where the expressions (2.24) and (2.25) for the Hamiltonian will also be re-derived.

As an example, consider the Lagrangian density

$$\mathscr{L} = \frac{1}{2}\left(\phi_{,\alpha}\phi^{,\alpha} - \mu^2\phi^2\right) \tag{2.26}$$

for a single real field $\phi(x)$, with μ a constant, which has the dimensions (length)$^{-1}$. In the next chapter we shall see that the quanta of this field are spinless neutral bosons with reduced Compton wavelength μ^{-1}, i.e. particles of mass $(\hbar\mu/c)$. The equation of motion (2.16) for this field is the Klein–Gordon equation

$$\left(\Box + \mu^2\right)\phi(x) = 0, \tag{2.27}$$

the conjugate field (2.22) is

$$\pi(x) = \frac{1}{c^2}\dot{\phi}(x) \tag{2.28}$$

and the Hamiltonian density (2.25) is

$$\mathscr{H}(x) = \frac{1}{2}\left[c^2\pi^2(x) + (\nabla\phi)^2 + \mu^2\phi^2\right]. \tag{2.29}$$

2.3   Quantized Lagrangian Field Theory

It is now easy to go from the classical to the quantum field theory by interpreting the conjugate coordinates and momenta of the discrete lattice approximation, Eqs. (2.17) and (2.19), as Heisenberg operators, and subjecting these to the usual canonical commutation relations:

$$\left.\begin{aligned}[\phi_r(j,\ t),\ \pi_s(j',\ t)] &= i\hbar\frac{\delta_{rs}\delta_{jj'}}{\delta\mathbf{x}_j} \\ [\phi_r(j,\ t),\phi_s(j',\ t)] &= [\pi_r(j,\ t),\pi_s(j',\ t)] = 0\end{aligned}\right\}. \tag{2.30}$$

If we let the lattice spacing go to zero, Eqs. (2.30) go over into the commutation relations for the fields:

$$\left.\begin{aligned}[\phi_r(\mathbf{x},\ t),\pi_s(\mathbf{x}',\ t)] &= i\hbar\,\delta_{rs}\,\delta(\mathbf{x}-\mathbf{x}') \\ [\phi_r(\mathbf{x},\ t),\phi_s(\mathbf{x}',\ t)] &= [\pi_r(\mathbf{x},\ t),\pi_s(\mathbf{x}',t)] = 0\end{aligned}\right\}, \tag{2.31}$$

since in the limit, as $\delta\mathbf{x}_j \rightarrow 0$, $\delta_{jj'}/\delta\mathbf{x}_j$ becomes the three-dimensional Dirac delta function $\delta(\mathbf{x} - \mathbf{x}')$, the points $\mathbf{x}$ and $\mathbf{x}'$ lying in the jth and j'th cell, respectively. Note that the canonical commutation relations (2.31) involve the fields at the same time; they are equal-time commutation relations. In the next chapter we shall obtain the commutators of the fields at different times.

For the Klein–Gordon field (2.26), Eqs. (2.31) reduce to the commutation relations:

$$\left.\begin{aligned}[\phi(\mathbf{x},\ t),\dot{\phi}(\mathbf{x}',\ t)] &= i\hbar c^2\delta(\mathbf{x}-\mathbf{x}') \\ [\phi(\mathbf{x},\ t),\phi(\mathbf{x}',\ t)] &= [\dot{\phi}(\mathbf{x},\ t),\dot{\phi}(\mathbf{x}',\ t)] = 0\end{aligned}\right\}. \tag{2.32}$$

In the next chapter we shall study the Klein–Gordon field in detail.

2.4   Symmetries and Conservation Laws

It follows from the Heisenberg equation of motion of an operator $O(t)$

$$i\hbar\frac{dO(t)}{dt} = [O(t), H]$$

(we are not considering operators with explicit time dependence) that O is a constant of the motion provided

$$[O, H] = 0.$$

Constants of the motion generally stem from invariance properties of systems under groups of transformations, e.g. translational and rotational invariance lead to conservation of linear and angular momentum, respectively. Such transformations lead to equivalent descriptions of the system; for example, referred to two frames of reference related by a Lorentz transformation. Quantum-mechanically, two such descriptions must be related by a unitary transformation U under which states and operators transform according to

$$|\Psi\rangle \to |\Psi'\rangle = U|\Psi\rangle, \quad O \to O' = UOU^\dagger. \tag{2.33}$$

The unitarity of the transformation ensures two things. Firstly, operator equations are covariant, i.e. have the same form whether expressed in terms of the original or the transformed operators. In particular, this will be true of the commutation relations of the fields and of the equations of motion, e.g. Maxwell's equations will be covariant with respect to Lorentz transformations. Secondly, under a unitary transformation, amplitudes, and hence observable predictions, are invariant.

If one deals with continuous transformations, the unitary operator U can be written

$$U = e^{i\alpha T} \tag{2.34}$$

where $T = T^\dagger$ and α is a real continuously variable parameter. For $\alpha = 0$, U goes over into the unit operator. For an infinitesimal transformation

$$U \approx 1 + i\delta\alpha T$$

and Eq. (2.33) becomes

$$O' = O + \delta O = (1 + i\delta\alpha T)O(1 - i\delta\alpha T)$$

i.e.

$$\delta O = i\delta\alpha[T, O]. \tag{2.35}$$

If the theory is invariant under this transformation, the Hamiltonian H will be invariant, $\delta H = 0$, and taking $O = H$ in Eq. (2.35) we obtain $[T, H] = 0$, i.e. T is a constant of the motion.

For a field theory derived from a Lagrangian density $\mathscr{L}$, one can construct conserved quantities from the invariance of $\mathscr{L}$ under symmetry transformations. We shall show that for such a theory, the invariance of $\mathscr{L}$ leads to equations of the form

$$\frac{\delta f^\alpha}{\delta x^\alpha} = 0 \tag{2.36}$$

where the f^α are functions of the field operators and their derivatives. If we define

$$F^\alpha(t) = \int d^3x f^\alpha(\mathbf{x}, t), \tag{2.37}$$

where integration is over all space, then the continuity equation (2.36) gives

$$\frac{1}{c}\frac{dF^0(t)}{dt} = -\int d^3x \sum_{j=1}^{3} \frac{\partial}{\partial x^j} f^j(\mathbf{x}, t) = 0 \tag{2.36a}$$

where the last step follows by transforming the integral into a surface integral by means of Gauss's divergence theorem and assuming (as always) that the fields, and hence the f^j, tend to zero sufficiently fast at infinity.[3] Hence

$$F^0 = \int d^3x f^0(\mathbf{x}, t) \tag{2.38}$$

is a conserved quantity. With $T = F^0$, the corresponding unitary operator is then given by Eq. (2.34).

The interpretation of the four-vector f^α follows from Eqs. (2.36)–(2.38). f^0/c and f^j are the three-dimensional volume and current densities of the conserved quantity F^0/c. Eq. (2.36a), applied to a finite three-dimensional volume V bounded by a surface S, then states that the rate of decrease of F^0/c within V equals the current of F^0/c flowing out through S. Correspondingly, the four-vector $f^\alpha(x)$, satisfying the conservation equation (2.36), is called a conserved current. (Strictly speaking, one should call it a four-current density.) The result, that the invariance of the Lagrangian density $\mathcal{L}$ under a continuous one-parameter set of transformations implies a conserved quantity, is known as Noether's theorem.

We apply these ideas to the transformation

$$\phi_r(x) \to \phi'_r(x) = \phi_r(x) + \delta\phi_r(x) \tag{2.39}$$

of the fields. The change induced in $\mathcal{L}$ is given by

$$\delta\mathcal{L} = \frac{\partial\mathcal{L}}{\partial\phi_r}\delta\phi_r + \frac{\partial\mathcal{L}}{\partial\phi_{r,\alpha}}\delta\phi_{r,\alpha} = \frac{\partial}{\partial x^\alpha}\left(\frac{\partial\mathcal{L}}{\partial\phi_{r,\alpha}}\delta\phi_r\right),$$

where the last step follows since $\phi_r(x)$ satisfies the field equations (2.16), and summations over repeated indices r and α are implied as previously. If $\mathcal{L}$ is invariant under the transformation (2.39) so that $\delta\mathcal{L} = 0$, the last equation reduces to the continuity equation (2.36) with

$$f^\alpha = \frac{\partial\mathcal{L}}{\partial\phi_{r,\alpha}}\delta\phi_r$$

---

[3] If one employs a finite normalization volume for the system, as we did in the last chapter, the surface integral vanishes on account of the periodic boundary conditions.

and the constant of the motion, from Eqs. (2.38) and (2.22), is

$$F^0 = c \int d^3\mathbf{x}\, \pi_r(x)\delta\phi_r(x). \tag{2.40}$$

An important particular case of the above arises for complex fields ϕ_r, i.e. non-Hermitian operators in the quantized theory. ϕ_r and $\phi_r^\dagger$ are then treated as independent fields, as discussed earlier. Suppose $\mathscr{L}$ is invariant under the transformation

$$\left.\begin{array}{l} \phi_r \to \phi_r' = e^{i\varepsilon}\phi_r \approx (1+i\varepsilon)\phi_r \\ \phi_r^\dagger \to \phi_r^{\dagger\,\prime} = e^{-i\varepsilon}\phi_r^\dagger \approx (1-i\varepsilon)\phi_r^\dagger \end{array}\right\} \tag{2.41}$$

where ε is a real parameter, and the right-hand expressions result for very small ε. From Eqs. (2.41)

$$\delta\phi_r = i\varepsilon\phi_r, \quad \delta\phi_r^\dagger = -i\varepsilon\phi_r^\dagger,$$

and Eq. (2.40) becomes

$$F^0 = i\varepsilon c \int d^3\mathbf{x}\left[\pi_r(x)\phi_r(x) - \pi_r^\dagger(x)\phi_r^\dagger(x)\right].$$

Since F^0 multiplied by any constant is also conserved, we shall, instead of F^0, consider

$$Q = -\frac{iq}{\hbar} \int d^3\mathbf{x}\left[\pi_r(x)\phi_r(x) - \pi_r^\dagger(x)\phi_r^\dagger(x)\right], \tag{2.42}$$

where q is a constant to be determined later. The reason for this change is that $\pm q$ will turn out to be the electric charges of the particles represented by the fields.

We evaluate the commutator $[Q, \phi_r(x)]$. Since ϕ_r and $\phi_r^\dagger$ are independent fields, ϕ_r commutes with all fields except π_r [see Eq. (2.31)]. Hence taking $(x')^0 = x^0 = ct$,

$$[Q, \phi_r(x)] = -\frac{iq}{\hbar} \int d^3\mathbf{x}'\,[\pi_s(x'), \phi_r(x)]\phi_s(x'),$$

and using the commutation relations (2.31) one obtains

$$[Q, \phi_r(x)] = -q\phi_r(x). \tag{2.43}$$

From this relation one easily verifies that if $|Q'\rangle$ is an eigenstate of Q with the eigenvalue Q', then $\phi_r(x)|Q'\rangle$ is also an eigenstate of Q belonging to the eigenvalue $(Q' - q)$, and correspondingly $\phi_r^\dagger(x)|Q'\rangle$ belongs to $(Q' + q)$. In the next chapter we shall see that, consistent with these results, ϕ_r and $\phi_r^\dagger$ are linear in creation and absorption operators, with ϕ_r absorbing particles of charge $(+q)$ or creating particles of charge $(-q)$, while $\phi_r^\dagger$ absorbs particles of charge $(-q)$ or creates particles of charge $(+q)$. Hence, we interpret the operator Q, Eq. (2.42), as the charge operator. We have therefore shown that charge is conserved $(dQ/dt = 0, [Q, H] = 0)$, provided the Lagrangian density $\mathscr{L}$ is invariant with respect to the transformation (2.41), which is known as a global phase transformation [corresponding to the fact that the phase ε in Eq. (2.41) is independent of x] or as a gauge transformation of the first kind. We see from Eq. (2.42) that we require complex, i.e. non-Hermitian, fields to represent particles with charge. Real, i.e. Hermitian, fields represent uncharged particles. Interpreting Eq. (2.42) as an operator involves the usual ambiguity as to the order of factors. We shall have to choose these so that for the vacuum state $|0\rangle$, in which no particles are present, $Q|0\rangle=0$. We shall return to this point in the next chapter.

The unitary transformation corresponding to the phase transformation (2.41) can, from Eq. (2.34), be written

$$U = e^{i\alpha Q}. \tag{2.44}$$

Hence for infinitesimally small α, we obtain from Eq. (2.33)

$$\begin{aligned} \phi'_r &= e^{i\alpha Q}\phi_r e^{-i\alpha Q} \\ &= \phi_r + i\alpha[Q, \phi_r] = (1 - i\alpha q)\phi_r \end{aligned} \tag{2.45}$$

where the last line follows from Eq. (2.43). Comparing Eqs. (2.45) and (2.41), we see that they are consistent if we take $\varepsilon = -\alpha q$.

Although we have talked of electric charge, with which one is familiar, this analysis applies equally to other types of charge, such as hypercharge.

Conservation of energy and momentum and of angular momentum follows from the invariance of the Lagrangian density $\mathscr{L}$ under translations and rotations. Since these transformations form a continuous group we need only consider infinitesimal transformations. Any finite transformation can be built up through repeated infinitesimal transformations. In four dimensions these are given by

$$x_\alpha \rightarrow x'_\alpha \equiv x_\alpha + \delta x_\alpha = x_\alpha + \varepsilon_{\alpha\beta}x^\beta + \delta_\alpha, \tag{2.46}$$

where δ_α is an infinitesimal displacement and $\varepsilon_{\alpha\beta}$ is an infinitesimal antisymmetric tensor, $\varepsilon_{\alpha\beta} = -\varepsilon_{\beta\alpha}$, to ensure the invariance of $x_\alpha x^\alpha$ under homogeneous Lorentz transformations ($\delta_\alpha = 0$).

The transformation (2.46) will induce a transformation in the fields which we assume to be

$$\phi_r(x) \rightarrow \phi'_r(x') = \phi_r(x) + \frac{1}{2}\varepsilon_{\alpha\beta}S^{\alpha\beta}_{rs}\phi_s(x). \tag{2.47}$$

In this section, summation over repeated indices r, s, labelling fields, as well as over Lorentz indices α, β, is implied. Here x and x' label the *same* point in space–time referred to the two frames of reference, and ϕ_r and ϕ'_r are the field components referred to these two coordinate systems. The coefficients $S^{\alpha\beta}_{rs}$ in Eq. (2.47) are antisymmetric in α and β, like $\varepsilon_{\alpha\beta}$, and are determined by the transformation properties of the fields. For example, for the vector potential $A_\alpha(x)$, Eq. (2.47) reduces to the transformation law of a vector.

Invariance under the transformations (2.46) and (2.47) means that the Lagrangian density expressed in terms of the new coordinates and fields has the same functional form as when expressed in the original coordinates and fields:

$$\mathscr{L}(\phi_r(x), \phi_{r,\alpha}(x)) = \mathscr{L}(\phi'_r(x'), \phi'_{r,\alpha}(x')). \tag{2.48}$$

(Here $\phi'_{r,\alpha}(x') \equiv \partial\phi'_r(x')/\partial x'^\alpha$.) From Eq. (2.48) the covariance of the field equations, etc. follows; i.e. they will have the same form expressed in terms of either the original or the transformed coordinates and fields.

The conservation laws follow by expressing the right-hand side of Eq. (2.48) in terms of the original coordinates and fields by means of Eqs. (2.46) and (2.47). We shall first state and discuss these results, postponing their derivation to the end of this section.

For a translation (i.e. $\varepsilon_{\alpha\beta} = 0$), one obtains the four continuity equations

$$\frac{\partial \mathcal{T}^{\alpha\beta}}{\partial x^\alpha} = 0, \tag{2.49}$$

where

$$\mathcal{T}^{\alpha\beta} \equiv \frac{\partial \mathcal{L}}{\partial \phi_{r,\alpha}} \frac{\partial \phi_r}{\partial x_\beta} - \mathcal{L} g^{\alpha\beta}, \tag{2.50}$$

and the four conserved quantities are

$$cP^\alpha \equiv \int d^3\mathbf{x}\, \mathcal{T}^{0\alpha} = \int d^3\mathbf{x} \left\{ c\pi_r(x) \frac{\partial \phi_r(x)}{\partial x_\alpha} - \mathcal{L} g^{0\alpha} \right\}. \tag{2.51}$$

P^α is the energy-momentum four-vector, with

$$\left. \begin{aligned} cP^0 &= \int d^3\mathbf{x} \left\{ \pi_r(x)\dot{\phi}_r(x) - \mathcal{L}(\phi_r, \phi_{r,\alpha}) \right\} \\ &= \int d^3\mathbf{x}\, \mathcal{H} = H \end{aligned} \right\} \tag{2.51a}$$

being the Hamiltonian, Eqs. (2.24) and (2.25), and

$$P^j = \int d^3\mathbf{x}\, \pi_r(x) \frac{\partial \phi_r(x)}{\partial x_j} \tag{2.51b}$$

being the momentum components of the fields. This interpretation will be confirmed when we come to express these operators in the number representation. Correspondingly $\mathcal{T}^{\alpha\beta}$ is called the energy-momentum tensor.

For a rotation (i.e. $\delta_\alpha = 0$) Eqs. (2.46)–(2.48) give the continuity equations

$$\frac{\partial \mathcal{M}^{\alpha\beta\gamma}}{\partial x^\alpha} = 0, \tag{2.52}$$

where

$$\mathcal{M}^{\alpha\beta\gamma} \equiv \frac{\partial \mathcal{L}}{\partial \phi_{r,\alpha}} S_{rs}^{\beta\gamma} \phi_s(x) + \left[x^\beta \mathcal{T}^{\alpha\gamma} - x^\gamma \mathcal{T}^{\alpha\beta} \right], \tag{2.53}$$

and the six conserved quantities (note that $\mathcal{M}^{\alpha\beta\gamma} = - \mathcal{M}^{\alpha\gamma\beta}$) are

$$\begin{aligned} cM^{\alpha\beta} &= \int d^3\mathbf{x}\, \mathcal{M}^{0\alpha\beta} \\ &= \int d^3\mathbf{x} \left\{ \left[x^\alpha \mathcal{T}^{0\beta} - x^\beta \mathcal{T}^{0\alpha} \right] + c\pi_r(x) S_{rs}^{\alpha\beta} \phi_s(x) \right\}. \end{aligned} \tag{2.54}$$

For two space-like indices ($i, j = 1, 2, 3$), M^{ij} is the angular momentum operator of the field (M^{12} being the z-component, etc.). Remembering that $\mathcal{T}^{0i}/c$ is the momentum density of the field [see Eqs. (2.51)], we interpret the term in square brackets in Eq. (2.54) as the orbital angular momentum and the last term as the intrinsic spin angular momentum.

We return to the derivation of the continuity equations (2.49) and (2.52).[4] The variation of a function $\phi_r(x)$ with the argument unchanged was defined in Eq. (2.39) as

$$\delta\phi_r(x) \equiv \phi'_r(x) - \phi_r(x). \tag{2.55a}$$

In addition, we now define the variation

$$\delta_{\mathrm{T}}\phi(x) \equiv \phi'_r(x') - \phi_r(x'). \tag{2.55b}$$

which results from changes of both the form and the argument of the function. We can then write

$$\delta_{\mathrm{T}}\phi_r(x) = \left[\phi'_r(x') - \phi_r(x')\right] + \left[\phi_r(x') - \phi_r(x)\right]$$
$$= \delta\phi_r(x') + \frac{\partial\phi_r}{\partial x_\beta}\delta x_\beta, \tag{2.56}$$

where δx_β is given by Eq. (2.46). To first order in small quantities this can be written

$$\delta_{\mathrm{T}}\phi_r(x) = \delta\phi_r(x) + \frac{\partial\phi_r}{\partial x_\beta}\delta x_\beta. \tag{2.57}$$

We can similarly write Eq. (2.48) as

$$0 = \mathscr{L}\left(\phi'_r(x'), \phi_{r,\alpha}(x')\right) - \mathscr{L}\left(\phi_r(x), \phi'_r, \alpha(x)\right)$$
$$= \delta\mathscr{L} + \frac{\partial\mathscr{L}}{\partial x^\alpha}\delta x^\alpha. \tag{2.58}$$

For $\delta\mathscr{L}$ we obtain, since $\phi_r(x)$ satisfies the field equations (2.16),

$$\delta\mathscr{L} = \frac{\partial\mathscr{L}}{\partial\phi_r}\delta\phi_r + \frac{\partial\mathscr{L}}{\partial\phi_{r,\alpha}}\delta\phi_{r,\alpha}$$
$$= \frac{\partial}{\partial x^\alpha}\left\{\frac{\partial\mathscr{L}}{\partial\phi_{r,\alpha}}\delta\phi_r\right\} = \frac{\partial}{\partial x^\alpha}\left\{\frac{\partial\mathscr{L}}{\partial\phi_{r,\alpha}}\left[\delta_{\mathrm{T}}\phi_r - \frac{\partial\phi_r}{\partial x_\beta}\delta x_\beta\right]\right\}. \tag{2.59}$$

We combine Eqs. (2.58) and (2.59) to obtain the continuity equations

$$\frac{\partial f^\alpha}{\partial x^\alpha} = 0, \tag{2.60}$$

where

$$f^\alpha \equiv \frac{\partial\mathscr{L}}{\partial\phi_{r,\alpha}}\delta_{\mathrm{T}}\phi_r - \mathscr{T}^{\alpha\beta}\delta x_\beta \tag{2.61}$$

with $\mathscr{T}^{\alpha\beta}$ given by Eq. (2.50).

We first consider translations, i.e. $\varepsilon_{\alpha\beta}=0$, so that from Eqs. (2.46) and (2.47) $\delta x_\beta=\delta_\beta$ and $\delta_{\mathrm{T}}\phi_r=0$. Eq. (2.61) reduces to $f^\alpha=-\mathscr{T}^{\alpha\beta}\,\delta x_\beta$, and, since the four displacements δ_β

---

[4] The reader may omit the rest of this section as the details of this derivation will not be required later on.

are independent of each other, Eq. (2.60) reduces to the four continuity equations (2.49) for energy and momentum conservation.

Finally, we consider rotations, i.e. $\delta_\alpha = 0$. From Eqs. (2.46) and (2.47) for δx_β and $\delta_T \phi_r$, and the antisymmetry of $\varepsilon_{\alpha\beta}$, Eq. (2.61) becomes

$$f^\alpha = \frac{1}{2} \varepsilon_{\beta\gamma} \mathcal{M}^{\alpha\beta\gamma} \tag{2.62}$$

where $\mathcal{M}^{\alpha\beta\gamma}$ is the tensor defined in Eq. (2.53). Since the rotations $\varepsilon_{\beta\gamma}$ are independent of each other, Eq. (2.60) reduces to the continuity equations (2.52).

Problems

2.1. Show that replacing the Lagrangian density $\mathcal{L} = \mathcal{L}(\phi_r, \phi_{r,\alpha})$ by

$$\mathcal{L}' = \mathcal{L} + \partial_\alpha \Lambda^\alpha(x),$$

where $\Lambda^\alpha(x)$, $\alpha = 0, \dots 3$ are arbitrary functions of the fields $\phi_r(x)$, does not alter the equations of motion.

2.2. The real Klein–Gordon field is described by the Hamiltonian density (2.29). Use the commutation relations (2.31) to show that

$$[H, \phi(x)] = -i\hbar c^2 \pi(x), \quad [H, \pi(x)] = i\hbar(\mu^2 - \nabla^2)\phi(x),$$

where H is the Hamiltonian of the field.

From this result and the Heisenberg equations of motion for the operators $\phi(x)$ and $\pi(x)$, show that

$$\dot\phi(x) = c^2 \pi(x), \quad (\Box + \mu^2)\phi(x) = 0.$$

2.3. Show that the Lagrangian density

$$\mathcal{L} = -\frac{1}{2}[\partial_\alpha \phi_\beta(x)][\partial^\alpha \phi^\beta(x)] + \frac{1}{2}[\partial_\alpha \phi^\alpha(x)][\partial_\beta \phi^\beta(x)] + \frac{\mu^2}{2}\phi_\alpha(x)\phi^\alpha(x)$$

for the real vector field $\phi^\alpha(x)$ leads to the field equations

$$[g_{\alpha\beta}(\Box + \mu^2) - \partial_\alpha \partial_\beta]\phi^\beta(x) = 0,$$

and that the field $\phi^\alpha(x)$ satisfies the Lorentz condition

$$\partial_\alpha \phi^\alpha(x) = 0.$$

2.4. Use the commutation relations (2.31) to show that the momentum operator of the fields

$$P^j = \int d^3x \pi_r(x) \frac{\partial \phi_r(x)}{\partial x_j} \tag{2.51b}$$

satisfies the equations

$$[P^j, \phi_r(x)] = -i\hbar \frac{\partial \phi_r(x)}{\partial x_j}, \quad [P^j, \pi_r(x)] = -i\hbar \frac{\partial \pi_r(x)}{\partial x_j}.$$

Hence show that any operator $F(x) = F(\phi_r(x), \pi_r(x))$, which can be expanded in a power series in the field operators $\phi_r(x)$ and $\pi_r(x)$, satisfies

$$\left[P^j, F(x) \right] = -i\hbar \frac{\partial F(x)}{\partial x_j} .$$

Note that we can combine these equations with the Heisenberg equation of motion for the operator $F(x)$

$$\left[H, \ F(x) \right] = -i\hbar c \frac{\partial F(x)}{\partial x_0}$$

to obtain the covariant equations of motion

$$\left[P^\alpha, F(x) \right] = -i\hbar \frac{\partial F(x)}{\partial x_\alpha} ,$$

where $P^0 = H/c$.

2.5. Under a translation of coordinates

$$x_\alpha \rightarrow x'_\alpha = x_\alpha + \delta_\alpha \quad (\delta_\alpha = \text{a constant four-vector})$$

a scalar field $\phi(x_\alpha)$ remains invariant:

$$\phi'(x'_\alpha) = \phi(x_\alpha), \quad \text{i.e.} \phi'(x_\alpha) = \phi(x_\alpha - \delta_\alpha).$$

Show that the corresponding unitary transformation

$$\phi(x) \rightarrow \phi'(x) = U\phi(x)U^\dagger$$

is given by $U = \exp[-i\delta_\alpha P^\alpha/\hbar]$, where P^α is the energy-momentum four-vector of the field, Eqs. (2.51). (You may find the results of the previous problem helpful.)

3

The Klein–Gordon Field

In Chapter 1 we quantized the electromagnetic field by Fourier analysing it and imposing harmonic oscillator commutation relations on the Fourier expansion coefficients. This approach naturally led to photons. In the last chapter, a different procedure, the canonical quantization formalism, led directly to quantized field operators. We shall now Fourier analyse these field operators and we shall see that the Fourier coefficients, which are now also operators, satisfy the same commutation relations as the absorption and creation operators of the number representation. In this way the interpretation in terms of field quanta is regained.

In this chapter we shall consider relativistic material particles of spin 0. Photons, which are much more complicated on account of their transverse polarization will be treated in Chapter 5.

3.1   The Real Klein–Gordon Field

For particles of rest mass m, energy and momentum are related by

$$E^2 = m^2 c^4 + c^2 \mathbf{p}^2. \tag{3.1}$$

If the particles can be described by a single scalar wavefunction $\phi(x)$, the prescription of non-relativistic quantum mechanics

$$\mathbf{p} \rightarrow -i\hbar\nabla, \quad E \rightarrow i\hbar \, \partial/\partial t \tag{3.2}$$

leads to the Klein–Gordon equation (2.27):

$$\left(\Box + \mu^2\right)\phi(x) = 0, \tag{3.3}$$

Quantum Field Theory, Second Edition   Franz Mandl and Graham Shaw
© 2010 John Wiley & Sons, Ltd

where $(\mu \equiv mc/\hbar)$. The interpretation of Eq. (3.3) as a single-particle equation leads to difficulties. These are related to defining a positive-definite particle density and to the two signs of the energy E which result from Eq. (3.1). We shall not discuss these difficulties, but only mention that they are typical of relativistic single-particle equations. We shall see that such difficulties do not occur in the many-particle theories which result when the fields, such as the Klein–Gordon field $\phi(x)$, are quantized.[1]

We know from Eq. (2.54), for the angular momentum of the field, that a single scalar field possesses orbital, but no spin, angular momentum, i.e. it represents particles of spin 0. Hence, the Klein–Gordon equation affords the appropriate description of π-mesons (pions) and K-mesons, both of which have spin 0.

We shall now consider a real scalar field $\phi(x)$, satisfying the Klein–Gordon equation (3.3). Such a field corresponds to electrically neutral particles. Charged particles, described by a complex field, will be dealt with in the next section.

We know from Section 2.2 that the Klein–Gordon equation (3.3) can be derived from the Lagrangian density

$$\mathcal{L} = \frac{1}{2}\left(\phi_{,\alpha}\phi^{,\alpha} - \mu^2\phi^2\right), \tag{3.4}$$

and that the field conjugate to ϕ is

$$\pi(x) = \frac{\partial\mathcal{L}}{\partial\dot{\phi}} = \frac{1}{c^2}\dot{\phi}(x). \tag{3.5}$$

On quantization, the real field ϕ becomes a Hermitian operator, $\phi^\dagger = \phi$, satisfying the equal-time commutation relations (2.32):

$$\left.\begin{array}{l} \left[\phi(\mathbf{x},\,t), \dot{\phi}\left(\mathbf{x}',\,t\right)\right] = i\hbar c^2 \delta\left(\mathbf{x} - \mathbf{x}'\right) \\ \left[\phi(\mathbf{x},\,t), \phi\left(\mathbf{x}',\,t\right)\right] = \left[\dot{\phi}(\mathbf{x},\,t), \dot{\phi}\left(\mathbf{x}',\,t\right)\right] = 0 \end{array}\right\}. \tag{3.6}$$

To establish contact with particles, we expand $\phi(x)$ in a complete set of solutions of the Klein–Gordon equation:

$$\phi(x) = \phi^+(x) + \phi^-(x) \tag{3.7a}$$

where

$$\phi^+(x) = \sum_{\mathbf{k}}\left(\frac{\hbar c^2}{2V\omega_{\mathbf{k}}}\right)^{1/2} a(\mathbf{k})e^{-ikx} \tag{3.7b}$$

and

$$\phi^-(x) = \sum_{\mathbf{k}}\left(\frac{\hbar c^2}{2V\omega_{\mathbf{k}}}\right)^{1/2} a^\dagger(\mathbf{k})e^{ikx}. \tag{3.7c}$$

---

[1] This is often referred to as second quantization, in contrast to the derivation of the single-particle wave equations by means of the substitution (3.2).

These equations are analogous to Eqs. (1.38) for the photon field. The summations are again over the wave vectors $\mathbf{k}$ allowed by the periodic boundary conditions, but k^0 and $\omega_{\mathbf{k}}$ are now given by

$$k^0 = \frac{1}{c}\omega_{\mathbf{k}} = +\left(\mu^2 + \mathbf{k}^2\right)^{1/2}, \tag{3.8a}$$

i.e. k is the wave four-vector of a particle of mass $m = \mu\hbar/c$, momentum $\hbar\mathbf{k}$ and energy

$$E = \hbar\omega_{\mathbf{k}} = +\left[m^2c^4 + c^2(\hbar\mathbf{k})^2\right]^{1/2}. \tag{3.8b}$$

The fact that each operator $a(\mathbf{k})$ occurs paired with its adjoint $a^\dagger(\mathbf{k})$ in Eqs. (3.7) ensures that ϕ is Hermitian.

From Eqs. (3.7) and the commutation relations (3.6), one easily obtains the commutation relations for the operators $a(\mathbf{k})$ and $a^\dagger(\mathbf{k})$. We shall leave the details for the reader (see Problem 3.1) and only quote the important result:

$$\left.\begin{array}{c} \left[a(\mathbf{k}), a^\dagger(\mathbf{k}')\right] = \delta_{\mathbf{k}\mathbf{k}'} \\ \left[a(\mathbf{k}), a(\mathbf{k}')\right] = \left[a^\dagger(\mathbf{k}), a^\dagger(\mathbf{k}')\right] = 0 \end{array}\right\}. \tag{3.9}$$

These are precisely the harmonic oscillator commutation relations [Eqs. (1.19) and (1.29)] and all the results can at once be taken over from Section 1.2. In particular, the operators

$$N(\mathbf{k}) = a^\dagger(\mathbf{k})a(\mathbf{k}) \tag{3.10}$$

have as their eigenvalues the occupation numbers

$$n(\mathbf{k}) = 0, 1, 2, \ldots, \tag{3.11}$$

and, correspondingly, $a(\mathbf{k})$ and $a^\dagger(\mathbf{k})$ are the annihilation and creation operators of particles with momentum $\hbar\mathbf{k}$ and energy $\hbar\omega_{\mathbf{k}}$, given by Eq. (3.8).

The Hamiltonian and momentum operators of the Klein–Gordon field are, from Eqs. (2.51), (3.4) and (3.5), given by

$$H = \int d^3x \frac{1}{2}\left[\frac{1}{c^2}\dot{\phi}^2 + (\nabla\phi)^2 + \mu^2\phi^2\right] \tag{3.12}$$

and

$$\mathbf{P} = -\int d^3x \frac{1}{c^2}\dot{\phi}\,\nabla\phi. \tag{3.13}$$

Substituting the expansion (3.7) in the last two equations one obtains

$$H = \sum_{\mathbf{k}} \hbar\omega_{\mathbf{k}}\left(a^\dagger(\mathbf{k})a(\mathbf{k}) + \frac{1}{2}\right), \tag{3.14}$$

$$\mathbf{P} = \sum_{\mathbf{k}} \hbar\mathbf{k}\left(a^\dagger(\mathbf{k})a(\mathbf{k}) + \frac{1}{2}\right), \tag{3.15}$$

confirming our interpretation of $[a^\dagger(\mathbf{k})a(\mathbf{k})]$ as the number operator for particles with wave vector $\mathbf{k}$. From the last two equations, one also sees directly that the momentum $\mathbf{P}$ is a constant of the motion for the free Klein–Gordon field. (Nevertheless we prefer the discussion of Section 2.4, because it reveals the fundamental and general connection between symmetries and conservation laws.)

From Eq. (3.14) we see that the state of lowest energy, the ground state, of the Klein–Gordon field is the vacuum state $|0\rangle$, in which no particles are present (all $n(\mathbf{k}) = 0$). We can also characterize this state by

$$a(\mathbf{k})|0\rangle = 0, \quad \text{all } \mathbf{k}, \tag{3.16a}$$

or, expressed in terms of the field operators (3.7), by

$$\phi^+(x)|0\rangle = 0, \quad \text{all } x. \tag{3.16b}$$

The vacuum has the infinite energy $\frac{1}{2}\sum_{\mathbf{k}}\hbar\omega_{\mathbf{k}}$. As discussed for the radiation field, only energy differences are observable. Hence, this infinite constant is harmless and easily removed by measuring all energies relative to the vacuum state.

One can avoid the explicit occurrence of such infinite constants by normal ordering of operators. In a *normal product*, all absorption operators stand to the right of all creation operators in each product of operators. Denoting the normal product by $N(\ldots)$ we have, for example,

$$N\big(a(\mathbf{k}_1)a(\mathbf{k}_2)a^\dagger(\mathbf{k}_3)\big) = a^\dagger(\mathbf{k}_3)a(\mathbf{k}_1)a(\mathbf{k}_2), \tag{3.17}$$

and

$$\begin{aligned}
N[\phi(x)\phi(y)] &= N[(\phi^+(x) + \phi^-(x))(\phi^+(y) + \phi^-(y))] \\
&= N[\phi^+(x)\phi^+(y)] + N[\phi^+(x)\phi^-(y)] \\
&\quad + N[\phi^-(x)\phi^+(y)] + N[\phi^-(x)\phi^-(y)] \\
&= \phi^+(x)\phi^+(y) + \phi^-(y)\phi^+(x) \\
&\quad + \phi^-(x)\phi^+(y) + \phi^-(x)\phi^-(y),
\end{aligned} \tag{3.18}$$

where the order of the factors has been interchanged in the second term, i.e. all positive frequency parts ϕ^+ (which contain only absorption operators) stand to the right of all negative frequency parts ϕ^- (which contain only creation operators).[2] Normal ordering does not fix the order of absorption or creation operators, each amongst themselves, but since each of these commute amongst themselves, such different ways of writing a normal product are equal; for example, expression (3.17) also equals $a^\dagger(\mathbf{k}_3)a(\mathbf{k}_2)a(\mathbf{k}_1)$. Hence in arranging a product of operators in normal order, one simply treats them as though all commutators vanish.

Clearly, the vacuum expectation value of any normal product vanishes. We redefine the Lagrangian density $\mathscr{L}$ and all observables, such as the energy–momentum or angular momentum of the field, or their densities, as normal products. We are free to do this, as it merely corresponds to a particular order of factors before quantization. With observables

---

[2] This definition of the normal product will be modified when fermions are introduced. Another notation commonly used for the normal product $N(AB\ldots L)$ is $: AB\ldots L:$.

defined as normal products, their vacuum expectation values vanish. In particular, Eqs. (3.14) and (3.15) become

$$P^\alpha = (H/c, \mathbf{P}) = \sum_{\mathbf{k}} \hbar k^\alpha a^\dagger(\mathbf{k}) a(\mathbf{k}). \qquad (3.19)$$

From the vacuum state $|0\rangle$, one constructs particle states in the same way as was done for photons in Section 1.2. For example, one-particle states are linear superpositions of

$$a^\dagger(\mathbf{k})|0\rangle, \quad \text{all } \mathbf{k}; \qquad (3.20a)$$

two-particle states are linear superpositions of

$$a^\dagger(\mathbf{k}) a^\dagger(\mathbf{k}')|0\rangle, \quad \text{all } \mathbf{k} \text{ and } \mathbf{k}' \neq \mathbf{k}, \qquad (3.20b)$$

and

$$\frac{1}{\sqrt{2}} \left[a^\dagger(\mathbf{k}) \right]^2 |0\rangle, \quad \text{all } \mathbf{k}, \qquad (3.20c)$$

and so on. With the vacuum state normalized, i.e. $\langle 0|0 \rangle = 1$, the states (3.20) are also normalized. That is the purpose of the factor $1/\sqrt{2}$ in Eq. (3.20c). Similar factors occur for more than two particles.

The particles of the Klein–Gordon field are bosons; the occupation numbers can take on any value $n(\mathbf{k}) = 0, 1, 2, \ldots$ Eq. (3.20b) illustrates another aspect of boson states: they are symmetric under interchange of particle labels. Since all creation operators commute with each other, we have

$$a^\dagger(\mathbf{k}) a^\dagger(\mathbf{k}')|0\rangle = a^\dagger(\mathbf{k}') a^\dagger(\mathbf{k})|0\rangle. \qquad (3.21)$$

3.2 The Complex Klein–Gordon Field

We shall now extend the treatment of the last section to the complex Klein–Gordon field. The new feature this introduces is, as we know from Section 2.4, that we can associate a conserved charge with the field. For the real field this was not possible. We shall concentrate on this aspect of a conserved charge. In other respects, the real and complex fields are very similar, and we shall only quote the main results leaving verification to the reader.

For the complex Klein–Gordon field, the Lagrangian density (3.4) is replaced by

$$\mathscr{L} = \mathrm{N}\left(\phi^\dagger_{,\alpha} \phi^{,\alpha} - \mu^2 \phi^\dagger \phi \right), \qquad (3.22)$$

where we have at once written the quantized operator as a normal product. The field ϕ and its adjoint $\phi^\dagger$, treated as independent fields, lead to the Klein–Gordon equations

$$(\Box + \mu^2)\phi(x) = 0, \quad (\Box + \mu^2)\phi^\dagger(x) = 0. \qquad (3.23)$$

The fields conjugate to ϕ and $\phi^\dagger$ are

$$\pi(x) = \frac{1}{c^2} \dot{\phi}^\dagger(x), \quad \pi^\dagger(x) = \frac{1}{c^2} \dot{\phi}(x), \qquad (3.24)$$

and the equal-time commutation relations (2.31) become

$$
\left.\begin{aligned}
\left[\phi(\mathbf{x},\,t),\dot\phi^{\,\dagger}(\mathbf{x}',\,t)\right] &= i\hbar c^2 \delta(\mathbf{x}-\mathbf{x}') \\
\left[\phi(\mathbf{x},\,t),\,\phi(\mathbf{x}',\,t)\right] &= \left[\phi(\mathbf{x},\,t),\,\phi^\dagger(\mathbf{x}',\,t)\right] = \left[\dot\phi(\mathbf{x},\,t),\,\dot\phi(\mathbf{x}',\,t)\right] \\
&= \left[\dot\phi(\mathbf{x},\,t),\dot\phi^{\,\dagger}(\mathbf{x}',\,t)\right] = \left[\phi(\mathbf{x},\,t),\,\dot\phi(\mathbf{x}',\,t)\right] = 0
\end{aligned}\right\}.
\tag{3.25}
$$

Analogously to Eqs. (3.7), we write the Fourier expansions of the fields as

$$
\phi(x) = \phi^+(x) + \phi^-(x) = \sum_{\mathbf{k}} \left(\frac{\hbar c^2}{2V\omega_{\mathbf{k}}}\right)^{1/2}\left[a(\mathbf{k})e^{-ikx} + b^\dagger(\mathbf{k})e^{ikx}\right]
\tag{3.26a}
$$

and

$$
\phi^\dagger(x) = \phi^{\dagger+}(x) + \phi^{\dagger-}(x) = \sum_{\mathbf{k}} \left(\frac{\hbar c^2}{2V\omega_{\mathbf{k}}}\right)^{1/2}\left[b(\mathbf{k})e^{-ikx} + a^\dagger(\mathbf{k})e^{ikx}\right].
\tag{3.26b}
$$

($\phi^{\dagger+}$ and $\phi^{\dagger-}$ are the positive and negative frequency parts of $\phi^\dagger$.)

From Eqs. (3.25) and (3.26a), one obtains the commutation relations

$$
\left[a(\mathbf{k}), a^\dagger(\mathbf{k}')\right] = \left[b(\mathbf{k}), b^\dagger(\mathbf{k}')\right] = \delta_{\mathbf{k}\mathbf{k}'},
\tag{3.27a}
$$

and the commutator of any other pair of operators vanishes, i.e.

$$
\left[a(\mathbf{k}), a(\mathbf{k}')\right] = \left[b(\mathbf{k}), b(\mathbf{k}')\right] = \left[a(\mathbf{k}), b(\mathbf{k}')\right] = \left[a^\dagger(\mathbf{k}), b(\mathbf{k}')\right] = 0.
\tag{3.27b}
$$

From the commutation relations (3.27), it follows that we can interpret $a(\mathbf{k})$ and $a^\dagger(\mathbf{k})$, and $b(\mathbf{k})$ and $b^\dagger(\mathbf{k})$, as absorption and creation operators of two types of particles – we shall call them a-particles and b-particles – and

$$
N_a(\mathbf{k}) = a^\dagger(\mathbf{k})a(\mathbf{k}), \qquad N_b(\mathbf{k}) = b^\dagger(\mathbf{k})b(\mathbf{k}),
\tag{3.28}
$$

as the corresponding number operators with eigenvalues $0, 1, 2, \ldots$. Hence a number representation can be set up as before, with states containing a- and b-particles generated by means of the creation operators $a^\dagger$ and $b^\dagger$ from the vacuum state $|0\rangle$, which is now defined by

$$
a(\mathbf{k})|0\rangle = b(\mathbf{k})|0\rangle = 0, \qquad \text{all } \mathbf{k},
\tag{3.29a}
$$

or equivalently by

$$
\phi^+(x)|0\rangle = \phi^{\dagger+}(x)|0\rangle = 0, \qquad \text{all } x.
\tag{3.29b}
$$

Expressed in terms of the absorption and creation operators, the energy-momentum operator (2.51) of the complex Klein–Gordon field assumes the form we expect

$$
P^\alpha = (H/c, \mathbf{P}) = \sum_{\mathbf{k}} \hbar k^\alpha (N_a(\mathbf{k}) + N_b(\mathbf{k})).
\tag{3.30}
$$

We now turn to the charge. From the invariance of the Lagrangian density (3.22) under the phase transformation (2.41) follows the conservation of charge Q, Eq. (2.42), which now takes the form

$$Q = \frac{-iq}{\hbar c^2} \int d^3\mathbf{x} N \left[\phi^\dagger(x)\phi(x) - \phi(x)\phi^\dagger(x) \right]. \tag{3.31}$$

The corresponding charge–current density is given by

$$s^\alpha(x) = (c\rho(x), \mathbf{j}(x)) = \frac{-iq}{\hbar} N \left[\frac{\partial \phi^\dagger}{\partial x_\alpha} \phi - \frac{\partial \phi}{\partial x_\alpha} \phi^\dagger \right], \tag{3.32}$$

which obviously satisfies the continuity equation

$$\frac{\partial s^\alpha(x)}{\partial x^\alpha} = 0. \tag{3.33}$$

Expressed in terms of creation and absorption operators, Eq. (3.31) becomes

$$Q = q \sum_{\mathbf{k}} [N_a(\mathbf{k}) - N_b(\mathbf{k})] \tag{3.34}$$

which clearly commutes with the Hamiltonian H, Eq. (3.30).

It follows from Eq. (3.34) that one must associate charges $+q$ and $-q$ with a- and b-particles, respectively. Apart from the sign of the charge, a- and b-particles have identical properties. Furthermore, the theory is completely symmetric between them, as one sees from Eqs. (3.26)–(3.34). Interchanging a and b merely changes the sign of Q. This result is not restricted to spin 0 bosons, but holds generally. The occurrence of antiparticles in association with all particles of non-zero charge is a fundamental feature of relativistic quantum field theory, which is fully vindicated by experiment.

An example of a particle–antiparticle pair is the pair of charged pi-mesons. Taking $q = e(> 0)$, one can identify the π^+- and π^--mesons with the a- and b-particles of the complex Klein–Gordon field. On the other hand, for a real field, the charge operator Q, Eqs. (3.31) or (3.34), is identically zero, and such a field corresponds to a neutral meson, such as the π^0.

The above considerations are not restricted to electric charge. The invariance of the Lagrangian density $\mathscr{L}$ under phase transformations would allow conservation of other additive quantities, which by analogy one would call some kind of charge other than electric. The above argument would lead to the occurrence of pairs of particles and antiparticles differing from each other in the sign of this new kind of charge. Because of this, even electrically neutral particles may possess antiparticles. This situation does occur in nature. The electrically neutral pseudo-scalar K^0-meson possesses an antiparticle, the $\overline{K}^0$-meson, which is also electrically neutral. K^0 and $\overline{K}^0$ possess opposite hypercharge, $Y = \pm 1$, and are represented by a complex Klein–Gordon field ϕ. Hypercharge is very nearly conserved (unlike electric charge, which is always exactly conserved), which is why it is a useful concept. To be specific, hypercharge is conserved in the strong interactions which are responsible for nuclear forces and associated production of strange particles, but it is not conserved in the weak interactions (about 10^{12} times weaker than the strong interactions) responsible for the decay of strange particles.

Instead of treating the complex Klein–Gordon field directly in terms of ϕ and $\phi^\dagger$ as independent fields, as we have done, one can define two real Klein–Gordon fields, ϕ_1 and ϕ_2, by

$$\phi = \frac{1}{\sqrt{2}}(\phi_1 + i\phi_2), \qquad \phi^\dagger = \frac{1}{\sqrt{2}}(\phi_1 - i\phi_2), \tag{3.35}$$

and use these as independent fields. We shall not give the development in terms of the real fields as the two approaches are closely related and very similar. Since the fields ϕ_1 and ϕ_2 are real, the creation and annihilation operators associated with them cannot describe charged particles, and it is only linear combinations of them, corresponding to the complex fields (3.35), which describe charged particles. Consequently, when dealing with conserved charges, it is, in general, more natural to work directly with the complex fields.

3.3   Covariant Commutation Relations

While the equations of motion obtained using the Lagrangian formalism are manifestly covariant, this is not so obvious for the field commutation relations derived by the canonical formalism, since these single out equal times. Taking the real Klein–Gordon field as a typical example, we shall illustrate the covariance of the commutation relations by calculating the commutator $[\phi(x),\ \phi(y)]$ for two arbitrary space–time points x and y. Since this commutator is a scalar, it must equal an invariant function.

Writing $\phi = \phi^+ + \phi^-$, we note that

$$[\phi^+(x), \phi^+(y)] = [\phi^-(x), \phi^-(y)] = 0, \tag{3.36}$$

since $\phi^+(\phi^-)$ contains only absorption (creation) operators. Hence

$$[\phi(x),\ \phi(y)] = [\phi^+(x),\ \phi^-(y)] + [\phi^-(x),\ \phi^+(y)], \tag{3.37}$$

and we need only evaluate the first commutator on the right-hand side of this equation. From Eqs. (3.7) one obtains

$$[\phi^+(x), \phi^-(y)] = \frac{\hbar c^2}{2V} \sum_{\mathbf{k}\mathbf{k}'} \frac{1}{(\omega_{\mathbf{k}}\omega_{\mathbf{k}'})^{1/2}} \left[a(\mathbf{k}), a^\dagger(\mathbf{k}')\right] e^{-ikx + ik'y}$$

$$= \frac{\hbar c^2}{2(2\pi)^3} \int \frac{d^3k}{\omega_{\mathbf{k}}} e^{-ik(x-y)} \tag{3.38}$$

where we have taken the limit $V \to \infty$ [see Eq. (1.48)], and in the last integral $k_0 = \omega_{\mathbf{k}}/c$. We introduce the definition

$$\Delta^+(x) \equiv \frac{-ic}{2(2\pi)^3} \int \frac{d^3k}{\omega_{\mathbf{k}}} e^{-ikx}, \qquad k_0 = \omega_{\mathbf{k}}/c, \tag{3.39}$$

since this and related functions will occur repeatedly.[3] Eq. (3.38) can then be written

$$[\phi^+(x), \phi^-(y)] = i\hbar c \Delta^+(x - y), \tag{3.40}$$

---

[3] There is no generally accepted definitions of these Δ-functions, with different definitions differing by constant factors, so care is required in using the literature.

and

$$[\phi^-(x), \phi^+(y)] = -i\hbar c\Delta^+(y-x) \equiv i\hbar c\Delta^-(x-y), \tag{3.41}$$

defining the function $\Delta^-(x)$. From Eqs. (3.40), (3.41) and (3.37) we obtain the commutation relation

$$[\phi(x), \phi(y)] = i\hbar c\Delta(x-y) \tag{3.42}$$

with $\Delta(x)$ defined by

$$\Delta(x) \equiv \Delta^+(x) + \Delta^-(x) = \frac{-c}{(2\pi)^3} \int \frac{d^3\mathbf{k}}{\omega_\mathbf{k}} \sin kx. \tag{3.43}$$

We see that $\Delta(x)$ is a real odd function, as required by the commutation relation (3.42), which (like $\Delta^\pm$) satisfies the Klein–Gordon equation

$$\left(\Box_x + \mu^2\right)\Delta(x-y) = 0. \tag{3.44}$$

The Δ-function (3.43) can be written

$$\Delta(x) = \frac{-i}{(2\pi)^3} \int d^4k\delta\left(k^2 - \mu^2\right)\varepsilon(k_0)e^{-ikx} \tag{3.45}$$

where $d^4k = d^3\mathbf{k}\,dk_0$, the k_0-integration is over $-\infty < k_0 < \infty$ and $\varepsilon(k_0)$ is defined by

$$\varepsilon(k_0) = \frac{k_0}{|k_0|} = \begin{cases} +1, & \text{if } k_0 > 0 \\ -1, & \text{if } k_0 < 0. \end{cases} \tag{3.46}$$

The equivalence of the definitions (3.43) and (3.45) is easily established; for example, by writing the δ-function in Eq. (3.45) as

$$\delta\left(k^2 - \mu^2\right) = \delta\left[k_0^2 - (\omega_\mathbf{k}/c)^2\right] = \frac{c}{2\omega_\mathbf{k}}\left[\delta\left(k_0 + \frac{\omega_\mathbf{k}}{c}\right) + \delta\left(k_0 - \frac{\omega_\mathbf{k}}{c}\right)\right] \tag{3.47}$$

and performing the k_0-integration.

The invariance of $\Delta(x)$ under proper Lorentz transformations (i.e. involving neither space nor time reflections) is obvious from Eq. (3.45), since each factor in the integrand is Lorentz-invariant [$\varepsilon(k_0)$ is invariant since proper Lorentz transformations do not interchange past and future].

The Lorentz-invariance of $\Delta(x)$ enables one to give a new interpretation to the equal-time commutation relation

$$[\phi(\mathbf{x}, t), \phi(\mathbf{y}, t)] = i\hbar c\Delta(\mathbf{x} - \mathbf{y}, 0) = 0 \tag{3.48}$$

which we had earlier [Eq. (3.6)].[4] The invariance of $\Delta(x-y)$ implies that

$$[\phi(x), \phi(y)] = i\hbar c\Delta(x-y) = 0, \quad \text{for}(x-y)^2 < 0, \tag{3.49}$$

i.e. the fields at any two points x and y, with space-like separation, commute. Hence if the field is a physical observable, measurements of the fields at two points with space-like

---

[4] This also follows directly from Eq. (3.43), since for $x^0 = 0$ the integrand is an odd function of k.

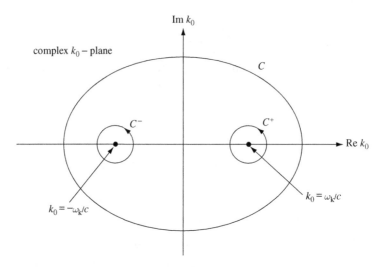

Figure 3.1   *Contours for the integral representation (3.50) of the functions $\Delta^\pm(x)$ and $\Delta(x)$*

separation must not interfere with each other. This is known as microcausality, since for a space-like separation, however small, a signal would have to travel with a velocity greater than the speed of light in order to cause interference, contrary to the special theory of relativity. When discussing the connnection between spin and statistics, at the end of Section 4.3, we shall see that the microcausality condition (3.49) is of fundamental importance, even if the field itself is not a physical observable.

A particularly useful way of representing Δ-functions is as contour integrals in the complex k_0-plane. The functions $\Delta^\pm(x)$ are given by

$$\Delta^\pm(x) = \frac{-1}{(2\pi)^4} \int_{C^\pm} \frac{d^4k\, e^{-ikx}}{k^2 - \mu^2} \tag{3.50}$$

with the contours C^+ and C^-, for Δ^+ and Δ^- respectively, shown in Fig. 3.1. Performing the contour integrations, one picks up the residues from one or other of the poles at $k_0 = \pm\omega_\mathbf{k}/c$, and Eq. (3.50) reduces to the definitions (3.39) and (3.41) of $\Delta^\pm(x)$. The function $\Delta(x)$, Eq. (3.43), is represented by the same integral (3.50) with the contour C shown in Fig. 3.1. Other Δ-functions are obtained by a different choice of contour.

3.4   The Meson Propagator

We shall now derive and discuss a Δ-function which is of great importance in quantum field theory. Its power and utility, particularly for the development of a systematic covariant perturbation theory, was first fully realized by Feynman. We shall again consider the real Klein–Gordon field.

To start with, we note that the Δ^+-function can be written as the vacuum expectation value of a product of two field operators. We have from Eq. (3.40) that

$$i\hbar c\Delta^+\left(x - x'\right) = \langle 0|\left[\phi^+(x), \phi^-\left(x'\right)\right]|0\rangle = \langle 0|\phi^+(x)\phi^-\left(x'\right)|0\rangle$$
$$= \langle 0|\phi(x)\phi\left(x'\right)|0\rangle. \tag{3.51}$$

We define the *time-ordered* or T-product by

$$T\left\{\phi(x)\phi(x')\right\} = \begin{cases} \phi(x)\phi(x'), & \text{if } t > t' \\ \phi(x')\phi(x), & \text{if } t' > t \end{cases} \tag{3.52}$$

($t \equiv x^0/c$, etc.), i.e. the operators are written in chronological order with time running from right to left: 'earlier' operators operate 'first'.[5] Using the step function

$$\theta(t) = \begin{cases} 1, & \text{if } t > 0 \\ 0, & \text{if } t < 0, \end{cases} \tag{3.53}$$

the T-product can be written

$$T\left\{\phi(x)\phi(x')\right\} = \theta(t - t')\phi(x)\phi(x') + \theta(t' - t)\phi(x')\phi(x). \tag{3.54}$$

The Feynman Δ-function Δ_F is defined by the vacuum expectation value of this T-product:

$$i\hbar c\, \Delta_F(x - x') \equiv \left\langle 0 \left| T\left\{\phi(x)\phi(x')\right\} \right| 0 \right\rangle. \tag{3.55}$$

From Eqs. (3.51) and (3.41), this leads to the explicit definition

$$\Delta_F(x) = \theta(t)\Delta^+(x) - \theta(-t)\Delta^-(x). \tag{3.56a}$$

Thus

$$\Delta_F(x) = \pm\Delta^\pm(x). \qquad \text{if } t \gtrless 0. \tag{3.56b}$$

We would like to be able to visualize the meaning of Δ_F, Eq. (3.55). For $t > t'$, this vacuum expectation value becomes $\langle 0|\phi(x)\phi(x')|0\rangle$. We can think of this expression as representing a meson being created at x', travelling to x, and being annihilated at x. The corresponding expression for $t' > t$, $\langle 0|\phi(x')\phi(x)|0\rangle$, admits a similar interpretation as a meson created at x, propagating to x', where it is absorbed. These two situations are illustrated schematically in Fig. 3.2. The dashed lines represent the propagation of the meson in the direction of the arrow, from x' to x or vice versa. Hence Δ_F, or the vacuum expectation value (3.55), is referred to as the *Feynman propagator* for the mesons of the Klein–Gordon field. We shall, briefly, call it the *meson propagator*, to distinguish it from the fermion and photon propagators to be introduced later.

To illustrate how these propagators arise, we shall consider qualitatively nucleon–nucleon scattering. In this process there will be two nucleons, but no mesons, present in the initial and final states (i.e. before and after the scattering). The scattering, i.e. the interaction, corresponds to the exchange of virtual mesons between the nucleons. The

---

[5] The definition (3.52) will be modified for fermions.

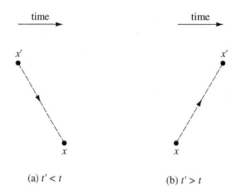

(a) $t' < t$ (b) $t' > t$

Figure 3.2   The meson propagator (3.55)

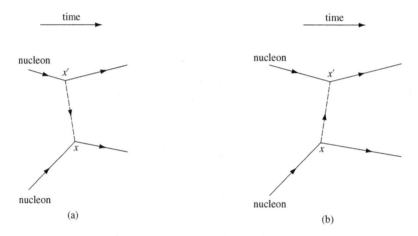

(a)   (b)

Figure 3.3 Contribution from one-meson exchange to nucleon–nucleon scattering, (a) $t' < t$; (b) $t' > t$

simplest such process is the one-meson exchange, schematically illustrated in Fig. 3.3. The continuous lines represent the nucleons, the dashed lines the mesons. As before, two situations arise according to whether $t > t'$ or $t' > t$. In the actual calculation, all values of x and x' are integrated over, corresponding to emission and absorption of the meson occurring at any two space–time points.

It is interesting to note that the division into the two types of process (a) and (b) of Fig. 3.3, depending on whether $t > t'$ or $t' > t$, is not Lorentz-invariant for $(x - x')$, a space-like separation. In this case what constitutes 'later' and what 'earlier' depends on the frame of reference. On the other hand, considering both cases *together* leads to the covariant Feynman propagator (3.55), which we represent by the single diagram in Fig. 3.4. No time-ordering is implied in this diagram and correspondingly there is no arrow on the meson line.

We have here introduced the ideas of *Feynman graphs* or diagrams. We shall deal with these fully later and shall see that they are a most useful way of picturing the mathematics.

nucleon

x'

nucleon

Figure 3.4 *Feynman graph for the one-meson contribution to nucleon–nucleon scattering*

But the reader must be warned not to take this pictorial description of the mathematics as a literal description of a process in space and time. For example, our naive interpretation of the meson propagator would imply that, for $(x - x')$, a space-like separation, the meson travels between the two points with a speed greater than the velocity of light. It is, however, possible to substantiate the above description if, instead of considering propagation between two points x and x', one calculates the probability for emission and absorption in two appropriately chosen four-dimensional regions.[6]

In the following we shall frequently require a representation of the meson propagator, not in coordinate, but in momentum space. This is given by the following integral representation, similar to Eq. (3.50) for $\Delta^{\pm}(x)$:

$$\Delta_F(x) = \frac{1}{(2\pi)^4} \int_{C_F} \frac{d^4 k \, e^{-ikx}}{k^2 - \mu^2}, \tag{3.57}$$

where the contour C_F is shown in Fig. 3.5. To verify Eq. (3.57), we evaluate the contour integral. For $x^0 > 0$, we must complete the contour C_F in the lower half k_0-plane [since $\exp(-ik_0 x^0) \to 0$ for $k_0 \to -i\infty$], and, comparing Eqs. (3.57) and (3.50), we obtain $\Delta_F(x) = \Delta^+(x)$, in agreement with Eq. (3.56b). For $x^0 < 0$, completion of the contour in the upper half k_0-plane similarly leads to agreement with Eq. (3.56b).

Instead of deforming the contour as in Fig. 3.5, we can move the poles an infinitesimal distance η off the real axis, as shown in Fig. 3.6, and perform the k_0-integration along the whole real axis, i.e. we replace Eq. (3.57) by

---

[6] See the article by G. Källén in *Encyclopedia of Physics*, vol. V, part 1, Springer, Berlin, 1958, Section 23. An English translation of this article, entitled *Quantum Electrodynamics*, has been published by Springer, New York, 1972, and by Allen & Unwin, London, 1972.

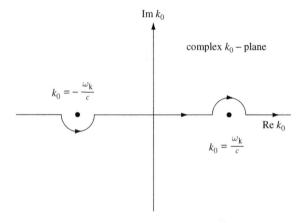

Figure 3.5   *The contour C_F for the meson propagator Δ_F, Eq. (3.57)*

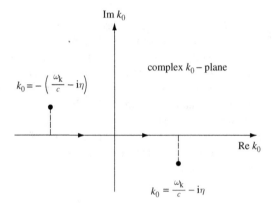

Figure 3.6   *Contour and displaced poles for the meson propagator Δ_F, Eq. (3.58)*

$$\Delta_F(x) = \frac{1}{(2\pi)^4} \int \frac{d^4k \, e^{-ikx}}{k_0^2 - \left(\dfrac{\omega_{\mathbf{k}}}{c} - i\eta\right)^2}$$

$$= \frac{1}{(2\pi)^4} \int \frac{d^4k \, e^{-ikx}}{k^2 - \mu^2 + i\varepsilon},$$

(3.58)

where $\varepsilon = 2\eta\omega_{\mathbf{k}}/c$ is a small positive number which we let tend to zero after integration. In Eq. (3.58), integration with respect to each of the four variables $k_0, \ldots, k_3$ is along the whole real axis $(-\infty,\infty)$.

The arguments of this section at once generalize to the case of the complex scalar field, discussed in Section 3.2. The charged meson propagator is now given by

$$\left\langle 0 \left| T\left\{ \phi(x)\phi^{\dagger}(x') \right\} \right| 0 \right\rangle = i\hbar c \Delta_F(x - x'),\tag{3.59}$$

where $\Delta_F(x)$ is the same function [Eqs. (3.56)–(3.58)] as for the real field. The interpretation of the vacuum expectation value (3.59) in terms of the emission, propagation and reabsorption of particles or antiparticles, depending on whether $t' < t$ or $t' > t$, is left to the reader.

Problems

3.1. From the expansion (3.7) for the real Klein–Gordon field $\phi(x)$, derive the following expression for the absorption operator $a(\mathbf{k})$:

$$a(\mathbf{k}) = \frac{1}{(2\hbar c^2 V \omega_{\mathbf{k}})^{1/2}} \int d^3 x e^{ikx} \left(i\dot{\phi}(x) + \omega_{\mathbf{k}} \phi(x) \right).$$

Hence derive the commutation relations (3.9) for the creation and annihilation operators from the commutation relations (3.6) for the fields.

3.2. With the complex Klein–Gordon fields $\phi(x)$ and $\phi^{\dagger}(x)$ expressed in terms of two independent real Klein–Gordon fields $\phi_1(x)$ and $\phi_2(x)$ by Eqs. (3.35), and with $\phi_r(x)$ expanded in the form

$$\phi_r(x) = \sum_{\mathbf{k}} \left(\frac{\hbar c^2}{2V\omega_{\mathbf{k}}} \right)^{1/2} [a_r(\mathbf{k})e^{-ikx} + a_r^{\dagger}(\mathbf{k})e^{ikx}], \qquad r = 1, 2,$$

show that

$$a(\mathbf{k}) = \frac{1}{\sqrt{2}} [a_1(\mathbf{k}) + ia_2(\mathbf{k})], \qquad b(\mathbf{k}) = \frac{1}{\sqrt{2}} [a_1(\mathbf{k}) - ia_2(\mathbf{k})].$$

Hence derive the commutation relations (3.25) from those for the real fields, and the commutation relations (3.27) from those for $a_r(\mathbf{k})$ and $a_r^{\dagger}(\mathbf{k}), r = 1, 2$.

3.3. From Eq. (3.58), or otherwise, show that the Feynman Δ-function satisfies the inhomogeneous Klein–Gordon equation

$$\left(\Box + \mu^2 \right) \Delta_F(x) = -\delta^{(4)}(x).$$

3.4. Derive Eq. (3.59) for the charged meson propagator, and interpret it in terms of emission and re-absorption of particles and antiparticles.

3.5. Charge conjugation for the complex Klein–Gordon field $\phi(x)$ is defined by

$$\phi(x) \rightarrow \mathscr{C}\phi(x)\mathscr{C}^{-1} = \eta_c \phi^{\dagger}(x)\tag{A}$$

where $\mathscr{C}$ is a unitary operator which leaves the vacuum invariant, $\mathscr{C}|0\rangle = |0\rangle$, and η_c is a phase factor.

Show that under the transformation (A) the Lagrangian density (3.22) is invariant and the charge-current density (3.32) changes sign.

Derive

$$\mathscr{C}a(\mathbf{k})\mathscr{C}^{-1} = \eta_c b(\mathbf{k}), \quad \mathscr{C}b(\mathbf{k})\mathscr{C}^{-1} = \eta_c^* a(\mathbf{k})$$

for the absorption operators, and hence show that

$$\mathscr{C}|a, \mathbf{k}\rangle = \eta_c^*|b, \mathbf{k}\rangle, \quad \mathscr{C}|b, \mathbf{k}\rangle = \eta_c|a, \mathbf{k}\rangle$$

where $|a, \mathbf{k}\rangle$ denotes the state with a single a-particle of momentum $\mathbf{k}$ present, etc. ($\mathscr{C}$ is called the charge conjugation operator. It interchanges particles and anti-particles: $a \leftrightarrow b$. The phase η_c is arbitrary and is usually set equal to unity, $\eta_c = 1$.)

3.6. The parity transformation (i.e. space inversion) of the Hermitian Klein–Gordon field $\phi(x)$ is defined by

$$\phi(\mathbf{x}, \ t) \ \rightarrow \ \mathscr{P}\,\phi(\mathbf{x}, \ t)\mathscr{P}^{-1} = \eta_p\phi(-\mathbf{x}, \ t) \tag{B}$$

where the parity operator $\mathscr{P}$ is a unitary operator which leaves the vacuum invariant, $\mathscr{P}|0\rangle = |0\rangle$, and $\eta_p = \pm 1$ is called the intrinsic parity of the field. Show that the parity transformation leaves the Lagrangian density (3.4) invariant.

Show that

$$\mathscr{P}|\mathbf{k}_1, \ \mathbf{k}_2, \ldots, \ \mathbf{k}_n\rangle = \eta_p^n| -\mathbf{k}_1, \ -\mathbf{k}_2, \ \ldots, \ -\mathbf{k}_n\rangle$$

for an arbitrary n-particle state.

For any operators A and B

$$e^{i\alpha A}Be^{-i\alpha A} = \sum_{n=0}^{\infty} \frac{(i\alpha)^n}{n!}B_n$$

where

$$B_0 = B, \quad B_n = [A, \ B_{n-1}], \quad n = 1, 2, \ldots,$$

holds identically. Hence prove that

$$\mathscr{P}_1\, a(\mathbf{k})\, \mathscr{P}_1^{-1} = ia(\mathbf{k}), \quad \mathscr{P}_2\, a(\mathbf{k})\, \mathscr{P}_2^{-1} = -i\eta_p a(-\mathbf{k}),$$

where the $a(\mathbf{k})$ are the annihilation operators of the field, and $\mathscr{P}_1$ and $\mathscr{P}_2$ are given by

$$\mathscr{P}_1 = \exp\left[-i\frac{\pi}{2}\sum_{\mathbf{k}} a^\dagger(\mathbf{k})a(\mathbf{k})\right], \quad \mathscr{P}_2 = \exp\left[i\frac{\pi}{2}\eta_p\sum_{\mathbf{k}} a^\dagger(\mathbf{k})a(-\mathbf{k})\right].$$

Hence, show that the operator $\mathscr{P} = \mathscr{P}_1\,\mathscr{P}_2$ is unitary and satisfies Eq. (B), i.e. it gives an explicit expression for the parity operator $\mathscr{P}$.

4

The Dirac Field

We now wish to consider systems of particles which satisfy the Pauli exclusion principle, i.e. which obey Fermi–Dirac statistics, so-called *fermions*. We saw in Chapter 2 that the canonical quantization formalism necessarily leads to bosons. On the other hand, the harmonic oscillator quantization, used in Chapter 1, allows an *ad hoc* modification which leads to Fermi–Dirac statistics. This modification was first introduced in 1928 by Jordan and Wigner and consists in replacing the commutation relations between absorption and creation operators by anticommutation relations. We shall develop this general formalism in Section 4.1.

In the remainder of this chapter, this formalism will be applied to the Dirac equation, i.e. to relativistic material particles of spin $\frac{1}{2}$. One of the distinctions between bosons and fermions is that the former always have integral spin $(0, 1, \ldots)$ whereas the latter must have half-integral spin $(\frac{1}{2}, \frac{3}{2}, \ldots)$. We shall see that this connection between spin and statistics is an essential feature of relativistic quantum field theory.

4.1 The Number Representation for Fermions

In Sections 1.2.2 and 1.2.3 we derived a number representation for bosons from the quantization of a system of independent harmonic oscillators. We shall now modify this formalism so as to obtain a number representation for fermions.

The essence of our earlier treatment can be stated as follows. We had operators a_r, $a_r^\dagger$, $r = 1, 2, \ldots$, satisfying the commutation relations

$$\left[a_r, a_s^\dagger\right] = \delta_{rs}, \quad \left[a_r, a_s\right] = \left[a_r^\dagger, a_s^\dagger\right] = 0, \tag{4.1}$$

and defined operators

$$N_r = a_r^\dagger a_r. \tag{4.2}$$

It then follows from the operator identity

$$[AB, \ C] = A[B, \ C] + [A, \ C]B \tag{4.3}$$

that

$$[N_r, \ a_s] = -\delta_{rs}a_s, \quad [N_r, \ a_s^\dagger] = \delta_{rs}a_s^\dagger. \tag{4.4}$$

The interpretation of a_r, $a_r^\dagger$ and N_r as absorption, creation and number operators, follows from Eqs. (4.2) and (4.4). In particular, N_r has the eigenvalues $n_r = 0, 1, 2\ldots$ The vacuum state $|0\rangle$ is defined by

$$a_r \ |0\rangle = 0, \ \text{all} \ r, \tag{4.5}$$

and other states are built up from the vacuum state as linear superpositions of states of the form

$$\left(a_{r_1}^\dagger\right)^{n_1} \left(a_{r_2}^\dagger\right)^{n_2} \ \ldots \ |0\rangle. \tag{4.6}$$

It is a remarkable fact that there is an alternative way of deriving the relations (4.4). Define the *anticommutator* of two operators A and B by

$$[A, \ B]_+ \equiv AB + BA. \tag{4.7}$$

We then have a second operator identity, analogous to (4.3),

$$[AB, \ C] = A[B, \ C]_+ - [A, \ C]_+B. \tag{4.8}$$

Suppose now that the operators a_r, $a_r^\dagger$, $r = 1, 2, \ \ldots,$ instead of satisfying the commutation relations (4.1) satisfy the anticommutation relations

$$\left[a_r, \ a_s^\dagger\right]_+ = \delta_{rs}, \quad [a_r, \ a_s]_+ = \left[a_r^\dagger, \ a_s^\dagger\right]_+ = 0; \tag{4.9}$$

in particular

$$(a_r)^2 = \left(a_r^\dagger\right)^2 = 0. \tag{4.9a}$$

One verifies from Eqs. (4.2), (4.8) and (4.9) that for the anticommuting operators [i.e. satisfying Eqs. (4.9)] the *same* commutation relations (4.4) hold which were previously derived for the commuting operators [i.e. satisfying Eqs. (4.1)]. This again leads to the interpretation of a_r, $a_r^\dagger$ and N_r as absorption, creation and number operators, but, from Eqs. (4.9), we now have

$$N_r^2 = a_r^\dagger a_r a_r^\dagger a_r = a_r^\dagger\left(1 - a_r^\dagger a_r\right)a_r = N_r, \tag{4.10}$$

whence

$$N_r(N_r - 1) = 0; \tag{4.10a}$$

i.e., for the anticommuting creation and absorption operators, the number operator N_r has the eigenvalues $n_r = 0$ and $n_r = 1$ only, i.e. we are dealing with Fermi–Dirac statistics.

The vacuum state $|0\rangle$ is again defined by Eq. (4.5). The state in which one particle is in the state r is

$$|1_r\rangle = a_r^\dagger \ |0\rangle. \tag{4.11}$$

For the two-particle states we have from the anticommutation relations (4.9) that for $r \neq s$

$$|1_r 1_s\rangle = a_r^\dagger a_s^\dagger |0\rangle = -a_s^\dagger a_r^\dagger |0\rangle = -|1_s 1_r\rangle \tag{4.12}$$

i.e. the state is antisymmetric under interchange of particle labels, as required for fermions. For $r = s$, we have

$$|2_r\rangle = \left(a_r^\dagger\right)^2 |0\rangle = 0, \tag{4.13}$$

thus regaining the earlier result that two particles cannot be in the same single-particle state.

In conclusion, we would like to note the fundamental difference, in spite of their superficial similarity, in the derivation of the boson and fermion results of this section. The boson commutation relations (4.1) are a direct consequence of the canonical commutation relations of non-relativistic quantum mechanics [compare the derivation of Eq. (1.19)]. We have no such foundation for the fermion anticommutation relations (4.9).

4.2   The Dirac Equation

We shall now consider the classical field theory of the Dirac equation, in preparation for going over to the quantized field theory in the next section.[1] The Dirac equation describes material particles of spin $\frac{1}{2}$. We shall see that in the quantum field theory, antiparticles again necessarily occur, e.g. for electrons these are the positrons. Because of our later applications to quantum electrodynamics, we shall for definiteness speak of electrons and positrons in this chapter, but the theory is equally applicable to other spin $\frac{1}{2}$ material particles such as muons and quarks.

The Dirac equation for particles of rest mass m

$$i\hbar \frac{\partial \psi(x)}{\partial t} = \left[c\boldsymbol{\alpha} \cdot (-i\hbar\boldsymbol{\nabla}) + \beta mc^2\right]\psi(x)$$

can be written

$$i\hbar\gamma^\mu \frac{\partial \psi(x)}{\partial x^\mu} - mc\psi(x) = 0 \tag{4.14}$$

where

$$\gamma^0 = \beta, \quad \gamma^i = \beta\alpha_i, \quad i = 1, 2, 3,$$

are Dirac 4×4 matrices which satisfy the anticommutation relations

$$[\gamma^\mu, \gamma^\nu]_+ = 2g^{\mu\nu} \tag{4.15}$$

---

[1] The reader is assumed to be familiar with the elementary theory of the Dirac equation, as discussed in, for example, L. I. Schiff, *Quantum Mechanics*, 3rd edn, McGraw-Hill, New York, 1968, pp. 472–488. Further results of the Dirac theory, which will be needed later, are derived and summarized in Appendix A.

and the Hermiticity conditions $\gamma^{0\dagger} = \gamma^0$ and $\gamma^{j\dagger} = -\gamma^j$ for $j = 1, 2, 3$, which can be combined into[2]

$$\gamma^{\mu\dagger} = \gamma^0 \gamma^\mu \gamma^0. \tag{4.16}$$

Correspondingly, $\psi(x)$ is a spinor wavefunction with four components $\psi_\alpha(x)$, $\alpha = 1, \ldots, 4$. The indices labelling spinor components and matrix elements will usually be suppressed.[3] Although it is at times convenient to use a particular matrix representation, this is generally not necessary. We shall formulate the theory in a representation-free way and only assume that the γ-matrices satisfy the anticommutation and Hermiticity relations (4.15) and (4.16). This will facilitate use of the most convenient representation in a given situation.

The adjoint field $\overline{\psi}(x)$ is defined by

$$\overline{\psi}(x) = \psi^\dagger(x)\gamma^0 \tag{4.18}$$

and satisfies the adjoint Dirac equation

$$i\hbar \frac{\partial \overline{\psi}(x)}{\partial x^\mu} \gamma^\mu + mc\overline{\psi}(x) = 0. \tag{4.19}$$

The Dirac equations (4.14) and (4.19) can be derived from the Lagrangian density

$$\mathscr{L} = c\overline{\psi}(x) \left[i\hbar\gamma^\mu \frac{\partial}{\partial x^\mu} - mc \right] \psi(x) \tag{4.20}$$

by varying the action integral (2.11) independently with respect to the fields ψ_α and $\overline{\psi}_\alpha$. From Eq. (4.20) one obtains for the conjugate fields of ψ_α and $\overline{\psi}_\alpha$

$$\pi_\alpha(x) = \frac{\partial \mathscr{L}}{\partial \dot{\psi}_\alpha} = i\hbar\psi_\alpha^\dagger, \quad \overline{\pi}_\alpha(x) = \frac{\partial \mathscr{L}}{\partial \dot{\overline{\psi}}_\alpha} \equiv 0. \tag{4.21}$$

The Hamiltonian and the momentum of the Dirac field are, from Eqs. (2.51), (4.20) and (4.21), given by

$$H = \int d^3x \overline{\psi}(x) \left[-i\hbar c\gamma^j \frac{\partial}{\partial x^j} + mc^2 \right] \psi(x) \tag{4.22}$$

and

$$\mathbf{P} = -i\hbar \int d^3x \psi^\dagger(x) \boldsymbol{\nabla}\psi(x). \tag{4.23}$$

---

[2] These conditions ensure the Hermiticity of the Dirac Hamiltonian

$$H = c\gamma^0\gamma^j(-i\hbar\,\partial/\partial x^j) + mc^2\gamma^0. \tag{4.17}$$

[3] In case of doubt, the reader should write the indices out explicitly; e.g. Eq. (4.14) becomes

$$\sum_{\beta=1}^{4} i\hbar\gamma^\mu_{\alpha\beta} \frac{\partial \psi_\beta(x)}{\partial x^\mu} - mc\psi_\alpha(x) = 0, \quad \alpha = 1, \ldots, 4. \tag{4.14a}$$

Eq. (4.22), of course, also follows from the usual definition of the Hamiltonian density (2.25) applied to the present case.

The angular momentum of the Dirac field follows similarly from Eq. (2.54). The transformation of the field under an infinitesimal Lorentz transformation, i.e. Eq. (2.47), is, in the case of the Dirac field, given by

$$\psi_\alpha(x) \rightarrow \psi'_\alpha(x') = \psi_\alpha(x) - \frac{i}{4}\varepsilon_{\mu\nu}\sigma^{\mu\nu}_{\alpha\beta}\psi_\beta(x), \tag{4.24}$$

where summation over $\mu, \nu = 0, \ldots, 3$ and $\beta = 1, \ldots, 4$ is implied, and where $\sigma^{\mu\nu}_{\alpha\beta}$ is the (α, β) matrix element of the 4×4 matrix

$$\sigma^{\mu\nu} \equiv \frac{i}{2}[\gamma^\mu, \gamma^\nu]. \tag{4.25a}$$

Eq. (4.24) is derived in Appendix A, Eq. (A.60). Eq. (2.54) now gives for the angular momentum of the Dirac field

$$\mathbf{M} = \int d^3\mathbf{x}\psi^\dagger(x)[\mathbf{x} \wedge(-i\hbar\nabla)]\psi(x) + \int d^3\mathbf{x}\psi^\dagger(x)\left(\frac{\hbar}{2}\boldsymbol{\sigma}\right)\psi(x), \tag{4.26}$$

where the 4×4 matrices

$$\boldsymbol{\sigma} = (\sigma^{23}, \sigma^{31}, \sigma^{12}) \tag{4.25b}$$

are the generalizations for the Dirac theory of the 2×2 Pauli spin matrices. We see that the two terms in Eq. (4.26) represent the orbital and spin angular momenta of particles of spin $\frac{1}{2}$.

The Lagrangian density (4.20) is invariant under the phase transformation (2.41). Hence Eq. (2.42) leads to the conserved charge

$$Q = q\int d^3\mathbf{x}\psi^\dagger(x)\psi(x), \tag{4.27}$$

and the charge–current density

$$s^\alpha(x) = (c\rho(x), \mathbf{j}(x)) = cq\bar{\psi}(x)\gamma^\alpha\psi(x) \tag{4.28}$$

satisfies the continuity equation (i.e. conservation equation)

$$\frac{\partial s^\alpha}{\partial x^\alpha} = 0. \tag{4.29}$$

The continuity equation also follows directly from the Dirac equations (4.14) and (4.19).

In order to quantize the Dirac field in the next section, we shall expand it in a complete set of solutions of the Dirac equation and then impose appropriate anticommutation relations on the expansion coefficients. To conclude this section, we shall therefore specify a complete orthonormal set of solutions of the Dirac equation (4.14).

We shall again consider a cubic enclosure, of volume V, with periodic boundary conditions. A complete set of plane wave states can then be defined as follows. For each momentum $\mathbf{p}$, allowed by the periodic boundary conditions, and positive energy

$$cp_0 = E_\mathbf{p} = +(m^2c^4 + c^2\mathbf{p}^2)^{1/2}, \tag{4.30}$$

the Dirac equation (4.14) possesses four independent solutions. These will be written

$$u_r(\mathbf{p})\frac{e^{-ipx/\hbar}}{\sqrt{V}}, \quad v_r(\mathbf{p})\frac{e^{ipx/\hbar}}{\sqrt{V}}, \quad r = 1, 2 \tag{4.31}$$

i.e. $u_r(\mathbf{p})$ and $v_r(\mathbf{p})$ are constant spinors satisfying the equations

$$(\not{p} - mc)u_r(\mathbf{p}) = 0, \quad (\not{p} + mc)v_r(\mathbf{p}) = 0, \quad r = 1, 2. \tag{4.32}$$

Here we introduce the very convenient notation $\not{A}$ (called A slash), which is defined for any four-vector A_μ by

$$\not{A} \equiv \gamma^\mu A_\mu. \tag{4.33}$$

Because of their time dependence, the solutions (4.31) involving u_r and v_r are referred to as positive and negative energy solutions respectively. We shall use these terms merely as a way of labelling the u- and v-solutions. We shall not follow up their interpretation in the single-particle theory, the resulting difficulties and the reinterpretation in terms of the hole theory. We shall see that the second quantization of the theory (i.e. when ψ and $\psi^\dagger$ become operators) leads directly to the interpretation in terms of particles and antiparticles without the intellectual contortions of the hole theory.[4]

The two-fold degeneracies of the two positive and the two negative energy solutions for a given momentum $\mathbf{p}$ result from the possible spin orientations. For the Dirac equation, only the longitudinal spin components (i.e. parallel to $\pm\mathbf{p}$) are constants of the motion, and we shall choose these spin eigenstates for the solutions (4.31). With

$$\sigma_\mathbf{p} = \frac{\boldsymbol{\sigma} \cdot \mathbf{p}}{|\mathbf{p}|}, \tag{4.34}$$

where $\boldsymbol{\sigma}$ is defined in Eqs. (4.25a) and (4.25b), we then choose the spinors in Eqs. (4.31) so that

$$\sigma_\mathbf{p} u_r(\mathbf{p}) = (-1)^{r+1}u_r(\mathbf{p}), \quad \sigma_\mathbf{p} v_r(\mathbf{p}) = (-1)^r v_r(\mathbf{p}), \quad r = 1, 2. \tag{4.35}$$

The asymmetry in labelling u- and v-spinors will be convenient for labelling the spin properties of particles and antiparticles.

We normalize the spinors u_r and v_r so that

$$u_r^\dagger(\mathbf{p})u_r(\mathbf{p}) = v_r^\dagger(\mathbf{p})v_r(\mathbf{p}) = \frac{E_\mathbf{p}}{mc^2}. \tag{4.36}$$

They then satisfy the orthonormality relations

$$\left. \begin{array}{c} u_r^\dagger(\mathbf{p})u_s(\mathbf{p}) = v_r^\dagger(\mathbf{p})v_s(\mathbf{p}) = \dfrac{E_\mathbf{p}}{mc^2}\delta_{rs} \\ u_r^\dagger(\mathbf{p})v_s(-\mathbf{p}) = 0 \end{array} \right\} \tag{4.37}$$

and the states (4.31) form a complete orthonormal set of solutions of the free-particle Dirac equation, normalized to $E_\mathbf{p}/mc^2$ in a volume V. These and other properties of the plane wave solutions (4.31) are discussed further in Appendix A.

---

[4] This remark is in no way meant to denigrate Dirac's tremendous intellectual achievement of inventing the hole theory originally.

4.3   Second Quantization

In order to quantize the Dirac field, we expand it in terms of the complete set of plane wave states (4.31):

$$\psi(x) = \psi^+(x) + \psi^-(x)$$
$$= \sum_{r\mathbf{p}} \left(\frac{mc^2}{VE_\mathbf{p}}\right)^{1/2} [c_r(\mathbf{p})u_r(\mathbf{p})\, e^{-ipx/\hbar} + d_r^\dagger(\mathbf{p})v_r(\mathbf{p})\, e^{ipx/\hbar}] \qquad (4.38a)$$

and hence the conjugate field $\overline{\psi} = \psi^\dagger \gamma^0$ has the expansion

$$\overline{\psi}(x) = \overline{\psi}^+(x) + \overline{\psi}^-(x)$$
$$= \sum_{r\mathbf{p}} \left(\frac{mc^2}{VE_\mathbf{p}}\right)^{1/2} [d_r(\mathbf{p})\overline{v}_r(\mathbf{p})\, e^{-ipx/\hbar} + c_r^\dagger(\mathbf{p})\overline{u}_r(\mathbf{p})\, e^{ipx/\hbar}] \qquad (4.38b)$$

where $\overline{u}_r = u_r^\dagger \gamma^0$, etc. The summations in Eqs. (4.38) are over the allowed momenta $\mathbf{p}$ and the spin states, labelled by $r = 1, 2$.[5] The factors $(mc^2/VE_\mathbf{p})^{1/2}$ will be convenient for the subsequent interpretation of the expansion coefficients. We have written $c_r^\dagger$ and $d_r^\dagger$ for two of these, anticipating that they will become operators on second quantization.

Eqs. (4.38) are closely analogous to the expansions of the complex Klein–Gordon field, Eqs. (3.26). However, the Dirac equation describes spin $\frac{1}{2}$ particles, such as electrons, which obey the Pauli principle and Fermi–Dirac statistics. Following the treatment in Section 4.1, we shall therefore impose the following *anti*commutation relations on the expansion coefficients:

$$[c_r(\mathbf{p}), c_s^\dagger(\mathbf{p}')]_+ = [d_r(\mathbf{p}), d_s^\dagger(\mathbf{p}')]_+ = \delta_{rs}\delta_{\mathbf{pp}'}, \qquad (4.39a)$$

and all other anticommutators vanish, i.e. with $c_r \equiv c_r(\mathbf{p})$, $c_s \equiv c_s(\mathbf{p}')$, etc.:

$$\left.\begin{array}{l} [c_r, c_s]_+ = [c_r^\dagger, c_s^\dagger]_+ = [d_r, d_s]_+ = [d_r^\dagger, d_s^\dagger]_+ = 0 \\ [c_r, d_s]_+ = [c_r, d_s^\dagger]_+ = [c_r^\dagger, d_s]_+ = [c_r^\dagger, d_s^\dagger]_+ = 0 \end{array}\right\} . \qquad (4.39b)$$

If we define the operators

$$N_r(\mathbf{p}) = c_r^\dagger(\mathbf{p})c_r(\mathbf{p}), \quad \overline{N}_r(\mathbf{p}) = d_r^\dagger(\mathbf{p})d_r(\mathbf{p}), \qquad (4.40)$$

the interpretation of $c_r, c_r^\dagger, N_r$ and $d_r, d_r^\dagger, \overline{N}_r$ as absorption, creation and number operators of two kinds of particles, both fermions, follows from the anticommutation relations (4.39), analogously to the development in Section 4.1.

The vacuum state $|0\rangle$ is defined by

$$c_r(\mathbf{p})|0\rangle = d_r(\mathbf{p})|0\rangle = 0, \quad \text{all } \mathbf{p}, \quad \text{and } r = 1, 2, \qquad (4.41)$$

or, equivalently, by

$$\psi^+(x)|0\rangle = \overline{\psi}^+(x)|0\rangle = 0, \quad \text{all } x. \qquad (4.42)$$

---

[5] We have chosen particular spin states u_r and v_r, but it should be clear to the reader that one may equally well use any other orthonormal spin states. The following arguments remain valid; only the interpretation of the spin properties of the states has to be modified.

States containing particles are generated from the vacuum state by means of the creation operators. As in Section 4.1, one sees that these states have all the properties characteristic of fermions [i.e. equations analogous to Eqs. (4.12) and (4.13) hold].

To obtain the physical properties of the particles associated with the c- and d-operators, we express the constants of the motion in terms of them. (The reader should be able to have a good guess at most of these properties.) In Section 4.2, we derived expressions for the energy, momentum, angular momentum and charge of the Dirac field [see Eqs. (4.22), (4.23), (4.26) and (4.27)]. However, these operators do not necessarily have the value zero for the vacuum state. We found a similar situation for the Klein–Gordon field [see Eqs. (3.15) and (3.16)]. As in the latter case, we automatically measure quantities relative to the vacuum state if we redefine the expressions for the constants of the motion with the operators ordered as normal products (i.e. in any product, absorption operators occur to the right of creation operators) so that vacuum values necessarily vanish.

For fermions we must modify our earlier definition of the normal product. In arranging a product of boson operators in normal order, one treats them as though all commutators vanish [see Eqs. (3.18) and (3.19)]. For fermion operators, one treats them as though all *anticommutators* vanish, e.g. with $\psi_\alpha \equiv \psi_\alpha(x)$ and $\psi_\beta \equiv \psi_\beta(x')$, etc., one has

$$N\left(\psi_\alpha \psi_\beta\right) = N[(\psi_\alpha^+ + \psi_\alpha^-)(\psi_\beta^+ + \psi_\beta^-)]$$
$$= \psi_\alpha^+ \psi_\beta^+ - \psi_\beta^- \psi_\alpha^+ + \psi_\alpha^- \psi_\beta^+ + \psi_\alpha^- \psi_\beta^- \qquad (4.43)$$

which should be compared with Eq. (3.19) for bosons. Similar results hold if in Eq. (4.43) one or both operators are replaced by their adjoint operators, $\overline{\psi}$, or for products of more than two fields.[6]

With the expressions for the constants of the motion, i.e. Eqs. (4.22), (4.23) and (4.26)–(4.28), modified to be normal products, e.g.

$$H = \int d^3 x\, N\left\{\overline{\psi}(x)\left[-i\hbar c \gamma^j \frac{\partial}{\partial x^j} + mc^2\right]\psi(x)\right\}, \qquad (4.22a)$$

etc., we substitute the expansions (4.38) for ψ and $\overline{\psi}$. Using the orthonormality properties of the single-particle states (4.31) and, in the calculation of H, that they are solutions of the Dirac equation, we obtain for the energy, momentum and charge operators

$$H = \sum_{r\mathbf{p}} E_\mathbf{p}[N_r(\mathbf{p}) + \overline{N}_r(\mathbf{p})] \qquad (4.44)$$

$$\mathbf{P} = \sum_{r\mathbf{p}} \mathbf{p}[N_r(\mathbf{p}) + \overline{N}_r(\mathbf{p})] \qquad (4.45)$$

---

[6] The fermion operators ψ and $\overline{\psi}$ are non-commuting quantities, not only through their dependence on the absorption and creation operators, but also as four-component spinors. Hence care is required in changing the order of operators; e.g. if O is a 4 × 4 matrix (such as a product of γ-matrices) then

$$N(\overline{\psi} O \psi) = \overline{\psi}_\alpha^+ O_{\alpha\beta} \psi_\beta^+ - \psi_\beta^- O_{\alpha\beta} \overline{\psi}_\alpha^+ + \overline{\psi}_\alpha^- O_{\alpha\beta} \psi_\beta^+ + \overline{\psi}_\alpha^- O_{\alpha\beta} \psi_\beta^-, \qquad (4.43a)$$

i.e. suppressing the spinor indices the second term on the right-hand side is $-\psi^- O^\mathrm{T} \overline{\psi}^+$, where O^T is the transposed matrix: $O_{\beta\alpha}^\mathrm{T} = O_{\alpha\beta}$ In case of doubt, the reader should write the spinor indices out explicitly.

$$Q = -e \sum_{r\mathbf{p}} [N_r(\mathbf{p}) - \overline{N}_r(\mathbf{p})]. \tag{4.46}$$

In the last equation we have taken the parameter q to be the charge of the electron: $q = -e < 0$. Hence identifying the mass m in the Dirac equation with the mass of the electron, we can interpret the particles associated with the c- and d-operators as electrons and positrons, respectively.

To identify the spin properties, we calculate the spin angular momentum in the states $c_r^\dagger(\mathbf{p})|0\rangle$ and $d_r^\dagger(\mathbf{p})|0\rangle$, containing one electron or one positron of momentum $\mathbf{p}$. From Eqs. (4.26) and (4.34) we define the longitudinal spin operator, i.e. in the direction of motion $\mathbf{p}$, by

$$S_{\mathbf{p}} = \frac{\hbar}{2} \int d^3\mathbf{x} N[\psi^\dagger(x)\sigma_{\mathbf{p}}\psi(x)]. \tag{4.47}$$

It is left to the reader to verify that

$$\begin{aligned} S_{\mathbf{p}} c_r^\dagger(\mathbf{p})|0\rangle &= (-1)^{r+1} \frac{\hbar}{2} c_r^\dagger(\mathbf{p})|0\rangle, \\ S_{\mathbf{p}} d_r^\dagger(\mathbf{p})|0\rangle &= (-1)^{r+1} \frac{\hbar}{2} d_r^\dagger(\mathbf{p})|0\rangle, \quad r = 1, 2. \end{aligned} \tag{4.48}$$

We see from Eqs. (4.48) that in both the electron state $c_r^\dagger(\mathbf{p})|0\rangle$ and the positron state $d_r^\dagger(\mathbf{p})|0\rangle$, the spin component in the direction of motion has the value $+\hbar/2$ for $r = 1$, and the value $-\hbar/2$ for $r = 2$. We refer to these two spin states, i.e. spin parallel and antiparallel to the direction of motion, as having positive (right-handed) and negative (left-handed) *helicity*, respectively. (Right- and left-handed here specifies the screw sense of the spin in the direction of motion.) We shall call $S_{\mathbf{p}}$ the helicity operator of a spin $\frac{1}{2}$ particle (whether electron or positron) with momentum $\mathbf{p}$.

It follows from Eqs. (4.44)–(4.46) and (4.48) that, as for the complex Klein–Gordon field, the theory is completely symmetric between particles (electrons) and antiparticles (positrons). These have the same properties, except for the reversal of the sign of the electric charge. (As a result, other electromagnetic properties, such as the magnetic moments, have opposite signs.)

The symmetry of the theory between particles and antiparticles is not obvious from the expansions of the field operators ψ and $\overline{\psi}$, Eqs. (4.38). This is due to the fact that we have not chosen a specific spinor representation and in most representations the positive and negative energy spinors will look very different. The expansions (4.38) only manifest the particle–antiparticle symmetry for representations of a particular kind, known as Majorana representations. Labelling the γ-matrices in a Majorana representation with the subscript M, the defining property of a Majorana representation is that

$$\gamma_M^{\mu*} = -\gamma_M^{\mu}, \quad \mu = 0, \ldots, 3, \tag{4.49}$$

where the asterisk denotes complex conjugation, i.e. all four γ-matrices are pure imaginary. A particular Majorana representation is given in Appendix A, Eqs. (A.79). Here we only require the defining property (4.49).

We see from Eq. (4.49) that in a Majorana representation, the operator

$$\left(i\hbar\gamma^\mu_M \frac{\partial}{\partial x^\mu} - mc\right) \tag{4.49a}$$

is real. Hence if ψ_M is a solution of the Dirac equation in a Majorana representation, so is its complex conjugate ψ^*_M. It follows that if we denote the positive energy solutions (4.31) by

$$u_{Mr}(\mathbf{p})\frac{e^{-ipx/\hbar}}{\sqrt{V}}, \quad r = 1, 2, \tag{4.50a}$$

in a Majorana representation, then the corresponding negative energy solutions are

$$u^*_{Mr}(\mathbf{p})\frac{e^{ipx/\hbar}}{\sqrt{V}}, \quad r = 1, 2. \tag{4.50b}$$

Hence the expansions (4.38) become, in a Majorana representation,

$$\left. \begin{aligned} \psi_M(x) &= \sum_{r\mathbf{p}} \left(\frac{mc^2}{VE_\mathbf{p}}\right)^{1/2} [c_r(\mathbf{p})u_{Mr}(\mathbf{p})\, e^{-ipx/\hbar} + d^\dagger_r(\mathbf{p})u^*_{Mr}(\mathbf{p})\, e^{ipx/\hbar}] \\ \psi^{\dagger T}_M(x) &= \sum_{r\mathbf{p}} \left(\frac{mc^2}{VE_\mathbf{p}}\right)^{1/2} [d_r(\mathbf{p})u_{Mr}(\mathbf{p})\, e^{-ipx/\hbar} + c^\dagger_r(\mathbf{p})u^*_{Mr}(\mathbf{p})\, e^{ipx/\hbar}] \end{aligned} \right\}. \tag{4.51}$$

In the last equation we gave the expansion for $\psi^{\dagger T}$ rather than $\bar\psi$ to bring out the complete symmetry between particles and antiparticles. The absorption operators $c_r(\mathbf{p})$ and $d_r(\mathbf{p})$ are multiplied by the same single-particle wavefunctions and thus are absorption operators of particles and antiparticles in the same single-particle state, i.e. with the same momentum, energy and helicity. The same is true of the creation operators. Having used a Majorana representation to manifest the particle–antiparticle symmetry of the field operators, we shall now revert to the representation-free formulation of Eqs. (4.31) and (4.38), in which this symmetry is masked.

The anticommutation relations (4.39) for the creation and absorption operators imply anticommutation relations for the Dirac field operators ψ and $\bar\psi$. From Eqs. (4.39) and the expansions (4.38) of the fields, one obtains

$$[\psi_\alpha(x), \psi_\beta(y)]_+ = [\bar\psi_\alpha(x), \bar\psi_\beta(y)]_+ = 0, \tag{4.52a}$$

$$[\psi^\pm_\alpha(x), \bar\psi^\mp_\beta(y)]_+ = i\left(i\gamma^\mu \frac{\partial}{\partial x^\mu} + \frac{mc}{\hbar}\right)_{\alpha\beta} \Delta^\pm(x - y) \tag{4.52b}$$

where $\Delta^\pm(x)$ are the invariant Δ-functions introduced for the Klein–Gordon equation, Eqs. (3.39) and (3.41). Eqs. (4.52a) are obvious. The derivation of Eqs. (4.52b) is left as an exercise for the reader.

Omitting suffixes, i.e. considered as a 4×4 matrix equation, we can write Eqs. (4.52b) as

$$[\psi^\pm(x), \bar\psi^\mp(y)]_+ = iS^\pm(x - y) \tag{4.53a}$$

where the 4×4 matrix functions $S^{\pm}(x)$ are defined by

$$S^{\pm}(x) = \left(i\gamma^{\mu}\frac{\partial}{\partial x^{\mu}} + \frac{mc}{\hbar}\right)\Delta^{\pm}(x). \tag{4.54a}$$

From the last two equations

$$[\psi(x), \overline{\psi}(y)]_{+} = iS(x-y) \tag{4.53b}$$

where, analogously to $\Delta(x) = \Delta^{+}(x) + \Delta^{-}(x)$, we defined

$$S(x) = S^{+}(x) + S^{-}(x) = \left(i\gamma^{\mu}\frac{\partial}{\partial x^{\mu}} + \frac{mc}{\hbar}\right)\Delta(x). \tag{4.54b}$$

From the representations of the various Δ-functions, obtained in Chapter 3, Eqs. (4.54) provide representations of the corresponding S-functions. For example, from the integral representation (3.50) for $\Delta^{\pm}(x)$, we can write Eqs. (4.54a) as

$$S^{\pm}(x) = \frac{-\hbar}{(2\pi\hbar)^4}\int_{C^{\pm}} d^4p \; e^{-ipx/\hbar}\frac{\not{p}+mc}{p^2-m^2c^2}, \tag{4.55a}$$

where the contours $C^{\pm}$ in the complex p_0-plane are anticlockwise closed paths enclosing the poles at $p_0 = \pm(E_{\mathbf{p}}/c)$, corresponding to Fig. 3.1 for the complex $k_0(= p_0/c)$-plane. Since

$$(\not{p}\pm mc)(\not{p}\mp mc) = p^2 - m^2c^2,$$

the last equation is often abbreviated into the symbolic form

$$S^{\pm}(x) = \frac{-\hbar}{(2\pi\hbar)^4}\int_{C^{\pm}} d^4p \; \frac{e^{-ipx/\hbar}}{\not{p}-mc}. \tag{4.55b}$$

4.3.1 The spin-statistics theorem

We conclude Section 4.3 with a brief discussion on the connection between spin and statistics of particles. So far, we have quantized the Dirac equation according to the anticommutation relations (4.39) in order to obtain Fermi–Dirac statistics for electrons. It is interesting to ask what the consequences would be if we quantize the Dirac equation according to Bose–Einstein statistics, i.e. by replacing all the anticommutators in Eqs. (4.39) by commutators. With this change, the energy of the field, again calculated from Eq. (4.22a), is not given by Eq. (4.44), but by

$$H = \sum_{r\mathbf{p}} E_{\mathbf{p}}[N_r(\mathbf{p}) - \overline{N}_r(\mathbf{p})]. \tag{4.56}$$

We are now dealing with Bose–Einstein statistics, and the occupation number operators $N_r(\mathbf{p})$ and $\overline{N}_r(\mathbf{p})$ can take on all values $0, 1, 2, \ldots$. Hence, the Hamiltonian (4.56) does not possess a lower bound. If we demand the existence of a state of lowest energy (i.e. a stable ground state), we must quantize the Dirac equation according to Fermi–Dirac statistics.

One may similarly ask what the consequences are of quantizing the Klein–Gordon field according to Fermi–Dirac statistics. In Section 3.3 we referred to microcausality, i.e. the requirement that two observables $A(x)$ and $B(y)$ must be compatible if $(x - y)$ is a space-like interval, i.e.

$$[A(x), B(y)] = 0, \quad \text{for } (x - y)^2 < 0. \tag{4.57}$$

We have seen that the observables of the fields, such as the energy–momentum densities or the charge–current densities, are bilinear in the field operators [see, for example, Eqs. (3.12), (3.13), (3.32), (4.22), (4.23) and (4.28)]. Using the operator identities (4.3) and (4.8), one can show that for Eq. (4.57) to hold for such bilinear observables, the fields themselves must either commute or anticommute for $(x - y)$ a space-like interval. For the real Klein–Gordon field we must have either

$$[\phi(x), \phi(y)] = 0, \quad \text{for } (x - y)^2 < 0,$$

or

$$[\phi(x), \phi(y)]_+ = 0, \quad \text{for } (x - y)^2 < 0.$$

We know that the first of these relations holds if the Klein–Gordon field is quantized according to Bose-Einstein statistics [compare Eq. (3.50)]. It is easy to show that neither relation holds if we quantize according to Fermi–Dirac statistics, i.e. replace the commutators by anticommutators in the commutation relations (3.9). Hence, the requirement of microcausality forces us to quantize the Klein–Gordon field according to Bose–Einstein statistics.

These conclusions generalize to interacting particles and other spin values. Particles with integral spin must be quantized according to Bose–Einstein statistics, particles with half-integral spin according to Fermi–Dirac statistics. The 'wrong' spin-statistics connections lead to the two types of difficulties we found above. This spin-statistics theorem, to which no exception is known in nature, represents an impressive success for relativistic quantum field theory.

4.4   The Fermion Propagator

In Section 3.4 we introduced the meson propagator. Corresponding to Eq. (3.55), we now define the Feynman fermion propagator as

$$\left\langle 0 | T\left\{ \psi(x)\overline{\psi}(x') \right\} | 0 \right\rangle \tag{4.58}$$

where spinor indices have again been suppressed. For fermion fields, the time-ordered product is defined by

$$\begin{aligned} T\{\psi(x)\overline{\psi}(x')\} &= \theta\left(t - t'\right)\psi(x)\overline{\psi}(x') - \theta(t' - t)\overline{\psi}(x')\psi(x) \\ &= \begin{cases} \psi(x)\overline{\psi}(x'), & \text{if } t > t' \\ -\overline{\psi}(x')\psi(x), & \text{if } t' > t \end{cases} \end{aligned} \tag{4.59}$$

(where $t = x^0/c$, etc.). This definition differs by a factor (-1) in the $t' > t$ term from the corresponding boson definition, Eqs. (3.52) and (3.54). This change in sign reflects the anticommutation property of fermion fields. (A similar difference occurred in the definition of the normal products of boson and fermion fields.)

In order to calculate the fermion propagator (4.58), using Eq. (4.59), we note that

$$\langle 0|\psi(x)\overline{\psi}(x')|0\rangle = \langle 0|\psi^+(x)\overline{\psi}^-(x')|0\rangle$$
$$= \langle 0|[\psi^+(x), \overline{\psi}^-(x')]_+|0\rangle = iS^+(x - x'),$$

(4.60a)

where we used Eq. (4.53a); similarly

$$\langle 0|\overline{\psi}(x')\psi(x)|0\rangle = iS^-(x - x').$$

(4.60b)

Combining Eqs. (4.58)–(4.60), we obtain the fermion propagator:

$$\langle 0|T\{\psi(x)\overline{\psi}(x')\}|0\rangle = iS_F(x - x')$$

(4.61)

where $S_F(x)$ is defined, analogously to Eqs. (3.56) and (4.54), by

$$S_F(x) = \theta(t)S^+(x) - \theta(-t)S^-(x) = \left(i\gamma^\mu \frac{\partial}{\partial x^\mu} + \frac{mc}{\hbar}\right)\Delta_F(x).$$

(4.62)

Corresponding to the integral representation (3.58) for $\Delta_F(x)$, $S_F(x)$ can be written

$$S_F(x) = \frac{\hbar}{(2\pi\hbar)^4} \int d^4p\, e^{-ipx/\hbar} \frac{\not{p} + mc}{p^2 - m^2c^2 + i\varepsilon},$$

(4.63)

where the integration in the complex p_0-plane is along the whole real axis: $-\infty < p_0 < \infty$. (Compare Fig. 3.6.)

As for the meson propagator, it is useful to visualize the fermion propagator in terms of Feynman diagrams. (As mentioned before, one must not take this interpretation too literally.)

For $t' < t$, the contribution to the fermion propagator (4.61) stems from the term (4.60a) and thus leads to the interpretation of (4.61) as creation of an electron at x', its propagation to x and its annihilation at x. On the other hand, for $t < t'$, the contribution to (4.61) comes from (4.60b) and is pictured as the emission of a positron at x, and its propagation to x' where it is annihilated. The two cases are illustrated in Fig. 4.1. Note that in both diagrams the arrow on the fermion line points *from* the vertex associated with the $\overline{\psi}$-field (x') *to* the vertex associated with the ψ-field (x), i.e. the arrow runs in the same direction as time for electrons, in the opposite direction for positrons.

These ideas are illustrated in Fig. 4.2, which shows two of the leading contributions, in lowest order of perturbation theory, to Compton scattering by electrons. Fig. 4.2(a) represents an electron propagating in the direction of the arrow on the fermion line, emitting the final photon at the vertex x' and absorbing the initial photon at x. Fig. 4.2(b) represents the corresponding process for $t < t'$. The initial photon is annihilated at x, creating an electron–positron pair, i.e. the final electron and a positron which propagates to x, where it annihilates with the initial electron to produce the final-state photon. Note that the arrow is always in the same sense along a fermion line, and in *both* diagrams is from the x'-vertex (associated with the $\overline{\psi}$ operator) to the x-vertex (associated

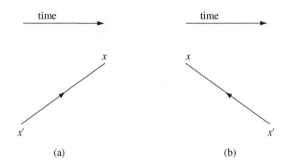

Figure 4.1   (a) $t' < t$: electron propagated from x' to x;
(b) $t' > t$: positron propagated from x to x'

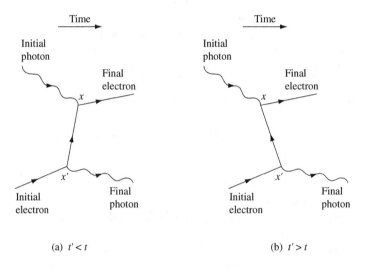

Figure 4.2   Contributions to Compton scattering: time-ordered graphs

with ψ) on the internal fermion line. Thus the two diagrams are topologically equivalent, i.e. they can be continuously deformed into each other.

The fermion propagator (4.61) includes both the contributions from Figs. 4.2(a) and (b), and we represent it by the single Feynman diagram in Fig. 4.3, in which no time ordering of the vertices x and x' is implied, and consequently no time direction is attached to the internal fermion line joining x' and x. The orientation of the line $x'x$ is of no significance in this diagram. However, we shall continue to interpret external lines (i.e. lines entering or leaving a diagram from outside) in the same way for Feynman graphs as for time-ordered graphs. A line entering a diagram from the left-hand side will be interpreted as a particle present initially, one leaving a diagram on the right-hand side as one present finally. With these conventions, Figs. 4.4(a) and (b) represent Compton scattering by electrons and positrons, without further labelling being necessary. The

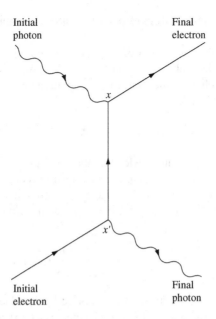

Initial
photon

Final
electron

x

x'

Initial
electron

Final
photon

Figure 4.3   *Contribution to Compton scattering: Feynman graph corresponding to the time-ordered graphs of Figure 4.2*

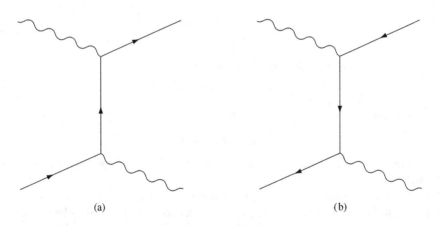

(a)   (b)

Figure 4.4   *Compton scattering:* (a) *by electrons;* (b) *by positrons*

arrows on the fermion lines are required to distinguish electrons (arrows on external lines from left to right) from positrons (arrows on external lines from right to left). Although no arrows are required on the photon lines, we shall at times use arrows on external photon lines to emphasize initially and finally present quanta. In the following we shall frequently use such Feynman diagrams.

4.5   The Electromagnetic Interaction and Gauge Invariance

We shall now consider the interaction of relativistic electrons with an electromagnetic field, specified by the scalar and vector potentials $\phi(x)$ and $A(x)$. For this purpose we shall take over the procedure which is successful in non-relativistic quantum mechanics. In the latter case, making the substitution

$$i\hbar \frac{\partial}{\partial t} \to i\hbar \frac{\partial}{\partial t} - q\phi(x), \qquad -i\hbar\nabla \to -i\hbar\nabla - \frac{q}{c}\mathbf{A}(x), \tag{4.64a}$$

in the free-particle Schrödinger equation leads to the correct wave equation for a particle of charge q in this field. [The corresponding classical result is contained in Eq. (1.59).]

The substitution (4.64a) is usually referred to as the 'minimal substitution'. In terms of the four-vector potential $A^\mu(x) = (\phi, A)$, the minimal substitution takes the explicitly covariant form

$$\partial_\mu \equiv \frac{\partial}{\partial x^\mu} \to D_\mu = \left[\partial_\mu + \frac{iq}{\hbar c}A_\mu(x)\right]. \tag{4.64b}$$

We shall assume that this substitution also correctly introduces the electromagnetic interaction into the Dirac equation. With the replacement (4.64b), and $q = -e$ for electrons, the Dirac equation (4.14) and the Lagrangian density (4.20) become

$$(i\hbar\gamma^\mu\partial_\mu - mc)\psi(x) = -\frac{e}{c}\gamma^\mu A_\mu(x)\psi(x), \tag{4.65}$$

and

$$\left. \begin{aligned} \mathscr{L} &= c\bar\psi(x)(i\hbar\gamma^\mu D_\mu - mc)\psi(x) \\ &= \mathscr{L}_0 + \mathscr{L}_1 \end{aligned} \right\} \tag{4.66}$$

where $\mathscr{L}_0$ is the Lagrangian density of the free Dirac field, i.e.

$$\mathscr{L}_0 = c\bar\psi(x)(i\hbar\gamma^\mu\partial_\mu - mc)\psi(x), \tag{4.67}$$

and $\mathscr{L}_1$ is the interaction Lagrangian density

$$\mathscr{L}_1 = e\bar\psi(x)\gamma^\mu\psi(x)A_\mu(x), \tag{4.68}$$

which couples the conserved current $s^\mu(x) = c(-e)\bar\psi\gamma^\mu\psi$, Eq. (4.28), to the electromagnetic field.

To obtain the complete Lagrangian density for electrodynamics, we must add to Eq. (4.66) the Lagrangian density $\mathscr{L}_{\mathrm{rad}}$ of the radiation field, i.e. of the electromagnetic field in the absence of charges. This division is analogous to that of the Hamiltonian in Chapter 1, Eqs. (1.61)–(1.63). $\mathscr{L}_{\mathrm{rad}}$ depends on the potential $A_\mu(x)$ only, and we shall study it in the next chapter.

We know that it is only the electromagnetic fields E and **B** which have physical significance, not the potential A_μ itself, i.e. the theory must be invariant under the gauge transformations of the potentials, Eqs. (1.3). The latter can be written in the covariant form

$$A_\mu(x) \to A'_\mu(x) = A_\mu(x) + \partial_\mu f(x) \tag{4.69a}$$

where $f(x)$ is an arbitrary function. The invariance of the theory under gauge transformations follows from that of the Lagrangian density. $\mathcal{L}_{rad}$ has this invariance property, as we shall see in the next chapter. However, applying (4.69a) to Eq. (4.66), we obtain

$$\mathcal{L} \rightarrow \mathcal{L}' = \mathcal{L} + e\overline{\psi}(x)\gamma^{\mu}\psi(x)\partial_{\mu}f(x) \tag{4.70}$$

i.e. $\mathcal{L}$ is not gauge invariant with respect to (4.69a). We can restore gauge invariance by demanding that coupled with the gauge transformation (4.69a) of the electromagnetic potentials, the Dirac fields transform according to

$$\left. \begin{array}{l} \psi(x) \rightarrow \psi'(x) = \psi(x)\, e^{ief(x)/\hbar c} \\ \overline{\psi}(x) \rightarrow \overline{\psi}'(x) = \overline{\psi}(x)\, e^{-ief(x)/\hbar c} \end{array} \right\} . \tag{4.69b}$$

Under the coupled transformations (4.69a) and (4.69b), the Lagrangian densities (4.67) and (4.68) transform according to

$$\mathcal{L}_0 \rightarrow \mathcal{L}'_0 = \mathcal{L}_0 - e\overline{\psi}(x)\gamma^{\mu}\psi(x)\partial_{\mu}f(x) \tag{4.71a}$$

$$\mathcal{L}_{I} \rightarrow \mathcal{L}'_{I} = \mathcal{L}_{I} + e\overline{\psi}(x)\gamma^{\mu}\psi(x)\partial_{\mu}f(x). \tag{4.71b}$$

Consequently, $\mathcal{L}=\mathcal{L}_0 + \mathcal{L}_{I}$ remains invariant under the coupled transformations.

Eqs. (4.69b) are called a *local* phase transformation, since the phase factors depend on x. In the special case that $f(x) = $ const., Eqs. (4.69b) reduce to a *global* phase transformation, considered in Section 2.4, where we saw that invariance under a global phase transformation leads to a conserved charge. We have now seen that gauge invariance of the theory requires invariance when simultaneously transforming the electromagnetic potentials according to the gauge transformation (4.69a) and the Dirac fields according to the local phase transformation (4.69b), and we shall in future refer to these coupled transformations as gauge transformations. It can be shown that the application of Noether's theorem to the invariance with respect to these (coupled) gauge transformations does not lead to a new conservation law, but only reproduces the conservation of charge.

In what follows we shall assume that Eq. (4.68) gives the correct interaction of quantum electrodynamics. One could try and add other gauge-invariant and Lorentz-invariant local interaction terms, but these can be excluded if we impose another restriction – renomalizability of the theory – as shown in Section 11.3.2. The ultimate justification for taking Eq. (4.68) as the correct interaction lies, of course, in the complete agreement between some of physics' most precise experiments and theoretical predictions based on this interaction.

Problems

4.1. From Eq. (4.53b), or otherwise, derive the equal-time anticommutation relation

$$[\psi(x), \overline{\psi}(y)]_{+}|_{x_0 = y_0} = \gamma^0\delta(\mathbf{x} - \mathbf{y}).$$

4.2. Show that the functions $S(x)$ and $S_F(x)$ are solutions of the homogeneous Dirac equation and of an inhomogeneous Dirac equation respectively.

4.3. Show that the charge–current density operator

$$s^\mu(x) = -ec\bar\psi(x)\gamma^\mu\psi(x)$$

of the Dirac equation satisfies the relation

$$[s^\mu(x), s^\nu(y)] = 0, \quad \text{for } (x-y)^2 < 0.$$

This relation shows that the charge–current densities, which are observable quantities, at two different space–time points x and y, are compatible, provided the interval $(x-y)$ is space-like, as required by microcausality.

4.4. Show that if in the expansion (3.7) of the real Klein–Gordon field $\phi(x)$ we impose the anticommutation relations

$$[a(\mathbf{k}),\, a^\dagger(\mathbf{k}')]_+ = \delta_{\mathbf{kk}'}, \quad [a(\mathbf{k}),\, a(\mathbf{k}')]_+ = [a^\dagger(\mathbf{k}),\, a^\dagger(\mathbf{k}')]_+ = 0,$$

then for $(x-y)$, a space-like interval:

$$[\phi(x),\phi(y)] \neq 0 \quad \text{and} \quad [\phi(x),\phi(y)]_+ \neq 0.$$

[We know from the discussion at the end of Section 4.3 that either the commutator or the anticommutator of $\phi(x)$ and $\phi(y)$ must vanish for $(x-y)$ space-like, if the bilinear observables constructed from ϕ are to satisfy the microcausality condition (4.57).]

4.5. For a Dirac field, the transformations

$$\psi(x) \to \psi'(x) = \exp(i\alpha\gamma_5)\psi(x), \quad \psi^\dagger(x) \to \psi^{\dagger'}(x) = \psi^\dagger(x)\exp(-i\alpha\gamma_5),$$

where α is an arbitrary real parameter, are called chiral phase transformations.

Show that the Lagrangian density (4.20) is invariant under chiral phase transformations in the zero-mass limit $m = 0$ only, and that the corresponding conserved current in this limit is the axial vector current $J_A^\alpha(x) \equiv \bar\psi(x)\gamma^\alpha\gamma_5\psi(x)$.

Deduce the equations of motion for the fields

$$\psi_L(x) \equiv \frac{1}{2}(1-\gamma_5)\psi(x), \quad \psi_R(x) \equiv \frac{1}{2}(1+\gamma_5)\psi(x)$$

for non-vanishing mass, and show that they decouple in the limit $m = 0$. Hence show that the Lagrangian density

$$\mathscr{L}(x) = i\hbar c\bar\psi_L(x)\gamma^\mu\partial_\mu\psi_L(x)$$

describes zero-mass fermions with negative helicity only, and zero-mass antifermions with positive helicity only. (This field is called the Weyl field and can be used to describe the neutrinos in weak interactions in the approximation of zero mass.)

5

Photons: Covariant Theory

In our discussion of the electromagnetic field in Chapter 1, we saw that only the transverse radiation field corresponds to independent dynamical degrees of freedom, and we only quantized this transverse field. On the other hand, the instantaneous Coulomb interaction between charges is fully determined by the charge distribution and, in the formulation of Chapter 1, is treated as a classical potential. This formulation of quantum electrodynamics is closely related to the classical theory and so facilitates interpretation in familiar terms. However, the decomposition of the fields into transverse and longitudinal components is clearly frame dependent and so hides the Lorentz invariance of the theory.

An explicitly Lorentz-covariant formulation of the theory is essential for a complete development of quantum electrodynamics. This is required to establish the renormalizability of the theory, i.e. the possibility of carrying out calculations to all orders of perturbation theory with finite self-consistent results, and it is very helpful in practice in calculating such higher-order radiative corrections.

We shall therefore in this chapter develop a covariant theory starting, in Section 5.1, from an explicitly covariant formulation of classical electrodynamics in which all four components of the four-vector potential $A^\mu(x) = (\phi, \mathbf{A})$ are treated on an equal footing. This corresponds to introducing more dynamical degrees of freedom than the system possesses and these will later have to be removed by imposing suitable constraints.

The quantized theory, derived in Section 5.2 by quantizing all four components of the four-vector potential $A^\mu(x)$, looks on the face of it very different from the theory of Chapter 1. However, the two formulations are equivalent, as we shall illustrate when discussing the photon propagator in Section 5.3.

5.1 The Classical Fields

To express Maxwell's equations in covariant form, we introduce the antisymmetric field tensor

Quantum Field Theory, Second Edition   Franz Mandl and Graham Shaw
© 2010 John Wiley & Sons, Ltd

$$
\begin{array}{cccccc}
\nu \to 0 & 1 & 2 & 3 & \mu \\
& & & & \downarrow
\end{array}
$$

$$
F^{\mu\nu}(x) = \begin{pmatrix}
0 & E_x & E_y & E_z \\
-E_x & 0 & B_z & -B_y \\
-E_y & -B_z & 0 & B_x \\
-E_z & B_y & -B_x & 0
\end{pmatrix}
\begin{array}{c}
0 \\ 1 \\ 2 \\ 3
\end{array}
\tag{5.1}
$$

In terms of $F^{\mu\nu}$ and the charge-current density $s^{\mu}(x) = (c\rho(x), \mathbf{j}(x))$, Maxwell's equations (1.1) become

$$
\partial_{\nu} F^{\mu\nu}(x) = \frac{1}{c} s^{\mu}(x)
\tag{5.2}
$$

$$
\partial^{\lambda} F^{\mu\nu}(x) + \partial^{\mu} F^{\nu\lambda}(x) + \partial^{\nu} F^{\lambda\mu}(x) = 0.
\tag{5.3}
$$

Since $F^{\mu\nu}$ is antisymmetric, Eq. (5.2) at once gives

$$
\partial_{\mu} s^{\mu}(x) = 0,
\tag{5.4}
$$

i.e. consistency requires conservation of the current to which the electromagnetic field is coupled.

The field $F^{\mu\nu}$ can be expressed in terms of the four-vector potential $A^{\mu}(x) = (\phi, \mathbf{A})$ by

$$
F^{\mu\nu}(x) = \partial^{\nu} A^{\mu}(x) - \partial^{\mu} A^{\nu}(x)
\tag{5.5}
$$

which is identical with Eqs. (1.2). In terms of the potentials, Eqs. (5.3) are satisfied identically, and Eqs. (5.2) become

$$
\Box A^{\mu}(x) - \partial^{\mu}(\partial_{\nu} A^{\nu}(x)) = \frac{1}{c} s^{\mu}(x).
\tag{5.6}
$$

These equations are Lorentz covariant, and they are also invariant under the gauge transformation

$$
A^{\mu}(x) \to A'^{\mu}(x) = A^{\mu}(x) + \partial^{\mu} f(x).
\tag{5.7}
$$

The field equations (5.6) can be derived from the Lagrangian density

$$
\mathscr{L} = -\frac{1}{4} F_{\mu\nu}(x) F^{\mu\nu}(x) - \frac{1}{c} s_{\mu}(x) A^{\mu}(x)
\tag{5.8}
$$

by treating the four components $A^{\mu}(x)$ as the independent fields in the variational principle (2.11)–(2.14). The form of this Lagrangian density ensures the correct behaviour of the field equations (5.6) under Lorentz and gauge transformations.[1]

---

[1] $\mathscr{L}$ is clearly Lorentz invariant. Under the gauge transformation (5.7)

$$
\mathscr{L} \to \mathscr{L} - \frac{1}{c} s_{\mu}(x) \partial^{\mu} f(x) = \mathscr{L} - \frac{1}{c} \partial^{\mu} [s_{\mu}(x) f(x)],
\tag{5.9}
$$

on account of current conservation. Although $\mathscr{L}$ is not invariant, it follows from Eqs. (2.11)–(2.14) that adding a four-divergence to the Lagrangian density does not alter the field equations, i.e. their gauge invariance is ensured by the form of $\mathscr{L}$. (See Problem 2.1.)

Unfortunately, the Lagrangian density (5.8) is not suitable for carrying out the canonical quantization. Eq. (5.8) leads to the conjugate fields

$$\pi^\mu(x) = \frac{\partial \mathscr{L}}{\partial \dot{A}_\mu} = -\frac{1}{c} F^{\mu 0}(x).$$

The antisymmetry of $F^{\mu\nu}$ then implies $\pi^0(x) \equiv 0$, and this is plainly incompatible with the canonical commutation relations (2.31) which we wish to impose.

A Lagrangian density which is suitable for quantization, first proposed by Fermi, is

$$\mathscr{L} = -\frac{1}{2} (\partial_\nu A_\mu(x))(\partial^\nu A^\mu(x)) - \frac{1}{c} s_\mu(x) A^\mu(x). \tag{5.10}$$

From Eq. (5.10) one obtains the conjugate fields

$$\pi^\mu(x) = \frac{\partial \mathscr{L}}{\partial \dot{A}_\mu} = -\frac{1}{c^2} \dot{A}^\mu(x) \tag{5.11}$$

which are now all non-vanishing so that the canonical quantization formalism can be applied.

The Lagrangian density (5.10) leads to the field equations

$$\Box A^\mu(x) = \frac{1}{c} s^\mu(x). \tag{5.12}$$

Comparison with Eqs. (5.6) shows that Eqs. (5.12) are only equivalent to Maxwell's equations if the potential $A^\mu(x)$ satisfies the constraint

$$\partial_\mu A^\mu(x) = 0. \tag{5.13}$$

Hence, to carry out the quantization, but end up with Maxwell's equations, we must in the first place quantize the theory for the general Lagrangian density (5.10), ignoring the constraint (5.13), and after quantization impose Eq. (5.13) or an equivalent constraint as a subsidiary condition. We shall consider this point in detail in the next section.

In the classical theory, starting from potentials $A^\mu(x)$ in an arbitrary gauge, we can always perform a gauge transformation (5.7) so that the transformed potentials $A'^\mu(x)$ satisfy the subsidiary condition (5.13). We achieve this by choosing the function $f(x)$ in Eq. (5.7) as a solution of

$$\partial_\mu A^\mu(x) + \Box f(x) = 0. \tag{5.14}$$

The subsidiary condition (5.13) does not specify the potentials uniquely. If the potentials $A^\mu(x)$ satisfy Eq. (5.13), so will any potentials $A'^\mu(x)$ obtained by the gauge transformation (5.7), provided the gauge function $f(x)$ satisfies

$$\Box f(x) = 0. \tag{5.15}$$

The subsidiary condition (5.13) is called the Lorentz condition. Its imposition represents a restriction on the choice of gauge. Any gauge in which Eq. (5.13) holds is called a Lorentz guage.

Using a Lorentz gauge has some important advantages. Firstly, the Lorentz condition (5.13) is a Lorentz-covariant constraint. This is in contrast to the condition for the Coulomb gauge, Eq. (1.6),

$$\nabla \cdot \mathbf{A} = 0,$$

which decomposes fields into transverse and longitudinal components and so is manifestly frame dependent. Secondly, the field equations (5.12) in a Lorentz gauge are much simpler than the corresponding Eqs. (5.6) in a general gauge. In particular, in the free field case ($s^\mu(x) = 0$) Eqs. (5.12) reduce to

$$\Box A^\mu(x) = 0. \tag{5.16}$$

Eq. (5.16) is the limit of the Klein–Gordon equation (3.3) for particles with mass zero. This will enable us to adapt many of our earlier results when considering the covariant quantization of the electromagnetic field.

Eq. (5.16) enables us to expand the free electromagnetic field $A^\mu(x)$ in a complete set of solutions of the wave equation, in close analogy to the expansion (3.7) for the Klein–Gordon field:

$$A^\mu(x) = A^{\mu+}(x) + A^{\mu-}(x) \tag{5.16a}$$

where

$$A^{\mu+}(x) = \sum_{r\mathbf{k}} \left(\frac{\hbar c^2}{2V\omega_\mathbf{k}} \right)^{1/2} \varepsilon_r^\mu(\mathbf{k}) a_r(\mathbf{k}) \, e^{-ikx} \tag{5.16b}$$

and

$$A^{\mu-}(x) = \sum_{r\mathbf{k}} \left(\frac{\hbar c^2}{2V\omega_\mathbf{k}} \right)^{1/2} \varepsilon_r^\mu(\mathbf{k}) a_r^\dagger(\mathbf{k}) \, e^{ikx}. \tag{5.16c}$$

The summations in these equations are over wave vectors $\mathbf{k}$ allowed by the periodic boundary conditions, and

$$k^0 = \frac{1}{c}\omega_\mathbf{k} = |\mathbf{k}|. \tag{5.17}$$

The summation over r, from $r = 0$ to $r = 3$, corresponds to the fact that for the four-vector field $A^\mu(x)$ there exist, for each $\mathbf{k}$, four linearly independent polarization states. These are described by the polarization vectors $\varepsilon_r^\mu(\mathbf{k})$, $r = 0, \ldots, 3$, which we choose to be real, and which satisfy the orthonormality and completeness relations

$$\varepsilon_r(\mathbf{k})\varepsilon_s(\mathbf{k}) = \varepsilon_{r\mu}(\mathbf{k})\varepsilon_s^\mu(\mathbf{k}) = -\zeta_r \, \delta_{rs}, \quad r, s = 0, \ldots, 3, \tag{5.18}$$

$$\sum_r \zeta_r \varepsilon_r^\mu(\mathbf{k})\varepsilon_r^\nu(\mathbf{k}) = -g^{\mu\nu}, \tag{5.19}$$

where

$$\zeta_0 = -1, \quad \zeta_1 = \zeta_2 = \zeta_3 = 1. \tag{5.20}$$

The classical potentials $A^\mu(x)$, $\mu = 0, \ldots 3$, are of course real quantities. Anticipating their interpretation in the quantized theory as operators, we have denoted the expansion coefficients in Eqs. (5.16) by a_r and $a_r^\dagger$.

Eqs. (5.16) should be compared with Eqs. (1.38). The latter expand the radiation field in terms of two transverse polarization states for each value of $\mathbf{k}$ and, in addition, we had the instantaneous Coulomb interaction between charges. Eqs. (5.16) give an expansion of the *total* field $A^\mu(x)$ in terms of four polarization states for each value of $\mathbf{k}$. We shall see in Section 5.3 that the two extra polarization states provide a covariant description of the instantaneous Coulomb interaction.

For many purposes one only requires the properties (5.18) and (5.19) of the polarization vectors. However a specific choice of polarization vectors in one given frame of reference often facilitates the interpretation. We shall choose these vectors as

$$\varepsilon_0^\mu(\mathbf{k}) = n^\mu \equiv (1, 0, 0, 0),
\tag{5.21a}$$

$$\varepsilon_r^\mu(\mathbf{k}) = (0, \boldsymbol{\varepsilon}_r(\mathbf{k})), \quad r = 1, 2, 3,
\tag{5.21b}$$

where $\boldsymbol{\varepsilon}_1(\mathbf{k})$ and $\boldsymbol{\varepsilon}_2(\mathbf{k})$ are mutually orthogonal unit vectors, which are also orthogonal to $\mathbf{k}$, and

$$\boldsymbol{\varepsilon}_3(\mathbf{k}) = \mathbf{k}/|\mathbf{k}|,
\tag{5.22a}$$

i.e.

$$\mathbf{k} \cdot \boldsymbol{\varepsilon}_r(\mathbf{k}) = 0, \quad r = 1, 2; \quad \boldsymbol{\varepsilon}_r(\mathbf{k}) \cdot \boldsymbol{\varepsilon}_s(\mathbf{k}) = \delta_{rs}, \quad r, s = 1, 2, 3.
\tag{5.22b}$$

ε_1^μ and ε_2^μ are called transverse, ε_3^μ longitudinal polarizations, and ε_0^μ scalar or time-like polarization.

For later use we note that $\varepsilon_3^\mu(\mathbf{k})$ can be written in the covariant form

$$\varepsilon_3^\mu(\mathbf{k}) = \frac{k^\mu - (kn)n^\mu}{[(kn)^2 - k^2]^{1/2}}.
\tag{5.22c}$$

This expression comes about since $(kn)n^\mu$ subtracts off the time-like component of k^μ, and the denominator makes ε_3^μ a space-like unit vector. We have not set $k^2 = 0$ in Eq. (5.22c), as it would be for a real photon, since we shall later require the more general case $k^2 \neq 0$.

Real polarization vectors correspond to linear polarization. To describe circular or elliptic polarization would require complex polarization vectors and corresponding modifications of Eqs. (5.18) and (5.19).

5.2 Covariant Quantization

We now apply the canonical formalism of Chapter 2 to quantize the free electromagnetic field, using the Lagrangian density (5.10) with $s_\mu(x) = 0$ and, in the first place, ignoring the Lorentz condition (5.13). With the fields $\pi^\mu(x)$ conjugate to $A_\mu(x)$ given by Eqs. (5.11), the equal-time commutation relations (2.31) become

$$\left[A^\mu(\mathbf{x}, t), A^\nu(\mathbf{x}', t)\right] = 0, \quad \left[\dot{A}^\mu(\mathbf{x}, t), \dot{A}^\nu(\mathbf{x}', t)\right] = 0,$$
$$\left[A^\mu(\mathbf{x}, t), \dot{A}^\nu(\mathbf{x}', t)\right] = -i\hbar c^2 g^{\mu\nu} \delta(\mathbf{x} - \mathbf{x}')
\tag{5.23}$$

Apart from the factor $(-g^{\mu\nu})$, these equations are identical with the commutation relations (3.6) of four independent Klein–Gordon fields, and each component $A^\mu(x)$ satisfies the wave equation (5.16) which is the limit of the Klein–Gordon equation (3.3) for particles of mass zero. [Both these points can be appreciated by comparing the Lagrangian densities (5.10) and (3.4).] This similarity enables us to take over earlier mathematical results, although their physical interpretation will have to be re-examined, taking into account the factor $(-g^{\mu\nu})$.

In Section 3.3, we derived the covariant commutation relations (3.42) for the Klein–Gordon field. From these we can at once write down the covariant commutation relations for the $A^\mu(x)$:

$$\left[A^\mu(x),\, A^\nu(x')\right] = i\hbar c D^{\mu\nu}(x - x'),\tag{5.24}$$

where

$$D^{\mu\nu}(x) = \lim_{m\to 0}\ [-g^{\mu\nu}\Delta(x)],\tag{5.25}$$

and $\Delta(x)$ is the invariant Δ-function (3.43).

The Feynman photon propagator is similarly given by

$$\langle 0|T\left\{A^\mu(x)A^\nu(x')\right\}|0\rangle = i\hbar c D_F^{\mu\nu}(x - x'),\tag{5.26}$$

where

$$D_F^{\mu\nu}(x) = \lim_{m\to 0}\ [-g^{\mu\nu}\,\Delta_F(x)] = \frac{-g^{\mu\nu}}{(2\pi)^4}\int \frac{\mathrm{d}^4 k\, \mathrm{e}^{-ikx}}{k^2 + i\varepsilon},\tag{5.27}$$

as is seen from Eqs. (3.55) and (3.58). The photon propagator will be discussed fully in the next section.

To gain the photon interpretation of the quantized fields, we substitute the field expansions (5.16) in the commutation relations (5.23), with the result

$$\left.\begin{array}{l}\left[a_r(\mathbf{k}),\, a_s^\dagger(\mathbf{k}')\right] = \zeta_r\,\delta_{rs}\,\delta_{\mathbf{k}\mathbf{k}'}\\[4pt]\left[a_r(\mathbf{k}),\, a_s(\mathbf{k}')\right] = \left[a_r^\dagger(\mathbf{k}),\, a_s^\dagger(\mathbf{k}')\right] = 0\end{array}\right\}.\tag{5.28}$$

From Eq. (5.20) $\zeta_r = 1$ for $r = 1, 2, 3$ so that for these values of r, Eqs. (5.28) are the standard boson commutation relations (3.9) leading to the usual number representation for transverse photons ($r = 1, 2$) and longitudinal photons ($r = 3$). For $r = 0$ (scalar photons) $\zeta_0 = -1$, and it consequently looks as though the usual roles of absorption and creation operators must be interchanged for $a_0(\mathbf{k})$ and $a_0^\dagger(\mathbf{k})$. However, effecting only this change results in other difficulties, and the standard formalism must be modified more radically. Of the several procedures available, we shall follow that due to Gupta and to Bleuler.

In the Gupta–Bleuler theory, the operators $a_r(\mathbf{k}), r = 1, 2, 3$ *and* 0, are interpreted as absorption operators, $a_r^\dagger(\mathbf{k}), r = 1, 2, 3$ *and* 0, as creation operators for transverse, long-itudinal *and* scalar photons. The vacuum state $|0\rangle$ is defined as the state in which there are no photons of any kind present, i.e.

$$a_r(\mathbf{k})|0\rangle = 0, \quad \text{all } \mathbf{k}, \quad r = 0 \ldots, 3,\tag{5.29a}$$

or, equivalently,

$$A^{\mu +}(x)|0\rangle = 0, \quad \text{all } x, \quad \mu = 0, \dots, 3. \tag{5.29b}$$

The operators $a_r^\dagger(\mathbf{k})$ operating on the vacuum state $|0\rangle$ create the one-photon states

$$|1_{\mathbf{k}r}\rangle = a_r^\dagger(\mathbf{k})|0\rangle \tag{5.30}$$

in which one transverse $(r = 1, 2)$, longitudinal $(r = 3)$ or scalar $(r = 0)$ photon of momentum $\mathbf{k}$ is present.

To justify this interpretation of the operators a_r and $a_r^\dagger$, we consider the Hamiltonian operator of the field. From Eq. (2.51a) this is given by

$$H = \int d^3\mathbf{x} N[\pi^\mu(x)\dot{A}_\mu(x) - \mathscr{L}(x)], \tag{5.31}$$

which, as usual, is to be taken as a normal product. On substituting the free Lagrangian density corresponding to Eq. (5.10), Eq, (5.11) for $\pi^\mu(x)$ and the expansions (5.16) for the fields, Eq. (5.31) becomes

$$H = \sum_{r\mathbf{k}} \hbar\omega_{\mathbf{k}} \zeta_r a_r^\dagger(\mathbf{k}) a_r(\mathbf{k}). \tag{5.32}$$

Despite the minus sign $(\zeta_0 = -1)$ associated with the scalar photons in Eq. (5.32), this energy is positive definite. For example, for the one-photon states (5.30) one easily obtains, using the commutation relations (5.28),

$$\begin{aligned} H|1_{\mathbf{k}r}\rangle &= \sum_{\mathbf{q}s} \hbar\omega_{\mathbf{q}} \zeta_s a_s^\dagger(\mathbf{q}) a_s(\mathbf{q}) a_r^\dagger(\mathbf{k})|0\rangle \\ &= \hbar\omega_{\mathbf{k}} a_r^\dagger(\mathbf{k})|0\rangle, \quad r = 0, \dots, 3, \end{aligned}$$

i.e. the energy has the positive value $\hbar\omega_{\mathbf{k}}$ for transverse, longitudinal and scalar photons. Correspondingly, we must define the number operators by

$$N_r(\mathbf{k}) = \zeta_r a_r^\dagger(\mathbf{k}) a_r(\mathbf{k}), \tag{5.33}$$

and these definitions, together with the commutation relations (5.28), lead to consistent number representations for all types of photons.

Although the formalism, as far as we have developed it, seems satisfactory, there are some difficulties which show up if we calculate the normalization of photon states. For example, the norm of the state (5.30) is

$$\langle 1_{\mathbf{k}r}|1_{\mathbf{k}r}\rangle = \langle 0|a_r(\mathbf{k})a_r^\dagger(\mathbf{k})|0\rangle = \zeta_r\langle 0|0\rangle = \zeta_r$$

(if we normalize $|0\rangle$ to $\langle 0|0\rangle = 1$), and for a scalar photon this norm is negative. More generally, one can show that for any state containing an odd number of scalar photons the norm is negative. At first sight this looks like a serious difficulty, since the probability interpretation of quantum mechanics depends on states having positive norms. However, no scalar or longitudinal photons have ever been observed. Both these points are related to the fact that so far we have ignored the Lorentz condition (5.13), so that our theory is not yet equivalent to Maxwell's equations. We must now try and impose the Lorentz condition.

Unfortunately, we cannot simply take the Lorentz condition (5.13) as an operator identity. Eq. (5.13) is incompatible with the commutation relations (5.24), since

$$\left[\partial_\mu A^\mu(x), A^\nu(x')\right] = i\hbar c \partial_\mu D^{\mu\nu}(x - x')$$

and this is not identically zero.

This problem was resolved by Gupta and Bleuler by replacing the Lorentz condition (5.13) by the weaker condition

$$\partial_\mu A^{\mu+}(x)|\Psi\rangle = 0, \tag{5.34}$$

involving absorption operators only. Eq. (5.34) is a restriction on the states which are allowed by the theory. From Eq. (5.34) and its adjoint

$$\langle\Psi|\partial_\mu A^{\mu-}(x) = 0$$

it follows that the Lorentz condition holds for expectation values:

$$\langle\Psi|\partial_\mu A^\mu(x)|\Psi\rangle = \langle\Psi|\partial_\mu A^{\mu+}(x) + \partial_\mu A^{\mu-}(x)|\Psi\rangle = 0. \tag{5.35}$$

This ensures that the Lorentz condition and hence Maxwell's equations hold as the classical limit of this theory.

In order to understand the meaning of the subsidiary condition (5.34), we express it in momentum space. On substituting Eqs. (5.16b) and (5.21), (5.22) for $A_\mu^+(x)$ and $\varepsilon_r^\mu(\mathbf{k})$, we obtain the conditions

$$[a_3(\mathbf{k}) - a_0(\mathbf{k})]|\Psi\rangle = 0, \quad \text{all } \mathbf{k}. \tag{5.36}$$

This is a constraint on the linear combinations of longitudinal and scalar photons, for each value of $\mathbf{k}$, that may be present in a state. It places no restriction on the transverse photons that may be present.

The effect of the subsidiary condition (5.36) becomes apparent if we calculate the expectation value of the energy of an allowed state $|\Psi\rangle$. Since from Eq. (5.36) and its adjoint we have

$$\langle\Psi|a_3^\dagger(\mathbf{k})a_3(\mathbf{k}) - a_0^\dagger(\mathbf{k})a_0(\mathbf{k})|\Psi\rangle = \langle\Psi|a_3^\dagger(\mathbf{k})[a_3(\mathbf{k}) - a_0(\mathbf{k})]|\Psi\rangle = 0,$$

it follows from Eq. (5.32) that

$$\langle\Psi|H|\Psi\rangle = \langle\Psi|\sum_{\mathbf{k}}\sum_{r=1}^{2}\hbar\omega_k a_r^\dagger(\mathbf{k})a_r(\mathbf{k})|\Psi\rangle, \tag{5.37}$$

i.e. only the transverse photons contribute to the expectation value of the energy as a consequence of the subsidiary condition. The same is true for all other observables.

Thus, as a result of the subsidiary condition, in free space, observable quantities will involve transverse photons only. This explains our earlier assertion that longitudinal and scalar photons are not observed as free particles. Only transverse photons are so observed, corresponding to the two degrees of freedom (for each $\mathbf{k}$) of the radiation field, which we found in the non-covariant formalism of Chapter 1 where we worked in the Coulomb gauge. In the covariant treatment, although they don't show up as free particles, the presence of longitudinal and scalar photons is not ruled out altogether. Of the resulting additional two degrees of freedom (for each **k**), one is removed by the subsidiary condition (5.36). The other can be shown to correspond to the arbitrariness in choice of Lorentz gauge. More specifically, one can show that altering the allowed admixtures of longitudinal and scalar photons is equivalent to a gauge transformation between two potentials, both of which are in Lorentz gauges. (See Problems 5.2 and 5.3.)

For free fields (i.e. no charges present), it is then simplest to work in a gauge such that the vacuum is represented by the state $|0\rangle$ in which no photons of any kind are present [see Eq. (5.29a)]. But the vacuum could also be described by any state containing no transverse and only allowed admixtures of scalar and longitudinal photons. This description would merely correspond to a different choice of Lorentz gauge. The situation is entirely analogous for states containing transverse photons.

For the electromagnetic field in the presence of charges, the situation is more complicated. We can no longer ignore the longitudinal and scalar photons. When discussing the photon propagator in the next section, we shall see that longitudinal and scalar photons play an important role as virtual particles in intermediate states and provide a covariant description of the instantaneous Coulomb interaction of Chapter 1. However, in this case too, one need consider only transverse photons in initial and final states of scattering processes. This corresponds to a particular choice of gauge and the fact that one can consider particles initially and finally, when they are far apart, as free. In Section 6.2 we shall return to this idea of switching the interaction between colliding particles on and off adiabatically as they approach and as they move apart.

We have developed the Gupta–Bleuler formalism only to the limited extent to which it is needed in applications. It is possible to develop a more complete systematic formalism in which states with negative norm do not appear as a blemish in Hilbert space, but occur in a self-consistent manner in a function space with an indefinite metric. For most purposes, this complete formalism is not required.[2]

5.3 The Photon Propagator

In Section 3.4 we interpreted the Klein–Gordon propagator (3.56) as the exchange of a virtual meson in an intermediate state. We now expect a similar interpretation for the photon propagator (5.26) but, corresponding to the four-vector nature of the field $A^\mu(x)$ and the resulting four independent polarization states, we expect the exchange of four kinds of photons, two corresponding to transverse polarization and one each to long-itudinal and scalar polarization. This description differs markedly from that of Chapter 1, where only transverse radiation occurred, but no longitudinal or scalar radiation. Instead we had the instantaneous Coulomb interaction between charges. We shall see that these two descriptions are indeed equivalent.

[2] The interested reader is referred to S. N Gupta, *Quantum Electrodynamics*, Gordon and Breach, New York, 1977; G. Källén, *Quantum Electrodynamics*, Springer, New York, 1972, and Allen & Unwin, London, 1972; or J.M. Jauch and F. Rohrlich, *The Theory of Photons and Electrons*, 2nd edn, Springer, New York, 1976, Section 6.3.

To establish this interpretation in terms of photon exchange we consider the momentum space propagator $D_F^{\mu\nu}(k)$, related to the configuration space propagator $D_F^{\mu\nu}(x)$ Eq. (5.27), by

$$D_F^{\mu\nu}(x) = \frac{1}{(2\pi)^4} \int d^4k D_F^{\mu\nu}(k)\, e^{-ikx}. \tag{5.38}$$

From Eqs. (5.27) and (5.19) we obtain

$$D_F^{\mu\nu}(k) = \frac{-g^{\mu\nu}}{k^2 + i\varepsilon} = \frac{1}{k^2 + i\varepsilon} \sum_r \zeta_r \varepsilon_r^\mu(\mathbf{k})\varepsilon_r^\nu(\mathbf{k}). \tag{5.39}$$

In order to interpret this expression, we use the special frame of reference in which the polarization vectors $\varepsilon_r^\mu(\mathbf{k})$ are given by Eqs. (5.21) and (5.22). The last equation then becomes:

$$D_F^{\mu\nu}(k) = \frac{1}{k^2 + i\varepsilon} \left\{ \sum_{r=1}^{2} \varepsilon_r^\mu(\mathbf{k})\varepsilon_r^\nu(\mathbf{k}) + \frac{[k^\mu - (kn)n^\mu][k^\nu - (kn)n^\nu]}{(kn)^2 - k^2} + (-1)n^\mu n^\nu \right\}. \tag{5.40}$$

This equation exhibits the contributions to the photon propagator from transverse, longitudinal and scalar photons.

By analogy with the meson case, we interpret the first term in Eq. (5.40),

$$_T D_F^{\mu\nu}(k) \equiv \frac{1}{k^2 + i\varepsilon} \sum_{r=1}^{2} \varepsilon_r^\mu(\mathbf{k})\varepsilon_r^\nu(\mathbf{k}), \tag{5.41a}$$

as the exchange of transverse photons. In the language of Chapter 1, it corresponds to the interaction of charges via the transverse radiation field.

The interpretation of the remaining two terms in Eq. (5.40) follows not from considering longitudinal and scalar photons separately, but from combining them into a term proportional to $n^\mu n^\nu$ plus the remainder. Eq. (5.40) then becomes

$$D_F^{\mu\nu}(k) = {}_T D_F^{\mu\nu}(k) + {}_C D_F^{\mu\nu}(k) + {}_R D_F^{\mu\nu}(k), \tag{5.42}$$

where

$$_C D_F^{\mu\nu}(k) \equiv \frac{n^\mu n^\nu}{(kn)^2 - k^2}, \tag{5.41b}$$

$$_R D_F^{\mu\nu}(k) \equiv \frac{1}{k^2 + i\varepsilon} \left[\frac{k^\mu k^\nu - (kn)(k^\mu n^\nu + k^\nu n^\mu)}{(kn)^2 - k^2} \right], \tag{5.41c}$$

and it is these linear combinations, both of which involve longitudinal and scalar photons, which allow a simple interpretation.

We first consider Eq. (5.41b) in configuration space. From Eqs. (5.38) and (5.21a) we obtain

$$c D_F^{\mu\nu}(x) = \frac{g^{\mu 0} g^{\nu 0}}{(2\pi)^4} \int \frac{d^3k\, e^{i k \cdot x}}{|\mathbf{k}|^2} \int dk^0\, e^{-i k^0 x^0}$$

$$= g^{\mu 0}\, g^{\nu 0} \frac{1}{4\pi |\mathbf{x}|} \delta(x^0). \tag{5.43}$$

This expression has the time dependence $[\delta(x^0)]$ and the space dependence $[1/|\mathbf{x}|]$ characteristic of an instantaneous Coulomb potential. Thus we see that the exchange of longitudinal and scalar photons, represented by the term (5.43), corresponds to the instantaneous Coulomb interaction between charges. In Chapter 1, we quantized the transverse radiation field only and treated the instantaneous Coulomb interaction as a classical potential, corresponding to the fact that the instantaneous Coulomb field does not represent independent dynamical degrees of freedom, but is fully determined by the charges. In the present treatment, the longitudinal and scalar field components are also quantized and the instantaneous Coulomb interaction emerges as an exchange of longitudinal and scalar photons.

Finally, we must discuss the remainder term (5.41c). In Chapter 1, the complete electromagnetic interaction between charges was represented in terms of the interactions via the transverse radiation field and the instantaneous Coulomb fields. Both these have been accounted for in the present treatment, and for the two treatments to be equivalent, the contribution of the remainder term (5.41c) to all observable quantities must vanish. This is indeed the case, the basic reason being that the electromagnetic field only interacts with the conserved charge–current density $s^\mu(x)$, Eqs. (5.2) and (5.4). We shall illustrate this for a simple example.

We shall see in Section 7.1, Eq. (7.14), that the scattering of charges by each other is, in lowest order of perturbation theory, given by the matrix element of the operator

$$\int d^4x \int d^4y\, s_1^\mu(x) D_{F\mu\nu}(x-y) s_2^\nu(y). \tag{5.44}$$

Here $s_1^\mu(x)$ and $s_2^\nu(y)$ are the two interacting charge–current densities. It is clear from Eq. (5.43) that the contribution of $_c D_{F\mu\nu}(x-y)$ to (5.44) corresponds to the instantaneous Coulomb interaction between the charge densities $\rho_1(\mathbf{x}, x^0) = s_1^0(\mathbf{x}, x^0)/c$ and $\rho_2(\mathbf{y}, x^0) = s_2^0(\mathbf{y}, x^0)/c$. Similarly, the transverse propagator $_T D_{F\mu\nu}(x-y)$ accounts for the electromagnetic interaction between the current densities $\mathbf{j}_1(x) = \mathbf{s}_1(x)$ and $\mathbf{j}_2(y) = \mathbf{s}_2(y)$.

The contribution to (5.44) of the remainder term $_R D_{F\mu\nu}(x-y)$ is easily shown to vanish, on account of current conservation. Transforming this contribution to expression (5.44) into momentum space, one obtains

$$\frac{1}{(2\pi)^4} \int d^4k\, s_1^\mu(-k) {_R D_{F\mu\nu}}(k) s_2^\nu(k), \tag{5.45}$$

where the momentum transforms $s_r^\mu(k)$, $r = 1, 2$, are defined, analogously to Eq. (5.38), by

$$s_r^\mu(x) = \frac{1}{(2\pi)^4} \int d^4k\, s_r^\mu(k)\, e^{-ikx}, \qquad r = 1, 2. \tag{5.46}$$

The current conservation equations, $\partial_\mu s_r^\mu(x) = 0$, translated into momentum space, become

$$k_\mu s_r^\mu(k) = 0, \quad r = 1, 2. \tag{5.47}$$

We see from the explicit form (5.41c) that each term in $_R D_{F\mu\nu}(k)$ is proportional to either k_μ or k_ν or both. Hence it follows from Eq. (5.47) that the expression (5.45) vanishes.

This completes our discussion of the equivalence of the two formulations of quantum electrodynamics. In doing this, we employed a special frame of reference leading to a division of the fields into transverse, longitudinal and scalar parts. In general, such a division is not required, and we shall work with manifestly covariant expressions involving summations over all four polarization states. In particular, the photon propagator, Eqs. (5.38) and (5.39), which will be very important in the development of quantum electrodynamics, has this property.

Problems

5.1. Show that the Lagrangian density obtained from

$$\mathscr{L} = -\frac{1}{4} F_{\mu\nu}(x) F^{\mu\nu}(x)$$

by adding the term $-\frac{1}{2}(\partial_\mu A^\mu(x))(\partial_\nu A^\nu(x))$, i.e.

$$\mathscr{L} = -\frac{1}{4} F_{\mu\nu}(x) F^{\mu\nu}(x) - \frac{1}{2}(\partial_\mu A^\mu(x))(\partial_\nu A^\nu(x)),$$

is equivalent to the Lagrangian density, proposed by Fermi:

$$\mathscr{L} = -\frac{1}{2}(\partial_\nu A_\mu(x))(\partial^\nu A^\mu(x)). \tag{5.48}$$

5.2. From the commutation relations (5.28) show that

$$\left[a_3(\mathbf{k}) - a_0(\mathbf{k}), a_3^\dagger(\mathbf{k}) - a_0^\dagger(\mathbf{k}) \right] = 0.$$

Show that the most general state representing the physical vacuum, i.e. the state in which there are no transverse photons present, but which contains the most general allowed admixture of scalar and longitudinal photons, is given by

$$|\Psi_{\text{SL}}\rangle = \sum_{n_1=0}^{\infty} \sum_{n_2=0}^{\infty} \ldots c(n_1, n_2, \ldots) \prod_{i=1}^{\infty} (\alpha_i^\dagger)^{n_i} |0\rangle$$

where

$$\alpha_i^\dagger \equiv a_3^\dagger(\mathbf{k}_i) - a_0^\dagger(\mathbf{k}_i),$$

$\mathbf{k}_i$ are the allowed wave vectors [see Eq. (1.13)], and $|0\rangle$ is the vacuum state in which there are no photons of any kind present. Show that the norm of this state is given by

$$\langle \Psi_{\text{SL}} | \Psi_{\text{SL}} \rangle = |c(0, 0, \ldots)|^2.$$

What is the most general state in which there are a definite number of transverse photons, with definite momenta and polarization vectors, present?

5.3. $|\Psi_T\rangle$ is a state which contains transverse photons only. Let

$$|\Psi'_T\rangle = \left\{ 1 + a\left[a_3^\dagger(\mathbf{k}) - a_0^\dagger(\mathbf{k}) \right] \right\} |\Psi_T\rangle,$$

where a is a constant. Show that replacing $|\Psi_T\rangle$ by $|\Psi'_T\rangle$ corresponds to a gauge transformation, i.e.

$$\langle \Psi'_T | A^\mu(x) | \Psi'_T \rangle = \langle \Psi_T | A^\mu(x) + \partial^\mu \Lambda(x) | \Psi_T \rangle,$$

where

$$\Lambda(x) = \left(\frac{2\hbar c^2}{V\omega_{\mathbf{k}}^3} \right)^{1/2} \mathrm{Re}\left(ia\, e^{-ikx} \right).$$

5.4. By making the minimal substitution

$$\partial_\alpha \phi(x) \rightarrow D_\alpha \phi(x) = \left[\partial_\alpha + \frac{ie}{\hbar c} A_\alpha(x) \right] \phi(x)$$

$$\partial_\alpha \phi^\dagger(x) \rightarrow [D_\alpha \phi(x)]^\dagger = \left[\partial_\alpha - \frac{ie}{\hbar c} A_\alpha(x) \right] \phi^\dagger(x)$$

in the Lagrangian density (3.22) of the complex Klein–Gordon field $\phi(x)$, derive the Lagrangian density $\mathscr{L}_I(x)$ for the interaction of the charged bosons, described by the field $\phi(x)$, with the electromagnetic field $A^\alpha(x)$.

Assuming that this interaction is invariant under the charge conjugation transformation $\mathscr{C}$, show that

$$\mathscr{C} A^\alpha(x) \mathscr{C}^{-1} = -A^\alpha(x).$$

[The transformation properties of $\phi(x)$ and of $s^\alpha(x)$, Eq. (3.32), under charge conjugation were discussed in Problem 3.5.]

Hence show that a single-photon state $|\mathbf{k}, r\rangle$ is an eigenstate of $\mathscr{C}$ with eigenvalue -1.

6

The S-Matrix Expansion

We shall now progress from the discussion of the free fields to the realistic and much more interesting case of fields in interaction, in which particles can be scattered, created and destroyed. In essence, this requires solving the coupled non-linear field equations for given conditions. In quantum electrodynamics, for example, one must solve the inhomogeneous wave equation (5.12) with the Dirac current density (4.28) as source term. This is an extremely difficult problem, which has only been solved in perturbation theory, i.e. the Hamiltonian of the system is divided into that of the free fields plus an interaction term. The latter is treated as a perturbation, which is justifiable if the interaction is sufficiently weak. For quantum electrodynamics, where the coupling of photons and electrons is measured by the small dimensionless fine structure constant $\alpha \approx 1/137$, this approach is outstandingly successful, not only in calculating processes in lowest order of perturbation theory, but also in calculating higher-order corrections.

In the Heisenberg picture, which we have so far been using, this programme is still very complex, and it was decisive for the successful development of the theory to work instead in the interaction picture. In Section 6.2 we shall study the equations of motion of the interacting fields in the interaction picture and we shall obtain a perturbation series solution suitable for collision processes. This solution, known as the S-matrix expansion, is due to Dyson. The Dyson expansion of the S-matrix is of great importance, since it contains the complete information about all collision processes in a form suitable for extracting the transition amplitude for a specific process to any order of perturbation theory. A systematic procedure for doing this will be developed in Section 6.3.

Before proceeding with these topics, we shall, in Section 6.1, introduce natural units, which considerably simplify details of the following calculations.

Quantum Field Theory, Second Edition   Franz Mandl and Graham Shaw
© 2010 John Wiley & Sons, Ltd

6.1   Natural Dimensions and Units

We have so far used c.g.s. units, in which the fundamental dimensions, in terms of which quantities are expressed, are mass (M), length (L) and time (T). In relativistic quantum field theory, expressions and calculations are much simplified if one uses natural units (n.u.). In natural units one takes mass, action (A) and velocity (V) as fundamental dimensions and chooses $\hbar$ as unit of action and the velocity of light c as unit of velocity. Hence $\hbar = c = 1$ in natural units, and c.g.s. expressions are transformed into natural units by putting $\hbar = c = 1$. In such n.u. expressions, all quantities have the dimensions of a power of M. Since

$$L = \frac{A}{MV} \text{ and } T = \frac{A}{MV^2} \tag{6.1}$$

one has the general result that a quantity which has the c.g.s. dimensions

$$M^p L^q T^r = M^{p-q-r} A^{q+r} V^{-q-2r}, \tag{6.2}$$

has the n.u. dimensions M^{p-q-r}. In natural units, many quantities have the same dimension. For example, the momentum–energy relation for a particle of mass m becomes, in natural units,

$$E^2 = m^2 + \mathbf{p}^2 = m^2 + \mathbf{k}^2, \tag{6.3}$$

so that mass, momentum, energy and wave number all have the same natural dimension M. The c.g.s. expression for the dimensionless fine structure constant

$$\alpha = \frac{e^2}{4\pi\hbar c} = \frac{1}{137.04} \text{ (c.g.s.)} \tag{6.4a}$$

becomes

$$\alpha = \frac{e^2}{4\pi} = \frac{1}{137.04} \quad \text{(n.u.)} \tag{6.4b}$$

so that, in natural units, electric charge is dimensionless (M^0).

From the general relation (6.2), or by using particular equations, one easily derives the n.u. dimensions of all quantities, and some of the more important ones are listed in Table 6.1.

Working in natural units it is very easy to obtain numerical results in any system of units. A quantity in natural units will have a dimension M^n. To convert this quantity to whatever c.g.s. units are convenient, one merely multiplies it by such powers of $\hbar$ and c, expressed in the appropriate units, as to give it the correct c.g.s. dimensions. One frequently interprets M as an energy and measures it in MeV. The conversion factors

$$\hbar = 6.58 \times 10^{-22} \text{ MeV} \cdot \text{s} \tag{6.5a}$$

$$\hbar c = 1.973 \times 10^{-11} \text{ MeV} \cdot \text{cm} \tag{6.5b}$$

then enable one easily to express quantities in terms of MeV, centimetres and seconds. Two examples will illustrate this.

Table 6.1   The c.g.s dimension $M^p L^q T^r$ and the n.u. dimensions $M^n = M^{p-q-r}$ of some quantities

Quantity	c.g.s.			n.u.
	p	q	r	n
Action	1	2	−1	0
Velocity	0	1	−1	0
Mass	1	0	0	1
Length	0	1	0	−1
Time	0	0	1	−1
Lagrangian or Hamiltonian densities	1	−1	−2	4
Fine structure constant α	0	0	0	0
Electric charge	$\frac{1}{2}$	$\frac{3}{2}$	−1	0
Klein–Gordon field $\phi(x)^*$	$\frac{1}{2}$	$\frac{1}{2}$	−1	1
Electromagnetic field $A^\mu(x)^*$	$\frac{1}{2}$	$\frac{1}{2}$	−1	1
Dirac fields $\psi(x)$ and $\bar{\psi}(x)^*$	0	$-\frac{3}{2}$	0	$\frac{3}{2}$

* The dimensions of the fields can, for example, be obtained from the Lagrangian densities, Eqs. (3.4), (5.10) and (4.20)

The Thomson cross-section (1.72) becomes, in natural units,

$$\sigma = \frac{8\pi}{3} \frac{\alpha^2}{m^2}. \tag{6.6}$$

With $m = 0.511$ MeV, we convert the right-hand side of this equation to cm^2 by multiplying by $(\hbar c$ in MeV · cm$)^2$ which, from Eq. (6.5b), gives

$$\sigma = \frac{8\pi}{3} \alpha^2 \frac{(1.973 \times 10^{-11} \text{MeV} \cdot \text{cm})^2}{(0.511 \text{MeV})^2} = 6.65 \times 10^{-25} \text{cm}^2.$$

Secondly, we quote[1] the n.u. expression for the lifetime τ of the positronium ground state $1\,^1S_0$. It is given by

$$\tau = \frac{2}{\alpha^5} \frac{1}{m}, \tag{6.7}$$

where m is the mass of the electron. With m in MeV, we must multiply Eq. (6.7) by ($\hbar$ in MeV·s), Eq. (6.5a), to obtain

$$\tau = \frac{2}{\alpha^5} \frac{(6.58 \times 10^{-22} \text{MeV} \cdot \text{s})}{(0.511 \text{MeV})} = 1.24 \times 10^{-10} \text{ s}.$$

This conversion factor is of course the same for converting any lifetime τ from natural units to seconds, the essential points being that in natural units τ has the dimension M^{-1} and must be expressed in (MeV)$^{-1}$.

These examples illustrate how very easy it is to obtain numerical results in any c.g.s. units from equations expressed in natural units. No advantage is gained by tediously retaining

---

[1] See J. M. Jauch and F. Rohrlich, *The Theory of Photons and Electrons*, 2nd edn, Springer, New York, 1976, p. 286, Eq. (12-108).

factors of $\hbar$ and c throughout a calculation or by converting a n.u. equation into c.g.s. form by inserting the appropriate factors of $\hbar$ and c prior to substituting numerical values.

Although rarely required, the c.g.s. form of an equation is easily obtained from its n.u. form. In a sum of terms, one must multiply each term by appropriate powers of $\hbar$ and c to make all the terms have the same c.g.s. dimensions. [E.g. a factor $(E + k)$, with E interpreted as an energy and k a wave number, could be turned into $(E + c\hbar k)$ or into $(E/c + \hbar k)$, etc.] To obtain the correct c.g.s. dimensions for the whole expression, it must be multiplied by a factor $\hbar^a c^b$ with the exponents a and b determined from dimensional arguments. Usually they are easily guessed.

From now on we shall in general work in natural units.

6.2   The *S*-Matrix Expansion

So far we have mainly considered the free, i.e. non-interacting, fields, using the Heisenberg picture (H.P.), in which state vectors are constant in time and the operators carry the full time dependence.

We now turn to the study of the interacting fields. For example, in quantum electrodynamics (QED), the interacting electron–positron and electromagnetic fields are described by the Lagrangian density

$$\mathscr{L} = \mathscr{L}_0 + \mathscr{L}_1 \qquad (6.8)$$

with the free-field Lagrangian density

$$\mathscr{L}_0 = \mathrm{N}[\overline{\psi}(x)(i\gamma^\mu \partial_\mu - m)\psi(x) - \frac{1}{2}\,(\partial_\nu A_\mu(x))(\partial^\nu A^\mu(x))] \qquad (6.9)$$

and the interaction Lagrangian density

$$\mathscr{L}_1 = \mathrm{N}[-s^\mu(x)A_\mu(x)] = \mathrm{N}[e\overline{\psi}(x)A\!\!\!/(x)\psi(x)] \qquad (6.10)$$

[see Eqs. (4.66)–(4.68) and (5.10)]. In Eqs. (6.9) and (6.10) we have written the free-field and the interaction Lagrangian densities as normal products. This ensures, as for the free-field cases considered earlier, that the vacuum expectation values of all observables, e.g. energy or charge, vanish. Corresponding to the division (6.8), the complete Hamiltonian H of the system is split into the free-field Hamiltonian H_0 and the interaction Hamiltonian H_I:

$$H = H_0 + H_I. \qquad (6.11)$$

As discussed at the beginning of this chapter, we shall employ the interaction picture (I.P.) which leads to two essential simplifications.[2]

Firstly, in the I.P., the operators satisfy the Heisenberg-like equations of motion (1.87), but involving the free Hamiltonian H_0 only, not the complete Hamiltonian H.

Secondly, if the interaction Lagrangian density $\mathscr{L}_1$ does not involve derivatives (and we shall restrict ourselves to this case until Chapter 19), the fields canonically conjugate to the

---

[2] The interaction picture, and its relation to the Heisenberg and Schrödinger pictures, is discussed in the appendix to Chapter 1 (Section 1.5). The reader who is not intimately familiar with this material is advised to study this appendix in depth at this stage.

interacting fields and to the free fields are identical. (For example in QED $\partial \mathcal{L}/\partial \dot{\psi}_\alpha = \partial \mathcal{L}_0/\partial \dot{\psi}_\alpha$, etc.) Since the I.P. and the H.P. are related by a unitary transformation, it follows that in the I.P., the interacting fields satisfy the same commutation relations as the free fields.

Thus in the I.P., the interacting fields satisfy the same equations of motion and the same commutation relations as the free-field operators. Consequently, we can take over the many results derived for free fields (in Chapters 3–5), as also true for the interacting fields in the I.P. In particular, the complete sets of plane wave states which we obtained continue to be solutions of the equations of motion, resulting in the same plane wave expansions of the field operators as before, the same number representations and the same explicit forms for the Feynman propagators.

In the I.P., the system is described by a time-dependent state vector $|\Phi(t)\rangle$. According to Eqs. (1.88) and (1.89), $|\Phi(t)\rangle$ satisfies the equation of motion

$$i\frac{\mathrm{d}}{\mathrm{d}t}|\Phi(t)\rangle = H_I(t)|\Phi(t)\rangle, \tag{6.12}$$

where

$$H_I(t) = e^{iH_0(t-t_0)}H_I^S e^{-iH_0(t-t_0)} \tag{6.13}$$

is the interaction Hamiltonian in the I.P., with H_I^S and $H_0 = H_0^S$ being the interaction and free-field Hamiltonians in the Schrödinger picture (S.P.). $H_I(t)$ is obtained by replacing, in H_I^S, the S.P. field operators by the time-dependent free-field operators. In Eqs. (6.12) and (6.13) we have omitted the labels I, used in Eqs. (1.88) and (1.89) to distinguish the I.P., as we shall be working exclusively in the I.P. in what follows.

Eq. (6.12) is a Schrödinger-like equation with the time-dependent Hamiltonian $H_I(t)$. With the interaction 'switched off' (i.e. we put $H_I \equiv 0$), the state vector is constant in time. The interaction leads to the state $|\Phi(t)\rangle$ changing with time. Given that the system is in a state $|i\rangle$ at an initial time $t = t_i$, i.e.

$$|\Phi(t_i)\rangle = |i\rangle, \tag{6.14}$$

the solution of Eq. (6.12) with this initial condition gives the state $|\Phi(t)\rangle$ of the system at any other time t. It follows from the Hermiticity of the operator $H_I(t)$ that the time development of the state $|\Phi(t)\rangle$ according to Eq. (6.12) is a unitary transformation. Accordingly, it preserves the normalization of states,

$$\langle\Phi(t)|\Phi(t)\rangle = \text{const.}, \tag{6.15}$$

and, more generally, the scalar product.

Clearly the formalism which we are here developing is not appropriate for the description of bound states, but it is particularly suitable for scattering processes. In a collision process the state vector $|i\rangle$ will define an initial state, long before the scattering occurs ($t_i = -\infty$), by specifying a definite number of particles, with definite properties and far apart from each other so that they do not interact. (For example, in QED, $|i\rangle$ would specify a definite number of electrons, positrons and photons with given momenta, spins and polarizations.) In the scattering process, the particles will come close together, collide (i.e. interact) and fly apart again. Eq. (6.12) determines the state $|\Phi(\infty)\rangle$ into which the initial state

$$|\Phi(-\infty)\rangle = |i\rangle, \tag{6.14a}$$

evolves at $t = \infty$, long after the scattering is over and all particles are far apart again. The S-matrix relates $|\Phi(\infty)\rangle$ to $|\Phi(-\infty)\rangle$ and is defined by

$$|\Phi(\infty)\rangle = S|\Phi(-\infty)\rangle = S|i\rangle. \tag{6.16}$$

A collision can lead to many different final states $|f\rangle$, and all these possibilities are contained within $|\Phi(\infty)\rangle$. (For example, an electron–positron collision may result in elastic scattering, bremsstrahlung (i.e. emission of photons), pair annihilation, etc.) Each of these final states $|f\rangle$ is specified in a way analogous to $|i\rangle$.

The transition probability that after the collision (i.e. at $t = \infty$) the system is in the state $|f\rangle$ is given by

$$|\langle f|\Phi(\infty)\rangle|^2. \tag{6.17}$$

($|\Phi(\infty)\rangle$ and $|i\rangle$ are assumed normed to unity.) The corresponding probability amplitude is

$$\langle f|\Phi(\infty)\rangle = \langle f|S|i\rangle \equiv S_{fi}. \tag{6.18}$$

With the state $|\Phi(\infty)\rangle$ expanded in terms of a complete orthonormal set of states,

$$|\Phi(\infty)\rangle = \sum_f |f\rangle\langle f|\Phi(\infty)\rangle = \sum_f |f\rangle S_{fi}, \tag{6.19}$$

the unitarity of the S-matrix can be written

$$\sum_f |S_{fi}|^2 = 1. \tag{6.20}$$

Eq. (6.20) expresses the conservation of probability. It is more general than the corresponding conservation of particles in non-relativistic quantum mechanics, since now particles can be created or destroyed.

In order to calculate the S-matrix we must solve Eq. (6.12) for the initial condition (6.14a). These equations can be combined into the integral equation

$$|\Phi(t)\rangle = |i\rangle + (-i) \int_{-\infty}^{t} dt_1 H_I(t_1) |\Phi(t_1)\rangle. \tag{6.21}$$

This equation can only be solved iteratively. The resulting perturbation solution, as a series in powers of H_I, will only be useful if the interaction energy H_I is small. This is the case for QED, where the dimensionless coupling constant characterizing the photon–electron interaction is the fine structure constant $\alpha \approx 1/137$.

Solving Eq. (6.21) by iteration

$$|\Phi(t)\rangle = |i\rangle + (-i) \int_{-\infty}^{t} dt_1 H_I(t_1)|i\rangle$$

$$+ (-i)^2 \int_{-\infty}^{t} dt_1 \int_{-\infty}^{t_1} dt_2 H_I(t_1) H_I(t_2)|\Phi(t_2)\rangle,$$

and so on, we obtain, in the limit $t \to \infty$, the S-matrix

$$S = \sum_{n=0}^{\infty} (-i)^n \int_{-\infty}^{\infty} dt_1 \int_{-\infty}^{t_1} dt_2 \ldots \int_{-\infty}^{t_{n-1}} dt_n H_1(t_1) H_1(t_2) \ldots H_1(t_n) \qquad (6.22a)$$

$$= \sum_{n=0}^{\infty} \frac{(-i)^n}{n!} \int_{-\infty}^{\infty} dt_1 \int_{-\infty}^{\infty} dt_2 \ldots \int_{-\infty}^{\infty} dt_n T\{H_1(t_1) H_1(t_2) \ldots H_1(t_n)\}. \qquad (6.22b)$$

Here, the time-ordered product $T\{\ldots\}$ of n factors is the natural generalization of the definitions (3.52) and (4.59) for two factors, i.e. the factors are ordered so that later times stand to the left of earlier times, and all boson (fermion) fields are treated as though their commutators (anticommutators) vanish. The equivalence of the two forms (6.22a) and (6.22b) only holds if H_1 contains an even number of fermion factors (as in QED), so that the reordering process introduces no extra factors (-1). The equivalence of the two forms holds separately for each term of the series. Its verification is left as an exercise for the reader. Finally we rewrite Eq. (6.22b) in terms of the interaction Hamiltonian density $\mathscr{H}_1(x)$ to obtain the explicitly covariant result

$$S = \sum_{n=0}^{\infty} \frac{(-i)^n}{n!} \int \ldots \int d^4x_1 d^4x_2 \ldots d^4x_n T\{\mathscr{H}_1(x_1) \mathscr{H}_1(x_2) \ldots \mathscr{H}_1(x_n)\}, \qquad (6.23)$$

the integrations being over all space–time. This equation is the Dyson expansion of the S-matrix. It forms the starting point for the approach to perturbation theory used in this book.

We have seen that the amplitude for a particular transition $|i\rangle \to |f\rangle$ is given by $\langle f|S|i\rangle$. To pick out from the expansion (6.23) the parts which contribute to this matrix element is a complex problem to which we shall return in the next section, but we must first discuss the specification of the initial and final states $|i\rangle$ and $|f\rangle$.

In the above perturbation formalism the states $|i\rangle$ and $|f\rangle$ are, as usual, eigenstates of the unperturbed free-field Hamiltonian H_0, i.e. with the interaction switched off ($H_I = 0$). This description appears wrong, since the particles we are dealing with are real physical particles, even when far apart. An electron, even when far away from other electrons, is surrounded by its photon cloud; it is a real electron, not a bare electron without its own electromagnetic field. Hence, the use of bare particle states $|i\rangle$ and $|f\rangle$ requires justification. One possible procedure is to appeal to the adiabatic hypothesis in which the interaction $H_1(t)$ is replaced by $H_1(t) f(t)$. The function $f(t)$ is chosen so that $f(t) = 1$ for a sufficiently long interval $-T \le t \le T$, and $f(t) \to 0$ monotonically as $t \to \pm \infty$. [In QED, for example, we could replace the elementary charge e by the time-dependent coupling constant $ef(t)$.] In this way the initial and final states are described by bare particles. During the interval $-\infty < t \le -T$, the equation of motion (6.12), with $H_1(t)$ replaced by $H_1(t)f(t)$, generates the real physical particles from the bare particles, and during the interval $|t| \le T$ we are dealing with the physical particles and the full interaction $H_1(t)$. In particular, the full interaction is effective during the interval $-\tau \le t \le \tau$, while the particles are sufficiently close together to interact (i.e. we must choose $T \gg \tau$). The essence of the adiabatic hypothesis is that the scattering, which occurs during the interval

$|t| \leq \tau$, cannot depend on our description of the system a long time before the scattering $(t << -\tau)$ or a long time after the scattering $(t >> \tau)$. Only at the end of a calculation do we take the limit $T \to \infty$. Of course, if we calculate a process in lowest-order perturbation theory [i.e. we use only the term of lowest order n in Eq. (6.23) which gives a non-vanishing result] then the interaction is exclusively used to cause the transition and not also to convert bare into real particles. We may then take the limit $T \to \infty$ from the start of the calculation and work with the full interaction $H_1(t)$.

6.3   Wick's Theorem

We must now see how to obtain from the S-matrix expansion (6.23) the transition amplitude $\langle f|S|i \rangle$ for a particular transition $|i\rangle \to |f\rangle$ in a given order of perturbation theory. The Hamiltonian density $\mathscr{H}_1(x)$ in Eq. (6.23) involves the interacting fields, each linear in creation and absorption operators. Hence the expansion (6.23) will describe a large number of different processes. However, only certain terms of the S-matrix will contribute to a given transition $|i\rangle \to |f\rangle$. For these terms must contain just the right absorption operators to destroy the particles present in $|i\rangle$, and they must contain the right creation operators to emit the particles present in $|f\rangle$. They may also contain additional creation and absorption operators, which create particles which are subsequently re-absorbed. These particles are only present in intermediate states and are called virtual particles.

Calculations can be greatly simplified by avoiding the explicit introduction of virtual intermediate particles. This can be achieved by writing the S-matrix expansion as a sum of normal products, since, in a normal product, *all* absorption operators stand to the *right* of *all* creation operators. Such an operator first absorbs a certain number of particles and then emits some particles. It does not cause emission and re-absorption of intermediate particles. Each of these normal products will effect a particular transition $|i\rangle \to |f\rangle$, which can be represented by a Feynman graph, similar to those introduced in Chapters 3 and 4.

Consider, for example, Compton scattering $(e^- + \gamma \to e^- + \gamma)$. The QED interaction Hamiltonian density is, from Eq. (6.10),

$$\mathscr{H}_1(x) = -\mathscr{L}_1(x) = -e\mathrm{N}[\overline{\psi}(x)A\!\!\!/(x)\psi(x)]. \tag{6.24}$$

Since the negative (positive) frequency parts $A^-, \overline{\psi}^-, \psi^- (A^+, \psi^+ \overline{\psi}^+)$ are linear in creation (absorption) operators for photons, electrons and positrons respectively, the only normal product which contributes to Compton scattering is

$$\overline{\psi}^- A^- \psi^+ A^+.$$

The method for expanding the S-matrix as a sum of normal products, which we shall now describe, is due to Dyson and Wick.

We first of all summarize the general definition of a normal product. Let $Q, R, \ldots, W$ be operators of the type $\psi^\pm, A^\pm$, etc., i.e. each is linear in either creation or absorption operators, then

$$\mathrm{N}(QR \ldots W) = (-1)^P (Q'R' \ldots W'). \tag{6.25a}$$

Here $Q', R', \ldots, W'$ are the operators $Q, R, \ldots, W$ reordered, so that all absorption operators (i.e. positive frequency parts) stand to the right of all creation operators (i.e. negative frequency parts). The exponent P is the number of interchanges of neighbouring fermion operators required to change the order $(QR \ldots W)$ into $Q'R' \ldots W'$. We generalize the definition (6.25a) by requiring the normal product to obey the distributive law

$$N(RS \ldots + VW \ldots) = N(RS \ldots) + N(VW \ldots). \tag{6.25b}$$

The QED interaction (6.24) is a normal product of field operators. We shall find that in other cases too, the interaction Hamiltonian density can be written as a normal products, i.e.

$$\mathscr{H}_1(x) = N\{A(x)B(x) \ldots\}, \tag{6.26}$$

where each of the fields $A(x)$, $B(x)$, $\ldots$, is linear in creation and absorption operators. Hence, we must consider the expansion into a sum of normal products of 'mixed' T-product (i.e. a T-product whose factors are normal products), such as occurs in the S-matrix expansion (6.23).

From the definition of the normal product, we have, for two field operators $A \equiv A(x_1)$ and $B \equiv B(x_2)$, that

$$AB - N(AB) = \begin{cases} [A^+, B^-]_+, & \text{for two fermion fields} \\ [A^+, B^-], & \text{otherwise} \end{cases}. \tag{6.27}$$

For two fermion fields, the anticommutators, and in all other cases, the commutators, are c-numbers, i.e. they do not involve creation or annihilation operators. [We had examples in Eqs. (3.40) and (4.53a).] Hence, the right-hand side of Eq. (6.27) is always a c-number. It is given by $\langle 0|AB|0\rangle$, as follows by taking the vacuum expectation value of Eq. (6.27). Hence Eq. (6.27) becomes:

$$AB = N(AB) + \langle 0|AB|0\rangle. \tag{6.28}$$

Since

$$N(AB) = \pm N(BA), \tag{6.29}$$

the minus sign applying in the case of two fermion fields, the plus sign in all other cases, it follows from Eq. (6.28) that for $x_1^0 \neq x_2^0$

$$T\{A(x_1)B(x_2)\} = N\{A(x_1)B(x_2)\} + \langle 0|T\{A(x_1)B(x_2)\}|0\rangle. \tag{6.30}$$

The case of equal times, $x_1^0 = x_2^0$, will be considered below.

The special notation

$$\overline{A(x_1)B(x_2)} \equiv \langle 0|T\{A(x_1)B(x_2)\}|0\rangle \tag{6.31}$$

will be convenient for this vacuum expectation value, which will be called the *contraction* of $A(x_1)$ and $B(x_2)$. Being a vacuum expectation value, it will vanish unless one of the field operators A and B creates particles which the other absorbs. The non-vanishing contractions are, of course, just the Feynman propagators, e.g. Eqs. (3.55), (3.59), (4.61) and (5.26):

$$\phi(x_1)\phi(x_2) = i\Delta_F(x_1 - x_2) \tag{6.32a}$$

$$\phi(x_1)\phi^\dagger(x_2) = \phi^\dagger(x_2)\phi(x_1) = i\Delta_F(x_1 - x_2) \tag{6.32b}$$

$$\psi_\alpha(x_1)\overline{\psi}_\beta(x_2) = -\overline{\psi}_\beta(x_2)\psi_\alpha(x_1) = iS_{F\alpha\beta}(x_1 - x_2) \tag{6.32c}$$

$$A^\mu(x_1)A^\nu(x_2) = iD_F^{\mu\nu}(x_1 - x_2). \tag{6.32d}$$

To generalize Eq. (6.30) to several operators $A \equiv A(x_1)$, ..., $M \equiv M(x_m)$, ..., the generalized normal product is defined by

$$N(A\,BC\,DE\,F \ldots J\,K\,LM \ldots$$

$$=(-1)^P AK\ BC\ EL \ \ldots\ N(DF \ldots JM \ldots\ ) \tag{6.33}$$

where P is the number of interchanges of neighbouring fermion operators required to change the order $(ABC\ldots)$ to $(AKB\ldots)$; for example

$$N(\psi_\alpha(x_1)\ \psi_\beta(x_2)A^\mu(x_3)\overline{\psi}_\gamma(x_4)\overline{\psi}_\delta(x_5))$$

$$=(-1)\ \psi_\beta(x_2)\overline{\psi}_\delta(x_5)N(\psi_\alpha(x_1)A^\mu(x_3)\overline{\psi}_\gamma(x_4)). \tag{6.34}$$

For the case of unequal times (i.e. $x_i^0 \neq x_j^0$, for $i \neq j$), Wick has proved the following generalization of Eq. (6.30):

$$T(ABCD \ldots WXYZ) = N(ABCD \ldots WXYZ)$$
$$+ N(AB\ C \ldots YZ) + N(ABC \ldots YZ) + \cdots + N(ABC \ldots YZ)$$
$$+ N(AB\ CD \ldots YZ) + \cdots + N(AB \ldots WX\ YZ) + \cdots. \tag{6.35}$$

On the right-hand side of this equation appears the sum of all possible generalized normal products that can be formed from $(ABCD \ldots WXYZ)$, the first, second and third lines representing all terms with no, one and two contractions, and so on. Each term on the right-hand side of this equation contains all the factors in the same order in which they occur in the T-product on the left-hand side.

Eq. (6.35) states Wick's theorem. We shall not reproduce its proof, which is by induction, and so not very illuminating.[3]

---

[3] G. C. Wick, *Phys. Rev.* **80** (1950) 268.

With the interaction (6.26), the S-matrix expansion (6.23) contains the mixed T-products

$$T\{\mathscr{H}_1(x_1) \ldots \mathscr{H}_1(x_n)\} = T\{N(AB \ldots)_{x_1} \ldots N(AB \ldots)_{x_n}\}. \tag{6.36}$$

Wick extended the theorem (6.35) to include such mixed T-products. In each factor $N(AB \ldots)_{x_r}$ we replace $x_r = (x_r^0, \mathbf{x}_r)$ by $\xi_r = (x_r^0 \pm \varepsilon, \mathbf{x}_r)$, $(\varepsilon > 0)$, depending on whether the substitution is made in the creation or absorption part of the field. Hence

$$T\{N(AB \ldots)_{x_1} \ldots N(AB \ldots)_{x_n}\} = \lim_{\varepsilon \to 0} T\{(AB \ldots)_{\xi_1} \ldots (AB \ldots)_{\xi_n}\}, \tag{6.37}$$

the normal and chronological orderings within each group $(AB \ldots)_{\xi_r}$ being the same on account of the $\pm \varepsilon$ in ξ_r^0. On expanding the right-hand side of Eq. (6.37) by Wick's theorem *before* going to the limit $\varepsilon \to 0$, contractions within one group $(AB \ldots)_{\xi_r}$ (i.e. over equal-times operators when $\varepsilon \to 0$) vanish as the group is already in normal order. We thus have the desired result: the mixed T-product (6.36) can be expanded according to Eq. (6.35), provided contractions over equal times are omitted:

$$T\{N(AB \ldots)_{x_1} \ldots N(AB \ldots)_{x_n}\} = T\{(AB \ldots)_{x_1} \ldots (AB \ldots)_{x_n}\}_{\text{no e.t.c.}} \tag{6.38}$$

where 'no e.t.c.' stands for 'no equal-times contractions'.

Eqs. (6.35) and (6.38) represent the desired result, enabling us to expand each term in the S-matrix expansion (6.23) into a sum of generalized normal products. Each of these normal products corresponds to a definite process, characterized by the operators not contracted, which absorb and create the particles present in the initial and final states respectively. The non-vanishing contractions which occur in these generalized normal products are the Feynman propagators (6.32), corresponding to virtual particles being emitted and re-absorbed in intermediate states. In the next chapter we shall see how to evaluate these individual contributions to $\langle f|S|i \rangle$, which result from the application of Wick's theorem.

7

Feynman Diagrams and Rules in QED

In the last chapter we obtained the S-matrix expansion (6.23) and Wick's theorem for writing the terms in this expansion as a sum of normal products. In this chapter we shall show how to calculate the matrix element $\langle f|S|i \rangle$ for a transition from an initial state $|i\rangle$ to a final state $|f\rangle$ in a given order of perturbation theory. For definiteness, we shall give this development for the important case of QED. Once this case is understood, the corresponding formalism for others is easily derived.

In Section 7.1 we shall show how to pick out from the S-matrix expansion the terms which contribute to $\langle f|S|i \rangle$ in a given order of perturbation theory. These terms are easily identified. They are those normal products which contain the appropriate destruction and creation operators to destroy the particles present in the initial state $|i\rangle$ and create those present in the final state $|f\rangle$.

In Section 7.2 we shall evaluate the transition amplitude $\langle f|S|i \rangle$ in momentum space. This leads to Feynman diagrams as a way of interpreting the terms in the Wick expansion. There exists a one-to-one correspondence between the diagrams and the terms, which can be summarized in simple rules. These enable one to write down transition amplitudes directly from the Feynman graphs, rather than proceed *ab initio* from Wick's theorem. In Section 7.3 we shall state these rules, known as Feynman rules, for QED. We shall have obtained these rules from the Dyson–Wick formalism, but historically they were first derived by Feynman using a strongly intuitive approach.

In the first three sections of this chapter we shall consider QED as the interaction of the electron–positron field with the electromagnetic field. In the last section (Section 7.4) we shall extend QED to include, in addition to the electron–positron field, other leptons, namely the muon and tauon.

Quantum Field Theory, Second Edition   Franz Mandl and Graham Shaw
© 2010 John Wiley & Sons, Ltd

7.1   Feynman Diagrams in Configuration Space

The processes to which the individual terms in the *S*-matrix expansion (6.23), i.e. in

$$S = \sum_{n=0}^{\infty} S^{(n)} \equiv \sum_{n=0}^{\infty} \frac{(-i)^n}{n!} \int \ldots \int d^4x_1 \ldots d^4x_n T\{\mathscr{H}_I(x_1) \ldots \mathscr{H}_I(x_n)\}, \qquad (7.1)$$

contribute are of course determined by the nature of the interaction $\mathscr{H}_I(x)$. For QED this is given by Eq. (6.24):

$$\begin{aligned}
\mathscr{H}_I(x) &= -eN\{\bar{\psi}(x)A\!\!\!/(x)\psi(x)\} \\
&= -eN\left\{ \left(\bar{\psi}^+ + \bar{\psi}^-\right)\left(A\!\!\!/^+ + A\!\!\!/^-\right)\left(\psi^+ + \psi^-\right)\right\}_x.
\end{aligned} \qquad (7.2)$$

With $\psi^+ (\bar{\psi}^-)$, $\bar{\psi}^+ (\psi^-)$ and $A^+(A^-)$ being linear in absorption (creation) operators of electrons, positrons and photons, respectively, the interaction (7.2) gives rise to eight basic processes, e.g. the term $-eN(\bar{\psi}^+ A^- \psi^+)_x$ corresponds to the annihilation of an electron–positron pair with the creation of a photon.

Using the conventions for Feynman diagrams explained at the end of Section 4.4, we can represent these eight processes by the Feynman graphs of Fig. 7.1, which have been grouped into pairs. The graphs in each pair correspond to absorption or emission of a photon, together with: (a) the scattering of an electron, (b) the scattering of a positron, (c) pair annihilation or (d) pair creation. These diagrams illustrate the basic processes to which the QED interaction gives rise and will be referred to as the basic vertex part. All other QED Feynman diagrams are built up by combining such basic vertex parts.

The diagrams of Fig. 7.1 also represent the processes arising from the first-order term $S^{(1)}$ in Eq. (7.1). However, these are not real physical processes, i.e. for none of them can

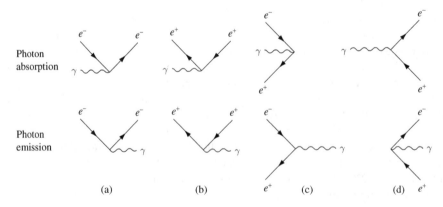

Figure 7.1   *The Feynman diagrams of the eight basic processes of the QED interaction $\mathscr{H}_I(x) = -eN(\bar{\psi}A\!\!\!/\psi)_x$. (a) e^- scattering; (b) e^+ scattering; (c) e^+e^- annihilation; (d) e^+e^- creation*

energy and momentum be conserved for real physical particles for which we must have $k^2 = 0$ for photons, and $p^2 = m^2$ for fermions. Consequently

$$\left\langle f \left| S^{(1)} \right| i \right\rangle = 0 \tag{7.3a}$$

for these transitions, as will be shown explicitly in the next section. More generally

$$\left\langle f \left| S^{(n)} \right| i \right\rangle = 0 \tag{7.3b}$$

for any unphysical process, i.e. for a transition between real physical states which violates a conservation law of the theory. This follows, since S generates a solution of the equations of motion, so that

$$\langle f | S | i \rangle = 0 \tag{7.3c}$$

for an unphysical process, and since Eq. (7.1) is a power series in the coupling constant e.

To obtain real processes, we must go at least to the second-order term $S^{(2)}$ in Eq. (7.1). This term contains two factors $\mathscr{H}_I$. Its expansion by Wick's theorem into a sum of normal products corresponds to all meaningful ways of joining two basic vertex parts into a Feynman diagram, as we shall now see.

Application of Wick's theorem, Eqs. (6.35) and (6.38), to $S^{(2)}$ leads to

$$S^{(2)} = \sum_{i=A}^{F} S_i^{(2)} \tag{7.4}$$

where

$$S_A^{(2)} = -\frac{e^2}{2!} \int d^4x_1 d^4x_2 N[(\overline{\psi}A\!\!\!/\psi)_{x_1} (\overline{\psi}A\!\!\!/\psi)_{x_2}] \tag{7.5a}$$

$$S_B^{(2)} = -\frac{e^2}{2!} \int d^4x_1 d^4x_2 \{ N[(\overline{\psi}A\!\!\!/\,\psi)_{x_1} (\overline{\psi}\,A\!\!\!/\psi)_{x_2}]$$

$$+ N[(\overline{\psi}A\!\!\!/\psi)_{x_1} (\overline{\psi}A\!\!\!/\psi)_{x_2}] \} \tag{7.5b}$$

$$S_C^{(2)} = -\frac{e^2}{2!} \int d^4x_1 d^4x_2 N[(\overline{\psi}\gamma^\alpha A_\alpha \psi)_{x_1} (\overline{\psi}\,\gamma^\beta A_\beta \,\psi)_{x_2}] \tag{7.5c}$$

$$S_D^{(2)} = -\frac{e^2}{2!} \int d^4x_1 d^4x_2 \{ N[(\overline{\psi}\gamma^\alpha A_\alpha \,\psi)_{x_1} (\overline{\psi}\,\gamma^\beta A_\beta \,\psi)_{x_2}]$$

$$+ N[(\overline{\psi}\gamma^\alpha A_\alpha \psi)_{x_1} (\overline{\psi}\gamma^\beta A_\beta \psi)_{x_2}] \} \tag{7.5d}$$

$$S_{\mathrm{E}}^{(2)} = -\frac{e^2}{2!} \int d^4x_1 d^4x_2 N[\overline{\psi} A\!\!\!/ \; \psi)_{x_1} (\overline{\psi} \; A\!\!\!/\psi)_{x_2}] \tag{7.5e}$$

$$S_{\mathrm{F}}^{(2)} = -\frac{e^2}{2!} \int d^4x_1 d^4x_2 (\overline{\psi} \; \gamma^\alpha A_\alpha \, \psi)_{x_1} (\overline{\psi}\gamma^\beta A_\beta \psi)_{x_2}. \tag{7.5f}$$

The first of these terms, $S_A^{(2)}$, Eq. (7.5a), is not very interesting. It corresponds to two processes of the kind illustrated in Fig. 7.1 going on independently of each other. Like $S^{(1)}$, this term does not lead to any real transitions.

The two terms in $S_B^{(2)}$, Eq. (7.5b), are identically equal to each other, as is seen by permuting the operators. This requires care, since the fermion fields are anticommuting operators and four-component spinors. Permuting the two groups $(\overline{\psi} A\!\!\!/\psi)$ involves an even permutation of fermion operators and the spinor indices of each group are self-contained.[1] Hence

$$N[(\overline{\psi} A\!\!\!/\psi)_{x_1} (\overline{\psi} A\!\!\!/\psi)_{x_2}] = N[(\overline{\psi}A\!\!\!/\psi)_{x_2} (\overline{\psi}A\!\!\!/ \; \psi)_{x_1}]. \tag{7.6}$$

Using this result and interchanging the integration variables $x_1 \leftrightarrow x_2$ in the second term of Eq. (7.5b), one obtains

$$S_B^{(2)} = -e^2 \int d^4x_1 d^4x_2 N[(\overline{\psi}A\!\!\!/ \; \psi)_{x_1} (\overline{\psi}A\!\!\!/ \; \psi)_{x_2}]. \tag{7.7}$$

This expression contains one fermion contraction. This is given by the fermion propagator (6.32c), which is a c-number and corresponds to a virtual intermediate fermion. For $t_2 < t_1$, we can think of it as a virtual electron propagating from x_2 to x_1, for $t_1 < t_2$ as a virtual positron propagating from x_1 to x_2. As explained in Section 4.4, no time-ordering is implied in the present formalism – indeed, all space–time points x_1 and x_2 are summed over – and these two cases are combined and jointly referred to as a virtual fermion propagating from x_2 (associated with $\overline{\psi}$) to x_1 (associated with ψ). In addition to this propagator, expression (7.7) contains two uncontracted fermion and two uncontracted photon operators. These absorb or create particles present initially or finally, so-called *external particles*. The operator $S_B^{(2)}$ contributes to many real processes. (To conserve energy and momentum, the initial and final states must each contain two particles.) Since the operators in $S_B^{(2)}$ are in normal order, it is easy to pick out the terms which contribute to a given process.

One of these processes is Compton scattering

$$\gamma + e^- \to \gamma + e^-, \tag{7.8}$$

---

[1] The reader can always resolve any cases of doubt by writing out explicitly the spinor indices.

already mentioned in Section 4.4. This process corresponds to selecting the positive frequency part $\psi^+(x_2)$ of $\psi(x_2)$ to absorb the initial electron, and the negative frequency part $\overline{\psi}^-(x_1)$ of $\overline{\psi}(x_1)$ to create the final electron. But either $A^+(x_1)$ or $A^+(x_2)$ can absorb the initial photon and correspondingly $A^-(x_2)$ or $A^-(x_1)$ must emit the final photon. Thus the part of Eq. (7.7) which causes Compton scattering is

$$S^{(2)}(\gamma e^- \to \gamma e^-) = S_a + S_b \tag{7.9}$$

where

$$S_a = -e^2 \int d^4x_1 d^4x_2 \overline{\psi}^-(x_1)\gamma^\alpha i S_F(x_1 - x_2)\gamma^\beta A_\alpha^-(x_1)A_\beta^+(x_2)\psi^+(x_2) \tag{7.10a}$$

$$S_b = -e^2 \int d^4x_1 d^4x_2 \overline{\psi}^-(x_1)\gamma^\alpha i S_F(x_1 - x_2)\gamma^\beta A_\beta^-(x_2)A_\alpha^+(x_1)\psi^+(x_2). \tag{7.10b}$$

In Eqs. (7.10), the operators have been put in a normal order and we have substituted Eq. (6.32c) for the fermion contraction.

The contributions S_a and S_b to Compton scattering are represented by the Feynman graphs in Figs. 7.2(a) and (b). The latter is the same as Fig. 4.3. (Remember that, except for the conventions about initial and final lines, there is no time ordering in Feynman graphs, so that the same graph can be drawn in many different ways.) In Fig. 7.2 we have attached the appropriate Lorentz indices (α, β) to vertices, and particle labels (γ, e^-) to external lines. We shall often omit these as redundant.

The other real processes described by Eq. (7.7) are Compton scattering by positrons, and the two-photon pair annihilation and creation processes, i.e.

$$\text{(i) } \gamma + e^+ \to \gamma + e^+, \quad \text{(ii) } e^+ + e^- \to \gamma + \gamma, \quad \text{(iii) } \gamma + \gamma \to e^+ + e^-. \tag{7.11}$$

The corresponding Feynman diagrams are shown in Figs. 7.3–7.5. We leave it to the reader to write down the operators for the first two of these processes. For the pair creation process one obtains

$$S^{(2)}(2\gamma \to e^+e^-) = -e^2 \int d^4x_1 d^4x_2 \overline{\psi}^-(x_1)\gamma^\alpha i S_F(x_1 - x_2)\gamma^\beta \psi^-(x_2)A_\alpha^+(x_1)A_\beta^+(x_2). \tag{7.12}$$

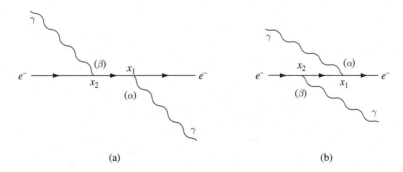

(a)   (b)

Figure 7.2 *The contributions S_a, S_b, Eqs. (7.10), to Compton scattering*

Although we have only shown one diagram in Fig. 7.5, Eq. (7.12) actually gives two contributions, since the operator $A_\beta^+(x_2)$ can absorb either of the initially present photons, with $A_\alpha^+(x_1)$ absorbing the other. A similar situation exists for the pair-annihilation process.

We next consider Eq. (7.5c). This term contains four uncontracted fermion operators. Accordingly, the real processes to which this term gives rise are fermion–fermion scattering: $e^- - e^-, e^+ - e^+$ or $e^- - e^+$ scattering, according to which positive and negative frequency parts are selected from the external fermion fields. The photon–photon contraction in Eq. (7.5c) describes the interaction between the charges as the exchange of transverse, longitudinal and scalar photons. This photon propagator occurs associated with two conserved current operators $s^\mu(x) = \left(\bar\psi\gamma^\mu\psi\right)_{x}$. As discussed in Section 5.3 [particularly the discussion of Eq. (5.44)], this covariant formulation is equivalent to the usual description of the interaction in terms of the instantaneous Coulomb interaction together with the exchange of transverse photons.

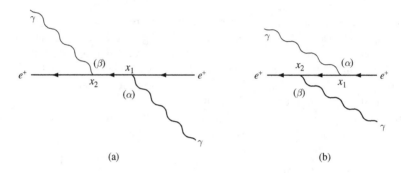

Figure 7.3   *The Feynman diagrams for Compton scattering by positrons*

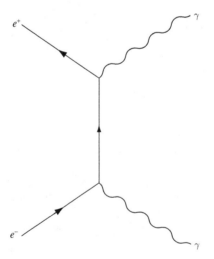

Figure 7.4   *The Feynman diagram for $e^+ + e^- \to \gamma + \gamma$*

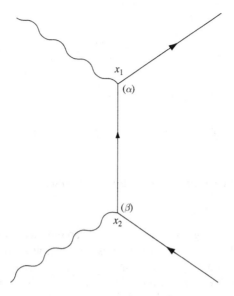

Figure 7.5 *The Feynman diagram for $\gamma + \gamma \to e^+ + e^-$, Eq. (7.12)*

We next consider electron–electron scattering,

$$e^- + e^- \to e^- + e^-, \qquad (7.13)$$

known as Møller scattering, in more detail. The part of the operator (7.5c) describing this process is

$$S^{(2)}(2e^- \to 2e^-)$$
$$= \frac{-e^2}{2!} \int d^4x_1 d^4x_2 N\left[(\overline{\psi}^-\gamma^\alpha\psi^+)_{x_1}(\overline{\psi}^-\gamma^\beta\psi^+)_{x_2}\right] iD_{F\alpha\beta}(x_1 - x_2), \qquad (7.14)$$

where we substituted Eq. (6.32d) for the photon contraction.

Let us label the initial and final electron states 1, 2 and $1'$, $2'$ respectively, i.e. with an obvious notation we are considering the transition

$$|i\rangle = c^\dagger(2)c^\dagger(1)|0\rangle \to |f\rangle = c^\dagger(2')c^\dagger(1')|0\rangle. \qquad (7.15)$$

Eq. (7.14) gives four contributions to the transition (7.15), since either initial electron can be absorbed by either ψ^+ operator, and either final electron can be emitted by either $\overline{\psi}^-$ operator. These four terms comprise two pairs, which differ only by the interchange of the integration variables $x_1 \leftrightarrow x_2$ in Eq. (7.14). We need consider only one of these pairs and multiply the result by a factor 2. The remaining two terms are represented by the Feynman graphs in Fig. 7.6.

We had another case of two identical contributions to a process, related to the interchange $x_1 \leftrightarrow x_2$, in connection with Eq. (7.5b). This represents a general result. The nth order term $S^{(n)}$ in the S-matrix expansion (7.1) contains a factor $1/n!$ and n integration variables $x_1, x_2, \ldots, x_n$. These are only summation variables and can be attached to the n

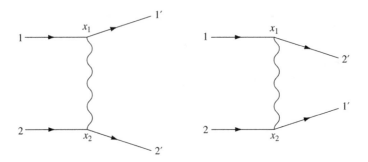

Figure 7.6   *The two diagrams for electron-electron scattering (Møller scattering)*

vertices of a given Feynman graph in $n!$ ways. We can omit the factor $1/n!$ if we consider only topologically different Feynman diagrams, i.e. diagrams which differ only in the labelling of vertices are considered the same. Some care is required in interpreting this statement. For example, the two diagrams of Fig. 7.6 are topologically different from each other because the two final electrons have different properties. (These were labelled $1'$ and $2'$. In practice they are the momenta and spins.) Permuting x_1 and x_2 does not interchange the two graphs of Fig. 7.6. As we shall see, their contributions occur with a relative minus sign and correspond to the 'direct minus exchange scattering', which the reader should recognize, from non-relativistic quantum mechanics, as characteristic of two identical fermions.

In order to obtain explicit expressions for these two contributions, let

$$\psi_j^+(x) = c(j)f_j(x), \quad \overline{\psi}_j^-(x) = c^\dagger(j)g_j(x) \tag{7.16}$$

be the parts of the operators $\psi^+(x)$ and $\overline{\psi}^-(x)$ proportional to $c(j)$ and $c^\dagger(j)$ respectively. $j = 1, 2, 1', 2'$ labels the electron states involved in the process. The part of the S-matrix operator (7.14) which effects the transition $|i\rangle \rightarrow |f\rangle$, Eq. (7.15), is then given by

$$S^{(2)}(e^-(1) + e^-(2) \rightarrow e^-(1') + e^-(2')) = S_a + S_b \tag{7.17a}$$

where S_a and S_b correspond to Figs. 7.6, and are given by

$$S_a = -e^2 \int d^4x_1 d^4x_2 N\left[(\overline{\psi}_{1'}^-\gamma^\alpha\psi_1^+)_{x_1}(\overline{\psi}_{2'}^-\gamma^\beta\psi_2^+)_{x_2}\right]iD_{F\alpha\beta}(x_1 - x_2) \tag{7.17b}$$

$$S_b = -e^2 \int d^4x_1 d^4x_2 N\left[(\overline{\psi}_{2'}^-\gamma^\alpha\psi_1^+)_{x_1}(\overline{\psi}_{1'}^-\gamma^\beta\psi_2^+)_{x_2}\right]iD_{F\alpha\beta}(x_1 - x_2). \tag{7.17c}$$

The relative minus sign of the two contributions is implied by the normal products in these equations. To arrange the creation and annihilation operators in both cases in the same order, e.g. as $c^\dagger(1')c^\dagger(2')c(1)c(2)$, requires the normal products in Eqs. (7.17b) and (7.17c) to be reordered equal to $-\overline{\psi}_{1'}^-(x_1)\overline{\psi}_{2'}^-(x_2)\psi_1^+(x_1)\psi_2^+(x_2)$ and $+\overline{\psi}_{1'}^-(x_2)\overline{\psi}_{2'}^-(x_1)\psi_1^+(x_1)\psi_2^+(x_2)$, respectively. Using Eqs. (7.16), we obtain from Eqs. (7.17) the transition amplitude

$$\langle f | S^{(2)}(2e^- \rightarrow 2e^-) | i \rangle$$
$$= \{-e^2 \int d^4x_1 \, d^4x_2 g_{1'}(x_1) \gamma^\alpha f_1(x_1) g_{2'}(x_2) \gamma^\beta f_2(x_2) i D_{F\alpha\beta}(x_1 - x_2)\} \qquad (7.18)$$
$$- \{1' \leftrightarrow 2'\} \qquad ,$$

where the term $\{1' \leftrightarrow 2'\}$ is just the first expression in braces with the labels $1'$ and $2'$ of the two final electron states interchanged. Our final result (7.18) has the desired form of a 'direct' amplitude minus an 'exchange' amplitude, the two amplitudes being transformed into each other by exchanging the single-particle states of the two electrons in the final state. In non-relativistic quantum mechanics, this result follows through the use of anti-symmetric wavefunctions according to Pauli's principle. In the above field-theoretic derivation, the anticommutativity of the fermion field operators is the crucial element.

These arguments generalize. Whenever the initial or final state contains several identical fermions, one obtains a completely antisymmetric transition amplitude $\langle f | S | i \rangle$. For example, if the initial state $|i\rangle$ contains s positrons in states $1, 2, \ldots, s$, the corresponding S-matrix operator will contain s uncontracted operators $N(\overline{\psi}(x_1)\overline{\psi}(x_2)\ldots\overline{\psi}(x_s))$. Any one of these operators $\overline{\psi}(x_1), \ldots, \overline{\psi}(x_s)$ can absorb the positron in state 1, and so on, giving $s!$ terms, whose sum is completely antisymmetric in the labels $1, 2, \ldots, s$, since the operators $\overline{\psi}(x_1), \ldots, \overline{\psi}(x_s)$ anticommute. An analogous argument holds for several identical final state fermions.

More curiously, the fact that the operator $\psi(x)$ can absorb an electron or create a positron implies that transition amplitudes are antisymmetric with respect to initial electron and final positron states. (A similar argument applies of course to $\overline{\psi}(x)$ and initial positrons and final electrons.) We have an example of this in electron–positron scattering,

$$e^+ + e^- \rightarrow e^+ + e^-,$$

known as Bhabha scattering. The part of the operator (7.5c) describing this process must contain the uncontracted operators $\overline{\psi}^+$, ψ^+, ψ^- and $\overline{\psi}^-$ to absorb and create the particles present initially and finally. As in the case of Møller scattering, four terms contribute, which again reduce to two by the general argument given above. It is left as an exercise to the reader to derive from Eq. (7.5c) the following expression for the S-matrix operator for Bhabha scattering:

$$S^{(2)}(e^+e^- \rightarrow e^+e^-) = S_a + S_b \qquad (7.19a)$$

where

$$S_a = -e^2 \int d^4x_1 d^4x_2 N\left[(\overline{\psi}^- \gamma^\alpha \psi^+)_{x_1}(\overline{\psi}^+ \gamma^\beta \psi^-)_{x_2}\right] i D_{F\alpha\beta}(x_1 - x_2), \qquad (7.19b)$$

$$S_b = -e^2 \int d^4x_1 d^4x_2 N\left[(\overline{\psi}^- \gamma^\alpha \psi^-)_{x_1}(\overline{\psi}^+ \gamma^\beta \psi^+)_{x_2}\right] i D_{F\alpha\beta}(x_1 - x_2). \qquad (7.19c)$$

The Feynman graph for S_a is shown in Fig. 7.7(a). It represents the scattering by photon exchange, as occurred for electron–electron scattering (Fig. 7.6). However, in the term S_b both initial particles are annihilated at x_2 and the final electron–positron pair is created at x_1. It corresponds to the *annihilation diagram* of Fig. 7.7(b). As in electron–electron scattering, there is a relative sign factor (-1) between the two contributions implicit in

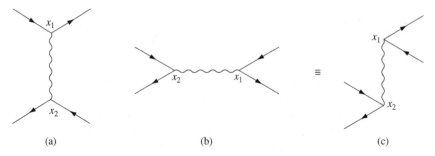

Figure 7.7 The contributions S_a and S_b to electron–positron scattering (Bhabha scattering): (a) represents photon exchange; (b) and (c) are equivalent ways of representing the pair-annihilation process

the normal products, which becomes explicit if the creation and annihilation operators are brought into the same normal order in both cases. That the diagrams of Figs. 7.7(a) and (b) are related by the interchange of an initial electron state and a final positron state is brought out by 'deforming' diagram 7.7(b) into diagram 7.7(c). Comparing diagrams 7.7(a) and (c) one sees that the latter is obtained from the former by interchanging the initial electron line at x_1 and the final positron line at x_2.

We shall now discuss briefly the remaining second-order terms $S_D^{(2)}$ to $S_F^{(2)}$, Eqs. (7.5d)–(7.5f).

Eq. (7.5d) contains two uncontracted fermion fields and gives rise to two processes according to whether the fermion present initially and finally is an electron or a positron. The two terms in Eq. (7.5d) are again equal to each other. For the electron case, this equation reduces to

$$S^{(2)}(e^- \rightarrow e^-)$$
$$= -e^2 \int d^4x_1\ d^4x_2 \overline{\psi}^-(x_1)\gamma^\alpha iS_F(x_1 - x_2)\gamma^\beta \psi^+(x_2)iD_{F\alpha\beta}(x_1 - x_2), \tag{7.20}$$

which corresponds to the diagram of Fig. 7.8. It represents a modification of the properties of a *bare* electron due to its interaction with the radiation field. It is one of the processes – in fact the simplest – which converts a bare electron into a *physical* electron, i.e. one surrounded by its photon cloud. This interaction changes the energy of the system, that is, the mass of the physical electron as compared with that of the bare electron. This is known as the *self-energy* of the electron, and Fig. 7.8 is called a self-energy diagram. Its evaluation leads to a divergent integral. These divergent self-energy effects can be eliminated by incorporating them in the properties of the physical electron. This is the process of *re-normalization*, which will be studied in Chapter 9.

Figure 7.8 The electron self-energy $S^{(2)}(e^- \rightarrow e^-)$, Eq. (7.20)

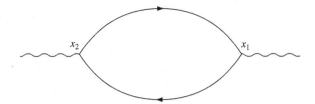

Figure 7.9 *The photon self-energy (vacuum polarization) $S^{(2)}(\gamma \to \gamma)$, Eq. (7.21)*

Fig. 7.9 similarly describes a *photon self-energy* arising from the term $S_E^{(2)}$, Eq. (7.5e). The interaction between the electromagnetic and the electron–positron fields enables the photon to create a virtual electron–positron pair, which subsequently annihilates again. An external electromagnetic field (for example the field of a heavy nucleus) will modify the distribution of these virtual electron–positron pairs, i.e. it will 'polarize the vacuum' in much the same way in which it would polarize a dielectric. For this reason, such photon self-energy graphs are called *vacuum polarization* diagrams. Like the electron self-energy, they lead to infinities, which are again eliminated by re-normalization (see Chapter 9).

Eq. (7.5e) for the photon self-energy can be written

$$S^{(2)}(\gamma \to \gamma) = -e^2 \int \mathrm{d}^4x_1\ \mathrm{d}^4x_2 \mathrm{N}[(\overline{\psi}A^-\psi)_{x_1}(\overline{\psi}\ A^+\psi)_{x_2}]. \tag{7.21}$$

Writing the spinor indices out explicitly, we can re-express the normal product in Eq. (7.21) as

$$\mathrm{N}[(\overline{\psi}_\lambda A^-_{\lambda\mu}\ \psi_\mu)_{x_1}(\overline{\psi}_\sigma\ A^+_{\sigma\tau}\psi_\tau)_{x_2}]$$

$$= (-1)\,\psi_\tau(x_2)\overline{\psi}_\lambda(x_1)A^-_{\lambda\mu}(x_1)\ \psi_\mu(x_1)\overline{\psi}_\sigma(x_2)A^+_{\sigma\tau}(x_2)$$

$$= (-1)\mathrm{Tr}[\mathrm{i}S_F(x_2 - x_1)A^-(x_1)\mathrm{i}S_F(x_1 - x_2)A^+(x_2)]. \tag{7.22}$$

(Here $A^-_{\lambda\mu}(x) \equiv \gamma^\alpha_{\lambda\mu}A^-_\alpha(x)$, etc.)

The minus sign in the last equation is characteristic of *closed fermion loops* (i.e. closed loops consisting of fermion lines only), which always involve the transposition of a single fermion operator from one end of a product of such factors to the other, i.e. an odd number of interchanges. The trace in Eq. (7.22) is equally characteristic. It corresponds to summing over all spin states of the virtual electron–positron pair. (The connection between spin sums and traces will be discussed in Section 8.2.)

Finally, Fig. 7.10 shows the graph representing Eq. (7.5f). This diagram has no external lines and consequently does not cause any transitions. One can show that such *vacuum diagrams* (i.e. diagrams without external lines) may be omitted altogether, at any rate in elementary applications.

This completes our initial analysis of the various terms which occur on decomposing $S^{(2)}$ into normal products. We have seen that the terms obtained correspond to specific

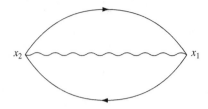

Figure 7.10   *The simplest vacuum diagram, Eq. (7.5f)*

processes and how Feynman diagrams greatly aid their interpretation. No new features occur for the higher-order terms $S^{(3)}$, ... The methods developed are sufficient to deal with such higher- order processes, although the complexity of the mathematics increases rapidly with order.

7.2   Feynman Diagrams in Momentum Space

In the last section we developed a technique for deriving the S-matrix operator which generates a particular transition $|i\rangle \rightarrow |f\rangle$ in a given order. In practice, one is usually interested in the corresponding matrix element $\langle f|S^{(n)}|i\rangle$. The states $|i\rangle$ and $|f\rangle$ are usually specified by the particles of known momenta, and spin and polarization properties present initially and finally. Explicit calculations of the matrix elements lead to a re-interpretation of the Feynman graphs as diagrams in momentum space. By studying some specific cases we shall see that these diagrams are closely related to the mathematical expressions they represent. It is possible to formulate a set of rules which enable one to write down the matrix elements directly from the Feynman diagrams without detailed calculations. These Feynman rules, which will be given in the next section, are the linchpin of practical calculations in perturbation theory.

Calculation of the matrix elements, with $|i\rangle$ and $|f\rangle$ specified as momentum eigenstates of the particles present, essentially corresponds to Fourier transforming the fields into momentum space in order to pick out the appropriate absorption and creation operators. For the propagators, these Fourier transforms are given, from Eqs. (6.32c), (6.32d), (4.63) and (5.27), by

$$\psi(x_1)\overline{\psi}(x_2) = iS_F(x_1 - x_2) = \frac{1}{(2\pi)^4} \int d^4p iS_F(p)e^{-ip(x_1-x_2)} \tag{7.23a}$$

$$A^\alpha(x_1)A^\beta(x_2) = iD_F^{\alpha\beta}(x_1 - x_2) = \frac{1}{(2\pi)^4} \int d^4k iD_F^{\alpha\beta}(k)e^{-ik(x_1-x_2)} \tag{7.23b}$$

where

$$S_F(p) = \frac{\not p + m}{p^2 - m^2 + i\varepsilon} \equiv \frac{1}{\not p - m + i\varepsilon} \tag{7.24a}$$

$$D_F^{\alpha\beta}(k) = \frac{-g^{\alpha\beta}}{k^2 + i\varepsilon}. \tag{7.24b}$$

The Fourier expansions of the uncontracted fields $\psi, \bar{\psi}$ and A_α are given by Eqs. (4.38) and (5.16). The effect of the uncontracted operators $\psi^+, \bar{\psi}^+$ and A_α^+, which occur in a term of the S-matrix expansion, acting on $|i\rangle$, is to give the vacuum state $|0\rangle$. For example, it follows from Eqs. (4.38) and (5.16) that

$$\psi^+(x)|e^-\mathbf{p}\rangle = |0\rangle \left(\frac{m}{VE_\mathbf{p}}\right)^{1/2} u(\mathbf{p})e^{-ipx} \tag{7.25a}$$

$$\bar{\psi}^+(x)|e^+\mathbf{p}\rangle = |0\rangle \left(\frac{m}{VE_\mathbf{p}}\right)^{1/2} \bar{v}(\mathbf{p})e^{-ipx} \tag{7.25b}$$

$$A_\alpha^+(x)|\gamma\mathbf{k}\rangle = |0\rangle \left(\frac{1}{2V\omega_\mathbf{k}}\right)^{1/2} \varepsilon_\alpha(\mathbf{k})e^{-ikx} \tag{7.25c}$$

Here we have suppressed the spin and polarization labels. For example, $|e^-\mathbf{p}\rangle$ and $|\gamma\mathbf{k}\rangle$ stand for the one-electron and one-photon states

$$|e^-\mathbf{p}\rangle \equiv |e^-\mathbf{p}r\rangle = c_r^\dagger(\mathbf{p})|0\rangle, \quad |\gamma\mathbf{k}\rangle \equiv |\gamma\mathbf{k}r\rangle = a_r^\dagger(\mathbf{k})|0\rangle, \quad r = 1, 2,$$

and $u(\mathbf{p})$ and $\varepsilon_\alpha(\mathbf{k})$ are short for $u_r(\mathbf{p})$ and $\varepsilon_{r\alpha}(\mathbf{k})$. In the following, we shall frequently simplify the notation in this way, writing $c(\mathbf{p})$ for $c_r(\mathbf{p})$, etc.

The effect of the uncontracted operators $\bar{\psi}^-, \psi^-$ and A_α^-, which occur in a term of the S-matrix expansion, acting on $|0\rangle$, is to produce the final state $|f\rangle$. In particular, we find from Eqs. (4.38) and (5.16) that

$$\bar{\psi}^-(x)|0\rangle = \sum |e^-\mathbf{p}\rangle \left(\frac{m}{VE_\mathbf{p}}\right)^{1/2} \bar{u}(\mathbf{p})e^{ipx} \tag{7.26a}$$

$$\psi^-(x)|0\rangle = \sum |e^+\mathbf{p}\rangle \left(\frac{m}{VE_\mathbf{p}}\right)^{1/2} v(\mathbf{p})e^{ipx} \tag{7.26b}$$

$$A_\alpha^-(x)|0\rangle = \sum |\gamma\mathbf{k}\rangle \left(\frac{1}{2V\omega_\mathbf{k}}\right)^{1/2} \varepsilon_\alpha(\mathbf{k})e^{ikx}, \tag{7.26c}$$

where the summations are over spin and polarization states, as well as momenta. It is straightforward to generalize the results (7.25) and (7.26) to states involving several particles.

Using Eqs. (7.23)–(7.26) it is easy to calculate S-matrix elements, as the following examples will show.

7.2.1   The first-order terms $S^{(1)}$

The Feynman graphs resulting from the first-order term

$$S^{(1)} = \mathrm{ie} \int \mathrm{d}^4 x \mathrm{N}(\bar{\psi}\slashed{A}\psi)_x \tag{7.27}$$

are just the basic vertex diagrams of Fig. 7.1. Let us calculate the matrix element $\langle f | S^{(1)} | i \rangle$ for one of these processes, namely for electron scattering with emission of a photon, illustrated in Fig. 7.11. In this figure, we state the energy–momentum four-vectors of the particles involved, but their spin and polarization labels have been suppressed, as discussed above. Fig. 7.11 represents the transition

$$|i\rangle = |e^- \mathbf{p}\rangle = c^\dagger(\mathbf{p})|0\rangle \rightarrow |f\rangle = |e^- \mathbf{p}'; \gamma \mathbf{k}'\rangle = c^\dagger(\mathbf{p}')a^\dagger(\mathbf{k}')|0\rangle, \tag{7.28}$$

i.e. $|i\rangle$ consists of an electron of momentum $\mathbf{p}$ (and spin state $s = 1, 2$), and $|f\rangle$ of an electron of momentum $\mathbf{p}'$ (and spin state $s' = 1, 2$) plus a photon of momentum $\mathbf{k}'$ (and polarization state $r' = 1, 2$). From Eqs. (7.25)–(7.28) we obtain

$$\langle f | S^{(1)} | i \rangle = \langle e^- \mathbf{p}'; \gamma \mathbf{k}' | \mathrm{ie} \int \mathrm{d}^4 x \bar{\psi}^-(x) \gamma^\alpha A_\alpha^-(x)\psi^+(x) | e^- \mathbf{p} \rangle$$

$$= \mathrm{ie} \int \mathrm{d}^4 x \left[\left(\frac{m}{V E_{\mathbf{p}'}} \right)^{1/2} \bar{u}(\mathbf{p}')\mathrm{e}^{\mathrm{i}p'x} \right] \gamma^\alpha \left[\left(\frac{1}{2V\omega_{\mathbf{k}'}} \right)^{1/2} \varepsilon_\alpha(\mathbf{k}')\mathrm{e}^{\mathrm{i}k'x} \right]$$

$$\times \left[\left(\frac{m}{V E_{\mathbf{p}}} \right)^{1/2} u(\mathbf{p})\mathrm{e}^{-\mathrm{i}px} \right]. \tag{7.29}$$

The x-dependent terms in this expression give

$$\int \mathrm{d}^4 x \exp\left[\mathrm{i}x(p' + k' - p)\right] = (2\pi)^4 \delta^{(4)}(p' + k' - p), \tag{7.30}$$

where we have anticipated going to the limits of an infinite volume, $V \rightarrow \infty$, and an infinite time interval during which the transition may occur. From Eqs. (7.29) and (7.30) we obtain

Figure 7.11   *The process $e^- \rightarrow e^- + \gamma$. The four-momenta of the particles are shown. The spin and polarization labels (s, s' and r) have been suppressed, as explained in the text*

$$\langle f|S^{(1)}|i\rangle = \left[(2\pi)^4 \delta^4 \left(p' + k' - p \right) \left(\frac{m}{VE_\mathbf{p}} \right)^{1/2} \left(\frac{m}{VE_{\mathbf{p}'}} \right)^{1/2} \left(\frac{1}{2V\omega_{\mathbf{k}'}} \right)^{1/2} \right] \mathcal{M} \quad (7.31)$$

where

$$\mathcal{M} = ie\bar{u}(\mathbf{p}')\slashed{\varepsilon}(\mathbf{k}' = \mathbf{p} - \mathbf{p}')u(\mathbf{p}). \quad (7.32)$$

Eqs. (7.31) and (7.32) are our final result. $\mathcal{M}$ is called the *Feynman amplitude* for the process represented by the Feynman graph in Fig. 7.11. Since this diagram is labelled by the momenta (and the implied spin and polarization labels) of the particles involved, it is called a Feynman diagram in momentum space, in contrast to the configuration-space diagrams of the last section, e.g. Fig. 7.1(a).

The δ-function in Eq. (7.31) arose from the x-integration in Eq. (7.29) over the three exponential functions associated with the two fermion lines and the photon line, which meet at the vertex x. This δ-function ensures conservation of energy and momentum for this process: $p = p' + k'$. [Correspondingly, the argument of the polarization vector $\varepsilon_\alpha(\mathbf{k}')$ in Eq. (7.32) was written $\mathbf{k}' = \mathbf{p} - \mathbf{p}'$.] We shall see that for more complicated Feynman diagrams, such a δ-function is obtained in this way for each vertex, ensuring energy–momentum conservation at each vertex and consequently for the process as a whole.

For the process $e^- \rightarrow e^- + \gamma$ and the other first-order processes, energy–momentum conservation is incompatible with the conditions for real particles $(p^2 = p'^2 = m^2,\ k'^2 = 0,$ in our case), so these are not real processes, as stated earlier.

7.2.2 Compton scattering

As a second example, we calculate the matrix element for Compton scattering, for which the S-matrix operator and Feynman graphs were given in Eqs. (7.9) and (7.10) and Fig. 7.2. Their counterparts in momentum space are shown in Fig. 7.12, corresponding to the transition

$$|i\rangle = c^\dagger(\mathbf{p})a^\dagger(\mathbf{k})|0\rangle \rightarrow |f\rangle = c^\dagger(\mathbf{p}')a^\dagger(\mathbf{k}')|0\rangle. \quad (7.33)$$

The S-matrix element for this transition is derived from Eqs. (7.9) and (7.10). Using Eqs. (7.23a), (7.25) and (7.26), one obtains

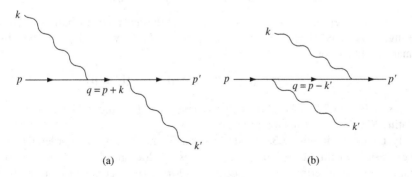

(a)   (b)

Figure 7.12 *Compton scattering by electrons*

$$\langle f|S_a|i\rangle = -e^2 \int d^4x_1 \, d^4x_2 \left[\left(\frac{m}{VE_{p'}}\right)^{1/2} \bar{u}(\mathbf{p}')e^{ip'x_1} \right] \left[\left(\frac{1}{2V\omega_{\mathbf{k}'}}\right)^{1/2} \not{\epsilon}(\mathbf{k}')e^{ik'x_1} \right]$$

$$\times \frac{1}{(2\pi)^4} \int d^4q \, iS_F(q)e^{-iq(x_1-x_2)}$$

$$\times \left[\left(\frac{1}{2V\omega_{\mathbf{k}}}\right)^{1/2} \not{\epsilon}(\mathbf{k})e^{-ikx_2} \right] \left[\left(\frac{m}{VE_\mathbf{p}}\right)^{1/2} u(\mathbf{p})e^{-ipx_2} \right].$$

$$(7.34)$$

Note that u and $\bar{u}$ are four-component spinors and that S_F and the factors $\not{\epsilon}$ are 4×4 matrices. The spinor indices are suppressed, but these quantities must always be written in the correct order of matrix algebra.

The x_1 and x_2 integrations in Eq. (7.34) give

$$\int d^4x_1 \exp[ix_1(p' + k' - q)] \int d^4x_2 \exp[ix_2(q - p - k)]$$

$$= (2\pi)^4 \delta^{(4)}(p' + k' - q)(2\pi)^4 \delta^{(4)}(q - p - k)$$

$$= (2\pi)^4 \delta^{(4)}(p' + k' - p - k)(2\pi)^4 \delta^{(4)}(q - p - k). \quad (7.35)$$

Hence energy and momentum are conserved at each vertex and overall for the process. In particular, the energy–momentum q of the virtual intermediate electron is fixed:

$$q = p + k = p' + k'. \quad (7.36)$$

Substituting Eq. (7.35) in (7.34) and carrying out the q integration, one obtains

$$\langle f|S_a|i\rangle = \left[(2\pi)^4 \delta^{(4)}(p' + k' - p - k) \right.$$

$$\left. \times \left(\frac{m}{VE_\mathbf{p}}\right)^{1/2} \left(\frac{m}{VE_{p'}}\right)^{1/2} \left(\frac{1}{2V\omega_{\mathbf{k}}}\right)^{1/2} \left(\frac{1}{2V\omega_{\mathbf{k}'}}\right)^{1/2} \right] \mathcal{M}_a \quad (7.37)$$

where $\mathcal{M}_a$, the Feynman amplitude associated with Fig. 7.12(a), is given by

$$\mathcal{M}_a = -e^2 \bar{u}(\mathbf{p}')\not{\epsilon}(\mathbf{k}')iS_F(q = p + k)\not{\epsilon}(\mathbf{k})u(\mathbf{p}). \quad (7.38a)$$

It is left as an exercise for the reader to show that the second contribution to Compton scattering, $\langle f|S_b|i\rangle$, is given by the same equation (7.37), with $\mathcal{M}_a$ replaced by the Feynman amplitude for Fig. 7.12(b):

$$\mathcal{M}_b = -e^2 \bar{u}(\mathbf{p}')\not{\epsilon}(\mathbf{k})iS_F(q = p - k')\not{\epsilon}(\mathbf{k}')u(\mathbf{p}). \quad (7.38b)$$

Our result, Eqs. (7.37) and (7.38), displays some general features which always occur in calculating S-matrix elements by these methods.

Firstly, the factors in Eqs. (7.38a) and (7.38b) are in the correct spinor order. Comparing these expressions with the Feynman graphs, Figs. 7.12(a) and (b), we can describe this order as: following a fermion line *in the sense of its arrows,* corresponds to writing the spinor factors *from right to left.*

Secondly, comparing these results with Eqs. (7.31) and (7.32), we note many common features. The square brackets in Eqs. (7.31) and (7.37) each contain a δ-function for overall energy–momentum conservation (multiplied by $(2\pi)^4$), and factors $(1/2V\omega_{\mathbf{k}})^{1/2}$ and $(m/VE_{\mathbf{p}})^{1/2}$ for each external photon and fermion line respectively. The Feynman amplitudes (7.32) and (7.38) contain a factor (ie), associated with each vertex in the related Feynman graphs, and factors $\bar{u}, u$ and $\slashed{\epsilon}$, associated in an obvious manner with external electron and photon lines. The one additional feature in Eqs. (7.38) is the presence of the factors $iS_{\mathrm{F}}(q)$ which correspond to the intermediate fermion lines in diagrams 7.12(a) and (b). These common features are examples of Feynman rules, which will be fully discussed in the next section.

Finally, we see that, for both Figs. 7.12(a) and (b), the intermediate particle cannot be a real particle: $q^2 \neq m^2$, since we cannot have energy–momentum conservation for three real particles at a vertex. This is in contrast to the non-covariant perturbation theory of non-relativistic quantum mechanics, where time and space coordinates (and consequently energy and three-momentum) are treated on different footings: particles in intermediate states satisfy the energy–momentum conditions of real particles (i.e. $p^2 = m^2$, $k^2 = 0$), but energy is not conserved in intermediate states although three-momentum is.

We briefly consider Compton scattering by positrons in order to establish some differences of detail which occur for positrons. The Feynman graphs in momentum space for this process are shown in Fig. 7.13. We leave it as exercises for the reader to show from first principles that the Feynman diagram 7.13(a) again leads to Eq. (7.37), with $\mathcal{M}_{\mathrm{a}}$ replaced by

$$\mathcal{M}'_{\mathrm{a}} = e^2 \bar{v}(\mathbf{p}) \slashed{\epsilon}(\mathbf{k}) iS_{\mathrm{F}}(q = -p - k) \slashed{\epsilon}(\mathbf{k}') v(\mathbf{p}') \tag{7.39}$$

and to obtain the corresponding result for diagram 7.13(b).

In Eq. (7.39), the spinor $v(\mathbf{p}')$ relates to the final-state positron, and the spinor $\bar{v}(\mathbf{p})$ to the initial-state positron. The order of the spinor factors in this equation corresponds to writing these factors *from right to left* as one follows the fermion line *in the sense of its arrows*. This is the same prescription as for electrons. Care is also needed in interpreting the momentum labels on Feynman diagrams. For *external* lines, the momenta shown are the *actual* four-momenta of the particles present initially and finally. This applies to electrons, positrons and photons. This means that on external *electron* lines, the flow of four-momentum is in the *same* sense as that of the arrows on the lines; on external *positron* lines it is in the sense *opposite* to that of the arrows. On *internal* fermion lines, on the other

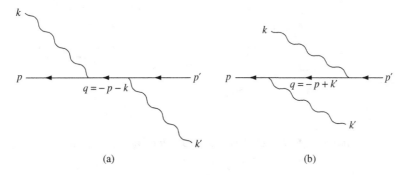

(a) (b)

Figure 7.13 *Compton scattering by positrons*

hand, the four-momentum labels on Feynman graphs *always* represent energy–momentum flow in the *same* direction as the arrows.

This completes our detailed analysis of Compton scattering. In the following examples, the detailed derivations will be left as exercises for the reader, and we shall concentrate on the remaining features of Feynman graphs not yet encountered.

7.2.3   Electron–electron scattering

The Feynman diagrams in configuration space for Møller scattering were shown in Fig. 7.6. The corresponding momentum space graphs are shown in Fig. 7.14. The S-matrix element for the transition

$$|i\rangle = c^\dagger(\mathbf{p}_2)c^\dagger(\mathbf{p}_1)|0\rangle \rightarrow |f\rangle = c^\dagger(\mathbf{p}_2')c^\dagger(\mathbf{p}_1')|0\rangle \tag{7.40}$$

is obtained from Eqs. (7.17). One finds

$$\langle f|S^{(2)}(2e^- \rightarrow 2e^-)|i\rangle$$
$$= \left[(2\pi)^4\delta^{(4)}(p_1' + p_2' - p_1 - p_2)\Pi\left(\frac{m}{VE_\mathbf{p}}\right)^{1/2}\right](\mathscr{M}_a + \mathscr{M}_b) \tag{7.41a}$$

where the meaning of $\Pi(m/VE_\mathbf{p})^{1/2}$ should be clear to the reader,[2] and where the Feynman amplitudes corresponding to Figs. 7.14(a) and (b) are given by

$$\mathscr{M}_a = -e^2\bar{u}(\mathbf{p}_1')\gamma^\alpha u(\mathbf{p}_1)iD_{F\alpha\beta}(k = p_2 - p_2')\bar{u}(\mathbf{p}_2')\gamma^\beta u(\mathbf{p}_2) \tag{7.41b}$$

$$\mathscr{M}_b = +e^2\bar{u}(\mathbf{p}_2')\gamma^\alpha u(\mathbf{p}_1)iD_{F\alpha\beta}(k = p_2 - p_1')\bar{u}(\mathbf{p}_1')\gamma^\beta u(\mathbf{p}_2). \tag{7.41c}$$

The last two equations exhibit explicitly the relative minus sign of the direct and exchange amplitudes, which reflects the exclusion principle, as discussed in the last section. The new feature in these equations is the appearance of the factors $iD_{F\alpha\beta}(k)$, corresponding to the

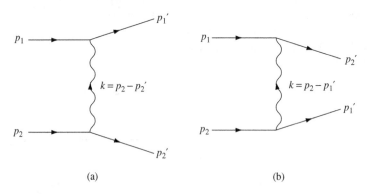

(a)   (b)

Figure 7.14   *Electron-electron scattering (Møller scattering)*

---

[2] i.e. even *without* detailed derivation, but Eqs. (7.41) should of course be derived by the reader.

internal photon lines in the Feynman graphs 7.14. Since from Eq. (7.24b) $D_{F\alpha\beta}(k) = D_{F\alpha\beta}(-k)$, the sense of k along an internal photon line is arbitrary. However, a definite direction must be chosen for k in order to assign consistent signs to k in the δ-functions associated with the vertices at the two ends of an internal photon line. For example, with the choice for k in Fig. 7.14(a), i.e. from the bottom to the top vertex, $p_2 = p'_2 + k$ and $p_1 + k = p'_1$, giving the correct overall energy–momentum conservation for the process.

7.2.4   Closed loops

A new feature occurs for Feynman diagrams containing closed loops of internal lines, such as the electron and photon self-energy diagrams, Figs. 7.8 and 7.9. For diagrams without loops, such as we have been considering in this section so far, energy–momentum conservation at the vertices determines the four-momenta of all internal lines completely. For loop diagrams this is not the case. Consider, as a typical example, the electron self-energy. Its Feynman graph in momentum space is shown in Fig. 7.15. Conservation of energy and momentum at the two vertices gives

$$p = q + k = p',$$ (7.42)

but this does not determine the internal momenta k and q separately. The intuitive response to this is that one must sum over all allowed values of k and q to find the total amplitude.

To see that this conjecture is correct, we require the matrix element of the electron self-energy operator $S^{(2)}(e^- \rightarrow e^-)$, Eq. (7.20), for the transition

$$|i\rangle = c^\dagger(\mathbf{p})|0\rangle \rightarrow |f\rangle = c^\dagger(\mathbf{p}')|0\rangle.$$ (7.43)

We leave it to the reader to obtain the result

$$\langle f|S^{(2)}(e^- \rightarrow e^-)|i\rangle$$

$$= -e^2 \left(\frac{m}{VE_p}\right)^{1/2} \left(\frac{m}{VE_{p'}}\right)^{1/2} \int d^4q\ d^4k \delta^{(4)}(p'-k-q)\delta^{(4)}(k+q-p)$$

$$\times\ iD_{F\alpha\beta}(k)\bar{u}(\mathbf{p}')\gamma^\alpha iS_F(q)\gamma^\beta u(\mathbf{p})$$ (7.44a)

$$= \left[(2\pi)^4\delta^{(4)}(p'-p)\left(\frac{m}{VE_p}\right)^{1/2}\left(\frac{m}{VE_{p'}}\right)^{1/2}\right]\mathcal{M}$$

Figure 7.15   *The electron self-energy*

where

$$\mathcal{M} = \frac{-e^2}{(2\pi)^4} \int d^4 k i D_{F\alpha\beta}(k) \bar{u}(\mathbf{p}) \gamma^\alpha i S_F(p-k) \gamma^\beta u(\mathbf{p}). \tag{7.44b}$$

As expected, this Feynman amplitude contains an integration over all internal photon momenta k, while for each value of k the internal fermion momentum has the value $q = p - k$, which corresponds to energy–momentum conservation at the vertices. This integration over an internal momentum is typical of a closed loop. (For another example, see Problem 7.2 on the photon self-energy.)

Eqs. (7.44) display the same structure which occurred in all our other examples. The origins of the individual factors in Eq. (7.44a) should be clear to the reader. The Feynman amplitude $\mathcal{M}$ contains factors iS_F and iD_F for fermion and photon propagators, spinors u and $\bar{u}$ for the initial and final external electron lines, and a γ-factor for each vertex. The spinor quantities are in the expected order as one follows the fermion line in the direction of its arrows. The remaining factor $(-e^2) = (ie)^2$ has its origin in the form of $S^{(2)}$ for QED, Eqs. (7.1) and (7.2).

7.3   Feynman Rules for QED

The S-matrix elements $\langle f|S|i \rangle$, which we have calculated for various processes, exhibit a definite structure, which allows one to identify individual factors and features with different aspects of the corresponding Feynman graphs. The same identification between the mathematical expressions and Feynman graphs is possible for all processes. Furthermore, no new features occur for other processes. This enables one to construct a set of rules for writing $\langle f|S|i \rangle$ down directly from the Feynman graphs.

In this section we shall state these rules for QED. They represent a generalization and a tidying-up of earlier results. Their origins should be clear to the reader. Where appropriate, we shall give explanations and cross-references, but we do not repeat everything from scratch; for example, we are assuming the conventions about arrows and momentum labels for Feynman graphs.

The expression for the S-matrix element for a transition can at once be written down by generalizing our earlier results. For the transition $|i\rangle \rightarrow |f\rangle$, where the initial and final states are specified by the momenta (and spin and polarization variables) of the particles present, the S-matrix element is given by

$$\langle f|S|i \rangle = \delta_{fi} + \left[(2\pi)^4 \delta^{(4)} (P_f - P_i) \prod_{\text{ext.}} \left(\frac{m}{VE} \right)^{1/2} \prod_{\text{ext.}} \left(\frac{1}{2V\omega} \right)^{1/2} \right] \mathcal{M}. \tag{7.45}$$

Here P_i and P_f are the total four-momenta in the initial and final states, and the products extend over all external fermions (e^- and e^+) and photons, E and ω being the energies of the individual external fermions and photons respectively.

The Feynman amplitude $\mathcal{M}$ is given by

$$\mathcal{M} = \sum_{n=1}^{\infty} \mathcal{M}^{(n)} \tag{7.46}$$

where the contribution $\mathcal{M}^{(n)}$ comes from the nth order perturbation term $S^{(n)}$. The Feynman amplitude $\mathcal{M}^{(n)}$ is obtained by drawing all topologically different, connected Feynman graphs in momentum space which contain n vertices and the correct external lines. The contribution to $\mathcal{M}^{(n)}$ from each graph is obtained from the following Feynman rules.

1. For each vertex, write a factor $ie\,\gamma^\alpha$ [see Eqs. (7.1) and (7.2)].
2. For each internal photon line, labelled by the momentum k, write a factor [see Eq. (7.24b)]

$$iD_{\mathrm{F}\alpha\beta}(k) = i\frac{-g_{\alpha\beta}}{k^2 + i\varepsilon}.\qquad (\alpha)\;\overset{k}{\sim\!\sim\!\sim\!\sim\!\sim\!\sim}\;(\beta)\qquad (7.47)$$

3. For each internal fermion line, labelled by the momentum p, write a factor [see Eq. (7.24a)]

$$iS_{\mathrm{F}}(p) = i\frac{1}{\not{p} - m + i\varepsilon}.\qquad \overset{p}{\bullet\!\!-\!\!\!\longrightarrow\!\!-\!\!\bullet}\qquad (7.48)$$

4. For each external line, write one of the following factors [see Eqs. (7.25) and (7.26)]:

   (a) for each initial electron:

   $$u_r(\mathbf{p})\quad p\;\bullet\!\!-\!\!\!\longrightarrow\!\!-\!\!\bullet\qquad (7.49\text{a})$$

   (b) for each final electron:

   $$\bar{u}_r(\mathbf{p})\quad \bullet\!\!-\!\!\!\longrightarrow\!\!-\!\!\bullet\;p\qquad (7.49\text{b})$$

   (c) for each initial positron:

   $$\bar{v}_r(\mathbf{p})\quad p\;\bullet\!\!-\!\!\!\longleftarrow\!\!-\!\!\bullet\qquad (7.49\text{c})$$

   (d) for each final positron:

   $$v_r(\mathbf{p})\quad \bullet\!\!-\!\!\!\longleftarrow\!\!-\!\!\bullet\;p\qquad (7.49\text{d})$$

   (e) for each initial photon:

   $$\varepsilon_{r\alpha}(\mathbf{k})\quad k\;\overset{(\alpha)}{\sim\!\sim\!\sim\!\sim\!\sim\!\sim}\bullet\qquad (7.49\text{e})$$

   (f) for each final photon[3]:

   $$\varepsilon_{r\alpha}(\mathbf{k})\quad \bullet\overset{(\alpha)}{\sim\!\sim\!\sim\!\sim\!\sim\!\sim}\;k\qquad (7.49\text{f})$$

In Eqs. (7.49) $\mathbf{p}$ and $\mathbf{k}$ denote the three-momenta of the external particles, and $r\,(=1, 2)$ labels their spin and polarization states.

5. The spinor factors (γ-matrices, S_{F}-functions, four-spinors) for each fermion line are ordered so that, reading from right to left, they occur in the same sequence as following the fermion line in the direction of its arrows.
6. For each closed fermion loop, take the trace and multiply by a factor (-1).

---

[3] For linear polarization states, which we are using, $\varepsilon_{r\alpha}(\mathbf{k})$ is real. In general it is complex (e.g. for circular polarization), and we must then replace $\varepsilon_{r\alpha}(\mathbf{k})$ by $\varepsilon_{r\alpha}^*(\mathbf{k})$ for a final-state photon.

This rule follows directly from the corresponding result in configuration space, derived in Section 7.1 [see Eq. (7.22)].

7. The four-momenta associated with the three lines meeting at each vertex satisfy energy–momentum conservation. For each four-momentum q which is not fixed by energy–momentum conservation, carry out the integration $(2\pi)^{-4} \int d^4q$. One such integration with respect to an internal momentum variable q occurs for each closed loop.

We had an example of this rule in Eq. (7.44b) for the electron self-energy. Inclusion of the factors $(2\pi)^{-4}$ in this rule is convenient, since all numerical factors (except for the phase factor of rule 8) are accounted for in this way, as will be shown below.

8. Multiply the expression by a phase factor δ_p, which is equal to $+1$ (-1) if an even (odd) number of interchanges of neighbouring fermion operators is required to write the fermion operators in the correct normal order.

In general this phase factor is only of significance when the contributions of several Feynman graphs are added, and only the relative signs matter. The situation most frequently met is the one we discussed for (e^-e^-)- and (e^-e^+)- scattering, involving contributions from diagrams which differ only by the interchange of external fermion lines associated with identical fermion operators. This corresponds to the interchange of either (i) two initial e^- (e^+) lines, or (ii) two final e^- (e^+) lines, or (iii) an initial $e^-(e^+)$ line with a final e^+ (e^-) line.

It remains to justify our assertion, made when discussing rule 7, that the above rules allow for all numerical factors. The only factors not taken into account so far are factors $(2\pi)^4$ which occur together with δ-functions or result from propagators. The x-integration at each vertex gives a factor $(2\pi)^4$ [see Eq. (7.30)], and the Fourier transform of each propagator gives a factor $(2\pi)^{-4}$ [see Eqs. (7.23)]. For a Feynman diagram containing n vertices and $f_i(b_i)$ internal fermion (photon) lines, the Feynman amplitude contains a factor

$$\left[(2\pi)^4\right]^{n-f_i-b_i-1} \tag{7.50}$$

where the exponent -1 allows for the factor $(2\pi)^4$, which was separated out in Eq. (7.45). It is left as a problem for the reader to show that for a Feynman diagram containing l closed loops

$$n - f_i - b_i - 1 = -l. \tag{7.51}$$

Since one momentum integral $\int d^4q$ occurs for each loop, we may omit the factor (7.50) from the Feynman amplitude expression, provided we replace each loop integral $\int d^4q$ by $(2\pi)^{-4} \int d^4q$.

This completes our discussion of Feynman rules for QED. The reader is recommended to use Feynman rules to re-derive the matrix elements $\langle f|S|i\rangle$, which we obtained from first principles earlier in this chapter. After a little practice, Feynman rules provide an extraordinarily simple method for obtaining even very complicated matrix elements, and for this reason they form the basis of most practical calculations. Similar diagrammatic techniques are also of great importance in many other fields, e.g. weak interactions (to be studied later in this book) and condensed matter physics, where analogous rules can be developed. In Appendix B, at the end of this book, we give a summary of these rules for QED, as well as those for the standard electro-weak theory, which will be derived later.

7.4   Leptons

So-far we have treated QED as the interaction of electrons and positrons with the electromagnetic field. More generally, QED is usually understood to include the interactions of all charged leptons with the electromagnetic field. In addition to electrons,[4] these are the muons ($\mu^\pm$) and the tauons ($\tau^\pm$). Both muons and tauons have spin $\frac{1}{2}$ and charge $\pm e$. Furthermore, within experimental accuracy, which is very high, they exhibit all the properties of particles whose interactions are identical with those of electrons, except for their masses: $m_\mu = 105.7$ MeV, $m_\tau = 1776.84 \pm 0.17$ MeV. This is referred to as e–μ–τ *universality*. This extended QED, which we shall now study, displays a new richness: processes involving more than one kind of lepton.

Assuming universality, the extension of the theory is almost trivial. Like the electron, we describe each kind of lepton by a Dirac spinor field: $\psi_l(x)$, where l labels the kind of lepton: $l = e, \mu, \tau$. The generalization of the free-field Lagrangian density for electrons, Eq. (4.67), is

$$\mathscr{L}_0 = \sum_l \overline{\psi}_l(x)(i\gamma^\alpha \partial_\alpha - m_l)\psi_l(x), \tag{7.52a}$$

and making the minimal substitution (4.64b) (with $q = -e$) leads to interaction Hamiltonian density

$$\mathscr{H}_1(x) = -\mathscr{L}_1(x) = -e \sum_l N\left[\overline{\psi}_l(x)\rlap{/}A(x)\psi_l(x)\right]. \tag{7.52b}$$

This equation describes a *local* interaction, since all field operators are evaluated at a single space–time point. This is appropriate for the interaction of the electromagnetic field with a point particle. While, within the limits of current experiment, leptons are point-like, hadrons have finite size. For example, the experimental value of the proton radius is of the order[5] 0.8×10^{-15} m. For this reason the electromagnetic interactions of charged hadrons cannot be described by expressions like (7.52b).

The second point to note about the interaction (7.52b) is that it consists of a sum of terms each of which involves *one* kind of lepton only. Hence, the interaction is described by basic vertex parts like those of Fig. 7.1, with *both* fermion lines at a vertex referring to the *same* kind of lepton. Instead of two electrons, as in Fig. 7.1, they could both be muons or both tauons. But we could not, for example, have one electron and one muon. The vertex part in Fig. 7.16 conserves charge, but it does not occur with the interaction (7.52b), since it would require an interaction term of the form $-e\overline{\psi}_\mu \rlap{/}A \psi_e$. Consequently, for any non-vanishing matrix element $\langle j | \mathscr{H}_1 | i \rangle$ the *electron number* $N(e)$, defined by

$$N(e) = N(e^-) - N(e^+) \tag{7.53a}$$

---

[4] Just as the names muon and tauon refer to both positively and negatively charged particles, so it is convenient to have a single word for electron and positron. It is usual to use electron for this purpose as well as for the negatively charged member of this pair. We shall follow this practice, adding the appropriate qualification when it is required to avoid ambiguity.

[5] The point-like nature of leptons has been tested to much shorter distances. (See Sections 8.4 and 8.5.)

Figure 7.16 *A basic vertex part that does NOT occur with the interaction (7.52b), which conserves electron and muon numbers at each vertex*

in an obvious notation, is conserved, as are the *muon and tauon numbers*[6]

$$N(\mu) = N(\mu^-) - N(\mu^+) \tag{7.53b}$$

$$N(\tau) = N(\tau^-) - N(\tau^+). \tag{7.53c}$$

Consequently processes like

$$e^- + \mu^+ \rightarrow e^+ + \mu^-, \tag{7.54}$$

although they conserve charge, are forbidden, and indeed are not observed.

The extension of the S-matrix formalism and of the Feynman rules to the QED interaction (7.52b) is now straightforward. Each term $(\bar{\psi}A\!\!\!/\psi)$ in our original interaction (7.2), which allowed for electrons only, is replaced by a sum $\Sigma_l(\bar{\psi}_l A\!\!\!/\psi_l)$, and the S-matrix expansion (7.1) leads to

$$S = \sum_{n=0}^{\infty} \frac{(ie)^n}{n!} \int \ldots \int d^4x_1 \ldots d^4x_n \sum_{l_1} \ldots \sum_{l_n}$$

$$\times T\{N(\bar{\psi}_{l_1}A\!\!\!/\psi_{l_1})_{x_1} \ldots N(\bar{\psi}_{l_n}A\!\!\!/\psi_{l_n})_{x_n}\}. \tag{7.55}$$

This expansion first of all contains terms involving one kind of lepton only. These are the terms we considered in the last two sections, but we could now be considering muons or tauons instead of electrons. $S^{(1)}$ is of this type and so are the terms in $S^{(2)}$ with $l_1 = l_2$ ($= e, \mu, \ldots$). The more interesting terms, for which new processes occur, are those involving more than one kind of lepton.

---

[6] These definitions will be modified when we consider weak interactions.

Consider, for example, the $l_1 = \mu$, $l_2 = e$ term in $S^{(2)}$, given by

$$S^{(2)}_{\mu e} = -e^2 \int d^4x_1 d^4x_2 T\{N(\overline{\psi}_\mu A\!\!\!/ \psi_\mu)_{x_1} N(\overline{\psi}_e A\!\!\!/ \psi_e)_{x_2}\}. \tag{7.56}$$

Using Wick's theorem [Eqs. (6.35) and (6.38)] to expand the T-product in terms of normal products, we obtain

$$S^{(2)}_{\mu e} = -e^2 \int d^4x_1 d^4x_2 N[(\overline{\psi}_\mu A\!\!\!/ \psi_\mu)_{x_1}(\overline{\psi}_e A\!\!\!/ \psi_e)_{x_2}]$$
$$- e^2 \int d^4x_1 d^4x_2 N[(\overline{\psi}_\mu A\!\!\!/ \,\psi_\mu)_{x_1}(\overline{\psi}_e A\!\!\!/ \,\psi_e)_{x_2}]. \tag{7.57}$$

All other unequal-time contractions vanish, as follows from the definition (6.31) of a contraction as a vacuum expectation value.[7]

The first term in Eq. (7.57), like the term $S^{(2)}_A$ Eq. (7.5a), corresponds to two independent unphysical processes of the kind shown in Fig. 7.1, except that now one refers to a muon instead of an electron.

The second term in Eq. (7.57) gives rise to processes which involve two external muons and two external electrons, and which must conserve charge, electron number and muon number. These include electron–muon scattering and, more interestingly, the process

$$e^+ + e^- \rightarrow \mu^+ + \mu^-, \tag{7.58}$$

i.e. the annihilation of an $(e^+ e^-)$ pair leading to the creation of a $(\mu^+ \mu^-)$ pair. The term in $S^{(2)}_{\mu e}$ responsible for this process is

$$S^{(2)}(e^+ e^- \rightarrow \mu^+ \mu^-)$$
$$= -e^2 \int d^4x_1 \, d^4x_2 N[(\overline{\psi}_\mu^- \gamma^\alpha \psi_\mu^-)_{x_1}(\overline{\psi}_e^+ \gamma^\beta \psi_e^+)_{x_2}] iD_{F\alpha\beta}(x_1 - x_2), \tag{7.59}$$

and from this operator one can calculate the transition matrix elements. For the transition shown in the Feynman graph in Fig. 7.17, i.e.

$$|i\rangle = |e^- \mathbf{p}_2; e^+ \mathbf{p}_1\rangle = c_e^\dagger(\mathbf{p}_2) d_e^\dagger(\mathbf{p}_1)|0\rangle$$
$$\rightarrow |f\rangle = |\mu^- \mathbf{p}_2'; \mu^+ \mathbf{p}_1'\rangle = c_\mu^\dagger(\mathbf{p}_2') d_\mu^\dagger(\mathbf{p}_1')|0\rangle, \tag{7.60}$$

one would in this way obtain the Feynman amplitude

$$\mathcal{M}^{(2)}(e^+ e^- \rightarrow \mu^+ \mu^-) = -ie^2 \overline{u}_\mu(\mathbf{p}_2') \gamma^\alpha v_\mu(\mathbf{p}_1') D_{F\alpha\beta}(p_1 + p_2) \overline{v}_e(\mathbf{p}_1) \gamma^\beta u_e(\mathbf{p}_2). \tag{7.61}$$

Here the labels e and μ, attached to the fermion lines in Fig. 7.17, to creation and absorption operators in Eq. (7.60), and to spinors in Eq. (7.61), distinguish electrons and muons.

We do not advocate deriving the Feynman amplitude (7.61) from first principles (other than as an exercise), since it is trivial to extend the rules for calculating amplitudes, which were given in the last section, to QED involving several leptons. The S-matrix operator

---

[7] When dealing with several fermion fields, we must assume that the field operators for different fermion fields anticommute. We continue to assume that fermion and boson field operators always commute with each other. (See J. D. Bjorken and S. D. Drell, *Relativistic Quantum Fields*, McGraw-Hill, New York, 1965, p. 98.)

Figure 7.17   *The process* $e^+ + e^- \to \mu^+ + \mu^-$

(7.59) differs from the operator S_b, Eq. (7.19c), for the annihilation diagram, Fig. 7.7(b), for (e^+e^-) scattering in that in the final state the electrons are replaced by muons. Hence the amplitude (7.61) can be derived in two steps.

Firstly, obtain the Feynman amplitude $\mathcal{M}_b$ corresponding to the operator S_b, Eq. (7.19c), for the transition analogous to (7.60), but with all particles electrons. $\mathcal{M}_b$ can be written down directly, using the Feynman rules of Section 7.3.

Secondly, in the expression for $\mathcal{M}_b$, replace all quantities referring to the final state electrons by the corresponding quantities for muons.

It is left as an exercise for the reader to verify that this procedure leads to Eq. (7.61).

There is one important difference between the $e^+ + e^- \to \mu^+ + \mu^-$ process and (e^+e^-) scattering. For the latter, a second contribution stems from S_a, Eq. (7.19b), corresponding to Fig. 7.7(a). For the former, as we have seen, Eq. (7.57) gives no such contribution. It would correspond to replacing the final electron lines in Fig. 7.7(a) by muon lines, so that each vertex would involve one electron line and one muon line (e.g. the vertex at the top would look like Fig. 7.16), violating the conservation of both electron number $N(e)$ and of muon number $N(\mu)$ at each vertex.

From this example, it is easy to see how to extend the rules of Section 7.3. For any process, one must draw all relevant Feynman diagrams which conserve $N(e)$, $N(\mu)$,... at each vertex, i.e. the two lepton lines entering and leaving a vertex must be of the same kind (both e or both μ, etc.). The Feynman amplitude corresponding to each of these diagrams is then written down directly using the Feynman rules of the last section.

Problems

7.1. Derive the lowest-order non-vanishing S-matrix element (7.19) and hence the corresponding Feynman amplitude for Bhabha scattering, i.e. the process
$$e^+(\mathbf{p}_1, r_1) + e^-(\mathbf{p}_2, r_2) \to e^+(\mathbf{p}'_1, s_1) + e^-(\mathbf{p}'_2, s_2).$$

7.2. Show that the Feynman amplitude for the photon self-energy diagram in Fig. 7.9 is given by
$$\mathcal{M} = \frac{-e^2}{(2\pi)^4} \int d^4p \, \mathrm{Tr}[\not{\epsilon}_r(\mathbf{k}) S_F(p+k) \not{\epsilon}_r(\mathbf{k}) S_F(p)],$$

where $\mathbf{k}$ and $\varepsilon_r(\mathbf{k})$ are the momentum and polarization vectors of the photon.

7.3. A real scalar field $\phi(x)$, associated with a spin-zero boson B, is described by the Lagrangian density

$$\mathscr{L}(x) = \mathscr{L}_0(x) + \mathscr{L}_1(x)$$

where $\mathscr{L}_0$ is the free-field density (3.4), and

$$\mathscr{L}_1(x) = g[\phi(x)]^4/4!$$

describes an interaction of the field with itself, with g a real coupling constant. (Normal ordering of operators is assumed throughout.)

Write down the S-matrix expansion, and pick out the normal ordered term that gives rise to the BB scattering process

$$B(\mathbf{k}_1) + B(\mathbf{k}_2) \rightarrow B(\mathbf{k}_3) + B(\mathbf{k}_4)$$

in first-order perturbation theory. Draw the Feynman diagram representing this term, and show that the corresponding S-matrix element is given by

$$\langle k_3, k_4 | S^{(1)} | k_1, k_2 \rangle = (2\pi)^4 \delta^{(4)}(k_3 + k_4 - k_1 - k_2) \prod_i \left(\frac{1}{2V\omega_i} \right)^{1/2} \mathscr{M}$$

with the Feynman amplitude $\mathscr{M} = ig$. [Note that $\mathscr{M}$ is independent of the boson four-momenta $k_i^\alpha \equiv (\omega_i, \mathbf{k}_i)$.]

7.4. Pseudo-scalar meson theory is defined by the Lagrangian density

$$\mathscr{L}(x) = \mathscr{L}_0(x) + \mathscr{L}_1(x)$$

where

$$\mathscr{L}_0(x) = \frac{1}{2}[\partial_\alpha \phi(x)\partial^\alpha \phi(x) - \mu^2 \phi^2(x)] + \overline{\psi}(x)(i\gamma^\alpha \partial_\alpha - m)\psi(x)$$

represents a free real spin 0 field $\phi(x)$ and a free fermion field $\psi(x)$, and

$$\mathscr{L}_1(x) = -ig\overline{\psi}(x)\gamma_5\psi(x)\phi(x)$$

describes their interaction.

The interaction Lagrangian density $\mathscr{L}_1(x)$ is similar to that of QED, except that $e\gamma^a$ is replaced by $(-ig\gamma_5)$, and the photon field $A_\alpha(x)$ is replaced by the meson field $\phi(x)$. Exploit this similarity to write down the Feynman rules for pseudo-scalar meson theory.

7.5. A real scalar field $\phi(x)$ is described by the Lagrangian density

$$\mathscr{L}(x) = \mathscr{L}_0(x) + \mu U(\mathbf{x})\phi^2(x),$$

where $\mathscr{L}_0$ is the free-field Lagrangian density (3.4), and $U(\mathbf{x})$ *is* a static external potential.

Derive the equation of motion

$$(\Box + \mu^2)\phi(x) = 2\mu U(\mathbf{x})\phi(x).$$

Show that, in lowest order, the S-matrix element for an incoming boson, with momentum $k_i = (\omega_i, \mathbf{k}_i)$, to be scattered to a state with momentum $k_f = (\omega_f, \mathbf{k}_f)$, is given by

$$\left\langle \mathbf{k}_f \left| S^{(1)} \right| \mathbf{k}_i \right\rangle = \frac{\mathrm{i}2\pi\delta(\omega_f - \omega_i)}{(2V\omega_i)^{1/2}(2V\omega_f)^{1/2}} 2\mu \, \tilde{U}(\mathbf{k}_f - \mathbf{k}_i)$$

where

$$\tilde{U}(\mathbf{q}) = \int \mathrm{d}^3\mathbf{x}\, U(\mathbf{x}) \mathrm{e}^{-\mathrm{i}q \cdot x}.$$

This type of problem, with a static external potential, will be considered further in Chapter 8.

8

QED Processes in Lowest Order

In the last chapter we established the Feynman rules for obtaining the matrix element S_{fi} for any collision process in QED. In this chapter we shall start by deriving from S_{fi} the experimentally observable quantities, i.e. the cross-sections. This is a straightforward generalization of the corresponding kinematical and phase–space arguments of non-relativistic collision theory.

The cross-sections obtained in this way are fully polarized, i.e. the photons and leptons present initially and finally are in definite polarization states. (As is customary, we use the term 'polarization state' for both photons and fermions, meaning a spin state in the latter case.) In most practical situations, the beams of colliding particles are unpolarized, and the polarizations of the particles produced in the collision are not observed. It then becomes necessary to average and sum over polarization states of initial and final particles, respectively. The very powerful and elegant techniques for performing these spin and polarization sums are developed in Sections 8.2 and 8.3. The corresponding formalism for analysing polarization properties is more complex, and we shall consider a simple example only.

In Sections 8.4–8.6, we shall illustrate our results by deriving the cross-sections, in lowest non-vanishing order of perturbation theory, for some of the processes considered in the previous chapter. By the end of this chapter, the reader should be able to deal in a similar way with any collision problem in QED. (A reader who tires of these applications should not be tempted also to omit Sections 8.7–8.9, which introduce some fundamental new ideas.)

We shall extend the S-matrix formalism to allow for the presence of an external electromagnetic field, i.e. of a field whose quantum fluctuations are negligible, so that it can be described by an unquantized classical field. As an application of these ideas, we shall consider the scattering of electrons by the Coulomb field of a nucleus, both elastic scattering (Section 8.7) and inelastic scattering, accompanied by emission of radiation, i.e. bremsstrahlung (Section 8.8).

In studying these Coulomb scattering processes, we shall encounter a new feature. There exists the possibility of the emission by a charged particle of one or more *very soft* photons (i.e. with very little energy). Experimentally, because of finite energy resolution, the distinction between elastic and inelastic scattering becomes blurred. This unrealistic separation into elastic and inelastic scattering events leads to the infrared divergence. In the last section of this chapter we shall see how this difficulty is resolved.

8.1   The Cross-Section

We consider a scattering process in which two particles, they may be leptons or photons, with four-momenta $p_i = (E_i, \mathbf{p}_i)$, $i = 1, 2$, collide and produce N final particles with momenta $p_f' = \left(E_f', \mathbf{p}_f'\right)$, $f = 1, \ldots, N$. Initial and final particles are assumed to be in definite polarization states. As in Chapter 7, the indices labelling these states will in general be suppressed. Eq. (7.45), defining the Feynman amplitude $\mathcal{M}$ for this process, can now be written

$$
S_{fi} = \delta_{fi} + (2\pi)^4 \delta^{(4)} \left(\sum p_f' - \sum p_i \right) \prod_i \left(\frac{1}{2VE_i} \right)^{1/2}
$$
$$
\times \prod_f \left(\frac{1}{2VE_f'} \right)^{1/2} \prod_l (2m_l)^{1/2} \mathcal{M}
$$
(8.1)

where the index l runs over all external leptons in the process.

Eq. (8.1) corresponds to the limit of an infinite time interval, $T \to \infty$, and an infinite volume, $V \to \infty$. For finite T and V, we would have obtained the same expression (8.1) with

$$
(2\pi)^4 \delta^{(4)} \left(\sum p_f' - \sum p_i \right) = \lim_{\substack{T \to \infty \\ V \to \infty}} \delta_{TV} \left(\sum p_f' - \sum p_i \right)
$$
$$
\equiv \lim_{\substack{T \to \infty \\ V \to \infty}} \int_{-T/2}^{T/2} dt \int_V d^3\mathbf{x} \, \exp\left[i x \left(\sum p_f' - \sum p_i \right) \right]
$$
(8.2)

replaced by $\delta_{TV} \left(\Sigma p_f' - \Sigma p_i \right)$. In deriving the cross-section, it will help to take T and V finite, to begin with. In this case the transition probability per unit time

$$
w = |S_{fi}|^2 / T
$$
(8.3)

involves the factor $\left[\delta_{TV} \left(\Sigma p_f' - \Sigma p_i \right) \right]^2$. For large values of T and V, we can then take

$$
\delta_{TV} \left(\sum p_f' - \sum p_i \right) = (2\pi)^4 \delta^{(4)} \left(\sum p_f' - \sum p_i \right)
$$
(8.4)

and

$$\left[\delta_{TV}\left(\sum p_f' - \sum p_i\right)\right]^2 = TV(2\pi)^4\delta^{(4)}\left(\sum p_f' - \sum p_i\right) \tag{8.5}$$

with errors which tend to zero as $T \to \infty$ and $V \to \infty$. Hence Eq. (8.3) becomes

$$w = V(2\pi)^4\delta^{(4)}\left(\sum p_f' - \sum p_i\right)\left(\prod_i \frac{1}{2VE_i}\right)\left(\prod_f \frac{1}{2VE_f'}\right)\left(\prod_l (2m_l)\right)|\mathcal{M}|^2. \tag{8.6}$$

Eq. (8.6) is the transition rate to one definite final state. To obtain the transition rate to a group of final states with momenta in the intervals $\left(\mathbf{p}_f', \mathbf{p}_f' + d\mathbf{p}_f'\right), f = 1, \ldots, N$, we must multiply w by the number of these states which is

$$\prod_f \frac{V d^3\mathbf{p}_f'}{(2\pi)^3}. \tag{8.7}$$

The differential cross-section is the transition rate into this group of final states for one scattering centre and unit incident flux. With our choice of normalization for the states, the volume V which we are considering contains one scattering centre, and the incident flux is v_{rel}/V, where v_{rel} is the relative velocity of the colliding particles.

Combining these results with Eq. (8.6), we obtain the required expression for the differential cross-section

$$d\sigma = w\frac{V}{v_{rel}}\prod_f \frac{V d^3\mathbf{p}_f'}{(2\pi)^3}$$

$$= (2\pi)^4\delta^{(4)}\left(\sum p_f' - \sum p_i\right)\frac{1}{4E_1E_2 v_{rel}}\left(\prod_l (2m_l)\right)\left(\prod_f \frac{d^3\mathbf{p}_f'}{(2\pi)^3 2E_f'}\right)|\mathcal{M}|^2. \tag{8.8}$$

Eq. (8.8) holds in any Lorentz frame in which the colliding particles move collinearly. In such a frame the relative velocity v_{rel} is given by the expression

$$E_1 E_2 v_{rel} = [(p_1 p_2)^2 - m_1^2 m_2^2]^{1/2}, \tag{8.9}$$

where m_1 and m_2 are the rest masses of the colliding particles. Two important examples of such frames are the centre-of-mass (CoM) system, and the laboratory (Lab) system. In the CoM system we have $\mathbf{p}_1 = -\mathbf{p}_2$, and hence

$$v_{rel} = \frac{|\mathbf{p}_1|}{E_1} + \frac{|\mathbf{p}_2|}{E_2} = |\mathbf{p}_1|\frac{E_1 + E_2}{E_1 E_2} \qquad \text{(CoM)}. \tag{8.10a}$$

In the laboratory system, the target particle (particle 2, say) is at rest, $\mathbf{p}_2 = 0$, and

$$v_{rel} = \frac{|\mathbf{p}_1|}{E_1} \qquad \text{(Lab)}. \tag{8.10b}$$

Eqs. (8.10) of course also follow from the general result (8.9).

The relativistic invariance of the cross-section formula (8.8) follows from Eq. (8.9) and from the Lorentz invariance of $d^3\mathbf{p}/2E$ for any four-vector $p = (E, \mathbf{p})$.[1]

Because of conservation of energy and momentum, the final-state momenta $\mathbf{p}'_1, \ldots, \mathbf{p}'_N$ are not all independent variables. In order to obtain a differential cross-section in the independent variables appropriate to a given situation, we integrate Eq. (8.8) with respect to the remaining variables. We illustrate this for the frequently occurring case of a process leading to a two-body final state. Eq. (8.8) now becomes

$$d\sigma = f\left(p'_1, p'_2\right) \delta^{(4)}\left(p'_1 + p'_2 - p_1 - p_2\right) d^3\mathbf{p}'_1 \, d^3\mathbf{p}'_2, \tag{8.12a}$$

where

$$f\left(p'_1, p'_2\right) \equiv \frac{1}{64\pi^2 v_{\text{rel}} E_1 E_2 E'_1 E'_2} \left(\prod_l (2m_l)\right) |\mathcal{M}|^2. \tag{8.12b}$$

Integration of Eq. (8.12a) with respect to $\mathbf{p}'_2$ gives

$$d\sigma = f\left(p'_1, p'_2\right) \delta\left(E'_1 + E'_2 - E_1 - E_2\right) |\mathbf{p}'_1|^2 d|\mathbf{p}'_1| d\Omega'_1, \tag{8.13}$$

where $\mathbf{p}'_2 = \mathbf{p}_1 + \mathbf{p}_2 - \mathbf{p}'_1$, and integrating Eq. (8.13) over $|\mathbf{p}'_1|$ we obtain[2]

$$d\sigma = f\left(p'_1, p'_2\right) |\mathbf{p}'_1|^2 d\Omega'_1 \left[\frac{\partial\left(E'_1 + E'_2\right)}{\partial|\mathbf{p}'_1|}\right]^{-1}, \tag{8.15}$$

where $\mathbf{p}'_2 = \mathbf{p}_1 + \mathbf{p}_2 - \mathbf{p}'_1$, and the partial derivative is evaluated with the polar angles θ'_1, ϕ'_1 of the vector $\mathbf{p}'_1$ constant.

To obtain the differential cross-section in the CoM system, we note that in the CoM system, $\mathbf{p}'_1 = -\mathbf{p}'_2$. From

$$\left(E'_f\right)^2 = \left(m'_f\right)^2 + |\mathbf{p}'_f|^2, \qquad f = 1, 2, \tag{8.16}$$

we find

$$\frac{\partial\left(E'_1 + E'_2\right)}{\partial|\mathbf{p}'_1|} = |\mathbf{p}'_1| \frac{E_1 + E_2}{E'_1 E'_2}, \tag{8.17}$$

---

[1] $d^3\mathbf{p}/2E$ can be written in the explicitly invariant form

$$\frac{d^3\mathbf{p}}{2E} = \int d^4p \, \delta(p^2 - m^2)\theta(p^0) \tag{8.11}$$

where $m^2 = E^2 - \mathbf{p}^2$, $\theta(p^0)$ is the step function (3.53), and the integration is with respect to p^0 over the range $-\infty < p^0 < \infty$.

[2] We are here using the general relation

$$\int f(x, y)\delta[g(x, y)] \, dx = \int f(x, y)\delta[g(x, y)] \left(\frac{\partial x}{\partial g}\right)_y dg$$

$$= \left[\frac{f(x, y)}{(\partial g/\partial x)_y}\right]_{g = 0} \tag{8.14}$$

and, combining Eqs. (8.15), (8.12b), (8.10a) and (8.17), we obtain the CoM differential cross-section

$$\left(\frac{d\sigma}{d\Omega_1'}\right)_{\text{CoM}} = \frac{1}{64\pi^2(E_1 + E_2)^2}\frac{|\mathbf{p}_1'|}{|\mathbf{p}_1|}\left(\prod_l 2m_l\right)|\mathcal{M}|^2. \tag{8.18}$$

Finally, we note that all the cross-section formulae which we have derived apply irrespective of whether identical particles are present or not. However, on calculating total cross-sections in cases where two or more final-state particles are identical, one must integrate only over those ranges of angles which correspond to physically distinguishable events. For example, if the CoM cross-section (8.18) refers to a process with two identical particles in the final state, then the scattering angles $(\theta_1', \phi_1') = (\alpha, \beta)$ and $(\theta_1', \phi_1') = (\pi - \alpha, \pi + \beta)$ describe the same process. Hence the total CoM cross-section is obtained by integrating Eq. (8.18) only over the forward hemisphere $0 \leq \theta_1' \leq \frac{1}{2}\pi$, i.e.

$$\sigma_{\text{CoM}}^{\text{tot}} = \int_0^1 d(\cos\theta_1') \int_0^{2\pi} d\phi_1'\left(\frac{d\sigma}{d\Omega_1'}\right)_{\text{CoM}} = \frac{1}{2}\int_{4\pi} d\Omega_1'\left(\frac{d\sigma}{d\Omega_1'}\right)_{\text{CoM}}, \tag{8.19}$$

where the last integral is over the complete solid angle 4π, as indicated.

8.2 Spin Sums

In the last section we considered a reaction in which the initial and final states are completely specified, including the polarization states of the leptons and photons present initially and finally. In many experiments, the colliding particles are unpolarized and the polarizations of the final-state particles are not detected. To obtain the corresponding unpolarized cross-section from Eq. (8.8), we must *average* $|\mathcal{M}|^2$ over all initial polarization states, and we must *sum* it over all final polarization states. In this section we shall show how to obtain these averages and sums over initial and final lepton spins. We shall find that the unpolarized cross-section can always be expressed in terms of traces of products of γ-matrices.

Consider a Feynman amplitude of the form

$$\mathcal{M} = \bar{u}_s(\mathbf{p}')\Gamma u_r(\mathbf{p}). \tag{8.20}$$

This occurs, for example, for Compton scattering [see Fig. 7.12 and Eqs. (7.38)]. Here the spinors $u_r(\mathbf{p})$ and $\bar{u}_s(\mathbf{p}')$ completely specify the momenta and spins of the electron in the initial and final states, and the operator Γ is a 4×4 matrix built up out of γ-matrices. Eq. (8.20) gives rise to an unpolarized cross-section proportional to

$$X \equiv \frac{1}{2}\sum_{r=1}^{2}\sum_{s=1}^{2}|\mathcal{M}|^2 \tag{8.21}$$

where we have averaged over initial spins $\left(\frac{1}{2}\sum_r\right)$ and summed over final spins $(\sum_s)$. Defining

$$\tilde{\Gamma} \equiv \gamma^0\Gamma^\dagger\gamma^0, \tag{8.22}$$

we can write Eq. (8.21) as

$$X = \frac{1}{2} \sum_r \sum_s (\bar{u}_s(\mathbf{p}')\Gamma u_r(\mathbf{p}))(\bar{u}_r(\mathbf{p})\tilde{\Gamma} u_s(\mathbf{p}')). \tag{8.23}$$

Writing out the spinor indices explicitly, this can be written

$$X = \frac{1}{2} \left(\sum_s u_{s\delta}(\mathbf{p}')\bar{u}_{s\alpha}(\mathbf{p}') \right) \Gamma_{\alpha\beta} \left(\sum_r u_{r\beta}(\mathbf{p})\bar{u}_{r\gamma}(\mathbf{p}) \right) \tilde{\Gamma}_{\gamma\delta}.$$

We introduce the positive energy projection operator [Eqs. (A.31) and (A.35)],

$$\Lambda^+_{\alpha\beta}(\mathbf{p}) = \left(\frac{\not{p}+m}{2m} \right)_{\alpha\beta} = \sum_{r=1}^{2} u_{r\alpha}(\mathbf{p})\bar{u}_{r\beta}(\mathbf{p}) \tag{8.24a}$$

in order to eliminate the sums over positive energy states.[3] This leads to our final result

$$\begin{aligned} X &= \frac{1}{2} \Lambda^+_{\delta\alpha}(\mathbf{p}')\Gamma_{\alpha\beta}\Lambda^+_{\beta\gamma}(\mathbf{p})\tilde{\Gamma}_{\gamma\delta} \\ &= \frac{1}{2}\text{Tr}\left[\Lambda^+(\mathbf{p}')\Gamma\Lambda^+(\mathbf{p})\tilde{\Gamma} \right] \\ &= \frac{1}{2}\text{Tr}\left[\frac{\not{p}'+m}{2m}\Gamma\frac{\not{p}+m}{2m}\tilde{\Gamma} \right]. \end{aligned} \tag{8.25}$$

In addition to the amplitude (8.20), involving the absorption and emission of an external negative lepton, there are also Feynman amplitudes of the form

$$\mathcal{M} = \bar{v}_s(\mathbf{p}')\Gamma v_r(\mathbf{p}) \tag{8.26a}$$

$$\mathcal{M} = \bar{u}_s(\mathbf{p}')\Gamma v_r(\mathbf{p}) \tag{8.26b}$$

$$\mathcal{M} = \bar{v}_s(\mathbf{p}')\Gamma u_r(\mathbf{p}). \tag{8.26c}$$

These represent: (a) absorption and emission of a positive lepton, as in Compton scattering by positrons [Eq. (7.39) and Fig. 7.13]; (b) creation of a lepton pair, as in $2\gamma \rightarrow e^+e^-$ (Fig. 7.5); and (c) annihilation of a lepton pair, as in $e^+e^- \rightarrow 2\gamma$ (Fig. 7.4).

The spin sums for these cases are performed as for the case which we considered in detail, but using the negative energy projection operator [Eqs. (A.31) and (A.35)]

$$\Lambda^-_{\alpha\beta}(\mathbf{p}) = -\left(\frac{\not{p}-m}{2m} \right)_{\alpha\beta} = -\sum_{r=1}^{2} v_{r\alpha}(\mathbf{p})\bar{v}_{r\beta}(\mathbf{p}) \tag{8.24b}$$

to eliminate sums over negative energy states. For example, Eq. (8.26b) leads to

$$\begin{aligned} \frac{1}{2}\sum_r \sum_s |\mathcal{M}|^2 &= -\frac{1}{2}\text{Tr}\left[\Lambda^+(\mathbf{p}')\Gamma\Lambda^-(\mathbf{p})\tilde{\Gamma} \right] \\ &= \frac{1}{2}\text{Tr}\left[\frac{\not{p}'+m}{2m}\Gamma\frac{\not{p}-m}{2m}\tilde{\Gamma} \right]. \end{aligned} \tag{8.27}$$

---

[3] Equation numbers (A. X) refer to equations in Appendix A at the end of the book. This appendix gives a self-contained account of the properties of Dirac spinors, etc. which we here require, and a reader not familiar with these is advised to study the appendix.

Spin sums, and consequently traces like Eqs. (8.25) and (8.27), frequently occur in practice. There exist simple techniques for calculating such traces. These use algebraic identities for γ-matrices (see Appendix A, Section A.2) and some general rules for calculating the traces of products of γ-matrices (see Section A.3). Later in this chapter we shall repeatedly apply these results and methods in calculating the unpolarized cross-sections for various processes in QED.

To conclude this section, we briefly discuss how to calculate the spin polarization properties of a process. This involves evaluating $|\mathcal{M}|^2$ for specific initial and final spin states. This can be done either by using a specific matrix representation for the spinors or by employing helicity or spin-projection operators to select the appropriate spin states. The latter technique again leads to traces and is usually the more convenient one.

We shall illustrate this method for the particular process resulting from the Feynman amplitude (8.20), in which the incident electron has positive helicity and the outgoing electron has negative helicity. The cross-section for this helicity flip process is proportional to

$$X = |\bar{u}_2(\mathbf{p'})\Gamma u_1(\mathbf{p})|^2$$
$$= (\bar{u}_2(\mathbf{p'})\Gamma u_1(\mathbf{p}))(\bar{u}_1(\mathbf{p})\tilde{\Gamma}u_2(\mathbf{p'})). \tag{8.28}$$

We introduce the helicity projection operators

$$\Pi^{\pm}(\mathbf{p}) = \frac{1}{2}(1 \pm \sigma_{\mathbf{p}}), \tag{A.37}$$

which have the properties

$$\Pi^+(\mathbf{p})u_r(\mathbf{p}) = \delta_{1r}u_r(\mathbf{p}), \qquad \Pi^-(\mathbf{p})u_r(\mathbf{p}) = \delta_{2r}u_r(\mathbf{p}). \tag{A.40}$$

Eq. (8.28) then becomes

$$X = (\bar{u}_2(\mathbf{p'})\Gamma\Pi^+(\mathbf{p})u_1(\mathbf{p}))(\bar{u}_1(\mathbf{p})\tilde{\Gamma}\Pi^-(\mathbf{p'})u_2(\mathbf{p'}))$$
$$= \sum_r \sum_s (\bar{u}_s(\mathbf{p'})\Gamma\Pi^+(\mathbf{p})u_r(\mathbf{p}))(\bar{u}_r(\mathbf{p})\tilde{\Gamma}\Pi^-(\mathbf{p'})u_s(\mathbf{p'}))$$
$$= \text{Tr}[\Lambda^+(\mathbf{p'})\Gamma\Pi^+(\mathbf{p})\Lambda^+(\mathbf{p})\tilde{\Gamma}\Pi^-(\mathbf{p'})], \tag{8.29}$$

where the last line follows from Eqs. (8.23) and (8.25) with Γ and $\tilde{\Gamma}$ replaced by $\Gamma\Pi^+(\mathbf{p})$ and $\tilde{\Gamma}\Pi^-(\mathbf{p'})$.

In the relativistic limit $E \gg m$, the helicity projection operators (A.40) simplify to

$$\Pi^{\pm}(\mathbf{p}) = \frac{1}{2}(1 \pm \gamma^5) \quad (E \gg m), \tag{A.43}$$

which also leads to a considerable simplification of Eq. (8.29) in the relativistic limit $E \gg m$, $E' \gg m$.

8.3 Photon Polarization Sums

In the last section, we showed how to perform spin sums in order to obtain unpolarized cross-sections. We now consider the corresponding photon polarization sums. We met an example of this for Thomson scattering in Section 1.4.4, where we first obtained the fully polarized cross-section, Eq. (1.69), and then explicitly performed the summing and averaging over final and initial polarizations by means of Eq. (1.71). An alternative covariant formalism exists for obtaining the unpolarized cross-section directly. This formalism depends on the gauge invariance of the theory, the consequences of which we shall now consider in more detail.

Gauge invariance of the theory implies the gauge invariance of the matrix elements, i.e. of the Feynman amplitudes. It is, of course, only the matrix element itself, corresponding to the sum of all possible Feynman graphs in a given order of perturbation theory, which must be gauge invariant. The contributions to the amplitude from individual Feynman graphs are, in general, not gauge invariant. For example, for Compton scattering, the individual amplitudes $\mathcal{M}_a$ and $\mathcal{M}_b$, Eqs. (7.38a) and (7.38b), are not gauge invariant, but their sum, $\lambda(\mathcal{M}_a + \mathcal{M}_b)$, is. (The verification of this statement, using the method to be developed in this section, is left as a problem for the reader (see Problem 8.7).)

For any process involving external photons, the Feynman amplitude $\mathcal{M}$ is of the form

$$\mathcal{M} = \varepsilon_{r_1}^{\alpha}(\mathbf{k}_1)\varepsilon_{r_2}^{\beta}(\mathbf{k}_2)\ldots \mathcal{M}_{\alpha\beta\ldots}(\mathbf{k}_1, \mathbf{k}_2, \ldots), \qquad (8.30)$$

with one polarization vector $\varepsilon(\mathbf{k})$ for each external photon, and the tensor amplitude $\mathcal{M}_{\alpha\beta\ldots}(\mathbf{k}_1, \mathbf{k}_2, \ldots)$ independent of these polarization vectors. [This follows from our fourth Feynman rule, Eqs. (7.49e) and (7.49f); we are again using real polarization vectors.]

The polarization vectors are of course gauge dependent. For example, for a free photon, described in a Lorentz gauge by the plane wave

$$A^{\mu}(x) = \text{const. } \varepsilon_r^{\mu}(\mathbf{k}) \, e^{\pm ikx},$$

the gauge transformation

$$A^{\mu}(x) \rightarrow A^{\mu}(x) + \partial^{\mu}f(x), \qquad \text{with } f(x) = \tilde{f}(k)e^{\pm ikx},$$

implies

$$\varepsilon_r^{\mu}(\mathbf{k})e^{\pm ikx} \rightarrow [\varepsilon_r^{\mu}(\mathbf{k}) \pm ik^{\mu} \tilde{f}(k)] \, e^{\pm ikx}. \qquad (8.31)$$

Invariance of the amplitude (8.30) under this transformation requires

$$k_1^{\alpha}\mathcal{M}_{\alpha\beta\ldots}(\mathbf{k}_1, \mathbf{k}_2, \ldots) = k_2^{\beta}\mathcal{M}_{\alpha\beta\ldots}(\mathbf{k}_1, \mathbf{k}_2, \ldots) = \cdots = 0, \qquad (8.32)$$

i.e. when any external photon polarization vector is replaced by the corresponding four-momentum, the amplitude must vanish.

To illustrate how Eq. (8.32) is used to calculate photon polarization sums, we consider, as a simple example, the matrix element

$$\mathcal{M}_r(\mathbf{k}) = \varepsilon_r^{\alpha}(\mathbf{k})\mathcal{M}_{\alpha}(\mathbf{k}),$$

corresponding to a process involving one external photon. The gauge invariance now implies

$$k^{\alpha}\mathcal{M}_{\alpha}(\mathbf{k}) = 0. \qquad (8.33)$$

The unpolarized cross-section for the process is proportional to

$$X = \sum_{r=1}^{2} |\mathcal{M}_r(\mathbf{k})|^2 = \mathcal{M}_\alpha(\mathbf{k})\mathcal{M}_\beta^*(\mathbf{k})\sum_{r=1}^{2} \varepsilon_r^\alpha(\mathbf{k})\varepsilon_r^\beta(\mathbf{k}). \qquad (8.34)$$

Using the relation

$$\sum_{r=1}^{2} \varepsilon_r^\alpha(\mathbf{k})\varepsilon_r^\beta(\mathbf{k}) = -g^{\alpha\beta} - \frac{1}{(kn)^2} [k^\alpha k^\beta - (kn)(k^\alpha n^\beta + k^\beta n^\alpha)], \qquad (8.35)$$

which follows from Eqs. (5.39) and (5.40) for a real photon $(k^2 = 0)$, and the gauge condition (8.33), we at once obtain from Eq. (8.34)

$$\sum_{r=1}^{2} |\mathcal{M}_r(\mathbf{k})|^2 = -\mathcal{M}^\alpha(\mathbf{k})\mathcal{M}_\alpha^*(\mathbf{k}). \qquad (8.36)$$

Eq. (8.36) is our desired result, and it is easily extended to transitions involving several external photons. This formalism necessitates working in a general Lorentz gauge, as the explicit gauge invariance of the matrix element may be lost in a particular gauge. (We shall meet an example of this when discussing Compton scattering in Section 8.6.) However, in practice, it may be advantageous to choose a particular gauge, which simplifies the algebra of the trace sums, and to carry out the photon polarization sums explicitly, as was done for Thomson scattering in Section 1.4.4. The use of both techniques will be illustrated later in this chapter (see Sections 8.6 and 8.8 on Compton scattering and on bremsstrahlung).

8.4   Lepton Pair Production in (e^+e^-) Collisions

As a first illustration of the use of the above methods in calculating processes to lowest non-vanishing order of perturbation theory, we shall consider the processes in which an electron–positron pair annihilates in collision, producing a charged lepton pair $(l^+ l^-)$. These processes are of considerable interest and have been studied experimentally over a wide range of energies. In this section, we shall take the final lepton pair to be muons or tauons, but not electrons. The case of Bhabha scattering (i.e. $e^+e^- \rightarrow e^+e^-$) will be considered in the next section.

We already considered the process

$$e^+(\mathbf{p}_1, r_1) + e^-(\mathbf{p}_2, r_2) \rightarrow l^+(\mathbf{p}_1', s_1) + l^-(\mathbf{p}_2', s_2) \qquad (8.37)$$

(where $l = \mu, \tau, \dots$) in Section 7.4. Its Feynman amplitude, corresponding to the Feynman graph of Fig. 7.17, is given by Eq. (7.61), which we now write in slightly modified notation as

$$\mathcal{M}(r_1, r_2, s_1, s_2) = ie^2 [\bar{u}_{s_2}(\mathbf{p}_2')\gamma_\alpha v_{s_1}(\mathbf{p}_1')]_{(l)} \frac{1}{(p_1 + p_2)^2} [\bar{v}_{r_1}(\mathbf{p}_1)\gamma^\alpha u_{r_2}(\mathbf{p}_2)]_{(e)}. \qquad (8.38)$$

The labels (l) and (e) distinguish quantities referring to leptons and to electrons. In Eq. (8.38) we have dropped the term $(+i\varepsilon)$ in the photon propagator. This term is only of significance at the pole of the propagator, and in the present case $(p_1 + p_2)^2 \geq 4m_e^2$ cannot vanish.

For the unpolarized cross-section we require

$$X = \frac{1}{4} \sum_{r_1} \sum_{r_2} \sum_{s_1} \sum_{s_2} |\mathcal{M}(r_1, r_2, s_1, s_2)|^2. \qquad (8.39)$$

Using the hermiticity condition $\gamma^{\alpha\dagger} = \gamma^0 \gamma^\alpha \gamma^0$ [Eq. (A.6)], Eq. (8.38) gives

$$\mathcal{M}^*(r_1, r_2, s_1, s_2) = -ie^2 \left[\bar{v}_{s_1}(\mathbf{p}_1') \gamma_\beta u_{s_2}(\mathbf{p}_2') \right]_{(l)} \frac{1}{(p_1 + p_2)^2} \left[\bar{u}_{r_2}(\mathbf{p}_2) \gamma^\beta v_{r_1}(\mathbf{p}_1) \right]_{(e)} \qquad (8.40)$$

and Eq. (8.39) becomes

$$X = \frac{e^4}{4 \left[(p_1 + p_2)^2 \right]^2} A_{(l)\alpha\beta} B_{(e)}^{\alpha\beta}. \qquad (8.41)$$

Here $A_{(l)\alpha\beta}$ is given by

$$
\begin{aligned}
A_{(l)\alpha\beta} &= \sum_{s_1} \sum_{s_2} \left[(\bar{u}_{s_2}(\mathbf{p}_2') \gamma_\alpha v_{s_1}(\mathbf{p}_1')) (\bar{v}_{s_1}(\mathbf{p}_1') \gamma_\beta u_{s_2}(\mathbf{p}_2')) \right]_{(l)} \\
&= \mathrm{Tr} \left[\frac{\not{p}_2' + m_l}{2m_l} \gamma_\alpha \frac{\not{p}_1' - m_l}{2m_l} \gamma_\beta \right],
\end{aligned}
\qquad (8.41a)
$$

where we used the energy projection operators (8.24a) and (8.24b). Similarly one obtains

$$B_{(e)}^{\alpha\beta} = \mathrm{Tr} \left[\frac{\not{p}_1 - m_e}{2m_e} \gamma^\alpha \frac{\not{p}_2 + m_e}{2m_e} \gamma^\beta \right]. \qquad (8.41b)$$

The traces (8.41a) and (8.41b) are easily evaluated using the results of Appendix A, Sections A.2 and A.3. Since the trace of a product of an odd number of γ-matrices vanishes [Eq. (A. 16)], Eq. (8.41a) becomes

$$A_{(l)\alpha\beta} = \frac{1}{4m_l^2} \left[\mathrm{Tr}(\not{p}_2' \gamma_\alpha \not{p}_1' \gamma_\beta) - m_l^2 \, \mathrm{Tr} \, (\gamma_\alpha \gamma_\beta) \right],$$

and from Eqs. (A.17) this gives

$$A_{(l)\alpha\beta} = \frac{1}{m_l^2} \left[p_{1\alpha}' p_{2\beta}' + p_{2\alpha}' p_{1\beta}' - (m_l^2 + p_1' p_2') g_{\alpha\beta} \right]. \qquad (8.42a)$$

Similarly one finds

$$B_{(e)}^{\alpha\beta} = \frac{1}{m_e^2} \left[p_1^\alpha p_2^\beta + p_2^\alpha p_1^\beta - (m_e^2 + p_1 p_2) g^{\alpha\beta} \right]. \qquad (8.42b)$$

Substituting Eqs. (8.42) into Eq. (8.41), one obtains

$$
\begin{aligned}
X = \frac{e^4}{2m_e^2 m_l^2 \left[(p_1 + p_2)^2 \right]^2} & \{ (p_1 p_1')(p_2 p_2') + (p_1 p_2')(p_2 p_1') \\
& + m_e^2 (p_1' p_2') + m_l^2 (p_1 p_2) + 2m_e^2 m_l^2 \}. \qquad (8.43)
\end{aligned}
$$

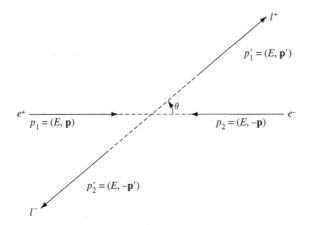

Figure 8.1 Kinematics for the process $e^+ e^- \rightarrow l^+ l^-$ in the CoM system

So far we have worked in an arbitrary reference frame. We now specialize to the CoM frame, as specified in Fig. 8.1. The kinematic factors occurring in Eq. (8.43) now take the form

$$
\left.
\begin{aligned}
p_1 p'_1 &= p_2 p'_2 = E^2 - pp' \cos \theta, & p_1 p'_2 &= p_2 p'_1 = E^2 + pp' \cos\theta \\
p_1 p_2 &= E^2 + p^2, & p'_1 p'_2 &= E^2 + p'^2 \\
& & (p_1 + p_2)^2 &= 4E^2
\end{aligned}
\right\}
\tag{8.44a}
$$

where

$$
p \equiv |\mathbf{p}|, \quad p' \equiv |\mathbf{p}'|.
\tag{8.44b}
$$

Furthermore, since $E \geq m_\mu \approx 207 m_e$, it is a very good approximation to take $p \equiv |\mathbf{p}| = E$, and to neglect terms proportional to m_e^2 inside the curly brackets in Eq. (8.43). On making these approximations and substituting Eqs. (8.43) and (8.44) in the CoM cross-section formula (8.18), we finally obtain

$$
\left(\frac{d\sigma}{d\Omega} \right)_{\text{CoM}} = \frac{\alpha^2}{16E^4} \left(\frac{p'}{E} \right) \left(E^2 + m_l^2 + p'^2 \cos^2 \theta \right)
\tag{8.45a}
$$

for the differential cross-section, and

$$
\sigma_{\text{tot}} = \frac{\pi \alpha^2}{4E^4} \left(\frac{p'}{E} \right) \left[E^2 + m_l^2 + \frac{1}{3} p'^2 \right]
\tag{8.45b}
$$

for the total cross-section. In the extreme relativistic limit, $E \gg m_l$, these formulae reduce to the much quoted

$$
\left.
\begin{aligned}
\left(\frac{d\sigma}{d\Omega} \right)_{\text{CoM}} &= \frac{\alpha^2}{16E^2} \left(1 + \cos^2 \theta \right) \\
\sigma_{\text{tot}} &= \frac{\pi \alpha^2}{3E^2}
\end{aligned}
\right\}
\quad (E \gg m_l).
\tag{8.46}
$$

Both the processes $e^+ e^- \rightarrow \mu^+ \mu^-$ and $e^+ e^- \rightarrow \tau^+ \tau^-$ have been extensively studied over a wide range of energies.

Figure 8.2 $E^2 \times \sigma_{tot}$ *(in arbitrary units) for the process* $e^+e^- \rightarrow \tau^+\tau^-$ *near threshold* $2E = 2m_\tau$. *[After W. Bacino et al., Phys. Rev. Lett. **41** (1978), 13.].* ⧍: *experimental data; curve: theoretical prediction (8.45b)*

Typical results are shown for $\tau^+\tau^-$ production near threshold and for both reactions at higher energies in Figs. 8.2 and 8.3 respectively. The latter results are of interest since they probe the interaction down to very small distances and so represent a severe test of QED. In the CoM system, the energy of the virtual photon in the process is $2E$, implying a timescale of order $\hbar/2E$ and a corresponding distance scale of order $\hbar c/2E$. For $E \approx 15$ GeV, this corresponds to a distance of the order 7×10^{-3} f. The agreement between theory and experiment implies that even at these very small distances, the electron, muon and tauon are adequately described as point charges. This small value should be compared with the experimental r.m.s. charge radius of the proton, which is of order 0.8 f.

Figure 8.3   *The total cross-sections (in nb) for the processes* $e^+e^- \rightarrow \mu^+\mu^-$ *and* $e^+e^- \rightarrow \tau^+\tau^-$ *at relativistic energies. [After D. P. Barber et al., Phys. Rev. Lett. **43**, (1979) 1915.]* ⧍: *experimental data; curves: theretical cross-section formula (8.46)*

Finally we note that at still higher energies, the contribution of the weak interaction to the cross sections must be taken into account. This will be discussed in the context of the unified theory of electromagnetic and weak interactions in Section 19.4.

8.5 Bhabha Scattering

We now consider elastic e^+e^- scattering. This process is a little more complicated than those considered in the last section since, in addition to the annihilation diagram [Figs. 7.17 or 7.7(b)], the scattering diagram [Fig. 7.7(a)] also contributes. The Feynman amplitude for the process

$$e^+(\mathbf{p}_1, r_1) + e^-(\mathbf{p}_2, r_2) \rightarrow e^+(\mathbf{p}'_1, s_1) + e^-(\mathbf{p}'_2, s_2) \qquad (8.47)$$

is

$$\mathcal{M} = \mathcal{M}_a + \mathcal{M}_b$$

where $\mathcal{M}_a$ and $\mathcal{M}_b$ correspond to the scattering and annihilation diagrams and are given by

$$\mathcal{M}_a = -ie^2 \left[\bar{u}(\mathbf{p}'_2)\gamma_\alpha u(\mathbf{p}_2)\right] \frac{1}{(p_1 - p'_1)^2} \left[\bar{v}(\mathbf{p}_1)\gamma^\alpha v(\mathbf{p}'_1)\right] \qquad (8.48a)$$

$$\mathcal{M}_b = ie^2 \left[\bar{u}(\mathbf{p}'_2)\gamma_\alpha v(\mathbf{p}'_1)\right] \frac{1}{(p_1 + p_2)^2} \left[\bar{v}(\mathbf{p}_1)\gamma^\alpha u(\mathbf{p}_2)\right]. \qquad (8.48b)$$

We have here suppressed the spin indices again. $\mathcal{M}_b$ is of course just the amplitude (8.38) with e^+e^- for the final state lepton pair. We again note the relative minus sign between these two terms corresponding to Feynman rule 8 (Section 7.3).

We shall evaluate the cross-section for this process in the CoM system, restricting ourselves for simplicity to the important case of the relativistic high-energy limit. The kinematics are now defined by Eqs. (8.44) with $p' = p$ and $E \gg m (\equiv m_e)$. The cross-section formula (8.18) now leads to

$$\left(\frac{d\sigma}{d\Omega}\right)_{\text{CoM}} = \frac{m^4}{16\pi^2 E^2} \left(X_{aa} + X_{bb} + X_{ab} + X^*_{ab}\right), \qquad (8.49)$$

where

$$X_{aa} = \frac{1}{4} \sum_{\text{spins}} |\mathcal{M}_a|^2 \qquad (8.50a)$$

$$X_{bb} = \frac{1}{4} \sum_{\text{spins}} |\mathcal{M}_b|^2 \qquad (8.50b)$$

$$X_{ab} = \frac{1}{4} \sum_{\text{spins}} \mathcal{M}_a \mathcal{M}^*_b, \qquad (8.50c)$$

the summations being over the spins of all four fermions.

The term X_{bb} follows at once as a special case of X, Eq. (8.43), with $m_l = m$ and $E \gg m$:

$$X_{bb} = \frac{e^4}{16m^4} \left[1 + \cos^2\theta + O\left(\frac{m^2}{E^2}\right)\right]. \qquad (8.51)$$

The evaluation of X_{aa} is essentially similar to that of X_{bb} and is left as an exercise for the reader. The result is

$$
\begin{aligned}
X_{aa} &= \frac{e^4}{2m^4 \left[(p_1 - p_1')^2 \right]^2} \left\{ (p_1 p_2)(p_1' p_2') + (p_1 p_2')(p_2 p_1') + O(E^2 m^2) \right\} \\
&= \frac{e^4}{8m^4 \sin^4(\theta/2)} \left[1 + \cos^4 \frac{\theta}{2} + O\left(\frac{m^2}{E^2} \right) \right].
\end{aligned}
\tag{8.52}
$$

The interference term X_{ab}, Eq. (8.50c), is more complicated, and we shall give its evaluation in some detail in order to illustrate how such more complicated spin sums can be handled simply. From Eqs. (8.50c) and (8.48),

$$
\begin{aligned}
X_{ab} &= \frac{-e^4}{4(p_1 - p_1')^2 (p_1 + p_2)^2} \sum_{\text{spins}} \left\{ [\bar{u}(\mathbf{p}_2')\gamma_\alpha u(\mathbf{p}_2)] [\bar{u}(\mathbf{p}_2)\gamma_\beta v(\mathbf{p}_1)] \right. \\
&\qquad\qquad \left. \times [\bar{v}(\mathbf{p}_1)\gamma^\alpha v(\mathbf{p}_1')] [\bar{v}(\mathbf{p}_1')\gamma^\beta u(\mathbf{p}_2')] \right. \\
&= \frac{-e^4}{4(p_1 - p_1')^2 (p_1 + p_2)^2} \text{Tr} \left\{ \frac{\slashed{p}_2' + m}{2m} \gamma_\alpha \frac{\slashed{p}_2 + m}{2m} \gamma_\beta \frac{\slashed{p}_1 - m}{2m} \gamma^\alpha \frac{\slashed{p}_1' - m}{2m} \gamma^\beta \right\} \\
&= \frac{-e^4}{64m^4 (p_1 - p_1')^2 (p_1 + p_2)^2} \left[\text{Tr}(\slashed{p}_2' \gamma_\alpha \slashed{p}_2 \gamma_\beta \slashed{p}_1 \gamma^\alpha \slashed{p}_1' \gamma^\beta) + O(E^2 m^2) \right].
\end{aligned}
$$

We must therefore evaluate the trace of a product of eight γ-matrices. It is usually possible and highly desirable to simplify such a product before taking the trace, rather than blindly use Eq. (A.18c). The contraction identities (Appendix A, Section A.2) are particularly useful in this respect. By means of

$$
\gamma_\lambda \gamma_\alpha \gamma_\beta \gamma_\gamma \gamma^\lambda = -2\gamma_\gamma \gamma_\beta \gamma_\alpha, \qquad \gamma_\lambda \gamma_\alpha \gamma_\beta \gamma^\lambda = 4g_{\alpha\beta} \tag{A14a}
$$

we find that the trace in X_{ab} equals

$$
-2 \, \text{Tr}(\slashed{p}_2' \slashed{p}_1 \gamma_\beta \slashed{p}_2 \slashed{p}_1' \gamma^\beta) = -8(p_2 p_1') \text{Tr}(\slashed{p}_2' \slashed{p}_1) = -32(p_2 p_1')(p_2' p_1),
$$

whence X_{ab} becomes

$$
\begin{aligned}
X_{ab} &= \frac{-e^4}{2m^4 (p_1 - p_1')^2 (p_1 + p_2)^2} \left[(p_1 p_2')(p_2 p_1') + O(E^2 m^2) \right] \\
&= \frac{-e^4}{8m^4 \sin^2(\theta/2)} \left[\cos^4 \frac{\theta}{2} + O\left(\frac{m^2}{E^2} \right) \right].
\end{aligned}
\tag{8.53}
$$

We see from this equation that X_{ab} is real. Hence substituting Eqs. (8.51)–(8.53) in Eq. (8.49), we obtain for the CoM differential cross-section in the high-energy limit ($E \gg m$)

$$
\left(\frac{d\sigma}{d\Omega} \right)_{\text{CoM}} = \frac{\alpha^2}{8E^2} \left[\frac{1 + \cos^4(\theta/2)}{\sin^4(\theta/2)} + \frac{1 + \cos^2\theta}{2} - \frac{2\cos^4(\theta/2)}{\sin^2(\theta/2)} \right].
\tag{8.54}
$$

The three terms in this equation correspond to the photon exchange diagram, Fig. 7.7(a), the annihilation diagram 7.7(b) and the interference term between them. It should be compared with the corresponding result (8.46) for the process $e^+e^- \rightarrow l^+l^-$ with $l \neq e$, when only the annihilation diagram is present.

At small angles, the exchange term dominates, giving rise to an infinite cross-section in the forward direction, $\theta = 0$, and an infinite total cross-section. These features are a consequence of the infinite range of the electromagnetic forces or, equivalently, of the zero mass of the photon. As $\theta \rightarrow 0$, the four-momentum $k^\alpha = (p_1 - p'_1)^\alpha$ of the exchanged photon tends to zero, and the factor

$$\frac{1}{k^2 + i\varepsilon} = \frac{1}{(p_1 - p'_1)^2 + i\varepsilon}$$

in the photon propagator diverges, from which the divergence of the amplitude (8.48a) and of the cross-section (8.54) follows.[4]

At large angles the photon-exchange term and the annihilation term are of comparable importance, and sensitive to the short-distance behaviour. For the annihilation term this was

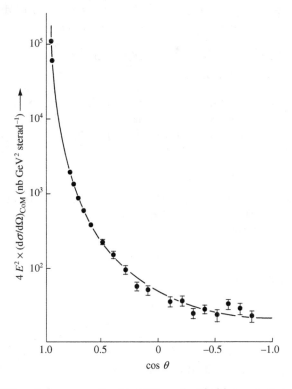

Figure 8.4 *The differential cross-section* $(d\sigma/d\Omega)_{CoM}$ *for Bhabha scattering,* $e^+e^- \rightarrow e^+e^-$, *at the total CoM energy* $2E = 34$ *GeV, [After H. J. Behrend et al., Phys. Lett,* **103B** *(1981), 148.]* $\updownarrow$: *experimental data; curve: QED cross-section formula (8.54)*

---

[4] As is often done, we suppressed the term $+ i\varepsilon$ in the photon propagator in Eq. (8.48a). This term is only relevant at the pole, $(p_1 - p'_1)^2 = 0$.

discussed in the last section. For the exchange diagram, the exchanged photon has the wave number $|\mathbf{k}| = |\mathbf{p}_1 - \mathbf{p}_1'| = 2E\sin(\theta/2)$, with an associated wavelength $\lambda = 2\pi/|\mathbf{k}|$.

Experimentally, the predicted behaviour has been confirmed over a wide range of energies and angles, and the interaction has been tested down to very short distances, comparable to those probed in the $e^+e^- \rightarrow \mu^+\mu^-$ and $e^+e^- \rightarrow \tau^+\tau^-$ experiments, which were discussed in the last section. Typical results for Bhabha scattering are shown in Fig. 8.4.

8.6   Compton Scattering

We shall now derive the cross-section for Compton scattering. In this process a photon is present in both the initial and final states, and we shall apply our earlier results to carry out the photon polarization sums, as well as the electron spin sums.

Suppose that in the initial state we have an electron with momentum $p = (E, \mathbf{p})$ in the spin state $u \equiv u_r(\mathbf{p})$, and a photon with momentum $k = (\omega, \mathbf{k})$ and polarization vector $\varepsilon \equiv \varepsilon_s(\mathbf{k})$, and that the corresponding quantities for the final state are $p' = (E', \mathbf{p}')$, $u' \equiv u_{r'}(\mathbf{p}')$, and $k' = (\omega', \mathbf{k}')$, $\varepsilon' \equiv \varepsilon_{s'}(\mathbf{k}')$. The differential cross-section for this process is given by Eqs. (8.15) and (8.12b) as

$$\frac{d\sigma}{d\Omega} = \frac{m^2\omega'}{16\pi^2 EE'\omega v_{\text{rel}}} \left[\left(\frac{\partial(E' + \omega')}{\partial\omega'} \right)_{\theta\phi} \right]^{-1} |\mathcal{M}|^2 \tag{8.55}$$

where $\mathcal{M}$ is the Feynman amplitude for this transition, (θ, ϕ) are the polar angles of $\mathbf{k}'$ and $d\Omega = \sin\theta\, d\theta\, d\phi$ is the corresponding element of solid angle. We shall take $\mathbf{k}$ as polar coordinate axis, so that θ is the photon scattering angle: $\mathbf{k} \cdot \mathbf{k}' = \omega\omega'\cos\theta$. In Eq. (8.55), initial and final momenta are related by the conservation laws

$$p + k = p' + k'. \tag{8.56}$$

In lowest order, the Feynman amplitude $\mathcal{M}$ results from the two Feynman graphs in Figs. 7.12(a) and 7.12(b), and the corresponding contributions to $\mathcal{M}$ are given by Eqs. (7.38a) and (7.38b). Defining

$$f_1 \equiv p + k, \quad f_2 \equiv p - k', \tag{8.57}$$

we obtain $\mathcal{M}$ from these equations as

$$\mathcal{M} = \mathcal{M}_a + \mathcal{M}_b, \tag{8.58}$$

where

$$\mathcal{M}_a = -ie^2 \frac{\bar{u}'\rlap{/}{\varepsilon}'(\rlap{/}{f}_1 + m)\rlap{/}{\varepsilon}u}{2(pk)}, \quad \mathcal{M}_b = ie^2 \frac{\bar{u}'\rlap{/}{\varepsilon}(\rlap{/}{f}_2 + m)\rlap{/}{\varepsilon}'u}{2(pk')}. \tag{8.59}$$

These results refer to a general reference frame. In most experiments, the photon beam is incident on a target of nearly stationary electrons. We shall now specialize Eq. (8.55) to the laboratory frame in which $p = (m, 0, 0, 0)$ and

$$\mathbf{p}' = \mathbf{k} - \mathbf{k}', \tag{8.60a}$$

$$E' = \left[m^2 + (\mathbf{k} - \mathbf{k'})^2\right]^{1/2} = \left[m^2 + \omega^2 + \omega'^2 - 2\omega\omega'\cos\theta\right]^{1/2}. \tag{8.60b}$$

From Eq. (8.56) we have generally

$$pk = p'k + k'k = pk' + k'k,$$

which in the laboratory system reduces to

$$\omega' = \frac{m\omega}{m + \omega(1 - \cos\theta)}. \tag{8.61}$$

This equation gives the energy shift of the scattered photon due to the recoil of the target electron. From Eq. (8.60b) we find

$$\left(\frac{\partial(E' + \omega')}{\partial\omega'}\right)_{\theta\phi} = \frac{m\omega}{E'\omega'}, \tag{8.62}$$

so that Eq. (8.55) gives for the differential cross-section in the laboratory frame

$$\left(\frac{d\sigma}{d\Omega}\right)_{\text{Lab}} = \frac{1}{(4\pi)^2}\left(\frac{\omega'}{\omega}\right)^2 |\mathcal{M}|^2. \tag{8.63}$$

The cross-sections (8.55) and (8.63) are fully polarized, i.e. both initial and final electrons and photons are in definite polarization states. To obtain the cross-section for Compton scattering by an unpolarized electron target, and with the spin of the final electron undetected, we must sum and average the above cross-section formulas over final and initial electron spins. To obtain the unpolarized cross-section, we must also sum and average over final and initial photon polarizations. We shall illustrate both the methods of Section 1.4.4 and of Section 8.3 for handling photon polarization.

To obtain the unpolarized cross-section directly, we use the covariant method of Section 8.3. Writing

$$\mathcal{M} \equiv \varepsilon_\alpha \varepsilon'_\beta \mathcal{M}^{\alpha\beta}, \tag{8.64}$$

we obtain

$$\frac{1}{4}\sum_{\text{pol}}\sum_{\text{spin}}|\mathcal{M}|^2 = \frac{1}{4}\sum_{\text{spin}}\mathcal{M}^{\alpha\beta}\mathcal{M}^*_{\alpha\beta}, \tag{8.65}$$

where the summations are over initial and final electron spins and photon polarizations. Eq. (8.65) is the analogue of Eq. (8.36) for the case of two external photons. Carrying out the spin summations, we obtain from Eqs. (8.65) and (8.58)–(8.59)

$$\frac{1}{4}\sum_{\text{pol}}\sum_{\text{spin}}|\mathcal{M}|^2 = \frac{1}{4}\sum_{\text{pol}}\sum_{\text{spin}}\left\{|\mathcal{M}_a|^2 + |\mathcal{M}_b|^2 + \mathcal{M}_a\mathcal{M}^*_b + \mathcal{M}_b\mathcal{M}^*_a\right\}$$

$$= \frac{e^4}{64m^2}\left\{\frac{X_{aa}}{(pk)^2} + \frac{X_{bb}}{(pk')^2} - \frac{X_{ab} + X_{ba}}{(pk)(pk')}\right\}, \tag{8.66}$$

where

$$X_{aa} = \text{Tr}\{\gamma^\beta (\rlap{/}f_1 + m)\gamma^\alpha (\rlap{/}p + m)\gamma_\alpha (\rlap{/}f_1 + m)\gamma_\beta (\rlap{/}p' + m)\} \tag{8.67a}$$

$$X_{bb} = \text{Tr}\{\gamma^\alpha (\rlap{/}f_2 + m)\gamma^\beta (\rlap{/}p + m)\gamma_\beta (\rlap{/}f_2 + m)\gamma_\alpha (\rlap{/}p' + m)\} \tag{8.67b}$$

$$X_{ab} = \text{Tr}\{\gamma^\beta (\rlap{/}f_1 + m)\gamma^\alpha (\rlap{/}p + m)\gamma_\beta (\rlap{/}f_2 + m)\gamma_\alpha (\rlap{/}p' + m)\} \tag{8.67c}$$

$$X_{ba} = \text{Tr}\{\gamma^\alpha (\rlap{/}f_2 + m)\gamma^\beta (\rlap{/}p + m)\gamma_\alpha (\rlap{/}f_1 + m)\gamma_\beta (\rlap{/}p' + m)\}. \tag{8.67d}$$

Note that the effect of the substitutions

$$k \leftrightarrow -k', \qquad \varepsilon \leftrightarrow \varepsilon', \tag{8.68a}$$

is to induce the transformations

$$f_1 \leftrightarrow f_2, \qquad \mathcal{M}_a \leftrightarrow \mathcal{M}_b \tag{8.68b}$$

and hence

$$X_{aa} \leftrightarrow X_{bb}, \qquad X_{ab} \leftrightarrow X_{ba}, \tag{8.68c}$$

We need therefore only calculate X_{aa} and X_{ab} from first principles. From Eq. (8.66), $X_{ba} = X_{ab}^*$. Furthermore, it follows from Eqs. (8.67c) and (8.67d) and the general property (A.20a) of γ-matrices, that $X_{ab} = X_{ba}$. Hence X_{ab} is real, and it is symmetric with respect to the transformation (8.68a), which provides two useful checks in its calculation.

The traces in Eqs. (8.67) involve products of up to eight γ-matrices. Their computation is much simplified by the use of the contraction identities (Appendix A, Section A.2) which eliminate four γ-matrices. We illustrate this for X_{aa}. The trace in Eq. (8.67a) contains the factor

$$Y \equiv \gamma^\beta (\rlap{/}f_1 + m)\gamma^\alpha (\rlap{/}p + m)\gamma_\alpha (\rlap{/}f_1 + m)\gamma_\beta$$

$$= \gamma^\beta (\rlap{/}f_1 + m)(-2\rlap{/}p + 4m)(\rlap{/}f_1 + m)\gamma_\beta$$

$$= 4\rlap{/}f_1 \rlap{/}p \rlap{/}f_1 + m\left[-16(pf_1) + 16f_1^2\right] + m^2\left(4\rlap{/}p - 16\rlap{/}f_1\right) + 16m^3.$$

Hence, using Eqs. (A.16) and (A.18a), (A.18b), one obtains directly

$$X_{aa} = \text{Tr}\{Y(\rlap{/}p' + m)\}$$

$$= 16\{2(f_1 p)(f_1 p') - f_1^2(pp') + m^2\left[-4(pf_1) + 4f_1^2\right]$$

$$\qquad + m^2\left[(pp') - 4(f_1 p')\right] + 4m^4\}. \tag{8.69}$$

If we express all quantities in terms of the three linearly independent scalars

$$p^2 = p'^2 = m^2, \quad pk = p'k', \quad pk' = p'k, \tag{8.70}$$

X_{aa} simplifies to

$$X_{aa} = 32\left[m^4 + m^2(pk) + (pk)(pk')\right]. \tag{8.71a}$$

From Eqs. (8.68) we have at once

$$X_{bb} = 32\left[m^4 - m^2(pk') + (pk)(pk')\right]. \tag{8.71b}$$

The interference term X_{ab}, Eq. (8.67c), is similarly computed with the result

$$X_{ab} = 16m^2 \left[2m^2 + (pk) - (pk') \right]. \tag{8.71c}$$

As expected, X_{ab} is real and symmetric, i.e. the substitution $k \leftrightarrow -k'$ transforms X_{ab} into

$$X_{ba} = X_{ab}. \tag{8.71d}$$

Substituting Eqs. (8.71a)–(8.71d) into Eq. (8.66), we obtain

$$\frac{1}{4} \sum_{\text{pol}} \sum_{\text{spin}} |\mathcal{M}|^2 = \frac{e^4}{2m^2} \left\{ \left(\frac{pk}{pk'} + \frac{pk'}{pk} \right) \right.$$
$$\left. + 2m^2 \left(\frac{1}{pk} - \frac{1}{pk'} \right) + m^4 \left(\frac{1}{pk} - \frac{1}{pk'} \right)^2 \right\}. \tag{8.72}$$

In the laboratory system, $pk = m\omega$, $pk' = m\omega'$, and from Eq. (8.61)

$$\frac{1}{\omega} - \frac{1}{\omega'} = \frac{1}{m}(\cos\theta - 1).$$

Hence Eq. (8.72) reduces to

$$\left[\frac{1}{4} \sum_{\text{pol}} \sum_{\text{spin}} |\mathcal{M}|^2 \right]_{\text{Lab}} = \frac{e^4}{2m^2} \left\{ \frac{\omega}{\omega'} + \frac{\omega'}{\omega} - \sin^2\theta \right\}, \tag{8.73}$$

and Eq. (8.63) gives the unpolarized cross-section

$$\left(\frac{d\sigma}{d\Omega} \right)_{\text{Lab}} = \frac{\alpha^2}{2m^2} \left(\frac{\omega'}{\omega} \right)^2 \left\{ \frac{\omega}{\omega'} + \frac{\omega'}{\omega} - \sin^2\theta \right\}. \tag{8.74}$$

By means of Eq. (8.61), ω' can of course be eliminated altogether from this equation. In the low-energy limit $\omega \ll m$, we have $\omega' \approx \omega$, i.e. the kinetic energy of the recoil electron is negligible, and Eq. (8.74) reduces to the Thomson scattering cross-section, Eq. (1.69a).

We shall now derive the cross-section for initial and final photons in states of definite polarization, i.e. summing and averaging over electron spins only. On also summing and averaging over photon polarizations, using the method of Section 1.4.4, we shall regain the result (8.74). In this case, the trace calculations cannot be simplified through use of the contraction identities. However, they are greatly facilitated by a suitable choice of gauge. In any reference frame, it is possible to find a Lorentz gauge in which the vacuum contains no longitudinal or scalar photons, and free photons are transverse (see Section 5.2). In this gauge, the polarization vectors of the external photons are of the form $\varepsilon = (0, \boldsymbol{\varepsilon})$, $\varepsilon' = (0, \boldsymbol{\varepsilon}')$, with

$$\varepsilon k = -\boldsymbol{\varepsilon} \cdot \mathbf{k} = 0, \qquad \varepsilon' k' = -\boldsymbol{\varepsilon}' \cdot \mathbf{k}' = 0. \tag{8.75a}$$

The analysis is further simplified if we work in the laboratory frame, $p = (m, 0, 0, 0)$, in which we also have

$$p\varepsilon = p\varepsilon' = 0. \tag{8.75b}$$

It follows from the anticommutation relations $\left[\gamma^\alpha, \gamma^\beta\right]_+ = 2g^{\alpha\beta}$ and the Dirac equation $(\not p - m)u(\mathbf{p}) = 0$, that

$$\not p \not\epsilon u = -m\not\epsilon u, \quad \not p \not\epsilon' \not u = -m\not\epsilon' u,$$

so that the matrix elements (8.59) simplify to

$$\mathcal{M}_{\mathrm a} = -\mathrm{i}e^2 \frac{\bar u\,'\not\epsilon'\not k'\not\epsilon' u}{2(pk)}, \quad \mathcal{M}_{\mathrm b} = -\mathrm{i}e^2 \frac{\bar u\,'\not\epsilon'\not k'\not\epsilon' u}{2(pk')}. \tag{8.76}$$

Note that Eq. (8.76) does not give a gauge-invariant expression for the matrix element $\mathcal{M} = \mathcal{M}_{\mathrm a} + \mathcal{M}_{\mathrm b}$. For example, the gauge transformation $\varepsilon \to \varepsilon + \lambda k$, where λ is a constant, leads to $\mathcal{M}_{\mathrm a} \to \mathcal{M}_{\mathrm a}$, (since $\not k \not k = k^2 = 0$) but $\mathcal{M}_{\mathrm b} \nrightarrow \mathcal{M}_{\mathrm b}$. This is of course due to the fact that we dropped the terms $p\varepsilon$ and $p\varepsilon'$ which are zero in our gauge.

Summing and averaging over electron spins now gives

$$\frac{1}{2}\sum_{\mathrm{spin}} |\mathcal{M}|^2 = \frac{e^4}{32m^2}\left\{ \frac{Y_{\mathrm{aa}}}{(pk)^2} + \frac{Y_{\mathrm{bb}}}{(pk')^2} + \frac{Y_{\mathrm{ab}} + Y_{\mathrm{ba}}}{(pk)(pk')} \right\} \tag{8.77}$$

where

$$Y_{\mathrm{aa}} = \mathrm{Tr}\{\not\epsilon'\not k\not\epsilon(\not p + m)\not\epsilon\not k\not\epsilon'(\not p' + m)\} \tag{8.78a}$$

$$Y_{\mathrm{bb}} = \mathrm{Tr}\{\not\epsilon\not k'\not\epsilon'(\not p + m)\not\epsilon'\not k'\not\epsilon(\not p' + m)\} \tag{8.78b}$$

$$Y_{\mathrm{ab}} = \mathrm{Tr}\{\not\epsilon'\not k\not\epsilon(\not p + m)\not\epsilon'\not k'\not\epsilon(\not p' + m)\} \tag{8.78c}$$

$$Y_{\mathrm{ba}} = \mathrm{Tr}\{\not\epsilon\not k'\not\epsilon'(\not p + m)\not\epsilon\not k\not\epsilon'(\not p' + m)\}. \tag{8.78d}$$

Substituting $k \leftrightarrow -k'$, $\varepsilon \leftrightarrow \varepsilon'$, again leads to $\mathcal{M}_{\mathrm a} \leftrightarrow \mathcal{M}_{\mathrm b}$, and

$$Y_{\mathrm{aa}} \leftrightarrow Y_{\mathrm{bb}}, \qquad Y_{\mathrm{ab}} \leftrightarrow Y_{\mathrm{ba}}, \tag{8.79}$$

and $Y_{\mathrm{ab}} = Y_{\mathrm{ba}} = Y_{\mathrm{ab}}^*$.

The traces in Eqs. (8.78) contain products of up to eight γ-matrices. We reduce this number using

$$\not A \not B = -\not B \not A + 2AB. \tag{8.80a}$$

For $A = B$, we have

$$\not A \not A = A^2, \tag{8.80b}$$

and, in particular,

$$\not p \not p = m^2, \quad \not k \not k = 0, \quad \not\epsilon \not\epsilon = \not\epsilon' \not\epsilon' = -1. \tag{8.80c}$$

For $AB = 0$, we have

$$\not A \not B = -\not B \not A, \tag{8.80d}$$

which will be particularly useful on account of Eqs. (8.75).

We illustrate the use of these tricks by computing Y_{aa}. Since $\not{k}\not{\varepsilon}'\not{k} = -\not{k}\not{k} = 0$ Eq. (8.78a), reduces to

$$Y_{aa} = \text{Tr}\{\not{\varepsilon}'\not{k}\not{\varepsilon}\not{p}\not{\varepsilon}\not{k}\not{\varepsilon}'\not{p}'\} = \text{Tr}\{\not{\varepsilon}'\not{k}\not{p}\not{k}\;\not{\varepsilon}'\not{p}'\}$$

since $\not{\varepsilon}\not{p}\not{\varepsilon} = -\not{p}\not{\varepsilon}\not{\varepsilon} = \not{p}$. Permuting $\not{p}$ and $\not{k}$, and using $\not{k}\not{k} = 0$, we obtain

$$
\begin{aligned}
Y_{aa} &= 2(pk)\text{Tr}\{\not{\varepsilon}'\not{k}\not{\varepsilon}'\not{p}'\} = 8(pk)[2(\varepsilon\prime k)(\varepsilon'p') + (kp')] \\
&= 8(pk)\left[2(\varepsilon'k)^2 + (pk')\right],
\end{aligned}
\tag{8.81a}
$$

since $p' - k = p - k'$ implies $\varepsilon'p' = \varepsilon'k$ and $kp' = pk'$. From Eq. (8.79) we have

$$Y_{bb} = -8(pk')\left[2(\varepsilon k')^2 - (pk)\right].\tag{8.81b}$$

The interference term Y_{ab}, Eq. (8.78c), is more complicated to evaluate. Its simplification depends essentially on writing $p' = p + k - k'$, so that the orthogonality relations (8.75) and Eqs. (8.80) can be used to the full. In this way one finds

$$Y_{ab} = 8(pk)(pk')\left[2(\varepsilon\varepsilon')^2 - 1\right] - 8(k\varepsilon')^2(pk') + 8(k'\varepsilon)^2(pk),\tag{8.81c}$$

which is real and symmetric (i.e. $Y_{ab} = Y_{ba}$). From Eqs. (8.81), (8.77) and (8.63), one obtains the differential cross-section for Compton scattering of polarized photons

$$\left(\frac{d\sigma}{d\Omega}\right)_{\text{Lab, pol}} = \frac{\alpha^2}{4m^2}\left(\frac{\omega'}{\omega}\right)^2\left\{\frac{\omega}{\omega'} + \frac{\omega'}{\omega} + 4(\varepsilon\varepsilon')^2 - 2\right\}.\tag{8.82}$$

Eq. (8.82) is known as the Klein–Nishina formula. From it one obtains the unpolarized cross-section by summing and averaging over final and initial photon polarizations. Since $\varepsilon\varepsilon' = -\boldsymbol{\varepsilon}\cdot\boldsymbol{\varepsilon}'$, one can write Eq. (1.71)

$$\frac{1}{2}\sum_{\text{pol}}(\varepsilon\varepsilon')^2 = \frac{1}{2}\left(1 + \cos^2\theta\right),\tag{8.83}$$

and applying this equation to Eq. (8.82) one at once regains the unpolarized cross-section formula (8.74).

8.7   Scattering by an External Field

So far, the electromagnetic field has been described by a quantized field, involving photon creation and annihilation operators. In some problems, where the quantum fluctuations of the field are unimportant, it may be adequate to describe the field as a purely classical function of the space–time coordinates. An example would be the scattering of electrons or positrons by an applied 'external' electromagnetic field $A_e^\alpha(x)$, such as the Coulomb field of a heavy nucleus.[5] More generally, one may have to consider both types of field,

---

[5] The meaning of 'external' in the present context and in the description of Feynman graphs (where we talk of external lines, particles, etc.) should not be confused.

replacing $A^\alpha(x)$ by the sum of the quantized and the classical fields, $A^\alpha(x) + A_e^\alpha(x)$. The S-matrix expansion of QED, Eqs. (7.1) and (7.2), then becomes

$$S = \sum_{n=0}^{\infty} \frac{(ie)^n}{n!} \int \cdots \int d^4x_1 \cdots d^4x_n T\left\{ N\left[\bar{\psi}(\not{A}+\not{A}_e)\psi\right]_{x_1} \cdots N\left[\bar{\psi}(\not{A}+\not{A}_e)\psi\right]_{x_n} \right\}.$$

(8.84)

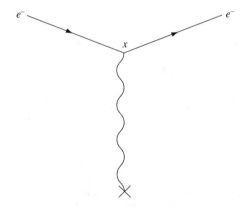

Figure 8.5 *Feynman graph in configuration space for electron scattering by an external source, marked by the cross*

As a simple example, we consider the scattering of an electron by a static external field

$$A_e^\alpha(x) = A_e^\alpha(\mathbf{x}) = \frac{1}{(2\pi)^3} \int d^3q \, e^{i\mathbf{q}\cdot\mathbf{x}} A_e^\alpha(\mathbf{q}),$$

(8.85)

where the Fourier transform of the field in momentum space, $A_e^\alpha(\mathbf{q})$, has been introduced for later convenience. In lowest order, the scattering arises from the first-order term in Eq. (8.84):

$$S_e^{(1)} = ie \int d^4x \, \bar{\psi}^-(x)\not{A}_e(x)\psi^+(x).$$

(8.86)

This is represented by the Feynman diagram in Fig. 8.5, in which the source of the classical field is represented by a cross. Consider the scattering of an electron from a state $|i\rangle$, with momentum $p = (E, \mathbf{p})$ and spinor $u_r(\mathbf{p})$, to a state $|f\rangle$, with momentum $p' = (E', \mathbf{p}')$ and spinor $u_s(\mathbf{p}')$. The evaluation of the matrix element $\langle f|S_e^{(1)}|i\rangle$ for this transition is similar to that in Section 7.2.1 for electron scattering by the quantized field. Going over to momentum space, one easily obtains

$$\langle f|S_e^{(1)}|i\rangle = \left[(2\pi)\delta(E'-E)\left(\frac{m}{VE}\right)^{1/2}\left(\frac{m}{VE'}\right)^{1/2}\right]\mathcal{M},$$

(8.87)

where

$$\mathcal{M} = ie\bar{u}_s(\mathbf{p}')\not{A}_e(\mathbf{q} = \mathbf{p}' - \mathbf{p})u_r(\mathbf{p}).$$

(8.88)

These equations should be compared with Eqs. (7.31) and (7.32) for the case of the quantized field. Unlike Eq. (7.31), Eq. (8.87) does not contain a momentum conserving δ-function, since we are ignoring the momentum of the source of the field, which experiences a recoil, $\mathbf{q} = \mathbf{p}' - \mathbf{p}$ being the momentum transferred to the particle by the field. On the other hand, the δ-function in Eq. (8.87) leads to conservation of the electron's energy, i.e. to elastic scattering ($|\mathbf{p}'| = |\mathbf{p}|$). It implies that the recoil energy of the source is negligible, and is a consequence of the static field assumption (8.85). The Feynman amplitude (8.88) is represented by the momentum space Feynman diagram of Fig. 8.6, in which it must be remembered that $|\mathbf{p}'| = |\mathbf{p}|$.

In addition to the replacement

$$(2\pi)^4 \delta^{(4)}(p' + k' - p) \rightarrow (2\pi)\delta(E' - E),$$

Eqs. (8.87) and (8.88) differ from Eqs. (7.31) and (7.32) in that the photon factors $(1/2V\omega_{\mathbf{k}'})^{1/2}\varepsilon_\alpha(\mathbf{k}')$ are replaced by the external field factor $A_{e\alpha}(\mathbf{q})$.

These results are easily generalized and lead to the following two changes in the Feynman rules, discussed in Section 7.3, in order to allow for the interaction with an external static field.

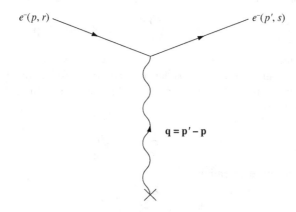

Figure 8.6 *Feynman graph in momentum space for electron scattering by an external source;* $|\mathbf{p}'| = |\mathbf{p}|$

(i) In Eq. (7.45), relating the Feynman amplitude $\mathcal{M}$ of a process to its S-matrix element $\langle f|S|i\rangle$, we must make the replacement

$$(2\pi)^4 \delta^{(4)}(P_f - P_i) \rightarrow (2\pi)\delta(E_f - E_i) \tag{8.89}$$

where E_i and E_f are the total energies of all particles present in the initial and final states.

(ii)   Corresponding to rule 4, Eqs. (7.49), for external lines (i.e. particles present initially or finally), we must add the following rule to the Feynman rules 1–8 of Section 7.3:

9. For each interaction of a charged particle with an external static field $A_e(\mathbf{x})$, write a factor

$$A_{e\alpha}(\mathbf{q}) = \int d^3x \, e^{-i\mathbf{q}\cdot\mathbf{x}} A_{e\alpha}(\mathbf{x}) \qquad (8.90)$$

Here $\mathbf{q}$ labels the momentum transferred from the field source (marked $\times$) to the particle. (The energy of the particle is conserved at the vertex.)

We next derive the cross-section for electron scattering by an arbitrary static external field in terms of the corresponding Feynman amplitude $\mathscr{M}$. The argument is very similar to that of Section 8.1 and we shall state it only briefly. Eq. (8.87) leads to the transition probability per unit time

$$w = \frac{1}{T}|\langle f|S_e^{(1)}|i\rangle|^2 = 2\pi\delta(E' - E)\left(\frac{m}{VE}\right)^2|\mathscr{M}|^2,$$

where, as in Eq. (8.6), we are considering a long but finite time interval T. Multiplying w by the density of final states

$$\frac{V\,d^3\mathbf{p}'}{(2\pi)^3} = \frac{V|\mathbf{p}'|^2 d|\mathbf{p}'|\,d\Omega'}{(2\pi)^3} = \frac{V|\mathbf{p}'|E'\,dE'\,d\Omega'}{(2\pi)^3},$$

and dividing by the incident electron flux $v/V = |\mathbf{p}|/(VE)$ gives the differential cross-section for electron scattering into an element of solid angle $d\Omega'$

$$\frac{d\sigma}{d\Omega'} = \left(\frac{m}{2\pi}\right)^2|\mathscr{M}|^2 = \left(\frac{me}{2\pi}\right)^2|\bar{u}_s(\mathbf{p}')\rlap{/}{A}_e(\mathbf{q})u_r(\mathbf{p})|^2, \qquad (8.91)$$

where we substituted Eq. (8.88) for $\mathscr{M}$, and $\mathbf{q} = \mathbf{p}' - \mathbf{p}$ and $|\mathbf{p}'| = |\mathbf{p}|$.

As an example of this result, we consider the scattering of electrons by the Coulomb field of a heavy nucleus, treated as a point charge (Mott scattering). In the Coulomb gauge, the potential is given by

$$A_e^\alpha(x) = \left(\frac{Ze}{4\pi|\mathbf{x}|}, 0, 0, 0\right), \qquad (8.92a)$$

with the momentum space potential

$$A_e^\alpha(\mathbf{q}) = \left(\frac{Ze}{|\mathbf{q}|^2}, 0, 0, 0\right). \qquad (8.92b)$$

On substituting Eq. (8.92b) in Eq. (8.91), and summing and averaging over electron spins, one obtains the unpolarized cross-section for Coulomb scattering

$$
\begin{aligned}
\frac{d\sigma}{d\Omega'} &= \frac{(2m\alpha Z)^2}{|\mathbf{q}|^4}\frac{1}{2}\sum_{r,s}|\bar{u}_s(\mathbf{p}')\gamma^0 u_r(\mathbf{p})|^2 \\
&= \frac{(\alpha Z)^2}{2|\mathbf{q}|^4}\mathrm{Tr}\{(\rlap{/}{p}'+m)\gamma^0(\rlap{/}{p}+m)\gamma^0\} \\
&= \frac{2(\alpha Z)^2}{|\mathbf{q}|^4}(E^2 + \mathbf{p}\cdot\mathbf{p}' + m^2).
\end{aligned}
\qquad (8.93a)
$$

Introducing the scattering angle θ, we have

$$\mathbf{p} \cdot \mathbf{p}' = |\mathbf{p}|^2 \cos\theta, \qquad |\mathbf{q}|^2 = |\mathbf{p}' - \mathbf{p}|^2 = 4|\mathbf{p}|^2 \sin^2(\theta/2), \qquad (8.94)$$

and, with $|\mathbf{p}| = Ev$, Eq. (8.93a) reduces to the Mott scattering formula

$$\frac{d\sigma}{d\Omega'} = \frac{(\alpha Z)^2}{4E^2 v^4 \sin^4(\theta/2)} \left[1 - v^2 \sin^2(\theta/2)\right] \qquad (8.93b)$$

for the scattering of relativistic electrons by a Coulomb field. In the non-relativistic limit, this reduces to the Rutherford scattering formula

$$\frac{d\sigma}{d\Omega'} = \frac{(\alpha Z)^2}{4m^2 v^4 \sin^4(\theta/2)}. \qquad (8.95)$$

We have here considered the nucleus as a point charge. We only mention that the treatment is easily modified to apply to the realistic case of a nucleus whose charge is distributed over a finite volume. For high-energy electrons, this leads to an important method of investigating the nuclear charge distribution. In particular, one obtains in this way the r.m.s. radius of the charge distribution, which has previously been referred to in connection with the finite size of the proton (see Problem 8.1).

The above analysis is easily extended to give the polarization properties of the electrons in Coulomb scattering.

At non-relativistic energies, the answer is, of course, well known from non-relativistic quantum mechanics. The Coulomb interaction (8.92) and the scattering amplitude are spin-independent, i.e. spin is conserved in Rutherford scattering. Interpreted in terms of helicity, this means that if the incident electron has positive helicity, then the outgoing electron has positive helicity for forward scattering, and negative helicity for backward scattering. At intermediate angles, it follows from the rotational properties of spinors that, for scattering through an angle θ, the probability of positive helicity (i.e no helicity flip) is $\cos^2(\theta/2)$, and of negative helicity (i.e. helicity flip) is $\sin^2(\theta/2)$. These results, of course, also follow from the matrix element in the first line of Eq. (8.93a). For the non-relativistic limit, it is most natural to use the Dirac–Pauli spinor representation [Eqs. (A.72) and (A.73) in Appendix A] for the spinors $u_r(\mathbf{p})$ and $u_s(\mathbf{p}')$. The indices r and s then label the spin components referred to a fixed axis in space. It follows that in the non-relativistic limit the leading term of the matrix element in Eq. (8.93a) is proportional to

$$\lim_{\substack{\mathbf{p} \to 0 \\ \mathbf{p}' \to 0}} \left(\bar{u}_s(\mathbf{p}')\gamma^0 u_r(\mathbf{p})\right) = u_s^\dagger(0)u_r(0) = \delta_{sr},$$

i.e. spin is conserved.

At relativistic energies, the scattering is spin-dependent, due to the interaction of the electron's magnetic moment with the magnetic field, which the electron sees in its own rest frame. [The velocity-dependent term $-v^2 \sin^2(\theta/2)$ in the Mott formula (8.93b) similarly results from this magnetic scattering.] To obtain the polarization properties, it is now most convenient to use the helicity states and the helicity projection operator formalism of Section 8.2. We consider a transition in which the incident electron has positive helicity [i.e. it is in the spin state $u_1(\mathbf{p})$]. The probability that the scattered electron has positive or negative helicity is, from Eq. (8.93a), proportional to

$$X_s = \left|\bar{u}_s(\mathbf{p}')\gamma^0 u_1(\mathbf{p})\right|^2, \quad \begin{cases} s = 1, & \text{positive helicity} \\ s = 2, & \text{negative helicity.} \end{cases} \tag{8.96}$$

From Eqs. (8.28) and (8.29) we see that this can be written

$$X_s = \mathrm{Tr}\left\{\Lambda^+(\mathbf{p}')\gamma^0\Pi^+(\mathbf{p})\Lambda^+(\mathbf{p})\gamma^0\Pi^\pm(\mathbf{p}')\right\}, \tag{8.97}$$

where the plus and minus signs on $\Pi^\pm(\mathbf{p}')$ correspond to $s = 1$ (no helicity flip) and $s = 2$ (helicity flip), respectively.

In the extreme relativistic limit $E = E' \gg m$, the helicity projection operators simplify to

$$\Pi^\pm(\mathbf{p}) = \frac{1}{2}\left(1 \pm \gamma^5\right). \tag{A.43}$$

Correspondingly, Eq. (8.97) reduces to

$$X_s = \frac{1}{16m^2}\mathrm{Tr}\left\{(\not{p}'+m)\gamma^0\left(1+\gamma^5\right)(\not{p}+m)\gamma^0\left(1\pm\gamma^5\right)\right\}. \tag{8.98}$$

Since p and p' are of order E, we would expect the leading contributions to X_s to come from the part of the trace in Eq. (8.98) which contains both $\not{p}$ and $\not{p}'$ as factors, so that X_s would be of order $(E/m)^2$. It is easy to show that for X_1, i.e. no helicity flip, this is indeed the case, but that for X_2, i.e. helicity flip, the only non-vanishing term is the part of the trace in Eq. (8.98) which is proportional to m^2 (i.e. independent of p and p'), so that X_2 is of order unity.[6] Thus in the extreme relativistic limit $E \gg m$, the helicity flip amplitude vanishes,

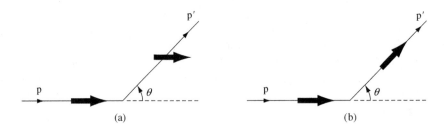

Figure 8.7   *The two extreme regimes for Coulomb scattering. (The broad arrows represent the electron spins.)* (a) *Non-relativistic energies,* $|\mathbf{p}| \ll E$: *spin is conserved.* (b) *Extreme relativistic energies,* $E \gg m$: *helicity is conserved*

---

[6] Using

$$[\gamma^5, \gamma^\alpha]_+ = 0, \quad (\gamma^5)^2 = 1, \quad \text{and}$$
$$\mathrm{Tr}\,\gamma^5 = \mathrm{Tr}(\gamma^5\gamma^\lambda) = \mathrm{Tr}(\gamma^5\gamma^\lambda\gamma^\mu) = \mathrm{Tr}(\gamma^5\gamma^\lambda\gamma^\mu\gamma^\nu) = 0, \tag{A.21}$$

one easily proves these points and calculates the traces in Eq. (8.98). (One finds that only the term containing p and p' makes a non-vanishing contribution to X_1.) The detailed proofs are left as exercises for the reader.

and helicity is conserved. This contrasts with the non-relativistic limit $|\mathbf{p}| \ll m$, where spin is conserved and the probability of helicity flip is $\sin^2(\theta/2)$. These two regimes of non-relativistic and extreme relativistic energies are schematically illustrated in Fig. 8.7. At intermediate energies, the traces in Eq. (8.97) must be evaluated exactly.

8.8 Bremsstrahlung

The deflection of an electron by the Coulomb field of a heavy nucleus, which we studied in the last section, implies that the electron must emit radiation and consequently be slowed down in its motion. More generally, the scattering of any charged particle leads to the emission of radiation. Both this process and the radiation are referred to as bremsstrahlung (literally translated, braking radiation). In this section we shall consider bremsstrahlung resulting from the scattering of electrons by the Coulomb field of a heavy nucleus. This process is of considerable practical importance. It is primarily responsible for the slowing down of fast electrons in their passage through matter, and it is widely used in electron accelerators to produce photon beams.

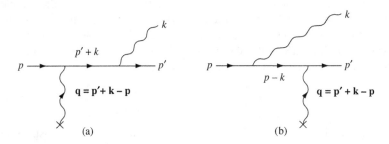

Figure 8.8 *The momentum space Feynman diagrams for bremsstrahlung*

Unlike elastic Coulomb scattering, we are now considering a process in which both the quantized field and the external field (i.e. the nuclear Coulomb field) play a role; the former to emit the radiation, the latter to ensure conservation of energy and momentum. In lowest order, the process is brought about by the second-order term ($n = 2$) in the S-matrix expansion (8.84) but, by this stage, the reader should have no difficulty in writing down the corresponding S-matrix element directly in momentum space. The two Feynman diagrams responsible for the transition are shown in Fig. 8.8, which also specifies the momenta of the particles. (The spin and polarization labels are again suppressed.) From the Feynman rules of Sections 7.3 and 8.7, the S-matrix element is given by

$$\langle f|S|i\rangle = 2\pi\delta(E' + \omega - E)\left(\frac{m}{VE}\right)^{1/2}\left(\frac{m}{VE'}\right)^{1/2}\left(\frac{1}{2V\omega}\right)^{1/2}\mathscr{M} \qquad (8.99)$$

where

$$\mathscr{M} = -e^2\bar{u}(\mathbf{p}')[\slashed{\varepsilon}(\mathbf{k})iS_F(p' + k)\slashed{A}_e(\mathbf{q}) + \slashed{A}_e(\mathbf{q})iS_F(p - k)\slashed{\varepsilon}(\mathbf{k})]u(\mathbf{p})$$

$$(8.100)$$

$$= -ie^2 \bar{u}(\mathbf{p}') \left[\not{\epsilon}(\mathbf{k}) \frac{\not{p}' + \not{k} + m}{2p'k} \not{A}_e(\mathbf{q}) + \not{A}_e(\mathbf{q}) \frac{\not{p} - \not{k} + m}{-2pk} \not{\epsilon}(\mathbf{k}) \right] u(\mathbf{p})$$

and $A_{e\alpha}(\mathbf{q} = \mathbf{p}' + \mathbf{k} - \mathbf{p})$ is the momentum space Coulomb potential, given in Eq. (8.92b).

The derivation of the cross-section formula for bremsstrahlung from Eq. (8.99) is very similar to that given in the last section for Coulomb scattering. Multiplying the transition rate $|\langle f|S|i\rangle|^2/T$, the density of final states

$$V\mathrm{d}^3\mathbf{p}'V\mathrm{d}^3\mathbf{k}/(2\pi)^6$$

and dividing by the incident electron flux $|\mathbf{p}|/(VE)$ leads to

$$\mathrm{d}\sigma = \frac{m^2}{(2\pi)^5 2\omega} \frac{|\mathbf{p}'|}{|\mathbf{p}|} |\mathcal{M}|^2 \mathrm{d}^3\mathbf{k}\,\mathrm{d}\Omega', \qquad (8.101)$$

where $\mathrm{d}\Omega'$ is the element of solid angle in the direction $\mathbf{p}'$ of the scattered electron.

Summing and averaging Eq. (8.101) over final and initial electron spins is a straightforward, but lengthy, calculation which leads to the Bethe–Heitler cross-section formula for bremsstrahlung in a Coulomb field.[7] We shall restrict ourselves to the simpler, but interesting, situation in which the emitted photon has very low energy, the so-called 'soft photon' limit $\omega \approx 0$. In this limit $\mathbf{q} = \mathbf{p}' - \mathbf{p}$ and $|\mathbf{p}'| = |\mathbf{p}|$, as for elastic scattering. In the Feynman amplitude (8.100), we neglect the $\not{k}$ terms in the numerators of the electron propagators and, using the Dirac equation, easily obtain

$$\mathcal{M} = -ie^2 \bar{u}(\mathbf{p}') \not{A}_e(\mathbf{q}) u(\mathbf{p}) \left[\frac{p'\varepsilon}{p'k} - \frac{p\varepsilon}{pk} \right] = -e\mathcal{M}_0 \left[\frac{p'\varepsilon}{p'k} - \frac{p\varepsilon}{pk} \right] \qquad (\omega \approx 0) \qquad (8.102)$$

where $\varepsilon \equiv \varepsilon(\mathbf{k})$ and $\mathcal{M}_0$ is the Feynman amplitude, Eq. (8.88), for elastic scattering without photon emission. Substituting Eq. (8.102) in Eq. (8.101) and comparing with Eq. (8.91), we can write the cross-section for soft bremsstrahlung in the form

$$\left(\frac{\mathrm{d}\sigma}{\mathrm{d}\Omega'} \right)_B = \left(\frac{\mathrm{d}\sigma}{\mathrm{d}\Omega'} \right)_0 \frac{\alpha}{(2\pi)^2} \left[\frac{p'\varepsilon}{p'k} - \frac{p\varepsilon}{pk} \right]^2 \frac{\mathrm{d}^3\mathbf{k}}{\omega} \qquad (\omega \approx 0) \qquad (8.103)$$

where $(\mathrm{d}\sigma/\mathrm{d}\Omega')_0$ is the cross-section for elastic scattering without photon emission, Eq. (8.91). In deriving Eqs. (8.102) and (8.103), we have not used the explicit form of $A_{e\alpha}(\mathbf{q})$, so that these equations hold for soft bremsstrahlung in an arbitrary static external field, not just for a Coulomb field.

Eqs. (8.102) and (8.103) contain two interesting features. Firstly, they each factorize into the corresponding quantity for elastic scattering without photon emission, multiplied by a factor which relates to the soft photon. Secondly, both the amplitude $\mathcal{M}$ and the cross-section (8.103) are singular in the infrared limit $\omega \to 0$. This infrared singularity arises because for $k = 0$ we have $p^2 = p'^2 = m^2$, so that the intermediate-state electrons in the

---

[7] The formula is derived in, for example, C. Itzykson and J. B. Zuber, *Quantum Field Theory*, McGraw-Hill, New York, 1980, Section 5-2-4, and in J. M. Jauch and F. Rohrlich, *The Theory of Photons and Electrons*, 2nd edn, Springer, New York, 1976, Section 15-6.

Feynman diagrams in Figs. 8.8(a) and (b) possess the four-momenta of a *real* electron, and correspondingly the propagators in Eq. (8.100) diverge.[8] Both the factorization property and the infrared singularity are characteristic features of soft photon-emission processes. We shall discuss them further in the next section.

It remains to sum over the polarization of the emitted photons, assuming that this is not observed. We shall do this by the gauge-invariant method of Section 8.3. Applying Eq. (8.36) to Eq. (8.103), we at once obtain the cross-section for the emission of soft bremsstrahlung in electron scattering by a static external field

$$\left(\frac{d\sigma}{d\Omega'}\right)_B = \left(\frac{d\sigma}{d\Omega'}\right)_0 \frac{(-\alpha)}{(2\pi)^2}\left[\frac{p'}{p'k} - \frac{p}{pk}\right]^2 \frac{d^3k}{\omega} \quad (\omega \approx 0). \tag{8.104}$$

8.9 The Infrared Divergence

The above results on soft photon emission in electron scattering have important implications for experiments on elastic electron scattering. In such an experiment, a photon may be emitted which is too soft to be detected, and it is the energy resolution ΔE of the apparatus which determines whether such a photon emission event is recorded as elastic or inelastic scattering. Consequently, the *experimental* cross-section is the sum of the elastic cross-section and the cross-section for bremsstrahlung of energy less than ΔE, i.e. it should be written

$$\left(\frac{d\sigma}{d\Omega'}\right)_{Exp} = \left(\frac{d\sigma}{d\Omega'}\right)_{El} + \left(\frac{d\sigma}{d\Omega'}\right)_B. \tag{8.105}$$

Here $(d\sigma/d\Omega')_{El}$ is the elastic cross-section and $(d\sigma/d\Omega')_B$ is the soft bremsstrahlung cross-section (8.104) integrated over the range of photon energy $0 \leq \omega \leq \Delta E$:

$$\left(\frac{d\sigma}{d\Omega'}\right)_B = \left(\frac{d\sigma}{d\Omega'}\right)_0 \alpha B, \tag{8.106}$$

where $(d\sigma/d\Omega')_0$ is the elastic cross-section in lowest order of perturbation theory and

$$B = \frac{-1}{(2\pi)^2} \int_{0 \leq |\mathbf{k}| \leq \Delta E} \frac{d^3k}{\omega}\left[\frac{p'}{p'k} - \frac{p}{pk}\right]^2. \tag{8.107}$$

We are assuming that ΔE is sufficiently small so that the soft photon result (8.104) is valid for $\omega \leq \Delta E$.

Unfortunately the integrand in Eq. (8.107) behaves like $1/\omega$ for small ω, so that the integral is logarithmically divergent at the lower limit of integration. This divergence is known as the *infrared catastrophe*. It is a consequence of the zero mass of the photon, and one way of dealing with this problem, due to Feynman, is to assign a fictitious small mass $\lambda(\neq 0)$ to the photon and at the end of the calculation take the limit $\lambda \to 0$ to regain QED.

---

[8] In a concise, but horrible, jargon, this situation is often described by saying that, as $k \to 0$, the internal electron lines in the Feynman diagrams 8.8(a) and (b) go on the mass shell; or, even more concisely and horribly, that they go on shell.

Introducing a non-zero photon mass leads to modifications of Eq. (8.107). The amplitude $\mathcal{M}$, Eq. (8.100), is modified since in the denominators of the electron propagators we must now put $k^2 = \lambda^2$, instead of $k^2 = 0$. It is easy to show that Eq. (8.103) is replaced by

$$\frac{d\sigma}{d\Omega'} = \left(\frac{d\sigma}{d\Omega'}\right)_0 \frac{\alpha}{(2\pi)^2} \left[\frac{2p'\varepsilon}{2p'k + \lambda^2} + \frac{2p\varepsilon}{-2pk + \lambda^2}\right]^2 \frac{d^3k}{\omega}. \qquad (8.108)$$

With non-zero mass, photons can be longitudinally as well as transversely polarized. The photon polarization sum is now effected by means of the formula for massive spin 1 bosons[9]

$$\sum_{r=1}^{3} \varepsilon_{r\alpha}\varepsilon_{r\beta} = -g_{\alpha\beta} + \frac{k_\alpha k_\beta}{\lambda^2}. \qquad (8.109)$$

The term $k_\alpha k_\beta / \lambda^2$ gives no contribution to the unpolarized cross-section in the limit when $\lambda \to 0$, hence the unpolarized soft bremsstrahlung cross-section, integrated over the photon energy range $\lambda \le \omega \le \Delta E$, is given by Eq. (8.106) with B, Eq. (8.107), now replaced by

$$B(\lambda) = \frac{-1}{(2\pi)^2} \int \frac{d^3k}{\omega_\lambda} \left[\frac{2p'}{2p'k + \lambda^2} + \frac{2p}{-2pk + \lambda^2}\right]^2 \qquad (8.110)$$

where $\omega_\lambda \equiv \left(\lambda^2 + \mathbf{k}^2\right)^{1/2}$ and the integration is over wave vectors $\mathbf{k}$ such that $\lambda \le \omega_\lambda \le \Delta E$. For $\lambda > 0$, $B(\lambda)$ is finite and well-defined. For $\lambda \to 0$, $B(\lambda)$ goes over into Eq. (8.107) and diverges again.

Returning to Eqs. (8.105) and (8.106), we see that the bremsstrahlung cross-section (8.106) is of order α relative to the lowest-order elastic cross-section $(d\sigma/d\Omega')_0$. Consequently it would be inconsistent to use the latter for $(d\sigma/d\Omega')_{\text{El}}$ in Eq. (8.105). Instead, we must include corrections of order α to the elastic cross-section, which arise from the next order in perturbation theory. Fig. 8.9 shows a Feynman diagram responsible for such a correction to the lowest-order graph, Fig. 8.6. We shall consider these corrections in detail in Section 9.7. For the moment we shall write

$$\left(\frac{d\sigma}{d\Omega'}\right)_{\text{El}} = \left(\frac{d\sigma}{d\Omega'}\right)_0 [1 + \alpha R(\lambda)] \qquad (8.111)$$

with the correction term $R(\lambda)$ also a function of the photon mass λ. Combining Eqs. (8.105), (8.106) and (8.111), we obtain

$$\left(\frac{d\sigma}{d\Omega'}\right)_{\text{Exp}} = \left(\frac{d\sigma}{d\Omega'}\right)_0 \{1 + \alpha[B(\lambda) + R(\lambda)] + O(\alpha^2)\}, \qquad (8.112)$$

where the term $O(\alpha^2)$ is to remind the reader that there exist higher-order corrections.

We have seen that $B(\lambda) \to \infty$ as $\lambda \to 0$. When calculating $R(\lambda)$ in Section 9.7, we shall find that $R(\lambda)$ is also singular at $\lambda = 0$, with $R(\lambda) \to -\infty$ as $\lambda \to 0$, in such a way that

---

[9] For a derivation of this formula, see Section 16.3 Eqs. (16.24)–(16.27).

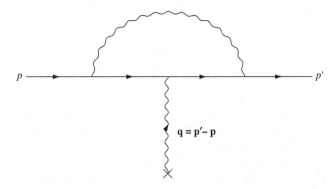

Figure 8.9 *A correction term to the elastic electron scattering by an external field*

the singularities in $B(\lambda)$ and $R(\lambda)$ exactly cancel and $[B(\lambda) + R(\lambda)]$ is well-defined and finite. Hence we can easily take the limit $\lambda \to 0$ in Eq. (8.112), obtaining a finite correction of order α to the lowest-order elastic cross-section $(d\sigma/d\Omega')_0$. Such corrections are called *radiative corrections*.

The experimental cross-section (8.112) depends on the energy resolution ΔE which occurs in Eqs. (8.107) and (8.110), and so depends on the details of the experimental set-up. Because of this, the radiative corrections are sometimes calculated and subtracted from the experimental data to give 'radiatively corrected cross-sections'. In this way the results of different experiments can be compared with each other and with theoretical predictions for $(d\sigma/d\Omega')_0$.

We can now understand the origin of the infrared catastrophe better. It arises through treating soft bremsstrahlung and elastic scattering as separate processes in perturbation theory. This separation is artificial, as one always observes elastic scattering together with some soft bremsstrahlung, and the cross-section for this is finite.

We have so far considered the emission of one soft photon and the infrared divergence to which this gives rise. For high-resolution experiments, the emission of many soft photons may become important, resulting in infrared divergences of higher order in α. We must then also modify Eq. (8.111) to take into account higher-order corrections to the elastic scattering. The statement that the infrared divergences cancel exactly to all orders of perturbation theory, leaving a finite radiative correction of order α, is the famous Bloch–Nordsieck theorem. In high-energy experiments, where radiative corrections can become as large as 50% in some kinematic regions, these multiphoton contributions must be incorporated. It can be shown[10] that the most important radiative correction is given by the generalization of Eq. (8.112) to all orders and obtained from it by exponentiation, i.e.

$$\left(\frac{d\sigma}{d\Omega'}\right)_{\text{Exp}} = \left(\frac{d\sigma}{d\Omega'}\right)_0 e^{\alpha[B(\lambda)+R(\lambda)]}. \tag{8.113}$$

---

[10] See the comprehensive article on the infra-red problem by D. R. Yennie, S. C. Frautschi and H. Suura, *Ann. Phys.* (N.Y.), **13** (1961), 379.

Correspondingly, the elastic cross-section, calculated to all orders of perturbation theory, is given by

$$\left(\frac{d\sigma}{d\Omega'}\right)_{El} = \left(\frac{d\sigma}{d\Omega'}\right)_{0} e^{\alpha R(\lambda)}. \tag{8.114}$$

Since $R(\lambda) \rightarrow -\infty$ as $\lambda \rightarrow 0$, it follows that the pure elastic cross-section, without photon emission, vanishes: there is no truly elastic scattering; all observed scattering is accompanied by emission of radiation. This corresponds to the result of classical electro-dynamics that an accelerated charge must radiate.

Any process involving charged particles initially or finally can be accompanied by the emission of soft photons, and infrared divergences occur in all these processes. Our conclusions, reached for the case of elastic electron scattering by an external field, hold generally. In particular, the Bloch–Nordsieck theorem applies, i.e. for all processes in QED the infrared divergences cancel exactly to all orders of perturbation theory, leaving finite radiative corrections of order α. In the next two chapters we shall go on to calculate these corrections.

Problems

8.1. In our discussion of electron scattering by an infinitely heavy nucleus, Eqs. (8.92)–(8.95), we treated the nucleus as a point charge. More realistically, we could treat the nucleus as a spherical charge distribution $Ze\rho(r)$, where

$$\int d^3\mathbf{r}\rho(r) = 1.$$

Show that the elastic scattering cross-section is now given by

$$\frac{d\sigma}{d\Omega'} = \left(\frac{d\sigma}{d\Omega'}\right)_M |F(\mathbf{q})|^2,$$

where $(d\sigma/d\Omega')_M$ is the Mott cross-section (8.93b), $ZeF(\mathbf{q})$ is the Fourier transform of the charge distribution and $\mathbf{q} = \mathbf{p}' - \mathbf{p}$ ($\mathbf{p}$ and $\mathbf{p}'$ are the initial and final momenta of the electron).

$F(\mathbf{q})$ is called the form factor of the charge distribution. Show that it is a function of $|\mathbf{q}|^2$ only, and that the root-mean-square radius r_m of the charge distribution is given by

$$r_m^2 = -6 \left.\frac{dF(\mathbf{q})}{d(|\mathbf{q}|^2)}\right|_{|\mathbf{q}| = 0}.$$

8.2. Consider elastic $e^- - \mu^-$ scattering at energies sufficiently high so that the mass of the electron can be neglected throughout. Show that the differential cross-section for this process, in the frame of reference in which the muon is initially at rest, is given by

$$\frac{d\sigma}{d\Omega'} = \left(\frac{d\sigma}{d\Omega'}\right)_M \left[1 + \frac{2E}{m_\mu}\sin^2\frac{\theta}{2}\right]^{-1} \left[1 - \frac{q^2}{2m_\mu^2}\tan^2\frac{\theta}{2}\right].$$

Here E and E' are the initial and final energies of the electron, and θ is the angle through which the electron is scattered. q is the four-momentum of the exchanged photon, whence

$$q^2 = -4EE'\sin^2(\theta/2),$$

and $(d\sigma/d\Omega')_M$ is the Mott cross-section (8.93b) in the extreme relativistic limit $v \to 1$ (and, of course, $Z = 1$).

8.3. In the last problem the interaction of the muon is characterized by the vertex factor $ie\,\gamma^\alpha$ (Feynman rule 1). The method of the last problem can be applied to electron–proton scattering by replacing the factor $ie\gamma^\alpha$ of the muon vertex by the factor

$$-ie\left[\gamma^\alpha F_1\left(q^2\right) + \frac{\kappa_p}{2m_p}F_2\left(q^2\right)i\sigma^{\alpha\beta}q_\beta\right]$$

where q is the momentum transfer four-vector from the electron, and κ_p is the anomalous magnetic moment of the proton, in units of the nuclear magneton. $F_1(q^2)$ and $F_2(q^2)$ are form factors representing the internal structure of the proton, with $F_1(0) = F_2(0) = 1$, so that for $\mathbf{q} \to 0$, i.e. a stationary proton interacting with static electric and magnetic fields, the proton has the correct electrostatic and magnetostatic interactions.

Show that in the laboratory frame in which the proton is initially at rest the differential cross-section for elastic scattering of electrons of energy $E(\gg m_e)$ is given by the Rosenbluth cross-section

$$\frac{d\sigma}{d\Omega'} = \left(\frac{d\sigma}{d\Omega'}\right)_M\left[1 + \frac{2E}{m_p}\sin^2\frac{\theta}{2}\right]^{-1}$$

$$\times\left\{\left[F_1^2(q^2) - \frac{\kappa_p^2}{4m_p^2}q^2F_2^2(q^2)\right] - \frac{q^2}{2m_p^2}\left[F_1(q^2) + \kappa_pF_2(q^2)\right]^2\tan^2\frac{\theta}{2}\right\}.$$

Here θ is the angle of scattering, and $(d\sigma/d\Omega')_M$ is the Mott cross-section (8.93b) with $Z = 1$ and $v = 1$.

8.4. Show that the probability of helicity flip occurring in Mott scattering is given by

$$\frac{m^2\sin^2(\theta/2)}{E^2\cos^2(\theta/2) + m^2\sin^2(\theta/2)}.$$

8.5. Show that the Mott scattering formula (8.93b) also gives the differential cross-section for the scattering of positrons by a heavy nucleus, treated as a point charge. (This equality of electron and positron scattering holds for the lowest-order calculations only.)

8.6. Show that the differential cross-section in the centre-of-mass system for electron–electron scattering in the high-energy limit $(E \gg m)$ is given by

$$\left(\frac{d\sigma}{d\Omega}\right)_{\text{CoM}} = \frac{\alpha^2}{8E^2}\left[\frac{1 + \cos^4(\theta/2)}{\sin^4(\theta/2)} + \frac{2}{\sin^2(\theta/2)\cos^2(\theta/2)} + \frac{1 + \sin^4(\theta/2)}{\cos^4(\theta/2)}\right],$$

where θ is the scattering angle and E is the energy of either electron in the CoM system.

8.7. Show that the Feynman amplitude for Compton scattering

$$\mathcal{M} = \mathcal{M}_a + \mathcal{M}_b \qquad\qquad (8.58 - 59)$$

is gauge invariant, although the individual contributions $\mathcal{M}_a$ and $\mathcal{M}_b$ are not, by considering the gauge transformation [compare Eq. (8.31)]

$$\varepsilon(\mathbf{k}) \rightarrow \varepsilon(\mathbf{k}) + \lambda k, \quad \varepsilon'(\mathbf{k}') \rightarrow \varepsilon'(\mathbf{k}') + \lambda' k'.$$

9

Radiative Corrections

In the last chapter we applied QED to calculate processes in lowest order of perturbation theory. On taking higher orders into account, one expects corrections of the order of the fine-structure constant α to the lowest-order results, known as radiative corrections. However, on doing such a calculation one encounters divergent integrals. The divergent electron self-energy term, Eq. (7.44b), corresponding to the Feynman diagram, Fig. 7.15, is a typical example.

In this chapter we shall show how to overcome these difficulties. This involves three steps. Firstly, one regularizes the theory, that is, one modifies it so that it remains finite and well-defined in all orders of perturbation theory. The second step originates from the recognition that the non-interacting leptons and photons from which perturbation theory starts are not the same thing as the real physical particles which interact. The interaction modifies the properties of the particles, e.g. the charge and mass of the electron, and the predictions of the theory must be expressed in terms of the properties of the physical particles, not of the non-interacting (or bare) particles. The second step, called renormalization, consists of relating the properties of the physical particles to those of the bare particles and expressing the predictions of the theory in terms of the masses and charges of the physical particles. The third step consists of reverting from the regularized theory back to QED. The original infinities of QED now appear in the relations between bare and physical particles. These relations, like the bare particles themselves, are totally unobservable. In contrast, the observable predictions of the theory, expressed in terms of the measured charges and masses of the particles, remain finite as QED is restored. In particular, the radiative corrections are finite and of order α.

The programme we have outlined can be carried through to all orders of perturbation theory, so that the radiative corrections can be calculated to extraordinarily high accuracy. The complete agreement of these predictions with equally precise experiments, for example for the anomalous magnetic moments of leptons and for the Lamb shift, constitutes one of the great triumphs of physics.

In this chapter we shall almost exclusively consider the calculations of radiative corrections in lowest order of perturbation theory. We shall develop the general methods for this in Sections 9.1–9.5, and we shall consider applications and comparison with experiment in Section 9.6. The more technical details of regularization are relegated to the next chapter. As a further application, in Section 9.7, we shall once more consider the infrared divergence completing the discussion which was started in Section 8.9. Finally, in Section 9.8, we shall briefly indicate how the considerations of this chapter can be generalized to yield finite radiative corrections to all orders of perturbation theory.

9.1   The Second-Order Radiative Corrections of QED

The radiative corrections to any process in QED are obtained, like the lowest-order matrix elements themselves, from the S-matrix expansion, Eq. (8.84), using the Feynman rules of Sections 7.3 and 8.7. The Feynman diagrams representing the radiative corrections to a process contain additional vertices, compared with the diagrams describing the process in lowest order of perturbation theory, corresponding to the emission and re-absorption of virtual photons. Restricting oneself to the Feynman diagrams which contain two extra vertices corresponds to calculating the radiative corrections in lowest order of perturbation theory, involving only one virtual photon. Thus these corrections are of second order in the electronic charge, i.e. first order in the fine-structure constant, relative to the lowest-order matrix element.

To introduce the basic ideas involved in calculating second-order radiative corrections, we shall consider the elastic scattering of electrons by a static external field $A_e^\mu(\mathbf{x})$. As we saw in Section 8.7, the lowest order of perturbation theory describes this process by the Feynman diagram of Fig. 9.1 and by the Feynman amplitude

$$\mathcal{M}^{(0)} = \mathrm{i}e_0 \bar{u}(\mathbf{p}') \slashed{A}_e(\mathbf{p}' - \mathbf{p})u(\mathbf{p}). \tag{9.1}$$

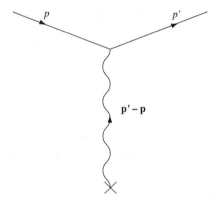

Figure 9.1 *The lowest-order contribution to the elastic scattering of electrons by a static external field*

Here, and from now on, the charge of the bare (i.e. non-interacting) electron is denoted by $(-e_0)$. Similarly, we shall denote the mass of the bare electron by m_0.

The radiative corrections to the amplitude (9.1) follow from the S-matrix expansion (8.84). We shall assume that the external field is weak, so that we need only retain terms linear in $A_e^\mu(\mathbf{x})$ in this expansion. It can then be written

$$S = 1 + \sum_{n=1}^{\infty} \frac{(ie_0)^n}{(n-1)!} \int \cdots \int d^4x_1 \cdots d^4x_n$$
$$\times T\{N(\bar{\psi} \slashed{A}_e \psi)_{x_1} N(\bar{\psi} \slashed{A} \psi)_{x_2} \cdots N(\bar{\psi} \slashed{A} \psi)_{x_n}\}. \tag{9.2}$$

The leading amplitude $\mathscr{M}^{(0)}$ stems from the $n=1$ term in Eq. (9.2). The $n=2$ term in Eq. (9.2) is linear in the quantized radiation field $A^\mu(x)$. Hence it necessarily involves emission or absorption of a photon and describes inelastic processes such as bremsstrahlung, discussed in Section 8.8. The second-order radiative correction follows from the $n=3$ term in Eq. (9.2).

The four contributions to this second-order correction are shown in Fig. 9.2. Each of these can be regarded as a modification of the lowest-order diagram 9.1, corresponding to one of the substitutions shown in Fig. 9.3. For example, the Feynman graph 9.2(a) is obtained by making the substitution of Fig. 9.3(a) in the incoming electron line of diagram 9.1, and so on.

The loop diagrams 9.3(a) and (b) we have met before (see Figs. 7.8, 7.9 and 7.15). They represent the self-energy parts of an electron and of a photon due to the interaction of the electron–positron field with the photon field. The electron self-energy loop in diagram 9.3(a) represents the lowest-order process which turns a bare electron into a real physical (i.e. interacting) electron. Similarly, the loop of diagram 9.3(b) represents the photon self-energy in lowest, i.e. second order only. The fermion–photon interaction here manifests itself in the creation and annihilation of a virtual electron–positron pair, and the loop diagram 9.3(b) is referred to as a vacuum polarization diagram. Lastly, the substitution shown in Fig. 9.3(c) represents the lowest-order modification of the basic vertex part, i.e. of the basic fermion–photon interaction $N(\bar{\psi} \slashed{A} \psi)$, due to the emission and re-absorption of a virtual photon during the process of interaction.

The Feynman amplitudes for the diagrams 9.2(a)–(d) follow from the Feynman rules and are given by

$$\mathscr{M}_a^{(2)} = ie_0 \bar{u}(\mathbf{p}') \slashed{A}_e(\mathbf{p}' - \mathbf{p}) iS_F(p) ie_0^2 \textstyle\sum(p) u(\mathbf{p}) \tag{9.3a}$$

$$\mathscr{M}_b^{(2)} = ie_0 \bar{u}(\mathbf{p}') ie_0^2 \textstyle\sum(p') iS_F(p') \slashed{A}_e(\mathbf{p}' - \mathbf{p}) u(\mathbf{p}) \tag{9.3b}$$

$$\mathscr{M}_c^{(2)} = ie_0 \bar{u}(\mathbf{p}') \gamma^\lambda u(\mathbf{p}) iD_{F\lambda\mu}(q) ie_0^2 \Pi^{\mu\nu}(q) A_{ev}(\mathbf{p}' - \mathbf{p}) \tag{9.3c}$$

$$\mathscr{M}_d^{(2)} = ie_0 \bar{u}(\mathbf{p}') e_0^2 \Lambda^\mu(p',p) u(\mathbf{p}) A_{e\mu}(\mathbf{p}' - \mathbf{p}), \tag{9.3d}$$

where

$$ie_0^2 \textstyle\sum(p) = \frac{(ie_0)^2}{(2\pi)^4} \int d^4k iD_{F\alpha\beta}(k) \gamma^\alpha iS_F(p-k) \gamma^\beta \tag{9.4}$$

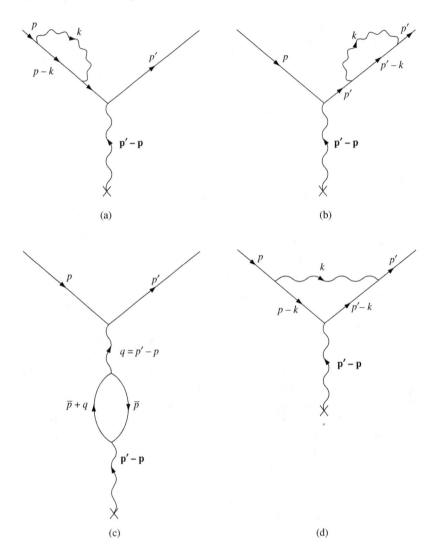

Figure 9.2 *The four contributions to the second-order radiative corrections to electron scattering*

$$ie_0^2 \Pi^{\mu\nu}(q) = \frac{(ie_0)^2}{(2\pi)^4}(-1)\mathrm{Tr} \int d^4\bar{p}\gamma^\mu iS_F(\bar{p}+q)\gamma^\nu iS_F(\bar{p}) \tag{9.5}$$

and

$$e_0^2 \Lambda^\mu(p', p) = \frac{(ie_0)^2}{(2\pi)^4} \int d^4k\gamma^\alpha iS_F(p'-k)\gamma^\mu iS_F(p-k)\gamma^\beta iD_{F\alpha\beta}(k). \tag{9.6}$$

We see that in order to calculate the second-order radiative correction to electron scattering we must evaluate the three loop integrals (9.4)–(9.6).

Unfortunately, on substituting the explicit expressions for the electron and photon propagators, all three integrals are found to be divergent for large values of the momentum

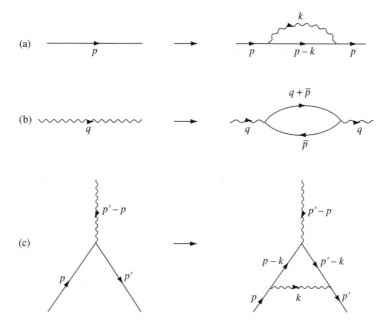

Figure 9.3 *The second-order corrections to fermion and photon lines and to the basic vertex part: (a) the fermion self-energy part* $ie_0^2\Sigma(p)$, *Eq.(9.4); (b) the photon self-energy part* $ie_0^2\Pi^{\mu\nu}(q)$, *Eq. (9.5); (c) the vertex part* $e_0^2\Lambda^\mu(p', p)$, *Eq.(9.6)*

variables of integration.[1] From dimensional arguments, one sees that for $k \to \infty$, the integrals (9.4) and (9.6) appear to be of order k and $\ln k$, respectively, while the integral (9.5) appears of order $\bar{p}^2$ as $\bar{p} \to \infty$.[2] In the next four sections of this chapter, we shall consider these three divergent integrals in detail. We shall show how the concepts of charge and mass renormalization enable one to extract unambiguously finite radiative corrections of order α, expressed in terms of the observed charge $(-e)$ and the observed mass m of the real physical electron, and not in terms of the unobservable charge $(-e_0)$ and the unobservable mass m_0 of the bare electron. The great importance of this analysis is due to the fact that in calculating the radiative corrections of lowest order to any process, these same three divergent integrals occur and no others. Consequently, once we have coped with these three integrals, the calculation of the second-order radiative correction to any process presents no difficulties of principle.

We illustrate this for Compton scattering, previously studied in Section 7.2.2. In lowest order, Compton scattering is described by the Feynman graphs of Fig. 9.4. The second-order radiative corrections arise from all connected Feynman graphs containing four vertices and the correct external lines.

---

[1] In addition, the integrals (9.4) and (9.6) also lead to infrared divergences at the lower limit $k \to 0$. These infrared divergences will also be dealt with later in this chapter.

[2] Of course, these dimensional arguments only give the maximum possible degree of divergence of each integral. If the coefficient of the leading divergence happens to vanish, the actual divergence is less severe.

Figure 9.4   *The two lowest-order Feynman graphs for Compton scattering*

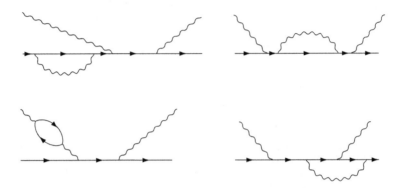

Figure 9.5   *Four of the fourth-order contributions to Compton scattering, obtained by inserting a self-energy or vertex modification in the Feynman graph 9.4(a)*

The substitutions of Fig. 9.3 in the diagrams 9.4 lead to 14 contributions, with insertions of the electron self-energy part (in internal or external electron lines), of the photon self-energy part and of the vertex correction being responsible for 6, 4 and 4 graphs, respectively. Four of these graphs are shown in Fig. 9.5. The Feynman amplitudes for these 14 graphs will contain the three divergent loop integrals (9.4)–(9.6) as factors, similar to the way they occur in the electron scattering amplitudes (9.3). After renormalization, the evaluation of these amplitudes presents no difficulties and leads to finite radiative corrections of order α.

In addition to these 14 Feynman diagrams, there are four fourth-order diagrams which are not obtained from the second-order diagrams 9.4 by inserting a self-energy or vertex correction. These are shown in Fig. 9.6(a)–(d).

Writing down the Feynman amplitudes for the diagrams 9.6(a) and (b), using the Feynman rules, one finds that these amplitudes are finite and well defined, yielding radiative corrections of order α.

Finally, we must consider the triangle graphs of Fig. 9.6(c) and (d). Their contributions differ only in sign and cancel each other exactly. This is most easily seen by replacing diagram 9.6(d) by the equivalent diagram 9.6(e). Diagrams 9.6(c) and (e) differ only in the directions of the arrows of the fermion propagators around the fermion loops, clockwise in graph 9.6(c) and anticlockwise in graph 9.6(e). We know from our earlier discussions that reversing the sense of a fermion propagator is equivalent to interchanging the virtual electron and positron states which the propagator represents. Hence reversing the arrows on the three propagators in the triangle of diagram 9.6(c) is equivalent to replacing e_0 by $(-e_0)$ at each of the three vertices of the triangle. Consequently the Feynman amplitudes of diagrams 9.6(c) and (e) differ only by a factor $(-1)^3$. (This result, of course, also follows

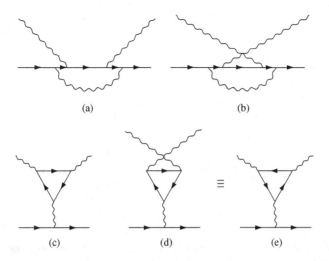

Figure 9.6 *The fourth-order contributions to Compton scattering which cannot be obtained by inserting a self-energy or vertex modification in the lowest-order Feynman graphs 9.4*

directly from the explicit expressions for the Feynman amplitudes of the two triangle graphs.) This result for the triangle diagrams is a particular case of Furry's theorem, which states that the contributions of diagrams which contain a closed fermion loop with an odd number of vertices, cancel. Such diagrams always occur in pairs differing only in the senses of the arrows on the fermion loops, and, by the same argument as above, their contributions will cancel.

The situation which has here been outlined for Compton scattering is characteristic of radiative corrections. The second-order radiative correction to any process is obtained by modifying the lowest-order graphs in all possible ways according to the substitutions of Fig. 9.3. After renormalization of the divergent loop integrals (9.4)–(9.6), these modified graphs give finite second-order radiative corrections. In general, there will be additional second-order radiative corrections from graphs which are not self-energy or vertex modifications of the lowest-order graphs. These contributions are finite and well-defined and so do not require renormalization.

9.2   The Photon Self-Energy

We shall first consider the effects of the photon self-energy insertion, Fig. 9.3(b), in the photon propagator. For example, in Møller scattering, the lowest-order diagram 9.7(a) is accompanied by the Feynman diagram 9.7(b) as one of the contributions to the second-order radiative correction. In the Feynman amplitudes, going from diagram 9.7(a) to 9.7(b) corresponds to the replacement

$$iD_{F\alpha\beta}(k) \rightarrow iD_{F\alpha\mu}(k)ie_0^2\Pi^{\mu\nu}(k)iD_{Fv\beta}(k), \tag{9.7}$$

where $ie_0^2\Pi^{\mu\nu}(k)$ is given by Eq. (9.5) which can be written

$$ie_0^2\Pi^{\mu\nu}(k) = \frac{-e_0^2}{(2\pi)^4}\int d^4p \frac{Tr[\gamma^\mu(\not{p}+\not{k}+m_0)\gamma^\nu(\not{p}+m_0)]}{\left[(p+k)^2 - m_0^2 + i\varepsilon\right]\left[p^2 - m_0^2 + i\varepsilon\right]}. \tag{9.8}$$

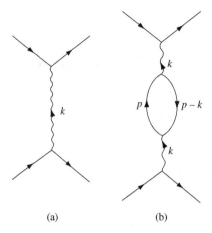

(a)   (b)

Figure 9.7 *Møller scattering: (b) represents the second-order photon self-energy correction to the lowest-order diagram (a)*

This integral is quadratically divergent for large p. In order to handle it, we must regularize it, that is, we must modify it so that it becomes a well-defined finite integral. For example, this could be achieved by multiplying the integrand in Eq. (9.8) by the convergence factor

$$f\left(p^2,\, \Lambda^2\right) = \left(\frac{-\Lambda^2}{p^2 - \Lambda^2}\right)^2. \tag{9.9}$$

Here, Λ is a cut-off parameter. For large, but finite, values of Λ, the integral now behaves like $\int \mathrm{d}^4 p / p^6$ for large p, and is well-defined and convergent. For $\Lambda \to \infty$, the factor $f\left(p^2,\, \Lambda^2\right)$ tends to unity, and the original theory is restored. One can think of such convergence factors either as a mathematical device, introduced to overcome a very unsatisfactory feature of QED, or as a genuine modification of QED at very high energies, i.e. at very small distances, which should show up in experiments at sufficiently high energies. We shall return to this point later.

The convergence factor (9.9) was introduced in order to illustrate the idea of regularization in a simple way. However, this factor does not provide a suitable regularization procedure for QED, as it does not ensure zero rest mass for the real physical photon, nor the related gauge invariance of the theory. It is most natural and desirable to employ a regularization procedure which ensures zero photon rest mass and gauge invariance for all values of the cut-off Λ and in each order of perturbation theory. Different regularization formalisms which satisfy both these requirements exist. In the limit in which the original theory is restored, the detailed form of the regularization procedure does not affect any physical results. In the next chapter, where we shall study regularization in detail, we shall see how this can be achieved. In the present chapter we shall assume the theory has been regularized in this way, so that all expressions are well-defined, finite and gauge invariant. Regularization will be implied; for example, we shall write $\Pi^{\mu\nu}(k)$ for the regularized form of the loop integral (9.8) rather than calling it $\Pi^{\mu\nu}(k, \Lambda)$.

Figure 9.8   *The modified photon propagator*

In order to interpret the effects of the radiative correction of Fig. 9.7(b), we shall consider it together with the lowest-order diagram, Fig. 9.7(a), from which it originates. Taking these diagrams together corresponds to the replacement shown in Fig. 9.8, i.e. the propagator modification

$$iD_{F\alpha\beta}(k) \rightarrow iD_{F\alpha\beta}(k) + iD_{F\alpha\mu}(k)ie_0^2\Pi^{\mu\nu}(k)iD_{Fv\beta}(k) \tag{9.10a}$$

or, more explicitly,

$$\frac{-ig_{\alpha\beta}}{k^2+i\varepsilon} \rightarrow \frac{-ig_{\alpha\beta}}{k^2+i\varepsilon} + \frac{-ig_{\alpha\mu}}{k^2+i\varepsilon} ie_0^2\Pi^{\mu\nu}(k)\frac{-ig_{v\beta}}{k^2+i\varepsilon}. \tag{9.10b}$$

We can simplify this expression. It follows from Lorentz invariance that $\Pi^{\mu\nu}(k)$ must be of the form

$$\Pi^{\mu\nu}(k) = -g^{\mu\nu}A(k^2) + k^\mu k^\nu B(k^2), \tag{9.11}$$

since this is the most general second-rank tensor which can be formed using only the four-vector k^μ. Furthermore, the photon propagator always occurs coupled to conserved currents, leading to expressions, analogous to Eq. (5.45), of the form

$$\int d^4k s_1^\alpha(-k)iD_{F\alpha\beta}(k)s_2^\beta(k).$$

If in this expression we make the replacement (9.10a) and substitute Eq. (9.11) for $\Pi^{\mu\nu}(k)$, it follows from current conservation, Eq. (5.47), that terms proportional to the photon momentum k give vanishing contributions. Hence, we can omit the term $k^\mu k^\nu B(k^2)$ from Eq. (9.11), on substituting this equation in Eqs. (9.10), with the result

$$\frac{-ig_{\alpha\beta}}{k^2+i\varepsilon} \rightarrow \frac{-ig_{\alpha\beta}}{k^2+i\varepsilon}\left[1 - e_0^2A(k^2)\frac{1}{k^2+i\varepsilon}\right]. \tag{9.12}$$

The right-hand expression in Eq. (9.12) represents the photon propagator, modified to include the second-order photon self-energy effects. We rewrite Eq. (9.12) in the form

$$\frac{-ig_{\alpha\beta}}{k^2+i\varepsilon} \rightarrow \frac{-ig_{\alpha\beta}}{k^2+i\varepsilon + e_0^2A(k^2)} + O(e_0^4). \tag{9.13}$$

The equivalence of Eqs. (9.12) and (9.13) follows, since after regularization, all quantities, and in particular $A(k^2)$, are finite. Hence in the spirit of perturbation theory, we can, for sufficiently small e_0^2, expand the right-hand side of Eq. (9.13) in powers of e_0^2, thereby regaining Eq. (9.12).

The left-hand side of Eq. (9.13) represents the photon propagator in lowest order of perturbation theory, i.e. the propagator of the bare, non-interacting photon. This propagator possesses a pole at $k^2 = 0$, corresponding to the bare photon having zero rest mass. [Quite generally, the propagator of a particle of rest mass m possesses a pole for four-momenta p such that $p^2 = m^2$. Eqs. (3.58) and (7.48) illustrate this for the meson and fermion propagators.] The right-hand side of Eq. (9.13) represents the photon propagator, including second-order self-energy corrections, i.e. it is the propagator of a real physical photon [albeit to $O(e_0^2)$ only]. If, as discussed previously, the real photon rest mass also vanishes, then the real photon propagator must also have its pole at $k^2 = 0$. From Eq. (9.13) this implies

$$A(0) \equiv A(k^2 = 0) = 0. \tag{9.14}$$

Hence $A(k^2)$ can be written

$$A(k^2) = k^2 A'(0) + k^2 \Pi_c(k^2) \tag{9.15}$$

where

$$A'(0) \equiv A'(k^2 = 0) = \frac{dA(k^2)}{d(k^2)}\bigg|_{k^2 = 0}$$

and $\Pi_c(k^2)$ vanishes linearly with k^2 at $k^2 = 0$. Substituting Eq. (9.15) into (9.12) and multiplying by e_0^2, we obtain

$$\frac{-ig_{\alpha\beta}}{k^2 + i\varepsilon} e_0^2 \rightarrow \frac{-ig_{\alpha\beta}}{k^2 + i\varepsilon} e_0^2 [1 - e_0^2 A'(0)] + \frac{ig_{\alpha\beta}}{k^2 + i\varepsilon} e_0^4 \Pi_c(k^2) \tag{9.16}$$

In formulating the Feynman rules, we associated a charge e_0 with each vertex. In writing Eq. (9.16), we have effectively incorporated into it the two factors e_0 which occur at the vertices at the ends of a photon propagator.

We must now interpret Eq. (9.16). The first term on the right-hand side is just the left-hand term multiplied by the constant $[1 - e_0^2 A'(0)]$. It is as though the magnitude of the two charges interacting through the photon propagator [for example, in the Møller scattering diagram 9.7(b)] is not e_0 but e, given by

$$e^2 = e_0^2 [1 - e_0^2 A'(0)]. \tag{9.17}$$

In this equation we have introduced the concept of *charge renormalization*. Eq. (9.17) defines the *renormalized electronic charge* $(-e)$, i.e. the charge of the physical, interacting electron, in contrast to the charge $(-e_0)$ of the bare non-interacting electron. We have considered the photon self-energy in second order only. There are of course higher-order corrections and we shall write Eq. (9.17) more completely as

$$e \equiv Z_3^{1/2} e_0 = e_0 \left[1 - \frac{1}{2} e_0^2 A'(0) + O(e_0^4) \right]. \tag{9.18}$$

The constant Z_3 relating the bare charge e_0 to the charge e of the real physical particle (to all orders in e_0) is called a *renormalization constant*. The right-hand side of Eq. (9.18) gives the explicit expression for this constant correct to the second order in e_0.

It is, of course, the charge $(-e)$ of the real physical electron which is observable, not the charge $(-e_0)$ which was introduced as the coupling constant of the free fields. Hence, all observable quantities, such as cross-sections, should be expressed in terms of the real charge e, not in terms of the bare charge e_0. From Eq. (9.18) we have

$$e_0 = e\left[1 + \mathrm{O}(e^2)\right]$$

so that Eq. (9.16) can be written

$$\frac{-\mathrm{i}g_{\alpha\beta}}{k^2 + \mathrm{i}\varepsilon}e_0^2 \rightarrow \frac{-\mathrm{i}g_{\alpha\beta}}{k^2 + \mathrm{i}\varepsilon}e^2 + \frac{\mathrm{i}g_{\alpha\beta}}{k^2 + \mathrm{i}\varepsilon}e^4\Pi_c\left(k^2\right) + \mathrm{O}(e^6). \tag{9.19}$$

Eq. (9.19) is our final result. It gives the photon propagator (times e^2), expressed in terms of the real charge e and accurate to terms in e^4. The first term on the right-hand side is the original photon propagator, but multiplied by the square of the renormalized charge e, instead of the bare charge e_0. The second term, of order α ($\equiv e^2/4\pi$) relative to the first term, will lead to an observable radiative correction of this order in any process involving the photon propagator $\mathrm{i}D_\mathrm{F}(k)$ in the lowest order of perturbation theory.

We have discussed renormalization of the regularized theory for a general value of the cut-off parameter Λ, i.e. before taking the appropriate limit [in our unrealistic example, Eq. (9.9), $\Lambda \rightarrow \infty$] which restores QED.[3] We must now consider what happens as we take this limit. In the next chapter, we shall consider regularization in detail. Here, we shall only state the results. Before proceeding to the limit, which restores the original theory, all quantities are well-defined and finite. As the limit is taken, the regularized integral $\Pi_c(k^2)$ tends to a well-defined finite limit which is independent of the detailed form of the regularization procedure. On the other hand, as the limit is taken, divergences reappear in the relation between e and e_0, Eq. (9.18) ($A'(0)$ and the renormalization constant Z_3 become infinite). But this is a relation between the observable charge e of a real physical particle and the bare charge e_0 of a non-interacting particle, which is a theoretical construct and completely unobservable. Thus Eq. (9.18) is itself not amenable to experimental tests, and it is only in such untestable equations that divergences appear. Looking back at our final result for the modified photon propagator, Eq. (9.19), we see that in the QED limit it only involves the measured elementary charge e and the well-defined and finite limit of the loop integral $\Pi_c(k^2)$. Thus, the modified propagator, and the Feynman amplitudes and the physical predictions to which it gives rise, are well-defined and finite, even in the QED limit of the regularized theory. An explicit expression for the loop integral $\Pi_c(k^2)$ will be derived in Section 10.4, and it will be employed later in this chapter in considering the radiative corrections to electron scattering by an external field.

---

[3] It should be stressed that even for well-behaved finite theories, renormalization is necessary in order to express theoretical predictions for cross-sections, etc., in terms of observable quantities like the charge e of real particles rather than in terms of unobservable quantities such as the bare charge e_0.

9.3 The Electron Self-Energy

We next consider the insertion of a fermion self-energy correction in a fermion propagator, shown in Fig. 9.3(a) and given by Eq. (9.4). After simplification by means of the contraction identities (A. 14), this equation can be written

$$
ie_0^2 \Sigma(p) = \frac{e_0^2}{(2\pi)^4} \int d^4k \, \frac{1}{k^2 + i\varepsilon} \frac{2\not{p} - 2\not{k} - 4m_0}{(p-k)^2 - m_0^2 + i\varepsilon}
\tag{9.20}
$$

This loop integral is ultraviolet divergent (i.e. in the limit $k \to \infty$). Its treatment is very similar to that of the photon self-energy in the last section, so that we can be brief, except where differences occur.

One of the differences is that the regularization and renormalization of Eq. (9.20) leads to integrals which are not only ultraviolet divergent, but also infrared divergent, i.e. they also diverge in the limit of zero photon momentum k. We have met infrared divergences in our analysis of elastic and inelastic electron scattering (Section 8.9) and shall discuss their significance further in Section 9.7. Here we conveniently remove infrared and ultraviolet divergences from Eq. (9.20) by the replacement

$$
\frac{1}{k^2 + i\varepsilon} \to \frac{1}{k^2 - \lambda^2 + i\varepsilon} - \frac{1}{k^2 - \Lambda^2 + i\varepsilon}
\tag{9.21}
$$

where λ is a small infrared cut-off parameter, which ultimately is set equal to zero. In effect, this corresponds to temporarily assigning a small non-zero rest mass λ to the photon. Λ is again an ultraviolet cut-off parameter, and taking the limit $\Lambda \to \infty$ restores QED. In the following we shall assume that the loop integral (9.20) has been regularized through the replacement (9.21).

Proceeding as in the last section, we shall consider the fermion self-energy loop together with the propagator into which it is inserted, as shown in Fig. 9.9 and given by

$$
\frac{i}{\not{p} - m_0 + i\varepsilon} \to \frac{i}{\not{p} - m_0 + i\varepsilon} + \frac{i}{\not{p} - m_0 + i\varepsilon} ie_0^2 \Sigma(p) \frac{i}{\not{p} - m_0 + i\varepsilon}.
\tag{9.22}
$$

We rewrite this expression using the identity

$$
\frac{1}{A - B} = \frac{1}{A} + \frac{1}{A} B \frac{1}{A} + \frac{1}{A} B \frac{1}{A} B \frac{1}{A} + \dots
\tag{9.23}
$$

which holds for any two operators A and B, not necessarily commuting. (The identity is easily verified by post-multiplying it by $A - B$.) By means of this identity, Eq. (9.22) can be written, correct to terms of order e_0^2,

$$
\frac{i}{\not{p} - m_0 + i\varepsilon} \to \frac{i}{\not{p} - m_0 + e_0^2 \Sigma(p) + i\varepsilon} + O(e_0^4).
\tag{9.24}
$$

The left-hand side of this equation is the propagator of the non-interacting fermion. As expected, it has a pole at $\not{p} = m_0$, corresponding to the bare rest mass m_0 of the non-interacting particle.[4] The expression on the right-hand side of Eq. (9.24) represents the propagator of the interacting physical fermion. Hence, we require its pole to be at $\not{p} = m$, where

$$m = m_0 + \delta m \tag{9.25}$$

is the real fermion rest mass; due to the interaction of the fermion field and the electro-magnetic field, the rest mass m of the real fermion differs from the rest mass m_0 of the non-interacting bare fermion. In our perturbation treatment, δm is a power series in e_0^2, and in the lowest order – to which we are working – $\delta m = \mathrm{O}(e_0^2)$. m is called the renormalized mass and the replacement of m_0 by m using Eq. (9.25) is known as *mass renormalization*. It is similar to and as essential as charge renormalization: the experimentally determined mass of the electron is m, not m_0, and predictions from theory must be expressed in terms of the observable properties of the real interacting particles.

To identify the pole of the modified propagator (9.24) at $\not{p} = m$, we rewrite its denomi-nator. It follows from Lorentz invariance that $\Sigma(p)$ can depend on the momentum vector p through $\not{p}$ and $p^2 (= \not{p}\not{p})$ only. It will be convenient to expand $\Sigma(p)$ in powers of $(\not{p} - m)$ in the form

$$\Sigma(p) = A + (\not{p} - m)B + (\not{p} - m)\Sigma_{\mathrm{c}}(p), \tag{9.26}$$

where A and B are constants (i.e. independent of p) and $\Sigma_{\mathrm{c}}(p)$ vanishes linearly with $(\not{p} - m)$ at $\not{p} = m$. In particular,

$$A = \Sigma(p)|_{\not{p}=m}. \tag{9.27}$$

Substituting Eqs. (9.25) and (9.26) in the modified propagator on the right-hand side of Eq. (9.24), we see that this propagator has a pole at $\not{p} = m$, provided

$$\delta m = -e_0^2 A. \tag{9.28}$$

Eq. (9.24) then reduces to

$$\frac{\mathrm{i}}{\not{p} - m_0 + \mathrm{i}\varepsilon} \rightarrow \frac{\mathrm{i}}{(\not{p} - m)(1 + e_0^2 B) + e_0^2(\not{p} - m)\Sigma_{\mathrm{c}}(p) + \mathrm{i}\varepsilon} + \mathrm{O}(e_0^4), \tag{9.29}$$

Figure 9.9   *The modified fermion propagator*

---

[4] This is merely a concise way of saying that

$$\frac{\mathrm{i}}{\not{p} - m_0 + \mathrm{i}\varepsilon} \equiv \frac{\mathrm{i}(\not{p} + m_0)}{p^2 - m_0^2 + \mathrm{i}\varepsilon}$$

has a pole at $p^2 = m_0^2$.

or, retaining only terms to $O(e_0^2)$,

$$\frac{i}{\not{p} - m_0 + i\varepsilon} \rightarrow \frac{i}{\not{p} - m + i\varepsilon} \left[(1 - e_0^2 B) - e_0^2 \Sigma_c(p) \right] + O(e_0^4). \tag{9.30}$$

An alternative way of carrying out the mass renormalization is to express the Hamiltonian of QED in terms of the real electron mass m, instead of the bare mass m_0,

$$\mathscr{H} = \mathscr{H}_0 + \mathscr{H}_1 \tag{9.31a}$$

where

$$\mathscr{H}_0 = -\dot{A}_v(x) \dot{A}^v(x) + \frac{1}{2} \left(\partial_\mu A_v(x) \right) \left(\partial^\mu A^v(x) \right) + i\psi^\dagger(x) \dot{\psi}(x) - \overline{\psi}(x)(i\not{\partial} - m)\psi(x), \tag{9.31b}$$

$$\mathscr{H}_1 = -e_0 \overline{\psi}(x) \not{A}(x) \psi(x) - \delta m \overline{\psi}(x)\psi(x). \tag{9.31c}$$

As implied by our notation, we shall treat $\mathscr{H}_0$ as the free-field Hamiltonian density and $\mathscr{H}_1$ as the interaction Hamiltonian density. The non-interacting fermion now has the physical mass m and satisfies the Dirac equation

$$(i\not{\partial} - m)\psi(x) = 0,$$

etc., and the fermion propagator becomes, in lowest order,

$$\frac{i}{\not{p} - m + i\varepsilon}.$$

The use of the real electron mass m for the non-interacting fermion is compensated for through the *mass counterterm* $-\delta m \overline{\psi}\psi$, appearing as an additional interaction term in Eq. (9.31c). Graphically this term is represented by Fig. 9.10 corresponding to a two-line vertex at which an electron (or positron) is destroyed and recreated.

If we use this division, Eqs. (9.31a)–(9.31c), of the Hamiltonian density in the S-matrix expansion (6.23), the Feynman rules which we obtained in Sections 7.3 and 8.7 must be modified in two respects.

Firstly, the bare mass m_0 is replaced by the real fermion mass m throughout; in particular, in the fermion propagator. [Eq. (7.48), as it stands, already states the modified rule 3, but we must now interpret m as the real mass; in Chapter 7 it denoted the bare mass.]

Secondly, using Eq. (9.31c) as the interaction, leads to extra contributions to the S-matrix expansion, represented by Feynman graphs containing the two-line vertex part, Fig. 9.10. We see from the S-matrix expansion, Eq. (6.23), that with Eq. (9.31c) as interaction, each two-line vertex in a graph leads to a factor $i\delta m$ in the Feynman amplitude, just as each basic vertex part leads to a factor $ie\gamma^\alpha$ (rule 1 of Section 7.3). For each two-line vertex, we must therefore write a factor

$$i\delta m \equiv -ie_0^2 A = -ie_0^2 \Sigma(p)|_{\not{p} = m}. \tag{9.32}$$

We see that as a consequence of the replacement

$$m_0 \rightarrow m \tag{9.33a}$$

Figure 9.10 *The two-line vertex graph representing the mass counterterm* $-\delta m \bar{\psi} \psi$

for the free fermion, each Feynman graph containing a fermion self-energy loop must be considered together with an identical graph, in which the self-energy loop has been replaced by a two-line vertex, Fig. 9.10. (These two graphs are of the same order in e_0.) The net effect on the Feynman amplitude of taking into account both graphs is the replacement

$$ie_0^2\sum(p) \rightarrow ie_0^2\sum(p) + i\delta m = ie_0^2(\not{p} - m)B + ie_0^2(\not{p} - m)\sum\nolimits_c(p), \qquad (9.33b)$$

where we used Eqs. (9.26) and (9.32). We see that, quite generally, the mass counterterm cancels the constant term A arising from the loop term $ie_0^2\sum(p)$.

Returning to Eq. (9.24), we see that if in this equation we make the replacements (9.33a) and (9.33b), we obtain

$$\frac{i}{\not{p} - m + i\varepsilon} \rightarrow \frac{i}{(\not{p} - m)(1 + e_0^2B) + e_0^2(\not{p} - m)\sum\nolimits_c(p) + i\varepsilon} + O(e_0^4). \qquad (9.34)$$

This equation agrees with our earlier result (9.29), as it must, since the two derivations differ only in the way the same total Hamiltonian is split into free-field and interaction parts.

The modified fermion propagator has so far been expressed in terms of the bare charge e_0. We now define a renormalized charge e by the relation

$$e^2 \equiv Z_2 e_0^2 = e_0^2(1 - e_0^2B) + O(e_0^6). \qquad (9.35)$$

The general interpretation of this charge renormalization is the same as we met in the last section, but it has, of course, a different origin; it is due to the fermion self-energy, not the photon self-energy. Multiplying Eq. (9.30) by e_0^2, in order to incorporate the charges e_0 which are associated with the two vertices at the ends of the propagator, and expressing e_0 in terms of e, we can write Eq. (9.30) as

$$\frac{ie_0^2}{\not{p} - m_0 + i\varepsilon} \rightarrow \frac{ie^2}{\not{p} - m + i\varepsilon}[1 - e^2\sum\nolimits_c(p)] + O(e^6). \qquad (9.36)$$

The right-hand side of Eq. (9.36) gives the renormalized fermion propagator (times e^2), correct to terms in e^4. The first term, $ie^2(\not{p} - m + i\varepsilon)^{-1}$, is simply the zeroth approximation, i.e. the bare fermion expression (in lowest order $e = e_0$ and $m = m_0$). The term containing $\sum_c$ is a radiative correction of order α to the zeroth approximation.

Finally, we must take the limit $\Lambda \rightarrow \infty$ in order to regain QED. From Eq. (9.20), the loop integral $\sum(p)$ appears to be linearly divergent in this limit. As is shown explicitly in Section 10.2, it is actually only logarithmically divergent, with A given by

$$A = -\frac{3m}{8\pi^2}\ln\frac{\Lambda}{m}. \qquad (9.37)$$

One finds that the constants B and Z_2 are also logarithmically divergent in the limit $\Lambda \to \infty$, while the correction term $\Sigma_c(p)$ remains well-defined and finite in this limit and independent of the details of the regularization procedure. It is from this term that measurable radiative corrections, of order α, to the lowest-order predictions are derived. In contrast, the divergent constants A, B and Z_2 occur only in untestable relations connecting physical and bare quantities.

9.4   External Line Renormalization

In the last two sections we considered the second-order self-energy insertions in photon and fermion propagators. When considering radiative corrections, these insertions must, of course, also be made in external lines. Their only effect now is a charge renormalization, but they do not lead to any finite radiative corrections. We shall derive these results for the case of an initially present electron.

Proceeding as in the last section, we consider the incident electron line together with its self-energy insertions, i.e. the replacement shown in Fig. 9.11. The corresponding replacement in the Feynman amplitudes is, from Eqs. (9.32) and (9.33), given by

$$u(\mathbf{p}) \to u(\mathbf{p}) + \frac{i}{\not{p} - m + i\varepsilon} \left[ie_0^2(\not{p} - m)B + ie_0^2(\not{p} - m)\Sigma_c(p) \right] u(\mathbf{p}). \tag{9.38}$$

Since $(\not{p} - m)u(\mathbf{p}) = 0$ and $(\not{p} - m)\Sigma_c(p)$ vanishes quadratically with $\not{p} - m$ as $\not{p} \to m$ [see Eq. (9.26)], we can drop the last term in Eq. (9.38) and write it

$$u(\mathbf{p}) \to \left[1 - \frac{e_0^2}{\not{p} - m + i\varepsilon}(\not{p} - m)B \right] u(\mathbf{p}). \tag{9.39}$$

Unfortunately, the term proportional to B in this equation is indeterminate, as it stands. This ambiguity is resolved by explicitly using the adiabatic hypothesis, discussed in Section 6.2, to describe how the self-energy effects convert the incident electron from a bare particle to a physical particle. In effect, the interaction is switched off as $t \to \pm\infty$ by multiplying the charge e_0 by a suitable factor $f(t)$, so that the interaction (9.31c) is replaced by

$$\mathscr{H}_I = -e_0 f(t)\overline{\psi}(x)\slashed{A}(x)\psi(x) - \delta m[f(t)]^2\overline{\psi}(x)\psi(x). \tag{9.40}$$

The precise form of the function $f(t)$ is not important. We require that $f(t) \to 0$ as $t \to \pm\infty$ and that $f(t)$ does not differ significantly from unity during a time interval T which is long compared to the duration of the scattering process considered. In terms of the Fourier transform

$$f(t) = \int_{-\infty}^{\infty} F(E)e^{iEt}dE = \int_{-\infty}^{\infty} F(E)e^{iqx}dE, \tag{9.41}$$

Figure 9.11   *The second-order modification of an external initial electron line*

where $q \equiv (E, \mathbf{0})$, we require the normalization

$$f(0) = \int_{-\infty}^{\infty} F(E)\, dE = 1 \tag{9.42}$$

and that $F(E)$ is almost a δ-function with a peak of width $1/T$ situated at $E = 0$. As $F(E) \to \delta(E)$, the original theory with $f(t) = 1$ is restored.

The modified interaction (9.40) behaves in many ways like an external field interaction. But whereas the static external field $A_e^\alpha(\mathbf{x})$, Eq. (8.85), conserves energy, but not three-momentum, the interaction (9.40) conserves three-momentum, but not energy. The effect of using the energy-non-conserving interaction (9.40) is to replace the original fermion self-energy insertion, Fig. 9.12(a), by the modified insertion shown in Fig. 9.12(b), where the vectors $q = (E, \mathbf{0})$ and $q' = (E', \mathbf{0})$ in the fermion propagators describe the non-conservation of energy at the vertices. Correspondingly, Eq. (9.39) is replaced by

$$u(\mathbf{p}) \rightarrow \left[1 - \int dE dE'\, F(E) F(E') \frac{e_0^2 B}{\not{p} - \not{q} - \not{q'} - m + i\varepsilon} (\not{p} - \not{q} - m) \right] u(\mathbf{p}). \tag{9.43}$$

The evaluation of this integral becomes trivial if in the numerator we make the replacement

$$(\not{p} - \not{q} - m) \rightarrow (\not{p} - \not{q} - m) - \frac{1}{2}(\not{p} - m) = \frac{1}{2}(\not{p} - 2\not{q} - m),$$

justified since $(\not{p} - m)u(p) = 0$. In the resulting integral we can make the further replacement

$$\frac{1}{2}(\not{p} - 2\not{q} - m) \rightarrow \frac{1}{2}(\not{p} - \not{q} - \not{q'} - m),$$

since apart from this factor the integral is symmetric in q and q'. With these replacements and the normalization condition (9.42) for $F(E)$, Eq. (9.43) reduces to

$$u(\mathbf{p}) \rightarrow \left(1 - \frac{1}{2} e_0^2 B \right) u(\mathbf{p}). \tag{9.44}$$

The last expression is independent of the adiabatic switch-off function $F(E)$. Hence, it already expresses our result in the limit $F(E) \to \delta(E)$, which restores our original theory with $f(t) = 1$.

Figure 9.12 *The fermion self-energy loop: (a) for the QED interaction (9.31c); (b) for the modified interaction (9.40); $q = (E, 0)$ and $q' = (E', 0)$ represent the energy non-conservation at the vertices*

Eq. (9.44) is a second-order result. We can rewrite it in the form

$$u(\mathbf{p}) \rightarrow Z_2^{1/2} u(\mathbf{p}), \qquad (9.45a)$$

where the renormalization constant Z_2, defined in Eq. (9.35), relates the bare charge e_0 to the physical charge e. Although we have only derived Eq. (9.45a) in second-order perturbation theory, it holds to all orders.

Similar arguments applied to self-energy insertions in other external fermion lines lead to the analogous results

$$\bar{u}(\mathbf{p}) \rightarrow Z_2^{1/2} \bar{u}(\mathbf{p}), \quad v(\mathbf{p}) \rightarrow Z_2^{1/2} v(\mathbf{p}), \quad \bar{v}(\mathbf{p}) \rightarrow Z_2^{1/2} \bar{v}(\mathbf{p}). \qquad (9.45b)$$

Similarly, the photon self-energy insertion of Fig. 9.3(b) in an external photon line leads to

$$\varepsilon^{\mu}(\mathbf{k}) \rightarrow Z_3^{1/2} \varepsilon^{\mu}(\mathbf{k}), \qquad (9.45c)$$

where the charge renormalization constant Z_3 is defined by Eq. (9.18).

The modifications (9.45) of the wave functions of the external particles due to self-energy effects are referred to as external line renormalization or wave function renormalization. When considering the modified photon and fermion propagators in the last two sections, we interpreted the parameters Z_3 and Z_2 as renormalization constants of the charges acting at the vertices at the ends of the photon and fermion propagators respectively [see Eqs. (9.18) and (9.35)]:

$$e_0 \rightarrow e = Z_3^{1/2} e_0, \quad e_0 \rightarrow e = Z_2^{1/2} e_0. \qquad (9.46)$$

We can equally interpret the wavefunction renormalizations (9.45) as charge renormalizations. For this purpose, we associate the factors $Z_3^{1/2}$ and $Z_2^{1/2}$ in Eqs. (9.45) with the charges acting at the vertices to which the external lines are attached. Eqs. (9.46) are now valid generally for each internal or external line attached to a vertex. For external lines, these charge renormalizations are the *only* self-energy effects (apart from the electron mass renormalization, $m_0 \rightarrow m$, allowed for automatically through the mass counterterm). This is in contrast to self-energy modifications of photon and fermion propagators [see Eqs. (9.19) and (9.36)], which lead to additional finite radiative corrections.

9.5   The Vertex Modification

We finally consider the second-order vertex modification shown in Fig. 9.13. This corresponds to the replacement

$$ie_0\gamma^{\mu} \rightarrow i\Gamma^{\mu}(p', p) = ie_0[\gamma^{\mu} + e_0^2 \Lambda^{\mu}(p', p)], \qquad (9.47)$$

where $\Lambda^{\mu}(p', p)$ is from Eq. (9.6) given by

$$\Lambda^{\mu}(p', p) = \frac{-i}{(2\pi)^4} \int \frac{d^4k}{k^2 + i\varepsilon} \gamma^{\alpha} \frac{1}{\not{p}' - \not{k} - m + i\varepsilon} \gamma^{\mu} \frac{1}{\not{p} - \not{k} - m + i\varepsilon} \gamma_{\alpha}. \qquad (9.48)$$

Figure 9.13 *The second-order vertex modification*

(m now of course denotes the real fermion mass.) $\Lambda^\mu(p', p)$ is both ultraviolet and infrared divergent. We regularize it by the same replacement (9.21) of the photon propagator which we used for the fermion self-energy loop.

When considering the fermion self-energy, we separated off the free-particle part $\Sigma(p)|_{\not p = m}$, Eq. (9.27). The troublesome part of $\Lambda^\mu(p', p)$, which diverges logarithmically as the cut-off parameter Λ tends to infinity and which we want to separate off, is again given by the free-particle value

$$\bar{u}(\mathbf{P})\Lambda^\mu(P, P)u(\mathbf{P}). \tag{9.49}$$

Here $u(\mathbf{P})$ is a free-particle spinor and P a free-particle momentum vector, i.e. $P^2 = m^2$. From Lorentz invariance, expression (9.49) must be of the form

$$a\bar{u}(\mathbf{P})\gamma^\mu u(\mathbf{P}) + bP^\mu\,\bar{u}(\mathbf{P})u(\mathbf{P}), \tag{9.50}$$

where a and b are scalar constants. But from Gordon's identity (see Problem A.2)

$$P^\mu\bar{u}(\mathbf{P})u(\mathbf{P}) = m\bar{u}(\mathbf{P})\gamma^\mu u(\mathbf{P}), \tag{9.51}$$

and combining the last three equations we can write

$$\bar{u}(\mathbf{P})\Lambda^\mu(P, P)u(\mathbf{P}) = L\bar{u}(\mathbf{P})\gamma^\mu u(\mathbf{P}), \tag{9.52}$$

where L is a scalar constant.

For general four-vectors p and p', we define $\Lambda_c^\mu(p', p)$ by

$$\Lambda^\mu(p', p) = L\gamma^\mu + \Lambda_c^\mu(p', p). \tag{9.53}$$

We see from Eq. (9.52) that for a free-particle four-momentum P

$$\bar{u}(\mathbf{P})\Lambda_c^\mu(P, P)u(\mathbf{P}) = 0. \tag{9.54}$$

The motivation for writing $\Lambda^\mu(p', p)$ in the form (9.53) is that in the limit $\Lambda \to \infty$, in which QED is restored, L diverges, but the second term $\Lambda_c^\mu(p', p)$ remains well defined and finite. This can be seen from the expression (9.48) as it stands. With the abbreviations

$$\Delta \equiv \not p - \not k - m + i\varepsilon, \quad q \equiv p - P, \quad q' \equiv p' - P,$$

we can use the identity (9.23) to expand the fermion propagators in Eq. (9.48) in powers of $\not{q}$ and $\not{q}'$:

$$\frac{1}{\not{p}' - \not{k} - m + i\varepsilon} \gamma^\mu \frac{1}{\not{p} - \not{k} - m + i\varepsilon} \equiv \frac{1}{\not{q}' + \Delta} \gamma^\mu \frac{1}{\not{q} + \Delta}$$
$$= \left(\frac{1}{\Delta} - \frac{1}{\Delta} \not{q}' \frac{1}{\Delta} + \cdots \right) \gamma^\mu \left(\frac{1}{\Delta} - \frac{1}{\Delta} \not{q} \frac{1}{\Delta} + \cdots \right). \tag{9.55}$$

Substituting this expansion in Eq. (9.48), we see that the leading term in $\Lambda^\mu(p', p)$ arises from $(1/\Delta)\gamma^\mu(1/\Delta)$ in Eq. (9.55) and is simply $\Lambda^\mu(P, P)$. This term may (and indeed does) diverge logarithmically as $k \rightarrow \infty$, since Δ is linear in k. All other terms necessarily converge as $k \rightarrow \infty$, since they contain additional factors Δ in the denominator.

Substituting Eq. (9.53) in Eq. (9.47), we obtain

$$ie_0\gamma^\mu \rightarrow i\Gamma^\mu(p', p) = ie_0\left[\gamma^\mu(1 + e_0^2 L) + e_0^2\Lambda_c^\mu(p', p)\right]. \tag{9.56}$$

The term proportional to γ^μ on the right-hand side of this equation is the original basic vertex part $ie_0\gamma^\mu$, but with a renormalized charge. We define a charge renormalization constant Z_1 by

$$e \equiv \frac{e_0}{Z_1} = e_0(1 + e_0^2 L) + O(e_0^5), \tag{9.57}$$

where $O(e_0^5)$ indicates that there are also higher-order contributions to the charge renormalization which results from all vertex modifications, whereas we have only considered the lowest, second-order correction. Expressing the right-hand side of Eq. (9.56) in terms of the renormalized charge e, we can write this equation

$$ie_0\gamma^\mu \rightarrow i\Gamma^\mu(p', p) = ie\left[\gamma^\mu + e^2\Lambda_c^\mu(p', p)\right] + O(e^5), \tag{9.58}$$

where $O(e^5)$ again indicates higher-order corrections.

Eq. (9.58) is our final result for the second-order vertex modification. It consists of the charge renormalization (9.57) and the correction term $\Lambda_c^\mu(p', p)$. In the regularized form, all quantities are well-defined and finite. In the limit $\Lambda \rightarrow \infty$, in which QED is restored, L (and Z_1) become infinite, but this only affects the unobservable relation (9.57). On the other hand, $\Lambda_c^\mu(p', p)$ tends to a well-defined finite limit, which is independent of the regularization procedure and which contributes to the lowest-order radiative correction of a process.

We now combine the charge renormalization (9.57) resulting from the vertex modification with the charge renormalizations (9.46) resulting from photon and fermion self-energy effects. Since each vertex has one photon line and two fermion lines attached, it follows from these equations that the net effect is the replacement of the bare charge e_0 at each vertex (i.e. everywhere in the theory) by the renormalized charge

$$e = e_0 Z_3^{1/2} Z_2 / Z_1. \tag{9.59}$$

This and all earlier results in this chapter have been derived in second-order perturbation theory only, but they can be shown to hold in all orders.

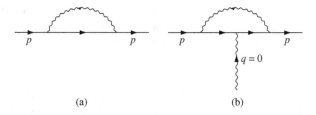

Figure 9.14 (a) *The second-order self-energy loop.* (b) *The vertex modification obtained by inserting a zero-energy photon in the fermion propagator of* (a)

The result (9.59) allows a remarkable simplification. This is due to the fact that the fermion self-energy insertion $\Sigma(p)$, Eq. (9.4), and the vertex insertion $\Lambda^\mu(p', p)$, Eq. (9.6), are related by Ward's identity which is given by

$$\frac{\partial \sum(p)}{\partial p_\mu} = \Lambda^\mu(p,p) \tag{9.60}$$

(see Problem 9.2). Ward's identity relates the fermion self-energy graph, Fig. 9.14(a), to the vertex modification obtained from it by insertion of a zero-energy photon in the intermediate fermion propagator, as shown in Fig. 9.14(b). Eq. (9.60) is a second-order result. However, Ward's identity can be generalized and holds in all orders of perturbation theory, allowing one to obtain higher-order vertex modifications by differentiation of higher-order fermion self-energy insertions. This greatly simplifies the calculation of higher-order radiative corrections.

Ward's identity also implies a relation between the charge renormalization constants Z_1 and Z_2, which we shall now derive. With $u(\mathbf{P})$ a free-particle spinor, we obtain from Eq. (9.60)

$$\bar{u}(\mathbf{P})\frac{\partial \sum(P)}{\partial P_\mu}\, u(\mathbf{P}) = \bar{u}(\mathbf{P})\Lambda^\mu(P,\, P)u(\mathbf{P}). \tag{9.61}$$

Using Eq. (9.26), we obtain for the left-hand side of Eq. (9.61)

$$B\bar{u}(\mathbf{P})\gamma^\mu u(\mathbf{P}),$$

since

$$\bar{u}(\mathbf{P})(\not{P} - m) = \sum\nolimits_c (P)u(\mathbf{P}) = 0.$$

From Eqs. (9.53) and (9.54), the right-hand side of Eq. (9.61) equals

$$L\bar{u}(\mathbf{P})\gamma^\mu u(\mathbf{P}),$$

and hence

$$B = L. \tag{9.62}$$

From Eqs. (9.35) and (9.57), this relation can be expressed in terms of the charge-renormalization constants Z_1 and Z_2 as

$$Z_2 = Z_1. \tag{9.63}$$

Although we have only derived Eq. (9.63) in second-order perturbation theory, it is an exact relation holding in all orders of perturbation theory.

As a consequence of the equality (9.63), Eq. (9.59) reduces to

$$e = e_0 Z_3^{1/2}. \tag{9.64}$$

Thus, the charge renormalization does not depend on fermion self-energy effects or vertex modifications, but originates solely from photon self-energy effects, i.e. from vacuum polarization. This has an interesting consequence when considering, not only electrons and positrons, but also other kinds of leptons, $\mu^{\pm}, \tau^{\pm}$. It is easy to see that Eqs. (9.63) and (9.64) continue to hold for each type of lepton, with the same constant Z_3 in each case.[5] Consequently the observed equality of the physical charges of particles implies the equality of their bare charges.

We have now completed our analysis of the second-order modifications of photon and fermion lines and of vertices, and we can summarize our results as follows. If we ascribe the physical masses $(m_e, m_\mu, \ldots)$ to the leptons and throughout replace the bare charge e_0 by the physical charge $e = e_0 Z_3^{1/2}$, then the only modifications due to second-order self-energy and vertex corrections in QED are the propagator modifications

$$\frac{-ig^{\alpha\beta}}{k^2 + i\varepsilon} \rightarrow \frac{-ig^{\alpha\beta}}{k^2 + i\varepsilon} \left[1 - e^2 \Pi_c(k^2) \right] + O(e^4) \tag{9.65a}$$

$$\frac{i}{\not{p} - m + i\varepsilon} \rightarrow \frac{i}{\not{p} - m + i\varepsilon} \left[1 - e^2 \Sigma_c(p) \right] + O(e^4) \tag{9.65b}$$

and the vertex modification

$$ie_0\gamma^\mu \rightarrow ie\left[\gamma^\mu + e^2 \Lambda_c^\mu(p', p) \right] + O(e^5). \tag{9.65c}$$

As stated earlier, in the limit $\Lambda \rightarrow \infty$ in which QED is restored, the regularized functions Π_c, Σ_c and Λ_c^μ tend to well-defined finite limits so that the modifications (9.65) lead to radiative corrections of order α. Instead of interpreting regularization as a mathematical device for coping with the divergences of the theory, we can keep the cut-off parameter Λ finite and interpret the regularized theory as a genuine modification of QED. We must then ask what limits are set on the value of Λ and the validity of QED by experiment. The most restrictive limits are obtained from the lepton pair processes $e^+ e^- \rightarrow l^+ l^-$. As discussed in Section 8.4, these processes probe the behavior of the photon propagator at $k^2 = (2E)^2$, where E is the electron energy in the centre-of-mass system. If the modified propagator (9.21), with Λ kept finite, is used, then agreement with the experimental data for electron energies E of order 15 GeV is only obtained if $\Lambda \gtrsim 150$ GeV, corresponding to distances of order $\Lambda^{-1} \lesssim 2 \times 10^{-3}$ f. We conclude that, at presently accessible energies, the observable predictions of QED are insensitive to modifications of the theory at distances much shorter than 10^{-3} f.

In contrast to the regularized functions Π_c, Σ_c and Λ_c^μ, one finds that δm diverges like $\ln \Lambda$ as $\Lambda \rightarrow \infty$, and indeed all divergent quantities of QED diverge logarithmically. In our

---

[5] In addition to the vacuum polarization loops formed from electron–positron pairs, the photon self-energy will now contain contributions from loops formed from $\mu^+ - \mu^-$ pairs, etc.

perturbation treatment, the leading divergent terms are of order α, so that appreciable differences between bare and physical quantities occur only for values of Λ which are enormously large. For example, we see from Eqs. (9.28) and (9.37) that for a significant electron mass correction δm to occur, i.e.$\delta m = \mathrm{O}(m)$, we require

$$\Lambda = \mathrm{O}\left(m\,e^{2\pi/3\alpha}\right) \approx 10^{121}\ \mathrm{GeV}.$$

On the other hand, for $\Lambda \ll 10^{121}$ GeV, one obtains $\delta m \ll m$, so that it seems reasonable to treat the mass correction δm in perturbation theory. Thus the reader who is rightly worried about treating large or infinite quantities in perturbation theory, even in unphysical relations, should think of Λ as finite, but much less than 10^{121} GeV. Furthermore, the physical predictions of the theory will not be measurably different from those obtained in the limit $\Lambda \rightarrow \infty$, provided Λ is much larger than the energy scale of the experiment considered.

9.6   Applications

We have so far shown how to calculate well-defined finite radiative corrections of order α. Applications of these results lead to some of the most spectacular successes of modern physics. In particular, for the anomalous magnetic moment of the electron and the muon, and for the energy levels of the hydrogen atom (the Lamb shift), comparison of theory with experiments leads to extraordinarily precise agreement. In this section we shall give these comparisons for both problems. For the magnetic moment we shall derive the radiative correction of order α to the Dirac value. The Lamb shift calculation is a bound-state problem and its proper treatment requires a fairly elaborate extension of the theory we have developed. We shall limit ourselves to giving a much simpler approximate non-relativistic derivation, due to Bethe, which calculates correctly the dominant contribution to the Lamb shift.

9.6.1   The anomalous magnetic moments

The magnetic moment of a particle shows up through the scattering of the particle by a magnetic field. For this reason we shall once more study the elastic scattering of an electron by a static external field. We considered this process in lowest order in Section 8.7 and found that it is represented by the Feynman graph in Fig. 9.15(a) and that its Feynman amplitude is given by Eq. (8.88):

$$ie\bar{u}(\mathbf{p}')A_e(\mathbf{q} = \mathbf{p}' - \mathbf{p})u(\mathbf{p}). \tag{9.66}$$

The radiative corrections of order α to this process stem from the Feynman graphs in Fig. 9.15(b)–(g). After renormalization, only diagrams (b) and (c) give contributions, with the Feynman amplitude to order e^3 given, from Eqs. (9.65), by

$$ie\bar{u}(\mathbf{p}')\gamma^\mu u(\mathbf{p})A_{e\mu}(\mathbf{q}) + ie\bar{u}(\mathbf{p}')\gamma^\mu u(\mathbf{p})[-e^2\Pi_c(q^2)]A_{e\mu}(\mathbf{q})$$

$$+ ie\bar{u}(\mathbf{p}')\left[e^2\Lambda_c^\mu(p', p)\right]u(\mathbf{p})A_{e\mu}(\mathbf{q}). \tag{9.67}$$

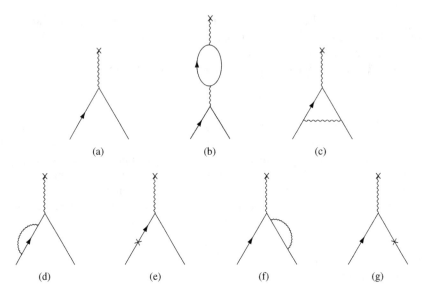

Figure 9.15   *Electron scattering by an external field: (a) the lowest-order graph; (b)–(g) the graphs of order e³. After re-normalization, only (b) and (c) contribute finite radiative corrections. (For greater clarity, in future we often mark only one arrow on each fermion line.)*

All other effects are absorbed into the mass and charge renormalizations. In particular, as we saw in Section 9.4, there are no observable radiative corrections associated with the external line diagrams (d)–(g). In connection with the charge renormalization resulting from the vacuum polarization graph (b), one point should be noted. When considering the vacuum polarization loop in Section 9.2, we associated a charge renormalization $e_0 \rightarrow e = e_0 Z_3^{1/2}$ with the charge acting at each end of the photon propagator. In the present context, i.e. for diagram (b), one of these factors $Z_3^{1/2}$ is absorbed into renormalizing the charges which act as source of the external field $A_{e\mu}(\mathbf{q})$.

To use Eq. (9.67) we need explicit expressions for $\Pi_c(q^2)$ and $\Lambda_c^\mu(p', p)$. In Section 10.4 we shall derive the result

$$e^2 \Pi_c(q^2) = -\frac{2\alpha}{\pi} \int_0^1 dz \, z(1-z) \ln\left[1 - \frac{q^2 z(1-z)}{m^2}\right]. \qquad (9.68)$$

For $q^2 \ll m^2$, the logarithm may be expanded to give

$$e^2 \Pi_c(q^2) = \frac{\alpha}{15\pi} \left(\frac{q^2}{m^2}\right) + \cdots \qquad (q^2 \ll m^2). \qquad (9.69)$$

In the next chapter, Section 10.5, we shall also show how to evaluate $\Lambda_c^\mu(p', p)$. In particular, we shall see that the third term in Eq. (9.67) contains the term

$$i e \bar{u}(\mathbf{p}') \left[\frac{i\alpha}{4\pi m} \sigma^{\mu\nu} q_\nu\right] u(\mathbf{p}) A_{e\mu}(\mathbf{q}). \qquad (9.70)$$

It is this contribution to the Feynman amplitude (9.67) that we wish to interpret. To do so, we use the Gordon identity (see Problem A.2) to rewrite the lowest-order scattering amplitude (9.66) as

$$\frac{ie}{2m}\bar{u}(\mathbf{p}')\left[(p'+p)^{\mu}+i\sigma^{\mu\nu}q_{\nu}\right]u(\mathbf{p})A_{e\mu}(\mathbf{q}). \tag{9.71}$$

In the non-relativistic limit of slowly moving particles and a static magnetic field, the second term in Eq. (9.71) is just the amplitude for the scattering of a spin $\frac{1}{2}$ particle with magnetic moment $(-e/2m)$, i.e. with gyromagnetic ratio $g=2$. The amplitude (9.70) is of the same form as the spin term in (9.71), i.e. it represents a correction to the value of the magnetic moment of the electron as given by the Dirac theory. This anomalous magnetic moment

$$-\frac{e}{2m}\left(1+\frac{\alpha}{2\pi}\right)$$

corresponds to a shift in the g-factor, usually quoted in the form

$$a_e \equiv \frac{g-2}{2} = \frac{\alpha}{2\pi} = 0.00116. \tag{9.72}$$

This result, first derived by Schwinger in 1948, is in excellent agreement with the first measurements by Kusch and Foley in 1947 and 1948, which gave the value

$$a_e = 0.00119 \pm 0.00005.$$

Subsequently, both theory and experiment have been greatly refined. Theoretically, the higher-order corrections of order α^2, α^3 and α^4 to a_e have been calculated. The result of these very heavy calculations (the α^4 term involves 891 Feynman graphs) is[6]

$$10^{12}a_e = 1159652183 \pm 8,$$

while the experimental value is

$$10^{12}a_e = 1159652181 \pm 7.$$

The agreement can only be described as remarkable.

A similar discussion applies to the anomalous magnetic moment of the muon. Since Eq. (9.72) is independent of the lepton mass, we obtain the same value for the muon in lowest order

$$a_{\mu} \equiv \frac{g_{\mu}-2}{2} = \frac{\alpha}{2\pi} = 0.00116,$$

where we have written g_{μ} to distinguish the muon g-factor from that of the electron. In higher order, differences arise, since vacuum polarization loops may involve any kind of lepton pair. This is illustrated in Fig. 9.16, where diagrams (a), (b) and (c), (d) show two e^4 contributions to a_e and a_{μ} respectively. Since the masses of the intermediate leptons occur

---

[6] See T. Aoyama et al., *Phys. Rev.* D77 (2008) 053012

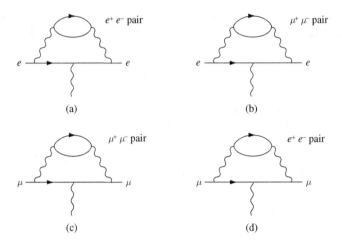

Figure 9.16 e^4 *vacuum polarization contributions to the g-factors of: (i) the electron: (a) and (b); (ii) the muon: (c) and (d)*

in the denominators of the corresponding propagators and since $m_\mu/m_e \approx 207$, the contribution of the muon pair diagram (b) is very small compared with that of the electron pair diagram (a). (Another way of putting this is to say that because of the much smaller electron mass, compared to m_μ, it is much easier to create a virtual electron–positron pair than a muon pair.) For the muon, on the other hand, the electron pair diagram (d) makes a large contribution to a_μ, in fact it is substantially bigger than that of the muon diagram (c). Similar conclusions hold generally for the contributions of vacuum polarization corrections.

High precision experiments of the muon magnetic moment have also been performed, leading to a value of[7]

$$10^{10}a_\mu = 11659208 \pm 6 \,,$$

whereas the QED prediction leads to a value of

$$10^{10}a_\mu = 11658472,$$

where the uncertainty on the latter is insignificant compared to the error on the experimental result. The last two numbers show a small discrepancy between theory and experiment. This can be attributed to the effect of strong interactions, e.g. vacuum polarization graphs involving $\pi^+\pi^-$ pairs and, to a lesser extent, weak interactions. The detailed analysis shows that these contributions are sufficiently large to close the gap between the theoretical and the experimental data for the muon. On the other hand, these effects are very small for the electron and do not upset the excellent agreement between theory and experiment.

---

[7] These results, for the muon magnetic moment are taken from the summary by A. Höcker and W. J Marciano, Particle Data Group (2008), loc. cit.

9.6.2   The Lamb shift

As a second important application we shall look at the radiative corrections to the energy levels of the hydrogen atom. Historically, the measurements by Lamb and Retherford in 1947 gave the main impetus to the development of modern QED. According to the Dirac theory, the $2s_{1/2}$ and $2p_{1/2}$ levels of hydrogen are degenerate. Lamb and Retherford's original experiment gave about 1000 MHz for the level splitting $E(2s_{1/2}) - E(2p_{1/2})$. This shift of the bound-state energy levels and the resulting splitting are known as the Lamb shift.

In the last sub-section we considered electron scattering by an external static potential, and we saw that the radiative corrections of order α stem from the Feynman graphs in Fig. 9.15(b)–(g). We can think of the same graphs as describing a bound-state level in hydrogen, if we interpret the electron lines and propagators, not in terms of free-particle states, but in terms of hydrogenic states. For the electron scattering case, we saw that the electron self-energy graphs 9.15(d)–(g) produce no observable radiative corrections [because $\Sigma_c(p)u(\mathbf{p}) = 0$], but contribute to the mass and charge renormalization only. For bound states, on the contrary, the electron self-energy graphs produce observable radiative corrections, and for s-states these graphs make the largest contribution to the level shift, with vacuum polarization and vertex corrections providing only a few % of the shift. It is therefore essential in calculating level shifts to take the bound-state aspect accurately into account. To do so, requires a lengthy analysis. The best approach is to use the bound interaction picture in which the nuclear Coulomb field in which the electron moves is included in the unperturbed Hamiltonian and only the remaining interaction constitutes the interaction Hamiltonian. Bethe in 1947 gave an approximate non-relativistic derivation of the Lamb shift, obtaining a surprisingly good result considering the nature of the calculation. We shall restrict ourselves to the Bethe approach, as it is much simpler and clearly exhibits the main origin of the Lamb shift.[8]

Bethe attributes the shift of a bound-state energy level to the self-energy of the electron in that bound state. However, a part of this self-energy effect has already been allowed for in using the physical electron mass in the calculation, and not the bare mass. Hence the true level shift is the difference between the self-energies of the bound and the free electron.

In Bethe's calculation, the hydrogen atom is treated non-relativistically, and second-order perturbation theory is used to calculate the interaction between the electron and the transverse photons. This is the formulation of QED which we gave in Chapter 1. The interaction Hamiltonian is, from Eq. (1.62),[9]

$$H_{\mathrm{I}} = -\frac{e}{m}\mathbf{A}(x)\boldsymbol{\cdot}\mathbf{p}. \tag{9.73}$$

---

[8] For a rigorous treatment, employing the bound interaction picture, the reader is referred to J. M. Jauch and F. Rohrlich, *The Theory of Photons and Electrons*, 2nd edn, Springer, New York, 1976, Sections 15-4 and S5-3. An alternative approach, which utilizes the Bethe derivation for the non-relativistic part of the calculation, is discussed in, for example, J. D. Bjorken and S. D. Drell, *Relativistic Quantum Mechanics*, McGraw-Hill, New York, 1964, Section 8.7, or C. Itzykson and J. B. Zuber, *Quantum Field Theory*, McGraw-Hill, New York, 1980, Section 7-3-2.

[9] The A^2 interaction term in Eq. (1.62) is independent of the electron momentum. Therefore it produces the same electron self-energy effects for the bound electron and the free electron, and so does not contribute to the level shift.

The level shift of a hydrogenic state $|nl\rangle \equiv \phi_{nl}(\mathbf{x})$ (where n and l are the principal and angular momentum quantum numbers) is then given by

$$\delta E(nl) = -\sum_\lambda \sum_{\mathbf{k}} \sum_{r=1,2} \frac{|\langle \lambda, n_r(\mathbf{k}) = 1|H_1|nl\rangle|^2}{E_\lambda + k - E_n}, \qquad (9.74)$$

where the intermediate state $|\lambda, n_r(\mathbf{k})=1\rangle$ consists of the hydrogen atom in one of the complete set of states $|\lambda\rangle \equiv \phi_\lambda(x)$ together with one transverse photon. (E_λ and E_n are the energy eigenvalues of $|\lambda\rangle$ and $|nl\rangle$.)

The matrix elements in Eq. 9.74 are given by Eq. (1.65). Using the dipole approximation, we replace the exponential in the matrix element (1.65) by unity.[10] Substituting for the matrix elements in Eq. 9.74, one obtains

$$\delta E(nl) = -\sum_\lambda \sum_{\mathbf{k}} \sum_{r=1,2} \left(\frac{e}{m}\right)^2 \frac{1}{2Vk} \frac{|\langle \lambda|\mathbf{\varepsilon}_r(\mathbf{k})\cdot\mathbf{p}|nl\rangle|^2}{E_\lambda + k - E_n} \qquad (9.75)$$

As in Section 1.3, we sum over photon polarizations ($r = 1, 2$) and, after converting the momentum sum into an integral, integrate over photon directions [see the corresponding analysis leading to Eq. (1.53)]. In this way, we obtain

$$\delta E(nl)= -\frac{1}{6\pi^2} \left(\frac{e}{m}\right)^2 \int_0^\infty k \; dk \; \sum_\lambda \frac{|\langle \lambda|\mathbf{p}|nl\rangle|^2}{E_\lambda + k - E_n} \qquad (9.76)$$

where

$$\langle \lambda|\mathbf{p}|nl\rangle = \int d^3x \phi_\lambda^*(\mathbf{x})(-i\nabla)\phi_{nl}(\mathbf{x}). \qquad (9.77)$$

The integral in Eq. (9.76) is linearly divergent as $k \to \infty$.

The corresponding self-energy $\delta E_f(\mathbf{p})$ for a free electron with momentum $\mathbf{p}$ is given by the same expression (9.76), where the matrix element (9.77) is now between plane-wave states and is diagonal, so that

$$\delta E_f(\mathbf{p}) = -\frac{1}{6\pi^2} \left(\frac{e}{m}\right)^2 \mathbf{p}^2 \int_0^\infty dk. \qquad (9.78)$$

This self-energy is proportional to the kinetic energy of the electron and can be interpreted in terms of a correction to the electron mass. Since the electron in the state $|nl\rangle$ has a momentum distribution, the corresponding self-energy is given by

$$\delta E_f(nl)= -\frac{1}{6\pi^2} \left(\frac{e}{m}\right)^2 \langle nl|\mathbf{p}^2|nl\rangle \int_0^\infty dk. \qquad (9.79)$$

This integral, like that in Eq. (9.76), is linearly divergent as $k \to \infty$.

---

[10] This is justified since the virtual photons which contribute significantly to the sum (9.74) have wavelengths which are large compared to the Bohr radius. See Eq. (9.88) below.

If one uses the physical mass of the electron in calculating the level shift of the state $|nl\rangle$, then the self-energy $\delta E_f(nl)$ has already been taken into account, and the observed level shift $\Delta E(nl)$ is given by

$$\Delta E(nl) = \delta E(nl) - \delta E_f(nl). \tag{9.80}$$

Since

$$\langle nl|\mathbf{p}^2|nl\rangle = \sum_\lambda |\langle \lambda|\mathbf{p}|nl\rangle|^2, \tag{9.81}$$

we obtain from Eqs. (9.76), (9.79) and (9.80)

$$\Delta E(nl) = \frac{1}{6\pi^2}\left(\frac{e}{m}\right)^2 \sum_\lambda |\langle \lambda|\mathbf{p}|nl\rangle|^2 \int_0^\infty dk \; \frac{E_\lambda - E_n}{E_\lambda - E_n + k} \cdots \tag{9.82}$$

This integral is only logarithmically divergent as $k \rightarrow \infty$. To make it converge, we replace the infinite upper limit by a finite cut-off $k = K \sim m$, i.e. we suppress contributions to the self-energy from virtual photons with energy $k \gtrsim m$. We may try to justify this cut-off as follows. In emitting a virtual photon, the electron experiences a recoil. If the non-relativistic treatment of the electron is meaningful, this recoil and hence the virtual photon energy k must be small compared to the electron rest mass. In other words, only transitions to non-relativistic hydrogenic states and virtual photons with energy $k \ll m$ may be important. Hence taking the upper limit of the integral in Eq. (9.82) as $k = K \sim m$, and assuming that in the sum over λ in this equation

$$|E_\lambda - E_n| \ll K$$

holds for the terms which matter, we obtain from Eq. (9.82)

$$\Delta E(nl) = \frac{1}{6\pi^2}\left(\frac{e}{m}\right)^2 \sum_\lambda |\langle \lambda|\mathbf{p}|nl\rangle|^2 (E_\lambda - E_n) \ln \frac{K}{|E_\lambda - E_n|}. \tag{9.83}$$

From this equation, the shift of any bound-state (ie energy) level is obtained. The fact that it depends only logarithmically on the cut-off K makes it insensitive to the exact value chosen for K.

To evaluate the λ-summation in Eq. (9.83), Bethe defines an average excitation energy $\langle E - E_n \rangle$ by the equation

$$\sum_\lambda |\langle \lambda|\mathbf{p}|nl\rangle|^2 (E_\lambda - E_n)\{\ln\langle E - E_n\rangle - \ln|E_\lambda - E_n|\} = 0. \tag{9.84}$$

Eq. (9.83) then becomes

$$\Delta E(nl) = \frac{1}{6\pi^2}\left(\frac{e}{m}\right)^2 \ln \frac{K}{\langle E - E_n \rangle} \sum_\lambda |\langle \lambda|\mathbf{p}|nl\rangle|^2 (E_\lambda - E_n). \tag{9.85}$$

The summation over λ in this equation can be performed, giving

$$\sum_{\lambda} |\langle\lambda|\mathbf{p}|nl\rangle|^2 (E_\lambda - E_n) = \frac{1}{2} e^2 |\phi_{nl}(0)|^2$$

$$= \begin{cases} \dfrac{e^2}{2\pi a^3 n^3}, & \text{if } l=0 \text{ (s-states)}, \\ 0, & \text{if } l \neq 0, \end{cases} \tag{9.86}$$

where $a = 4\pi/me^2$ is the Bohr radius.[11] Substitution of this result into Eq. (9.85) gives

$$\Delta E(nl) = \frac{8}{3\pi} \frac{\alpha^3}{n^3} \text{Ry} \ln \frac{K}{\langle E - E_n\rangle} \delta_{l0}, \tag{9.87}$$

where $\text{Ry} \equiv e^2/(8\pi a) = 13.6 \text{ eV}$ is the Rydberg energy.

Eq. (9.87) is the final result of Bethe's calculation. According to it, only s-states experience a level shift due to electron self-energy effects. For the 2s states of hydrogen, Bethe uses the value

$$\langle E - E_{2s}\rangle = 17.8 \text{ Ry}, \tag{9.88}$$

obtained by computation.[12] Thus the important intermediate states of the hydrogen atom are indeed non-relativistic, although they are highly excited continuum states. Using the value (9.87) and $K = m$, Bethe obtains from Eq. (9.87) the Lamb shift

$$E(2s_{1/2}) - E(2p_{1/2}) = 1040 \text{ MHz}, \tag{9.89}$$

in remarkable agreement with the experimental value of 1057.8 ± 0.1 MHz, obtained by Triebwasser, Dayhoff and Lamb in 1953 as the culmination of their measurements.

A proper relativistic calculation of the second-order radiative corrections leads to the value 1052.1 MHz for the $2s_{1/2} - 2p_{1/2}$ splitting.[13] Such a calculation, of course, contains no arbitrary cut-off parameter K, and in addition to the electron self-energy, it takes into account all e^2 radiative corrections, i.e. also the contributions from vacuum polarization and vertex corrections. States other than s-states now also experience level shifts, although these are much smaller, e.g. the $2p_{1/2}$ level is shifted downwards by 12.9 MHz.

We shall only consider the vacuum polarization contribution. We see from Eqs. (9.67) and (9.69) that the effect of the vacuum polarization on the scattering of non-relativistic electrons by a Coulomb potential (8.92) corresponds to modifying the Coulomb potential through the replacement

$$\frac{Ze}{|\mathbf{q}|^2} \rightarrow \frac{Ze}{|\mathbf{q}|^2} \left(1 + \frac{\alpha}{15\pi} \frac{|\mathbf{q}|^2}{m^2}\right), \tag{9.90a}$$

[11] A simple derivation of this result is given in J. J. Sakurai, *Advanced Quantum Mechanics*, Addison-Wesley, Reading, Mass., 1967, pp. 70–71.
[12] A later more accurate calculation gives the value 16.640 Ry.
[13] This and the other results we quote, together with detailed references, are given in the books by Jauch and Rohrlich and by Itzykson and Zuber which we list in the footnote on p. 187.

since for the static external field $q = (0, \mathbf{q})$. The corresponding modification in configuration space is, from Eq. (8.85),

$$\frac{Ze}{4\pi|\mathbf{x}|} \rightarrow \frac{Ze}{4\pi|\mathbf{x}|} + \frac{Ze\alpha}{15\pi m^2}\,\delta(\mathbf{x}). \tag{9.90b}$$

Since the bound states of the hydrogen atom are extremely non-relativistic, we can use the right-hand side of Eq. (9.90b) with $Z = 1$ as an effective potential to calculate the level shifts of the hydrogen atom due to vacuum polarization. In first-order perturbation theory, these shifts are given by

$$\Delta E_{nl}(\text{vac.pol.}) = \frac{-e^2\alpha}{15\pi m^2}|\phi_{nl}(0)|^2 = -\frac{8\alpha^3}{15\pi n^3}\text{Ry}\delta_{l0}. \tag{9.91}$$

For the 2s level of hydrogen this gives –27 MHz, a result which was first obtained by Uehling in 1935. We see from Eq. (9.91) that vacuum polarization only shifts s-levels, the shift being downwards. Qualitatively one can understand this in terms of the polarization by the external Coulomb field of the virtual electron–positron pairs, the electrons being attracted towards the nucleus, the positrons repelled. Thus the virtual electrons screen the nuclear charge. However, an s-state atomic electron will penetrate inside this screening and see the full nuclear charge, i.e. it will experience a more attractive potential, leading to stronger binding.

9.7   The Infrared Divergence

In Sections 8.8 and 8.9 we studied the scattering of electrons by an external field. Because of the finite energy resolution in any experiment, the observed elastic scattering cross-section always includes some bremsstrahlung, i.e. inelastic scattering with emission of a soft photon. We saw that the latter contribution is infrared divergent and asserted that this divergence is exactly cancelled by an infrared divergence in the radiative correction of order α to the elastic scattering. We shall now demonstrate this exact cancellation.

The Feynman amplitude, including the lowest-order radiative corrections, for elastic electron scattering is given by Eq. (9.67). The $O(\alpha)$ correction in the cross-section comes from the interference term of the lowest-order amplitude [the first term in Eq. (9.67)] with the radiative correction terms [the second and third terms in Eq. (9.67)]. Π_c is infrared finite, and the infrared divergence arises from the Λ_c^μ term, which we shall now study.

We regularize Eq. (9.48) for $\Lambda^\mu(p', p)$ by means of the replacement in Eq. (9.21), obtaining

$$e^2\Lambda^\mu(p',p) = \frac{-ie^2}{(2\pi)^4}\int \frac{d^4k}{k^2 - \lambda^2 + i\varepsilon}\,f(k)$$
$$\times \left\{ \frac{\gamma^\alpha(\not{p}' - \not{k} + m)\gamma^\mu(\not{p} - \not{k} + m)\gamma_\alpha}{\left[(p' - k)^2 - m^2 + i\varepsilon\right]\left[(p - k)^2 - m^2 + i\varepsilon\right]} \right\} \tag{9.92}$$

where

$$f(k) \equiv \frac{\lambda^2 - \Lambda^2}{k^2 - \Lambda^2 + i\varepsilon}. \tag{9.93}$$

Since we are now interested in the infrared divergence when $k \to 0$ and not in the ultraviolet divergence ($k \to \infty$), we shall omit the cut-off factor $f(k)$. We shall similarly drop terms linear in k and k^2 in the numerator and denominator in the expression in curly brackets in Eq. (9.92). Using $p^2 = p'^2 = m^2$ and the Dirac equation, we can simplify Eq. (9.92) to give

$$e^2 \bar{u}(\mathbf{p}')\Lambda^\mu(p', p)u(\mathbf{p}) = \frac{-ie^2}{(2\pi)^4}\bar{u}(\mathbf{p}')\gamma^\mu u(\mathbf{p})\left[\int \frac{d^4k}{k^2 - \lambda^2 + i\varepsilon} \frac{(p'p)}{(p'k)\,(pk)} + \cdots\right] \quad (9.94)$$

where, as throughout the following, the dots indicate terms which are finite in the limit $\lambda \to 0$ and which we therefore neglect. We evaluate the integral (9.94) by means of the identity

$$\frac{1}{k^2 - \lambda^2 + i\varepsilon} \equiv P\frac{1}{k^2 - \lambda^2} - i\pi\delta(k^2 - \lambda^2)$$

$$= P\frac{1}{k^2 - \lambda^2} - \frac{i\pi}{2\omega_\lambda}[\delta(k^0 - \omega_\lambda) + \delta(k^0 + \omega_\lambda)] \quad (9.95)$$

where $\omega_\lambda \equiv (\lambda^2 + \mathbf{k}^2)^{1/2}$. Performing the k^0-integration in (9.94) and omitting the infrared finite contribution from the principal value part of Eq. (9.95), one obtains

$$e^2\bar{u}(\mathbf{p}')\Lambda^\mu(p', p)u(\mathbf{p}) = e^2\bar{u}(\mathbf{p}')\gamma^\mu u(\mathbf{p})A(p', p) + \cdots, \quad (9.96a)$$

where

$$A(p', p) \equiv \frac{-1}{2(2\pi)^3}\int\frac{d^3\mathbf{k}}{\omega_\lambda}\frac{(p'p)}{(p'k)(pk)}. \quad (9.96b)$$

In Eq. (9.67) we require the renormalized part of Eq. (9.96), which is given by Eq. (9.53), i.e.

$$e^2\bar{u}(\mathbf{p}')\Lambda_c^\mu(p', p)u(\mathbf{p}) = e^2\bar{u}(\mathbf{p}')[\Lambda^\mu(p', p) - L\gamma^\mu]u(\mathbf{p}). \quad (9.97)$$

From Eqs. (9.53), (9.54) and (9.96) we obtain

$$e^2\bar{u}(\mathbf{p})\Lambda^\mu(p, p)u(\mathbf{p}) = e^2L\bar{u}(\mathbf{p})\gamma^\mu u(\mathbf{p}) = e^2\bar{u}(\mathbf{p})\gamma^\mu u(\mathbf{p})A(p, p) + \cdots$$

and a similar equation with p replaced by p', whence

$$L = A(p, p) + \cdots = A(p', p') + \cdots. \quad (9.98)$$

Combining Eqs. (9.96)–(9.98) leads to

$$e^2\bar{u}(\mathbf{p}')\Lambda_c^\mu(p', p)u(\mathbf{p})$$
$$= e^2\bar{u}(\mathbf{p}')\gamma^\mu u(\mathbf{p})\left\{A(p', p) - \frac{1}{2}A(p', p') - \frac{1}{2}A(p, p)\right\} + \cdots$$
$$= e^2\bar{u}(\mathbf{p}')\gamma^\mu u(\mathbf{p})\left\{\frac{1}{4(2\pi)^3}\int\frac{d^3\mathbf{k}}{\omega_\lambda}\left[\frac{p'}{p'k} - \frac{p}{pk}\right]^2\right\} + \cdots.$$

Substituting this expression in Eq. (9.67), we obtain the Feynman amplitude for elastic electron scattering

$$\mathcal{M} = \mathcal{M}_0\left\{1 + \frac{e^2}{4(2\pi)^3}\int\frac{d^3\mathbf{k}}{\omega_\lambda}\left[\frac{p'}{p'k} - \frac{p}{pk}\right]^2\right\} + \cdots \quad (9.99)$$

where $\mathcal{M}_0$ is the lowest-order elastic scattering amplitude

$$\mathcal{M}_0 = ie\bar{u}(\mathbf{p}')\mathcal{A}_e(\mathbf{p}' - \mathbf{p})u(\mathbf{p}).$$

Hence the elastic scattering cross-section (8.91) becomes

$$\left(\frac{d\sigma}{d\Omega}\right)_{\text{El}} = \left(\frac{d\sigma}{d\Omega}\right)_0 \left\{1 + \frac{\alpha}{(2\pi)^2} \int \frac{d^3k}{\omega_\lambda} \left[\frac{p'}{p'k} - \frac{p}{pk}\right]^2 \right\} + \cdots \qquad (9.100)$$

where $(d\sigma/d\Omega)_0$ is the lowest-order elastic scattering cross-section.

The cross-section 9.100 is infrared divergent in the limit as $\lambda \to 0$ and $\omega_\lambda \to |\mathbf{k}|$. On forming the experimentally observed cross-section (8.105), we see that this divergence in the elastic scattering is exactly cancelled by the infrared divergence which occurs in the soft bremsstrahlung cross-section, Eqs. (8.106) and (8.110), in the limit $\lambda \to 0$. Hence the experimental quantity remains finite as the limit $\lambda \to 0$ is taken, as asserted in Section 8.9. As stated in that section, this conclusion holds in all orders of perturbation theory: the infrared divergences which occur in the higher-order radiative corrections exactly cancel those in the inelastic processes involving emission of several soft photons.

9.8 Higher-Order Radiative Corrections

So far we have considered radiative corrections of order α only. The renormalization procedure we have developed can be extended and leads to finite radiative corrections in all orders of perturbation theory. The proof of this renormalizability of QED is of fundamental importance, but because of its complexity we confine ourselves to a qualitative discussion only.[14]

When studying radiative corrections of order α, we saw that these arise in two ways: from the e^2 modifications to propagators and basic vertex parts of the lowest-order Feynman graphs, and from higher-order graphs which cannot be obtained in this way. (This was illustrated for Compton scattering in Figs. 9.5 and 9.6.) This situation persists for higher-order corrections, and we shall first of all consider the higher-order modifications of propagators and vertices, starting with the electron propagator.

The e^2 correction to the electron propagator results from the insertion of the electron self-energy loop $ie_0^2\Sigma(p)$, Fig. 9.17, in the bare electron propagator, giving Fig. 9.9 and Eqs. (9.22). In higher orders, iterations of two, three or more such electron self-energy loops will occur. Their combined contributions produce the modification shown in Fig. 9.18 and given by

$$\frac{i}{\not{p} - m_0 + i\varepsilon} \to \frac{i}{\not{p} - m_0 + i\varepsilon} + \frac{i}{\not{p} - m_0 + i\varepsilon}\, ie_0^2\Sigma(p)\, \frac{i}{\not{p} - m_0 + i\varepsilon}$$

$$+ \frac{i}{\not{p} - m_0 + i\varepsilon}\, ie_0^2\Sigma(p) \frac{i}{\not{p} - m_0 + i\varepsilon}\, ie_0^2\Sigma(p)\frac{i}{\not{p} - m_0 + i\varepsilon} + \cdots \qquad (9.101a)$$

---

[14] For complete treatments, the reader is referred to the books listed in the footnote on p. 187 and to J. D. Bjorken and S. D. Drell, *Relativistic Quantum Fields*, McGraw-Hill, New York, 1965, Chapter 19.

Figure 9.17   *The second-order electron self-energy insertion* $ie_0^2\Sigma(p)$

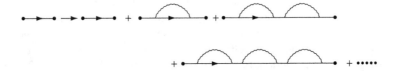

Figure 9.18   *The electron propagator obtained by including the electron self-energy insertion of order e^2 and its iterations*

$$= \frac{\text{i}}{\not{p} - m_0 + e_0^2\sum(p) + \text{i}\varepsilon}, \qquad (9.101\text{b})$$

where we used the identity (9.23) to obtain the last expression.

There are of course many other electron self-energy insertions. To include them all, we define a *proper* Feynman graph as a graph which cannot be split into two graphs by cutting a single internal line. Fig. 9.19(a) and (b) show some proper and improper electron self-energy graphs. We now re-define $ie_0^2\Sigma(p)$ as the sum of all proper electron self-energy insertions, as indicated in Fig. 9.20. With this interpretation of $ie_0^2\Sigma(p)$, the infinite series of terms on the right-hand side of Eq. 101a contains all electron self-energy graphs, so that the expression (9.101b) represents the complete (i.e. exact) electron propagator.

The expression (9.101b) for the complete electron propagator is of the same form as our earlier second-order result, Eq. (9.24), if in the latter equation we reinterpret $ie_0^2\Sigma(p)$ and

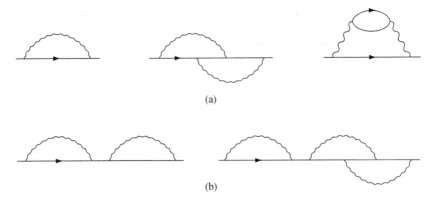

(a)

(b)

Figure 9.19   *Some electron self-energy graphs: (a) proper graphs; (b) improper graphs*

Figure 9.20 *The graphical representation of* $ie_0^2\Sigma(p)$, *redefined as the sum of all proper electron self-energy insertions*

omit the term $O(e_0^4)$. Hence, the discussion of mass renormalization, leading from Eq. (9.24) to Eq. (9.29), goes through exactly as before, giving

$$\frac{ie_0^2}{\not{p} - m_0 + i\varepsilon} \rightarrow \frac{ie_0^2}{(\not{p} - m)(1 + e_0^2 B) + e_0^2(\not{p} - m)\sum_c(p) + i\varepsilon}. \tag{9.102}$$

In Eq. (9.102) the bare and renormalized masses are again related by Eqs. (9.25), (9.27) and (9.28), which now hold to all orders, and B is again defined by Eq. (9.26). $ie_0^2\Sigma(p)$ in Eqs. (9.26) and (9.27) is the complete proper electron self-energy insertion, Fig. 9.20. Defining the renormalized charge e by

$$e^2 \equiv Z_2 e_0^2 = e_0^2/(1 + e_0^2 B), \tag{9.103}$$

we obtain from Eq. (9.102)

$$\frac{ie_0^2}{\not{p} - m_0 + i\varepsilon} \rightarrow \frac{ie^2}{(\not{p} - m) + e^2(\not{p} - m)\sum_c(p) + i\varepsilon}, \tag{9.104a}$$

which holds to all orders in e^2. In lowest order, Eqs. (9.103) and (9.104a) reduce to our previous results, Eqs. (9.35) and (9.36), where $\Sigma_c(p)$ is now calculated from just the first graph of Fig. 9.20.

Alternatively, and equivalently, one can introduce mass counterterms as in Eqs. (9.31). The analysis goes through exactly as before, with the counterterm (9.32) now defined in terms of the complete proper electron self-energy insertion. In this way one obtains, instead of Eq. (9.104a),

$$\frac{ie_0^2}{\not{p} - m + i\varepsilon} \rightarrow \frac{ie^2}{(\not{p} - m) + e^2(\not{p} - m)\sum_c(p) + i\varepsilon} \tag{9.104b}$$

in agreement with our earlier second-order results (9.34) and (9.35). We shall adopt this approach in what follows.

One can similarly deal with the photon propagator. We redefine $ie_0^2\Pi^{\mu\nu}(k)$ as the sum of all proper photon self-energy insertions, as indicated in Fig. 9.21. Iteration of this complete proper photon self-energy insertion leads, analogously to Eq. (9.101a), to the propagator modification

$$\frac{-ig_{\alpha\beta}}{k^2 + i\varepsilon} \rightarrow \frac{-ig_{\alpha\beta}}{k^2 + i\varepsilon} + \frac{-ig_{\alpha\mu}}{k^2 + i\varepsilon} ie_0^2\Pi^{\mu\nu}(k)\frac{-ig_{\nu\beta}}{k^2 + i\varepsilon}$$
$$+ \frac{-ig_{\alpha\mu}}{k^2 + i\varepsilon} ie_0^2\Pi^{\mu\nu}(k)\frac{-ig_{\nu\sigma}}{k^2 + i\varepsilon} ie_0^2\Pi^{\sigma\tau}(k)\frac{-ig_{\tau\beta}}{k^2 + i\varepsilon} + \dots, \tag{9.105a}$$

Figure 9.21   *The graphical representation of* $ie_0^2\Pi^{\mu\nu}(k)$, *redefined as the sum of all proper photon self-energy insertions*

which incorporates all photon self-energy terms, proper and improper. If we substitute Eq. (9.11) for $\Pi^{\mu\nu}(k)$ and omit terms proportional to $k^\mu k^\nu$, since the propagator is always coupled to conserved currents, we can sum the series in Eq. (9.105a) and obtain.

$$\frac{-ig_{\alpha\beta}}{k^2 + i\varepsilon} \rightarrow \frac{-ig_{\alpha\beta}}{k^2 + i\varepsilon + e_0^2 A(k^2)}. \qquad (9.105b)$$

This is just Eq. (9.13) which, with the redefinition of $ie_0^2\Pi^{\mu\nu}(k)$, is now exact to all orders in e^2, rather than to $O(e_0^4)$ only. We again demand $A(0) = 0$, Eq. (9.14). If we multiply Eq. (9.105b) by e_0^2 and express e_0^2 in terms of the renormalized charge

$$e \equiv e_0 Z_3^{1/2} = e_0 \left[1 + e_0^2 A'(0)\right]^{-1/2}, \qquad (9.106)$$

we finally obtain for the complete photon propagator

$$\frac{-ig_{\alpha\beta}}{k^2 + i\varepsilon} e_0^2 \rightarrow \frac{-ig_{\alpha\beta}}{k^2 + i\varepsilon + e^2 \Pi_c(k^2)} e^2 \qquad (9.107)$$

$A'(0)$ and $\Pi_c(k^2)$ are again defined in Eq. (9.15). In lowest order of perturbation theory, Eqs. (9.106) and (9.107) reduce to our previous results, Eqs. (9.18) and (9.19).

   In Section 9.4 we considered external line renormalization in lowest order of perturbation theory. We found that for external lines the only self-energy effects are mass and charge renormalizations. This result can be shown to hold in all orders of perturbation theory.

   Finally, we must consider the vertex function $i\Gamma^\mu(p', p)$. The second-order treatment of Section 9.5 is easily generalized by including all proper vertex modifications and redefining $ie_0^3\Lambda^\mu(p', p)$ as the sum of all such modifications, as indicated in Fig. 9.22. With this interpretation of $ie_0^3\Lambda^\mu(p', p)$, the basic results, Eqs. (9.57) and (9.58), remain unchanged, except that the qualifying terms $\left[O(e_0^5) \text{ and } O(e^5)\right]$ are absent. Since the Ward identity (9.60) can be shown to hold to all orders, it follows that the relations $Z_1 = Z_2$ and $e = e_0 Z_3^{1/2}$, Eqs. (9.63) and (9.64), are also exact.

Figure 9.22   *The graphical representation of* $ie_0^3\Lambda^\mu(p', p)$, *redefined as the sum of all proper vertex modifications*

We have now generalized the relations between bare and physical masses and charges to all orders. Correspondingly, the modifications of the bare propagators and the basic vertex part are given, to all orders, by

$$\frac{i}{\not{p} - m + i\varepsilon} \rightarrow \frac{i}{(\not{p} - m) + e^2(\not{p} - m)\Sigma_c(p) + i\varepsilon} \tag{9.108a}$$

$$\frac{-ig_{\alpha\beta}}{k^2 + i\varepsilon} \rightarrow \frac{-ig_{\alpha\beta}}{k^2 + i\varepsilon + e^2\Pi_c(k^2)} \tag{9.108b}$$

$$ie_0\gamma^\mu \rightarrow ie\left[\gamma^\mu + e^2\Lambda_c^\mu(p', p)\right]. \tag{9.108c}$$

Although the right-hand sides of these equations no longer depend explicitly on the bare charge e_0, they do depend implicitly on e_0 since Σ_c, Π_c and Λ_c^μ are expressed in terms of e_0. Dyson, Salam, Ward and others have shown that these quantities can be expressed in a consistent manner, order by order, in terms of the physical charge e. The right-hand sides of Eqs. (9.108) therefore represent the renormalized propagators and the renormalized vertex function.

Not all radiative corrections are due to self-energy and vertex modifications. To clarify the distinction, we define the process of *reduction* of a graph as removing self-energy and vertex modifications from it, i.e. as replacing them by bare propagators and basic vertex parts. Fig. 9.23 illustrates the process of reduction schematically. A graph from which all self-energy and vertex modifications have been removed, so that it cannot be reduced further, is called *irreducible* or alternatively a *skeleton graph*. For example, for Compton scattering the graphs in Figs. 9.4 and 9.6 are irreducible, whereas the graphs in Fig. 9.5 are reducible to the same skeleton graph, Fig. 9.4(a).

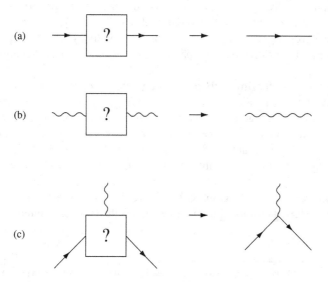

Figure 9.23 *Reduction of (a) an electron propagator; (b) a photon propagator; (c) a vertex part*

To calculate the Feynman amplitude for an arbitrary process to order n, we combine the above results and proceed as follows. Firstly, we draw all skeleton diagrams which contribute to the process (i.e. have the correct external lines) and have not more than n vertices. From these skeleton graphs, all graphs contributing to the process up to nth order are obtained by replacing the bare propagators and the basic vertex parts by the renormalized propagators and vertex functions, Eqs. (9.108), expanding these equations up to the appropriate powers in e^2. Thus, to calculate the amplitude for Compton scattering to fourth order, we require the second- and fourth-order skeleton graphs. These are shown in Figs. 9.4 and 9.6. (The triangle graphs 9.6(c) and (d) give zero contribution, as shown in Section 9.1, and should be omitted.) Substitution of Eqs. (9.108), expanded to $O(e^2)$, in the second-order skeleton graphs 9.4 generates the second-order amplitude, as well as the contributions of all reducible fourth-order graphs; for example, those shown in Fig. 9.5.[15] The skeleton graphs in Fig. 9.6 are already of fourth order. Hence, no additional fourth-order graphs are generated by the replacements (9.108), and in the amplitudes for these skeleton graphs we merely replace the bare charge e_0 by the physical charge e.

9.9   Renormalizability

We have now outlined a general method of calculating higher-order radiative corrections, expressed in terms of the mass and charge of the physical electron. In order to deal with well-defined finite quantities we had to regularize the theory through the introduction of suitable cut-off parameters. We must now consider whether the radiative corrections remain finite in the limit as we remove the cut-off parameters in order to restore QED.

In order to study the divergences of QED, it suffices to consider the *primitive divergences* of the theory, since all other divergences can be built up from these. A primitively divergent graph is a divergent graph which is converted to a convergent graph if any internal line is cut (i.e. replaced by two external lines). Clearly, a primitively divergent graph must be a proper graph, and it cannot contain any divergent sub-graphs.[16] Obvious examples of primitively divergent graphs are the second-order self-energy and vertex corrections of Fig. 9.3.

We want to identify all primitively divergent graphs. It follows from the definition of a primitively divergent graph that one obtains a convergent result if in the expression for the Feynman amplitude of the graph one keeps any one internal four-momentum fixed (this corresponds to cutting an internal line) and integrates over all others. Hence the degree of divergence of this Feynman amplitude can be obtained by naive dimensional arguments, i.e. by counting powers of momentum variables of integration in the numerators and denominators.

Let the primitively divergent graph G have n vertices, f_i(b_i) internal fermion (photon) lines and f_e(b_e) external fermion (photon) lines. If d is the number of internal momenta not

---

[15] The two graphs on the left-hand side of Fig. 9.5 contain modifications to external lines only. As we saw in Section 9.4, they do not contribute to the radiative corrections and so need not be considered explicitly.

[16] Any graph isolated from another graph G by cutting a finite number of internal lines is called a sub-graph of G

fixed by energy-momentum conservation at the vertices, then the degree of divergence of the Feynman amplitude of G is

$$K = 4d - f_i - 2b_i. \tag{9.109}$$

There are n δ-functions associated with the vertices of G. One of these δ-functions ensures overall conservation of energy and momentum and only involves external momenta. Hence of the $(f_i + b_i)$ internal momenta, only

$$d = f_i + b_i - (n - 1) \tag{9.110}$$

are independent variables. We also have the relations

$$2n = f_e + 2f_i, \quad n = b_e + 2b_i. \tag{9.111}$$

Combining Eqs. (9.109)–(9.111), we obtain

$$K = 4 - \frac{3}{2} f_e - b_e \geq 0 \tag{9.112}$$

as necessary condition for G to be a primitively divergent graph. $K = 0, 1, \ldots$ means at most a logarithmic, linear, ... divergence.

Eq. (9.112) is a most remarkable result because it depends only on the number of external fermion and photon lines of the graph and is independent of its internal structure. Furthermore it provides the vital information that the only graphs which may possibly be primitively divergent are those with $(f_e, b_e) = (0, 2), (0, 3), (0, 4), (2, 0)$ and $(2, 1)$, and that the divergences are at most quadratic. Two of these five types of graphs are, in fact, convergent. We met the simplest example of the type $(f_e, b_e) = (0, 3)$ in Section 9.1, Fig. 9.6, where we saw that the triangle graphs occur in pairs which exactly cancel. Since this result generalizes to all higher-order graphs of this type, one can omit such graphs altogether. Secondly, graphs with $(f_e, b_e) = (0, 4)$ do not lead to divergences. These graphs describe the scattering of light by light. The simplest Feynman graph for this process is shown in Fig. 9.24. From Eq. (9.112) such graphs could be logarithmically divergent. One can show that as a consequence of gauge invariance they are strongly convergent.

The remaining three types of graphs can be primitively divergent. They are just the electron and photon self-energy graphs and the vertex modifications. The only primitively divergent self-energy graphs are the second-order corrections, Fig. 9.3(a) and (b). The second-order vertex modification, Fig. 9.3(c), is also primitively divergent, but there exist infinitely many primitively divergent higher-order vertex modifications; for example, the third graph on the right-hand side of Fig. 9.22 is primitively divergent.

This exhausts the enumeration of primitively divergent graphs. In particular, the irreducible diagrams of any physical process. $(f_e + b_e \geq 4)$ are finite, and infinities can only arise through insertions of self-energy and vertex modifications in these graphs. Hence, if the residual modifications Σ_c, Π_c and Λ_c^μ, which result from renormalization and which occur in Eqs. (9.108), remain finite to all orders, then inserting them into the irreducible graphs cannot lead to divergences. The same approach which we used to analyse the primitively divergent second-order corrections (essentially a Taylor series expansion of the convergent integrands in the integrals representing the Feynman amplitudes) can be used to show that all primitively divergent contributions to Σ_c, Π_c and Λ_c^μ remain finite

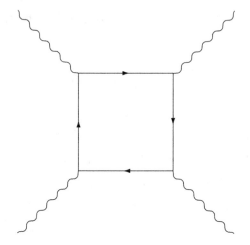

Figure 9.24   *The simplest light–light scattering diagram*

as the cut-off parameters are removed. This result can be extended to all proper self-energy and vertex modifications. Consequently, the predictions of QED, expressed in terms of the physical mass and charge of the electron, remain finite as all cut-off parameters are removed.

A field theory is called *renormalizable* if its predictions in terms of a finite number of parameters (i.e. masses and coupling constants) remain finite when all cut-offs are removed. QED is an example of a renormalizable theory. In such a theory, where the results are well defined and finite in the limit as the momentum cut-off Λ tends to infinity, the results are insensitive to the form of the cut-off, provided only that Λ is much greater than the momentum scale of the process under consideration. In other words, for QED, the discussion following Eqs. (9.65) goes through essentially unchanged to all orders, and theoretical predictions obtained with a finite Λ are not measurably different from those obtained in the limit $\Lambda \to \infty$, provided Λ is very much greater than the energy scale of the experiment. In contrast, a theory which is not renormalizable can still be made well-defined and finite by the introduction of suitable cut-off parameters Λ. However, in such a non-renormalizable theory the physical predictions diverge in the limit $\Lambda \to \infty$ and hence are inevitably sensitive to the form and magnitude of the cut-offs, even for very large Λ.

Problems

9.1. In Section 9.1 we used general arguments to show that the Feynman amplitudes of the two triangle graphs, Fig. 9.6(c) and (d), differ only in sign and exactly cancel each other. Derive this result from the explicit forms of these amplitudes. [*Note*: This proof does not require the evaluation of the traces to which these diagrams give rise, but only the use of the general properties of traces (see Section A.3 in Appendix A) to relate the two expressions to each other.]

9.2. From $[S_F(p)]^{-1} = \not{p} - m$ and $[S_F(p)][S_F(p)]^{-1} = 1$, derive

$$\partial S_F(p)/\partial p_\mu = - S_F(p)\gamma^\mu S_F(p).$$

Hence, derive the Ward identity

$$\frac{\partial \sum(p)}{\partial p_\mu} = \Lambda^\mu(p,p). \tag{9.60}$$

10

Regularization[1]

In the last chapter we saw that the calculations of radiative corrections in QED lead to divergent loop integrals. These divergences are removed by regularization, i.e. suitable modification of these integrals. After renormalization, those integrals which enter physical predictions remain finite when the regularization is removed, i.e. when the original theory is restored. There exist several regularization formalisms, and the regularized integrals depend on the formalism employed. However, in the limit in which the original theory is restored, the physical predictions become independent of the method of regularization used. At any rate, whenever different methods have been used, they have led to the same results. In this chapter we shall consider the explicit evaluation of single loop integrals using two regularization procedures.

Historically the oldest procedure is the cut-off method, which we discussed and used for illustration in Chapter 9. It has the advantage of relating the divergences to the short-distance and high-energy behavior of the theory. To illustrate this method, we shall, in Section 10.2, use it to calculate the electron mass shift δm. The cut-off method is difficult to apply in all but the simplest cases. In particular, using this method makes it difficult to ensure gauge invariance and the validity of the Ward identity to all orders of perturbation

---

[1] This chapter deals with the technical details of regularization procedures. They will only be required in Chapter 15, and readers may, if they wish, defer reading this chapter until then.

Quantum Field Theory, Second Edition   Franz Mandl and Graham Shaw
© 2010 John Wiley & Sons, Ltd

theory. In order to do so, one must adopt a clumsy and complicated cut-off procedure, such as the Pauli–Villars method.[2]

Subsequently an alternative method, known as dimensional regularization, was developed. Although less easy to interpret than the cut-off method, it is easier to apply, and it has the great advantage that it automatically ensures gauge invariance, and relatedly the validity of the Ward identity, to all orders of perturbation theory. Dimensional regularization is therefore of particular importance for QED and, even more so, for non-Abelian gauge theories, such as quantum chromodynamics and the Weinberg–Salam unified theory of weak and electromagnetic interactions. For non-Abelian gauge theories, gauge invariance implies several Ward identities. These are so restrictive that it is very difficult to satisfy them when employing a cut-off regularization procedure. In contrast, dimensional regularization automatically respects these identities and gauge invariance to all orders of perturbation theory, and for non-Abelian gauge theories this formalism has been almost exclusively used. In particular, the crucial proof that the Weinberg–Salam theory is renormalizable was carried out using dimensional regularization.

The formalism of dimensional regularization will be developed in Section 10.3. In the two succeeding sections it will be used to evaluate the vacuum polarization correction and the anomalous magnetic moment of the electron. Our development will again be restricted to the lowest order of perturbation theory. i.e. to single loop integrals, but the same methods can be extended to higher orders.[3]

The finite loop integrals which result after regularization can be evaluated by reducing them to certain standard forms using tricks invented for this purpose by Feynman. We shall obtain these results in Section 10.1.

10.1   Mathematical Preliminaries

10.1.1   Some standard integrals

We first list the standard integrals most frequently met in the evaluation of loop integrals. Below we comment on their derivation.

$$\int \frac{\mathrm{d}^4 k}{(k^2 - s + \mathrm{i}\varepsilon)^n} = \mathrm{i}\pi^2 (-1)^n \frac{\Gamma(n-2)}{\Gamma(n)} \frac{1}{s^{n-2}}, \quad n \geqslant 3, \tag{10.1}$$

$$\int \mathrm{d}^4 k \frac{k^\mu}{(k^2 - s + \mathrm{i}\varepsilon)^n} = 0, \quad n \geq 3, \tag{10.2}$$

---

[2] The Pauli–Villars formalism is discussed in, for example, J. M. Jauch and F. Rohrlich, *The Theory of Photons and Electrons*, 2nd edn, Springer, New York, 1976, Section 10-9, and in C. Itzykson and J, B. Zuber, *Quantum Field Theory*, McGraw-Hill, New York, 1980, Sections 7-1-1 and 8-4-2.

[3] For a full discussion of the technique of dimensional regularization, its applications and references, we refer the reader to G. Leibbrandt, *Rev. Mod. Phys.* **47** (1975) 849 G 'tHooft and M. Veltman, *Nucl. Phys.* **B44** (1972) 189, and C. Nash, *Relativistic Quantum Fields*, Academic Press, London, 1978.

$$\int d^4k \frac{k^\mu k^\nu}{(k^2 - s + i\varepsilon)^n} = i\pi^2 (-1)^{n+1} \frac{\Gamma(n-3)}{2\Gamma(n)} \frac{g^{\mu\nu}}{s^{n-3}}, \qquad n \geq 4, \tag{10.3}$$

$$\int \frac{d^4p}{(p^2 + 2pq + t + i\varepsilon)^n} = i\pi^2 \frac{\Gamma(n-2)}{\Gamma(n)} \frac{1}{(t - q^2)^{n-2}}, \qquad n \geq 3, \tag{10.4}$$

$$\int d^4p \frac{p^\mu}{(p^2 + 2pq + t + i\varepsilon)^n} = -i\pi^2 \frac{\Gamma(n-2)}{\Gamma(n)} \frac{q^\mu}{(t - q^2)^{n-2}}, \qquad n \geq 3, \tag{10.5}$$

$$\int d^4p \frac{p^\mu p^\nu}{(p^2 + 2pq + t + i\varepsilon)^n}$$

$$= i\pi^2 \frac{\Gamma(n-3)}{2\Gamma(n)} \frac{[2(n-3)q^\mu q^\nu + (t - q^2)g^{\mu\nu}]}{(t - q^2)^{n-2}}, \qquad n \geq 4. \tag{10.6}$$

In the right-hand expressions of Eqs. (10.1)–(10.6) we have put $\varepsilon = 0$, which is usually permissible. If subsequently this leads to ambiguities, then we must retain $(s - i\varepsilon)$ and $(t + i\varepsilon)$ in place of s and t in these expressions. In what follows, we shall usually anticipate the limit $\varepsilon \to 0$ in this way.

The formula (10.1) for the case $n = 3$ is obtained by performing the k^0 integration as a contour integral and the subsequent integration with respect to $\mathbf{k}$ using spherical polar coordinates.[4] The general result for $n \geq 3$ follows by repeated differentiation with respect to s. Eq. (10.2) is obvious from symmetry. Eqs. (10.4) and (10.5) follow from Eqs. (10.1) and (10.2) respectively by changing variables from k and s to

$$p = k - q, \quad t = q^2 - s. \tag{10.7}$$

Differentiating Eq. (10.5) with respect to q_ν leads to Eq. (10.6), and taking $q = 0$ in Eq. (10.6) gives Eq. (10.3).

Other integrals involving more complicated tensors in the numerator of the integrand are easily obtained from the above formulas by differentiation and changing variables, but the above results suffice for most purposes.

10.1.2 Feynman parameterization

The integrals (10.1)–(10.6) contain a single quadratic factor, raised to the power n, in the denominators, whereas usually one deals with integrals containing a product of several different quadratic factors in the denominator. These more general integrals are reduced to the desired form by means of an ingenious technique due to Feynman.

For a product of two quadratic factors a and b, one starts from the identity

$$\frac{1}{ab} = \frac{1}{b-a} \int_a^b \frac{dt}{t^2}. \tag{10.8}$$

---

[4] This derivation is given in J.J. Sakurai, *Advanced Quantum Mechanics*, Addison-Wesley, 1967, p. 315.

Defining the *Feynman parameter z* by

$$t = b + (a - b)z, \tag{10.9}$$

Eq. (10.8) can be written

$$\frac{1}{ab} = \int_0^1 \frac{dz}{[b + (a - b)z]^2}. \tag{10.10}$$

We see that by introducing the Feynman parameter z we have expressed $1/ab$ in terms of a single factor raised to the power 2. Although at this stage the integral in Eq. (10.10) may look like a complication, we shall see that Feynman parameterization allows us to evaluate all integrals straightforwardly.

The above method easily extends. For three factors, the alternative results

$$\frac{1}{abc} = 2\int_0^1 dx \int_0^x dy \frac{1}{[a + (b - a)x + (c - b)y]^3} \tag{10.11a}$$

$$= 2\int_0^1 dx \int_0^{1-x} dz \frac{1}{[a + (b - a)x + (c - a)z]^3} \tag{10.11b}$$

are proved by integrating with respect to y and z respectively and using Eq. (10.10). Eq. (10.11a) generalizes to an arbitrary number of factors, and the result

$$\frac{1}{a_0 a_1 a_2 \ldots a_n} = \Gamma(n+1) \int_0^1 dz_1 \int_0^{z_1} dz_2 \ldots \int_0^{z_{n-1}} dz_n$$

$$\times \frac{1}{[a_0 + (a_1 - a_0)z_1 + \cdots (a_n - a_{n-1})z_n]^{n+1}} \tag{10.12}$$

is established by induction.

Other useful results are obtained by differentiation with respect to one or more parameters. For example, differentiating Eq. (10.10) with respect to a gives

$$\frac{1}{a^2 b} = 2\int_0^1 dz \frac{z}{[b + (a - b)z]^3}. \tag{10.13}$$

Finally, we note that the modified photon propagator (9.21) can be written in the often useful form

$$\frac{1}{k^2 - \lambda^2 + i\varepsilon} - \frac{1}{k^2 - \Lambda^2 + i\varepsilon} = -\int_{\lambda^2}^{\Lambda^2} \frac{dt}{(k^2 - t + i\varepsilon)^2}, \tag{10.14}$$

which is the identity (10.8) again.

10.2   Cut-Off Regularization: The Electron Mass Shift

As an illustration of the cut-off method of regularization, we shall calculate the electron self-energy mass shift δm in second-order perturbation theory.

From Eqs. (9.27), (9.28) and (9.4), δm is given by

$$\delta m = i\bar{u}(\mathbf{p})\left\{\frac{-e_0^2}{(2\pi)^4}\int d^4k \frac{\gamma^\alpha(\not{p}-\not{k}+m)\gamma_\alpha}{(p-k)^2-m^2+i\varepsilon}\right.$$

$$\left. \times\left[\frac{1}{k^2-\lambda^2+i\varepsilon}-\frac{1}{k^2-\Lambda^2+i\varepsilon}\right]\right\}u(\mathbf{p}), \qquad (10.15)$$

where we have replaced the photon propagator in Eq. (9.4) by the modification (9.21) to avoid any difficulties which may arise from infrared divergences. Eq. (10.15) is simplified by using the contraction identities (A.14b) and by setting $\not{p}u(\mathbf{p})=mu(\mathbf{p})$ and $p^2=m^2$. If we also substitute Eq. (10.14), we obtain

$$\delta m = \frac{ie_0^2}{(2\pi)^4}\bar{u}(\mathbf{p})\left[\int d^4k \frac{2(\not{k}+m)}{k^2-2pk+i\varepsilon}\int_{\lambda^2}^{\Lambda^2}\frac{dt}{(k^2-t+i\varepsilon)^2}\right]u(\mathbf{p}), \qquad (10.16)$$

and applying Eq. (10.13) gives

$$\delta m = \frac{ie_0^2}{(2\pi)^4}\bar{u}(\mathbf{p})\left[\int_{\lambda^2}^{\Lambda^2}dt\int_0^1 dz\int d^4k\right.$$

$$\left. \times\frac{4(\not{k}+m)z}{[k^2-2pk(1-z)-tz+i\varepsilon]^3}\right]u(\mathbf{p}). \qquad (10.17)$$

The integral with respect to k in Eq. (10.17) is obtained from Eqs. (10.4) and (10.5), leading to

$$\delta m = \frac{me_0^2}{8\pi^2}\int_0^1 dz\int_{\lambda^2}^{\Lambda^2}dt\frac{2z-z^2}{tz+m^2(1-z)^2}$$

$$= \frac{m\alpha_0}{2\pi}\int_0^1 dz(2-z)\ln\frac{\Lambda^2z+m^2(1-z)^2}{\lambda^2z+m^2(1-z)^2}. \qquad (10.18)$$

This expression remains infrared finite in the limit $\lambda \to 0$, and we can therefore take $\lambda = 0$ in Eq. (10.18). As $\Lambda \to \infty$, the integral diverges logarithmically with the leading term given by

$$\delta m = \frac{m\alpha_0}{2\pi}\ln\frac{\Lambda^2}{m^2}\int_0^1 dz(2-z)+O(1)$$

$$= \frac{3m\alpha_0}{2\pi}\ln\frac{\Lambda}{m}+O(1). \qquad (10.19)$$

This is the result quoted in Section 9.3, Eqs. (9.37) and (9.28).

10.3   Dimensional Regularization

10.3.1   Introduction

The divergent loop integrals of field theory are four-dimensional integrals in energy-momentum space. Dimensional regularization consists in modifying the dimensionality of these integrals so that they become finite. In the first place, we generalize from a four-dimensional to a D-dimensional space, where D is a positive integer. The metric tensor $g^{\alpha\beta} = g_{\alpha\beta}$ of this space is defined by

$$\left. \begin{array}{ll} g^{00} = -g^{ii} = 1, & i = 1, 2, \ldots, D-1, \\ g^{\alpha\beta} = 0, & \alpha \neq \beta. \end{array} \right\} \tag{10.20}$$

Correspondingly, a four-vector k^α is replaced by a vector with D components

$$k^\alpha \equiv \left(k^0, \ k^1, \ \ldots, \ k^{D-1}\right), \tag{10.21}$$

and

$$k^2 = k_\alpha k^\alpha = \left(k^0\right)^2 - \sum_{i=1}^{D-1} \left(k^i\right)^2. \tag{10.22}$$

Loop integrals now become integrals in D dimensions with the volume element $d^D k = dk^0 dk^1 \ldots dk^{D-1}$. For example, Eq. (10.1) generalizes to[5]

$$\int \frac{d^D k}{(k^2 - s + i\varepsilon)^n} = i\pi^{D/2}(-1)^n \frac{\Gamma(n-D/2)}{\Gamma(n)} \frac{1}{s^{n-D/2}} \tag{10.23}$$

for integer values of $n > D/2$. For $n = D/2$ (e.g. for $n = 2$ when $D = 4$), the left-hand side of Eq. (10.23) is logarithmically divergent, and the right-hand side is also singular due to the pole of $\Gamma(z)$ at $z = 0$. However, for non-integer values of D, the right-hand side of Eq. (10.23) is perfectly well defined and finite. We can therefore use it to *define* a generalization of the integral on the left-hand side of Eq. (10.23) to D dimensions for non-integer values of D. In particular, we shall take $D = 4 - \eta$ where η is a small positive parameter. Restoring ordinary four-dimensional space (and, for example, QED) corresponds to the limit $\eta \to 0$.

Before going on to QED, we shall illustrate these ideas by a simple, but unrealistic, example. Suppose we were dealing with the divergent loop integral

$$\Pi(s) = \frac{1}{(2\pi)^4} \int \frac{d^4 k}{(k^2 - s + i\varepsilon)^2}. \tag{10.24}$$

In the cut-off method, we would multiply the integrand by, for example, $(-\Lambda^2)/(k^2 - \Lambda^2)$ and evaluate the resulting integral

$$\Pi_\Lambda(s) = \frac{1}{(2\pi)^4} \int \frac{d^4 k}{(k^2 - s + i\varepsilon)^2} \frac{-\Lambda^2}{k^2 - \Lambda^2} \tag{10.25}$$

---

[5] For the method of evaluation of such integrals, see the paper by 't Hooft and Veltman, quoted in the preamble to this chapter.

using the methods of Section 10.1. In the limit as $\Lambda \to \infty$, $\Pi_\Lambda(s)$ is of course logarithmically divergent. However, after renormalization, one would deal with a difference

$$\Pi_\Lambda(s) - \Pi_\Lambda(s_0)$$

and this has the well-defined finite limit

$$\lim_{\Lambda \to \infty} \{\Pi_\Lambda(s) - \Pi_\Lambda(s_0)\} = -\frac{i}{16\pi^2} \ln (s/s_0). \tag{10.26}$$

In the dimensional method, regularization is achieved by using Eq. (10.23) to extend the definition of $\Pi(s)$ to $D = 4 - \eta$ dimensions. Specifically, we write

$$\Pi(s) = \mu^{-\eta} \Pi_\eta(s) \tag{10.27a}$$

where

$$\Pi_\eta(s) = \frac{\mu^\eta}{(2\pi)^{4-\eta}} \int \frac{d^{4-\eta}k}{(k^2 - s + i\varepsilon)^2} = \frac{i}{16\pi^2} \frac{1}{(4\pi)^{-\eta/2}} \frac{\Gamma(\eta/2)}{\Gamma(2)} \left(\frac{s}{\mu^2}\right)^{-\eta/2} \tag{10.27b}$$

and μ is a mass scale which will be discussed below. The factor μ^η ensures that the natural dimension of $\Pi_\eta(s)$ is independent of η and is required in order to expand about four dimensions $\eta = 0$. For $\eta \to 0$, one has

$$x^{-\eta/2} = 1 - \frac{1}{2}\eta \ln x + \cdots \tag{10.28}$$

for arbitrary dimensionless x, and

$$\Gamma\left(\frac{\eta}{2}\right) = \frac{2}{\eta} - \gamma + \cdots \tag{10.29}$$

where $\gamma = 0.5772 \ldots$ is Euler's constant.[6] Hence, one obtains

$$\Pi_\eta(s) = \frac{i}{16\pi^2}\left[\frac{2}{\eta} - \gamma + \ln 4\pi\right] - \frac{i}{16\pi^2} \ln\left(\frac{s}{\mu^2}\right), \tag{10.30a}$$

whence

$$\lim_{\eta \to 0} \{\Pi_\eta(s) - \Pi_\eta(s_0)\} = -\frac{i}{16\pi^2} \ln(s/s_0), \tag{10.30b}$$

in agreement with the result (10.26) derived by the cut-off method, independent of the mass parameter μ.

This is a characteristic feature of dimensional regularization. In generalizing loop integrals to $D = 4 - \eta$ dimensions one must always introduce associated factors μ^η for consistency, as illustrated by the discussion of Eq. (10.27b) above. They are compensated by factors $\mu^{-\eta}$, as in Eq. (10.27a), which are usually absorbed into a redefinition of the associated coupling constants. For example, in QED, each loop integral is associated with a factor e_0^2, which is replaced by the coupling

$$\tilde{e}_0^2 = \mu^{-\eta} e_0^2, \tag{10.31}$$

---

[6] See, for example, M. Abramowitz and I. A. Stegun, *Handbook of Mathematical Functions*, Dover, New York, 1972, p. 255.

Table 10.1   *The n.u. dimensions of some quantities in $D = 4 - \eta$ space–time dimensions*

Quantity	Dimension
Action	0
Lagrangian density	$D = 4 - \eta$
Klein–Gordon field ϕ	$1 - \eta/2$
Electromagnetic field A^μ	$1 - \eta/2$
Dirac field ψ	$(3 - \eta)/2$
Electric charge e	$\eta/2$

which remains dimensionless independent of η. To see this we note that in $D = 4 - \eta$ dimensions the action is still given by

$$S = \int L \, dt,$$

while the Lagrangian density $\mathscr{L}$ is now defined by

$$L = \int d^{D-1} x \, \mathscr{L}$$

From this we see that the Lagrangian density $\mathscr{L}$ has dimension $D = 4 - \eta$, and the dimensions of the fields and the electric charge can be obtained from the Lagrangian densities (3.4), (5.10), (4.20) and (4.68). They are given in Table 10.1, from which one sees that $\tilde{e}_0^2$ remain dimensionless, while $\tilde{e}_0 \to e_0$ in the limit of four dimensions. In addition, the finite physical results obtained after renormalization are always independent of μ in the limit $D \to 4$, as we shall see explicitly in Sections 10.4 and 10.5.

10.3.2   General results

In order to apply dimensional regularization to QED, we must extend the above ideas in two ways. Firstly, we require other D-dimensional integrals, in addition to Eq. (10.23). Secondly, we must generalize expressions involving γ-matrices.

The relevant integrals are derived from Eq. (10.23) in much the same way in which the standard integrals (10.2)–(10.6) follow from Eq. (10.1). The only integrals we shall require, in addition to Eq. (10.23), are

$$\int d^D k \, \frac{k^\mu}{(k^2 - s + i\varepsilon)^n} = 0, \tag{10.32}$$

$$\int d^D k \, \frac{k^\mu k^\nu}{(k^2 - s + i\varepsilon)^n} = i\pi^{D/2} (-1)^{n+1} \frac{\Gamma(n - D/2 - 1)}{2\Gamma(n)} \frac{g^{\mu\nu}}{s^{n - D/2 - 1}}, \tag{10.33}$$

$$\int d^D k \, \frac{k^2}{(k^2 - s + i\varepsilon)^n} = i\pi^{D/2} (-1)^{n+1} \frac{\Gamma(n - D/2 - 1)}{2\Gamma(n)} \frac{D}{s^{n - D/2 - 1}}, \tag{10.34}$$

where Eq. (10.34) follows from Eq. (10.33), since

$$g_{\mu\nu}g^{\mu\nu} = D. \tag{10.35}$$

As regards the meaning of Eqs. (10.32)–(10.34), they are, in the first place, derived for integer values of D. For non-integer values, the integrals are defined by the expressions on the right-hand sides of these equations. We shall again write $D = 4 - \eta$ and shall require the limit $\eta \to 0$, i.e. $D \to 4$. The reader may feel some unease as to the meaning of $g^{\mu\nu}$ in the right-hand-side expression of Eq. (10.33) when D is not an integer. However, the singularity of this expression, in the limit $\eta \to 0$, arises from the factor $\Gamma(n - D/2 - 1)$, whereas $g^{\mu\nu}$ is non-singular in this limit. Hence only the value of $g^{\mu\nu}$ for $D=4$ enters the final results, and only this value will be required.

We must next see how to handle expressions involving γ-matrices. In the first place, we again consider general integer values of D and introduce a set of γ-matrices $\gamma^0, \gamma^1, \ldots, \gamma^{D-1}$, which satisfy the usual anticommutation relations

$$\gamma^\mu\gamma^\nu + \gamma^\nu\gamma^\mu = 2g^{\mu\nu}. \tag{10.36}$$

From these one derives contraction and trace relations, analogous to Eqs. (A.14)–(A.18). If the γ-matrices are $f(D) \times f(D)$ matrices, and I is the $f(D) \times f(D)$ unit matrix [i.e. $f(D=4) = 4$], one obtains the contraction identities

$$\left. \begin{array}{l} \gamma_\lambda\gamma^\lambda = DI \\ \gamma_\lambda\gamma^\alpha\gamma^\lambda = -(D-2)\gamma^\alpha \\ \gamma_\lambda\gamma^\alpha\gamma^\beta\gamma^\lambda = (D-4)\gamma^\alpha\gamma^\beta + 4g^{\alpha\beta} \end{array} \right\}, \tag{10.37}$$

etc., and the trace relations

$$\left. \begin{array}{l} \mathrm{Tr}\left(\gamma^\alpha\gamma^\beta\right) = f(D)g^{\alpha\beta} \\ \mathrm{Tr}\left(\gamma^\alpha\gamma^\beta\gamma^\gamma\gamma^\delta\right) = f(D)\left[g^{\alpha\beta}g^{\gamma\delta} - g^{\alpha\gamma}g^{\beta\delta} + g^{\alpha\delta}g^{\beta\gamma}\right] \\ \mathrm{Tr}\left(\gamma^\alpha\gamma^\beta \ldots \gamma^\mu\gamma^\nu\right) = 0 \end{array} \right\}, \tag{10.38}$$

where in the last relation $\left(\gamma^\alpha\gamma^\beta \ldots \gamma^\mu\gamma^\nu\right)$ contains an odd number of γ-matrices.

We shall now take over these γ-matrix relations uncritically to the case of $D = 4 - \eta$ dimensions, where η is a small positive number, even though the meaning and existence of γ-matrices in a non-integer number of dimensions is far from clear. In the limit $D \to 4$, the usual relations are restored, and since these relations (unlike the integrals) are non-singular in this limit, only their behavior at $D=4$ enters the final results of any calculation, as we shall see explicitly in the next two sections.

10.4   Vacuum Polarization

We shall now use dimensional regularization to derive the vacuum polarization expression

$$e^2\Pi_c\left(k^2\right) = -\frac{2\alpha}{\pi} \int_0^1 \mathrm{d}\,zz(1-z) \ln\left[1 - \frac{k^2z(1-z)}{m^2}\right], \tag{9.68}$$

which we only quoted in Chapter 9.

We take as starting point Eq. (9.8) for the photon self-energy loop which, after dimensional regularization, reads

$$
ie_0^2 \Pi^{\mu\nu}(k) = \frac{-\tilde{e}_0^2 \mu^{4-D}}{(2\pi)^D} \int d^D p \, \frac{N^{\mu\nu}(p,k)}{\left[(p+k)^2 - m^2 + i\varepsilon\right][p^2 - m^2 + i\varepsilon]}
\tag{10.39}
$$

where μ is the mass scale introduced in Section 10.3.1 and

$$
N^{\mu\nu}(p,k) \equiv \mathrm{Tr}[\gamma^\mu(\slashed{p} + \slashed{k} + m)\gamma^\nu(\slashed{p} + m)].
\tag{10.40}
$$

Evaluating the trace by means of Eqs. (10.38) gives

$$
N^{\mu\nu}(p,k) = f(D)\left\{(p^\mu + k^\mu)p^\nu + (p^\nu + k^\nu)p^\mu + \left[m^2 - p(p+k)\right]g^{\mu\nu}\right\}.
\tag{10.41}
$$

After Feynman parameterization and use of Eq. (10.10), we can write Eq. (10.39)

$$
ie_0^2 \Pi^{\mu\nu}(k) = \frac{-\tilde{e}_0^2 \mu^{4-D}}{(2\pi)^D} \int_0^1 dz \int d^D p \, \frac{N^{\mu\nu}(p,k)}{[p^2 - m^2 + (k^2 + 2pk)z + i\varepsilon]^2}.
\tag{10.42}
$$

If we introduce the new variable

$$
q^\mu = p^\mu + z\,k^\mu
\tag{10.43}
$$

Eq. (10.42) becomes

$$
ie_0^2 \Pi^{\mu\nu}(k) = \frac{-\tilde{e}_0^2 \mu^{4-D}}{(2\pi)^D} \int_0^1 dz \int d^D q \, \frac{N^{\mu\nu}(q - kz, k)}{[q^2 + k^2 z(1-z) - m^2 + i\varepsilon]^2}
\tag{10.44}
$$

with

$$
\begin{aligned}
N^{\mu\nu}(q - kz,\ k) = f(D)\Big\{&[2q^\mu q^\nu - q^2 g^{\mu\nu}] + [m^2 - k^2 z(1-z)]g^{\mu\nu} \\
&+ [-2z(1-z)(k^\mu k^\nu - k^2 g^{\mu\nu})] + \ldots \Big\}
\end{aligned}
\tag{10.45}
$$

where the dots indicate that terms linear in q have been omitted, since these terms vanish on integration, by Eq. (10.32). Combining the last two equations, we obtain

$$
ie_0^2 \Pi^{\mu\nu}(k) = \frac{-\tilde{e}_0^2 \mu^{4-D}}{(2\pi)^D} f(D) \int_0^1 dz \sum_{i=1}^3 I_i^{\mu\nu}(k,z),
\tag{10.46}
$$

where, using Eqs. (10.23) and (10.32)–(10.34), we have

$$
\begin{aligned}
I_1^{\mu\nu}(k,z) &\equiv \int d^D q \, \frac{[2q^\mu q^\nu - q^2 g^{\mu\nu}]}{[q^2 + k^2 z(1-z) - m^2 + i\varepsilon]^2} \\[2mm]
&= \frac{-ig^{\mu\nu}\pi^{D/2}\Gamma(1-D/2)}{[m^2 - k^2 z(1-z)]^{1-D/2}}(1 - D/2),
\end{aligned}
\tag{10.47a}
$$

$$I_2^{\mu\nu}(k,z) \equiv [m^2 - k^2 z(1-z)]g^{\mu\nu} \int d^D q \, \frac{1}{[q^2 + k^2 z(1-z) - m^2 + i\varepsilon]^2}$$

(10.47b)

$$= [m^2 - k^2 z(1-z)]g^{\mu\nu} \frac{i\pi^{D/2}\Gamma(2-D/2)}{[m^2 - k^2 z(1-z)]^{2-D/2}}$$

$$= -I_1^{\mu\nu}(k,z),$$

and

$$I_3^{\mu\nu}(k,z) \equiv [-2z(1-z)(k^\mu k^\nu - k^2 g^{\mu\nu})] \int d^D q \, \frac{1}{[q^2 + k^2 z(1-z) - m^2 + i\varepsilon]^2}$$

(10.47c)

$$= -2z(1-z)(k^\mu k^\nu - k^2 g^{\mu\nu}) \frac{i\pi^{D/2}\Gamma(2-D/2)}{[m^2 - k^2 z(1-z)]^{2-D/2}}.$$

Substituting Eqs. (10.47a)–(10.47c) into Eq. (10.46) gives

$$e_0^2 \Pi^{\mu\nu}(k) = (k^\mu k^\nu - k^2 g^{\mu\nu}) \, \tilde{e}_0^2 \Pi(k^2)$$

(10.48)

where

$$\Pi(k^2) = \frac{2\mu^{4-D} f(D)\Gamma(2-D/2)}{(4\pi)^{D/2}} \int_0^1 dz \, \frac{z(1-z)}{[m^2 - k^2 z(1-z)]^{2-D/2}}.$$

(10.49)

It follows from Eq. (10.48) that the gauge condition[7]

$$k_\mu \Pi^{\mu\nu}(k) = 0$$

(10.50)

holds for any four-vector k. Using dimensional regularization, gauge invariance is automatically satisfied. This is a general result and follows from the fact that dimensional regularization preserves local symmetries of the Lagrangian, such as gauge invariance.

Finally we set $D = 4 - \eta$ and take the limit $\eta \to 0$. Writing $f(4-\eta) = 4 - \eta f'(4) + \cdots$ and using Eqs. (10.28) and (10.29), we obtain from Eq. (10.49) in the limit as $\eta \to 0$

$$\Pi(k^2) = \frac{1}{12\pi^2}\left(\frac{2}{\eta} - \gamma - \frac{f'(4)}{2} + \ln 4\pi\right)$$

(10.52a)

$$- \frac{1}{2\pi^2} \int_0^1 dz \, z(1-z) \ln\left[\frac{m^2 - k^2 z(1-z)}{\mu^2}\right].$$

---

[7] That Eq. (10.50) is a consequence of the gauge invariance condition (8.32) can be seen as follows. Inserting a vacuum polarization loop in an external field line [for example, going from Fig. 9.1 to Fig. 9.2(c)] leads to the following modification of the corresponding Feynman amplitude

$$\mathcal{M}_\alpha(\mathbf{k})A_e^\alpha(\mathbf{k}) \to \mathcal{M}'_\alpha(\mathbf{k})A_e^\alpha(\mathbf{k}) = \left[\mathcal{M}_\alpha(\mathbf{k}) + \mathcal{M}_\lambda(\mathbf{k})iD_F^{\lambda\mu}(\mathbf{k})ie^2\Pi_{\mu\alpha}(\mathbf{k})\right]A_e^\alpha(\mathbf{k}).$$

(10.51)

If $\mathcal{M}$ and $\mathcal{M}'$ satisfy the gauge condition (8.32), then Eq. (10.50) must hold.

Comparing Eq. (10.48) with Eqs. (9.11) and (9.15) gives

$$\Pi(k^2) = A'(0) + \Pi_c(k^2), \tag{10.53}$$

and since $\Pi_c(0)=0$ we obtain from the last two equations

$$\begin{aligned}
e_0^2 \Pi_c(k^2) &= e_0^2[\Pi(k^2) - \Pi(0)] \\
&= -\frac{2\alpha_0}{\pi} \int_0^1 dz\, z(1-z) \ln\left[1 - \frac{k^2 z(1-z)}{m^2}\right]
\end{aligned} \tag{10.54}$$

in the limit $\eta \to 0$. Since $e_0 = e$ to lowest order, this is just the required result (9.68). We note that this result, and all other predictions of measurable quantities, are independent of $f'(4)$, whose value is arbitrary. From now on we shall follow the usual convention and set

$$f'(4) = 0,$$

when Eq. (10.52b) becomes

$$\begin{aligned}
\Pi(k^2) = \frac{1}{12\pi^2}\left(\frac{2}{\eta} - \gamma + \ln 4\pi\right) \\
- \frac{1}{2\pi^2}\int dz\, z(1-z) \ln\left[\frac{m^2 - k^2 z(1-z)}{\mu^2}\right].
\end{aligned} \tag{10.52b}$$

10.5   The Anomalous Magnetic Moment

As a second application of dimensional regularization we shall derive the anomalous magnetic moment of the electron to order α, Eq. (9.72).

We saw in Section 9.6.1 that this correction to the magnetic moment stems from the vertex correction, Fig. 9.15(c), given by Eq. (9.48). By dimensional regularization, retaining the infrared cut-off λ, we can rewrite this equation

$$\begin{aligned}
&e_0^2 \Lambda^\mu(p', p) \\
&= \frac{-i\tilde{e}_0^2 \mu^{4-D}}{(2\pi)^D}\int d^D k \frac{N^\mu(p',p,\,k)}{(k^2 - \lambda^2 + i\varepsilon)[(p'-k)^2 - m^2 + i\varepsilon][(p-k)^2 - m^2 + i\varepsilon]},
\end{aligned} \tag{10.55}$$

where μ is the mass scale introduced in Section 10.3.1 and

$$N^\mu(p', p,\, k) = \gamma^\alpha(\not{p}' - \not{k} + m)\gamma^\mu(\not{p} - \not{k} + m)\gamma_\alpha. \tag{10.56}$$

By means of Feynman parameterization [Eq. (10.11b)], we can write Eq. (10.55) as

$$e_0^2 \Lambda^\mu(p', p) = \frac{-i\tilde{e}_0^2 \mu^{4-D}}{(2\pi)^D}\int_0^1 dy \int_0^{1-y} dz \int d^D k \frac{2N^\mu(p',p,\,k)}{[k^2 - 2k(p'y + pz) - r + i\varepsilon]^3} \tag{10.57}$$

where

$$r \equiv \lambda^2(1-y-z) - y(p'^2 - m^2) - z(p^2 - m^2). \tag{10.58}$$

Expressed in terms of the new variable

$$t^\mu = k^\mu - a^\mu \equiv k^\mu - (p'y + pz)^\mu, \tag{10.59}$$

Eq. (10.57) becomes

$$e_0^2 \Lambda^\mu(p', p) = \frac{-i\tilde{e}_0^2 \mu^{4-D}}{(2\pi)^D} \int_0^1 dy \int_0^{1-y} dz \int d^D t \, \frac{2N^\mu(p', p, \; t+a)}{[t^2 - r - a^2 + i\varepsilon]^3}. \tag{10.60}$$

In order to carry out the t-integration, we rearrange $N^\mu(p', p, \; t+a)$ as a sum of terms proportional to different powers of t

$$N^\mu(p', p, \; t+a) = \sum_{i=0}^{2} N_i^\mu(p', p), \tag{10.61}$$

where

$$N_0^\mu(p', p) \equiv \gamma^\alpha(\not{p}' - \not{a} + m)\gamma^\mu(\not{p} - \not{a} + m)\gamma_\alpha \tag{10.62a}$$

$$N_1^\mu(p', p) \equiv -\gamma^\alpha \left[\not{t}\gamma^\mu(\not{p} - \not{a} + m) + (\not{p}' - \not{a} + m)\gamma^\mu \not{t} \right] \gamma_\alpha \tag{10.62b}$$

$$N_2^\mu(p', p) \equiv \gamma^\alpha \not{t}\gamma^\mu \not{t}\gamma_\alpha. \tag{10.62c}$$

(For simplicity we have written $N_i^\mu(p', p)$, although in addition to being functions of p' and p, N_0^μ depends on a, N_2^μ on t, and N_1^μ on both a and t.) Eq. (10.60) can then be written

$$e_0^2 \Lambda^\mu(p', p) = \sum_{i=0}^{2} e_0^2 \Lambda_{i\mu}(p', p), \tag{10.63a}$$

where

$$e_0^2 \Lambda_i^\mu(p', p) = \frac{-i\tilde{e}_0^2 \mu^{4-D}}{(2\pi)^D} \int_0^1 dy \int_0^{1-y} dz \int d^D t \, \frac{2N_i^\mu(p', p)}{[t^2 - r - a^2 + i\varepsilon]^3}. \tag{10.63b}$$

Of the three terms in Eq. (10.63a), Λ_1^μ vanishes, since the integrand is odd in t [see Eq. (10.32)], Λ_0^μ is non-vanishing and finite in the limit $D \to 4$, and Λ_2^μ diverges in this limit.

We first consider the divergent term Λ_2^μ. Substituting Eq. (10.62c) in Eq. (10.63b), we evaluate the t-integral by means of Eq. (10.33) and obtain

$$e_0^2 \Lambda_2^\mu(p', p) = \frac{\tilde{e}_0^2 \mu^{4-D}}{(2\pi)^D} \frac{\pi^{D/2} \Gamma(2 - D/2)}{\Gamma(3)} \int_0^1 dy \int_0^{1-y} dz \, \frac{\gamma^\alpha \gamma_\sigma \gamma^\mu \gamma^\sigma \gamma_\alpha}{[-(r+a^2)]^{2-D/2}}. \tag{10.64}$$

Setting $D = 4 - \eta$, one obtains from this equation, on account of Eqs. (10.37), (10.28) and (10.29),

$$e_0^2 \Lambda_2^\mu(p', p) = \gamma^\mu \frac{\tilde{e}_0^2}{8\pi^2} \int_0^1 dy \int_0^{1-y} dz \left\{ \left(\frac{2}{\eta} - \gamma + \ln 4\pi \right) \right.$$

$$\left. -2 - \ln \left[\frac{r + a^2}{\mu^2} \right] \right\}, \tag{10.65}$$

where we have neglected terms which vanish in the limit $\eta \to 0$.

In Chapter 9, the observable part $\Lambda_c^\mu(p', p)$ of the vertex correction was defined by the equation

$$\Lambda^\mu(p',p) = L\gamma^\mu + \Lambda_c^\mu(p',p). \tag{9.53}$$

We have now shown that the divergent part of $\Lambda^\mu(p',p)$ is given by the $1/\eta$ term in $\Lambda_2^\mu(p', p)$, Eq. (10.65). In identifying Eq. (9.53) with Eq. (10.63a), this $1/\eta$ term is incorporated in the term $L\gamma^\mu$ of Eq. (9.53). It follows that $\Lambda_c^\mu(p',p)$ is finite, as shown by a different approach in Section 9.5.

In order to derive the anomalous magnetic moment of the electron, we next consider the observable corrections which the vertex modification $e^2\Lambda^\mu(p',p)$ contributes to the scattering of electrons by an external static electromagnetic field. The Feynman amplitude for this is given by

$$\mathcal{M} = ie_0\bar{u}(\mathbf{p}')e_0^2\Lambda^\mu(p',p)u(\mathbf{p})A_{e\mu}(\mathbf{q}=\mathbf{p}'-\mathbf{p}). \tag{10.66}$$

The most general form for this amplitude, allowing for Lorentz invariance, the Lorentz gauge condition

$$q_\mu A_e^\mu(\mathbf{q}) = 0 \tag{10.67}$$

and the Gordon identity (see Problem A.2) is

$$\mathcal{M} = ie_0\bar{u}(\mathbf{p}')\left[\gamma^\mu F_1(q^2) + \frac{i}{2m}\sigma^{\mu\nu}q_\nu F_2(q^2)\right]u(\mathbf{p})A_{e\mu}(\mathbf{q}) \tag{10.68}$$

where F_1 and F_2 are arbitrary functions of q^2. In order to calculate the magnetic moment, we need only consider the second term in Eq. (10.68). Comparing Eqs. (10.68) and (10.65), we see that Λ_2^μ makes no contribution to $F_2(q^2)$, so that the correction to the magnetic moment arises entirely from the term $\Lambda_0^\mu(p',p)$, defined by Eqs. (10.62a) and (10.63b). The t-integration in the latter equation is easily carried out, using Eq. (10.23) with $D=4$, and leads to

$$\bar{u}(\mathbf{p}')e^2\Lambda_0^\mu(p', p)u(\mathbf{p}) = \frac{-\alpha}{4\pi}\int_0^1 dy \int_0^{1-y} dz \frac{\bar{u}(\mathbf{p}')N_0^\mu(p',p)u(\mathbf{p})}{(r+a^2)} \tag{10.69}$$

in the limit $\eta \to 0$, where we can set $e_0 = e$ to lowest order.

Eq. (10.69) is a well-defined finite expression. The remaining analysis required to obtain the anomalous magnetic moment is straightforward, but lengthy, and we shall omit it. It involves evaluation of the spinor matrix element $\bar{u}(\mathbf{p}')N_0^\mu(p',p)u(\mathbf{p})$, utilizing the commutation and contraction relations of the γ-matrices, the Dirac equation, the Gordon identity and the Lorentz condition (10.67). Retaining only terms which contribute to the magnetic moment [i.e. are of the form of the second term in Eq. (10.68)], one obtains

$$F_2(q^2) = \frac{m^2\alpha}{\pi}\int_0^1 dy \int_0^{1-y} dz \frac{(y+z)(1-y-z)}{\lambda^2(1-y-z)+(p'y+pz)^2}.$$

This integral is well-defined in the limit $\lambda \to 0$. Hence setting $\lambda = 0$ and $p' = p$, with $p^2 = m^2$, gives finally

$$F_2(0) = \frac{\alpha}{\pi}\int_0^1 dy \int_0^{1-y} dz \frac{1-y-z}{y+z} = \frac{\alpha}{2\pi}, \tag{10.70}$$

which is the second-order correction to the magnetic moment quoted earlier, Eq. (9.72).

Problems

10.1. Derive Eq. (10.26) from Eq. (10.25), i.e. by using the cut-off method of regularization.

10.2. In the dimensional regularization scheme, the electron self-energy (9.4) is given by

$$ie_0^2\sum(p) = \frac{-\tilde{e}_0^2\mu^{4-D}}{(2\pi)^D}\int d^D k \frac{\gamma^\alpha(\not{p}-\not{k}+m)\gamma^\beta}{(p-k)^2-m^2+i\varepsilon}\frac{g_{\alpha\beta}}{k^2-\lambda^2+i\varepsilon} \qquad (10.71)$$

in the limit as $D \to 4$, where μ is the mass scale introduced in Section 10.3.1 and we have also introduced a small cut-off parameter λ to guard against infrared divergences.

Show, using Feynman parameterization and the contraction identities (10.37) that this can be rewritten in the form

$$ie_0^2\sum(p) = \frac{-\tilde{e}_0^2}{(2\pi)^D}\mu^{4-D}\int_0^1 dz\int d^D q\frac{[Dm-(D-2)(\not{p}-\not{q}-\not{p}z)]}{(q^2-s)^2}, \qquad (10.72)$$

where

$$q = k - pz$$

and

$$s = m^2 z + \lambda^2(1-z) - p^2 z(1-z).$$

Finally, evaluate the integral over q using the standard integrals (10.23) and (10.32) and take the limits $\eta = D-4 \to 0$ and $\lambda \to 0$ to obtain the result[8]

$$e_0^2\sum(p) = \frac{\tilde{e}_0^2}{16\pi^2}(\not{p}-4m)\left[\frac{2}{\eta}-\gamma+\ln 4\pi\right] + \tilde{e}_0^2\sum_c(p) \qquad (10.73a)$$

where

$$16\pi^2\sum_c(p) = (2m-\not{p})$$
$$-2\int_0^1 dz[\not{p}(1-z)-4m]\ln\left(\frac{m^2z-p^2z(1-z)}{\mu^2}\right) \qquad (10.73b)$$

10.3. Compare the above result (10.73) with the expansion

$$e_0^2\sum(p) = -\delta m + (\not{p}-m)(1-Z_2) + (\not{p}-m)e_0^2\sum_c(p),$$

keeping only those terms which diverge in the limit $\eta = D-4 \to 0$. Hence show that the corresponding contributions to δm and Z_2 are given by

$$\delta m = \frac{1}{\eta}\left(\frac{3\alpha_0 m}{2\pi}\right) \qquad Z_2 = -\frac{1}{\eta}\left(\frac{\alpha_0}{2\pi}\right),$$

while $\sum_c(p)$ remains finite.

Evaluate the corresponding contribution to the vertex renormalization constant $Z_1 = (1 - e_0^2 L + \ldots)$ using Eqs. (9.53), (10.63) and (10.65). Hence, use Ward's identity to check the above value for Z_2.

---

[8] This result will be needed in the following problem and later in section 15.2.

11

Gauge Theories

So far we have focussed exclusively on quantum electrodynamics, where the form of the interaction is well known from the classical theory of Maxwell and Lorentz. For the equally fundamental 'strong' and 'weak' forces between elementary particles, there are no classical theories to guide us. The only way forward is to postulate suitable forms for the interactions and test their predictions against experiment. Fortunately these forms are severely restricted by general theoretical and experimental requirements. Theoretically, we will require the theory to be Lorentz invariant, local and renormalizable. By local we mean that the interaction Lagrangian density $\mathscr{L}_I(x)$ only involves products of fields evaluated at the same space–time point x^μ, in order to avoid problems with action at a distance. By renormalizable we mean that its physical predictions in terms of a finite number of free parameters (i.e. masses and coupling constants) remain finite and well defined when all cut-offs are removed, as discussed in Section 9.9 for QED. In addition, the postulated interactions should respect the particle spectrum, symmetries and conservation laws suggested by experiment, in so far as these are known.

It remains to find Lagrangians compatible with the above principles which successfully describe the strong and weak interactions. Historically, this was achieved by an approach which emphasized the roles of renormalizability and gauge invariance. These requirements completely determine the form of the interaction in QED, as we shall see. In 1954, this led Yang and Mills to propose generalizations of the electromagnetic gauge transformations (4.69) in order to construct other possible 'gauge theories'. At first, this approach seemed hopeless, since it predicted the existence of zero-mass bosons, which had not been observed. However, this problem was eventually overcome and the modern theories of strong, weak and electromagnetic interactions are all examples of gauge theories.

In this chapter, we concentrate on the roles of renormalizability and gauge invariance in determining the forms of the electromagnetic and strong interactions. The rest of the book will then be divided into two parts, which can be read independently of each other, as follows.

Quantum Field Theory, Second Edition   Franz Mandl and Graham Shaw
© 2010 John Wiley & Sons, Ltd

The modern theory of strong interactions is characterized by an apparently simple interaction Lagrangian, as we shall shortly see. However, apart from a few general features, the physical properties of the theory are difficult to ascertain because of technical problems in formulating perturbation theory, and because low-order perturbation theory is not in general a useful approximation. In Chapters 12–15. we shall introduce the formal methods required to deal with these problems and use them to obtain some of the more important properties of the strong interactions.

Chapters 16–19 are devoted to the unified theory of electromagnetic and weak interactions. In contrast to the theory of strong interactions, this is characterized by an extremely complicated interaction Lagrangian density. However, having finally obtained the interaction, the main physical properties of the theory are easily obtained using the perturbative methods already introduced in the context of QED.

11.1   The Simplest Gauge Theory: QED

In Section 4.5 we introduced the electromagnetic interaction into the free-fermion Lagrangian density

$$\mathcal{L}_0 = \overline{\psi}(x)(i\gamma^\mu \partial_\mu - m)\psi(x) \tag{11.1}$$

through the minimal substitution

$$\partial_\mu \rightarrow D_\mu = [\partial_\mu + iqA_\mu(x)], \tag{4.64b}$$

where q is the charge of the particle annihilated by the field $\psi(x)$. We required invariance of the resulting theory, i.e. of the Lagrangian density

$$\mathcal{L} = \overline{\psi}(x)(i\gamma^\mu D_\mu - m)\psi(x) = \mathcal{L}_0 - q\overline{\psi}(x)\gamma^\mu \psi(x)A_\mu(x),$$

under gauge transformations of the electromagnetic field

$$A_\mu(x) \rightarrow A'_\mu(x) = A_\mu(x) + \partial_\mu f(x), \tag{11.2a}$$

where $f(x)$ is a real differentiable function. This invariance was ensured if, coupled with the transformation (11.2a), the Dirac fields $\psi(x)$ and $\overline{\psi}(x)$ underwent the transformations

$$\left. \begin{array}{l} \psi(x) \rightarrow \psi'(x) = \psi(x)e^{-iqf(x)} \\ \overline{\psi}(x) \rightarrow \overline{\psi}'(x) = \overline{\psi}(x)e^{iqf(x)} \end{array} \right\}. \tag{11.2b}$$

Eqs. (11.2b) have the form of a *local* phase transformation. We shall refer to the *coupled* transformation (11.2a) and (11.2b) as a gauge transformation and to any theory which is invariant under such coupled transformations as a gauge theory. QED is the simplest example of a gauge theory.

In the above discussion, we started from the form of the electromagnetic interaction. The invariance of the theory under the gauge transformations (11.2a) of the electromagnetic potentials then required the Dirac fields ψ and $\overline{\psi}$ simultaneously to undergo the local phase transformation (11.2b). We can try and reverse this argument and start from the invariance of the free Lagrangian density $\mathscr{L}_0$ under the *global* phase transformation

$$\left. \begin{array}{l} \psi(x) \rightarrow \psi'(x) = \psi(x)e^{-i\alpha} \\ \overline{\psi}(x) \rightarrow \overline{\psi}'(x) = \overline{\psi}(x)e^{i\alpha} \end{array} \right\}, \tag{11.3}$$

when α is a real number. We know from the discussion following Eq. (2.40) and from Eqs. (4.27) to (4.29) that this invariance ensures 'current conservation', i.e. the current

$$s^\mu(x) = q\overline{\psi}(x)\gamma^\mu\psi(x) \tag{11.4}$$

satisfies $\partial_\mu s^\mu(x) = 0$, so that the charge

$$Q = q \int d^3x \psi^\dagger(x)\psi(x) \tag{11.5}$$

is conserved.

Invariance under the global phase transformation (11.3) allows us to change the phase of the field by the same amount at each space–time point. This appears unnecessarily restrictive in a local field theory. We shall therefore demand invariance with respect to the more general local phase transformations (11.2b). Under these transformations, the free-field Lagrangian density $\mathscr{L}_0$ becomes

$$\mathscr{L}_0 \rightarrow \mathscr{L}_0' = \mathscr{L}_0 + q\overline{\psi}(x)\gamma^\mu\psi(x)\partial_\mu f(x), \tag{11.6}$$

i.e. $\mathscr{L}_0$ is not invariant. Invariance of the theory is then restored if we can augment $\mathscr{L}_0$ by a term $\mathscr{L}_I$ such that the new Lagrangian density $\mathscr{L}=\mathscr{L}_0 + \mathscr{L}_I$ is invariant. This can be achieved by associating with the 'matter field' $\psi(x)$ a 'gauge field' $A_\mu(x)$ which transforms according to the gauge transformation (11.2a). The interaction between matter and gauge fields is then specified by making the minimal substitution (4.64b) in the free-field Lagrangian density $\mathscr{L}_0$ i.e. by replacing the ordinary derivative $\partial_\mu\psi(x)$ by the 'covariant derivative'

$$D_\mu\psi(x) = [\partial_\mu + iqA_\mu(x)]\psi(x). \tag{11.7}$$

$\mathscr{L}_0$ thus goes over into

$$\mathscr{L} = \overline{\psi}(x)(i\gamma^\mu D_\mu - m)\,\psi(x) = \mathscr{L}_0 + \mathscr{L}_I \tag{11.8a}$$

where

$$\mathscr{L}_I = -q\overline{\psi}(x)\gamma^\mu\psi(x)A_\mu(x). \tag{11.8b}$$

Under the coupled gauge transformation (11.2a) and (11.2b), the covariant derivative $D_\mu\psi(x)$ undergoes the transformation

$$D_\mu\psi(x) \rightarrow e^{-iqf(x)}D_\mu\psi(x), \tag{11.9}$$

i.e. it transforms in the same way as the field $\psi(x)$ itself, Eq. (11.2b). Hence $\mathscr{L}$ is invariant under gauge transformations, as required.

To summarize this approach: if one takes, as the basic requirement, invariance with respect to local phase transformations of the matter field, one is led to introduce a gauge field coupled to the matter field through the replacement of the ordinary derivative $\partial_\mu \psi$ by the covariant derivative $D_\mu \psi$, resulting in the familiar interaction (11.8b). This is called a *minimal gauge interaction* because it can, in principle, be modified by adding further, more complicated, terms, which are themselves gauge invariant. For the moment we will ignore such *non-minimal* interactions because they are not needed in QED and because they lead to non-renormalizable theories, as we shall show later.

11.2   Quantum Chromodynamics

In QED, the 'gauge theory' approach we have outlined is an interesting way of looking at an interaction whose form was originally determined in other ways. In strong interactions it was used to infer the detailed forms of interactions which were previously unknown. The resulting theory is called quantum chromodynamics (QCD) because of the central role played by the so-called colour charges. In this section, we will first introduce the idea of colour in the context of the simple quark model. We will then formulate QCD as a gauge theory in analogy to QED, but starting from the conservation of colour rather than electric charge.

11.2.1   Colour and confinement

Nucleons, pions and other hadrons are bound states of more fundamental spin-$\frac{1}{2}$ fermions called quarks.[1] There are now six known types of quarks, called *flavours*, characterized by different masses and by their electric charges. The d, s and b quarks have negative charge $-e/3$ while the u, c and t quarks have positive charge $2e/3$. In the simple quark model, the observed baryons are assumed to be bound states of three quarks, while the mesons are assumed to be bound states of a quark and an antiquark. This model gives a remarkably successful description of the observed hadron spectrum, but has two paradoxical features. Firstly free quarks and other fractionally charged states, like two quark bound states, are never observed. Secondly, the combined space and spin wavefunctions of the baryons are symmetric under interchange of quarks of the same flavour, in apparent contradiction to the fundamental spin-statistics theorem discussed in Section 4.3.1. This problem is especially acute for baryons like the $\Omega^- \equiv sss$ baryon, which seem to contain three identical spin $\frac{1}{2}$ quarks in a totally symmetric wavefunction.

Both the above phenomena are naturally explained by the theory of colour, developed by Han, Nambu and Greenberg in 1964–1965. The basic assumption is that in addition to space and spin degrees of freedom, quarks (but not leptons) possess another attribute called colour. Specifically it is assumed that any quark $q = u, d, s \dots$ can exist in three different 'colour states' – an assumption which has subsequently been directly verified by experiment.[2] Correspondingly quark wavefunctions

$$\Psi = \psi \chi^c$$

---

[1] No detailed knowledge of the quark model is required beyond what is summarized in this paragraph. For general background reading on the quark model, see B. R. Martin and G. Shaw, *Particle Physics*, 3rd edn., John Wiley & Sons, Ltd, Chichester (2008); in particular, Chapters 3 and 6.
[2] See for example, Section 7.2.3 of Martin and Shaw, cited above.

are written as the product of a space/spin part ψ and a colour wavefunction χ^c. For a single quark, the three independent colour wavefunctions are usually denoted by $\chi^c = r, g, b$ and are conveniently represented by the 'colour spinors'

$$
r = \begin{pmatrix} 1 \\ 0 \\ 0 \end{pmatrix} \quad g = \begin{pmatrix} 0 \\ 1 \\ 0 \end{pmatrix} \quad b = \begin{pmatrix} 0 \\ 0 \\ 1 \end{pmatrix} \tag{11.10}
$$

in the same way that the two non-relativistic spin wavefunctions $\chi = \alpha, \beta$ of a spin $\frac{1}{2}$ particle can be represented by the Pauli spinors

$$
\alpha = \begin{pmatrix} 1 \\ 0 \end{pmatrix} \quad \beta = \begin{pmatrix} 0 \\ 1 \end{pmatrix}.
$$

Just as the spin wavefunctions are acted on by spin operators which can be represented by two-dimensional Hermitian matrices, the colour wavefunctions are acted on by 'colour operators' which can be represented by three-dimensional Hermitian matrices. Apart from the unit matrix, there are eight linearly independent Hermitian matrices in three dimensions. These are conveniently chosen to be

$$
\hat{F}_i = \frac{1}{2}\lambda_i \quad (i = 1, 2, \ldots, 8) \tag{11.11a}
$$

where the matrices

$$
\lambda_1 = \begin{pmatrix} 0 & 1 & 0 \\ 1 & 0 & 0 \\ 0 & 0 & 0 \end{pmatrix} \quad \lambda_2 = \begin{pmatrix} 0 & -i & 0 \\ i & 0 & 0 \\ 0 & 0 & 0 \end{pmatrix} \quad \lambda_3 = \begin{pmatrix} 1 & 0 & 0 \\ 0 & -1 & 0 \\ 0 & 0 & 0 \end{pmatrix}
$$

$$
\lambda_4 = \begin{pmatrix} 0 & 0 & 1 \\ 0 & 0 & 0 \\ 1 & 0 & 0 \end{pmatrix} \quad \lambda_5 = \begin{pmatrix} 0 & 0 & -i \\ 0 & 0 & 0 \\ i & 0 & 0 \end{pmatrix} \quad \lambda_6 = \begin{pmatrix} 0 & 0 & 0 \\ 0 & 0 & 1 \\ 0 & 1 & 0 \end{pmatrix}
$$

$$
\lambda_7 = \begin{pmatrix} 0 & 0 & 0 \\ 0 & 0 & -i \\ 0 & i & 0 \end{pmatrix} \quad \lambda_8 = \frac{1}{\sqrt{3}}\begin{pmatrix} 1 & 0 & 0 \\ 0 & 1 & 0 \\ 0 & 0 & -2 \end{pmatrix}
$$

$$
\tag{11.11b}
$$

are three-dimensional analogs of the Pauli matrices (A.62).[3] The operators $\hat{F}_i (i = 1, 2, \ldots, 8)$ are called colour charges and satisfy commutation relations of the form

$$
[\hat{F}_i, \hat{F}_j] = i f_{ijk} \hat{F}_k, \tag{11.12a}
$$

where a sum over repeated colour indices $k = 1, 2, \ldots, 8$ is understood here and in what follows. The *structure constants* f_{ijk} are numbers which are easily determined using Eq. (11.11). They are totally antisymmetric

$$
f_{ijk} = -f_{jik} = f_{jki} = -f_{kji} = f_{kij} = -f_{ikj}, \tag{11.12b}
$$

---

[3] In both cases, they are traceless Hermitian matrices and the most general Hermitian matrix in two or three dimensions can be written as a linear combination of the unit matrix and the σ_i or λ_i respectively.

Table 11.1 *Some of the coefficients in the commutation relations (11.12) of the colour charges $\hat{F}_i (i=1, 2, \ldots, 8)$. All the other non-zero coefficients may be deduced from these by using the fact that f_{ijk} is totally antisymmetric*

ijk	123	147	156	246	257	345	367	458	678
f_{ijk}	1	$\frac{1}{2}$	$-\frac{1}{2}$	$\frac{1}{2}$	$\frac{1}{2}$	$\frac{1}{2}$	$-\frac{1}{2}$	$\frac{\sqrt{3}}{2}$	$\frac{\sqrt{3}}{2}$

so that f_{ijk} vanishes if any two of the indices i, j, k are equal. Some of the non-zero coefficients are given in Table 11.1, and the rest can be deduced from them using Eq. (11.12b).

In the colour theory, the colour charges are assumed to be exactly conserved. However they cannot all be required to have simultaneous eigenvalues since they do not commute with each other. Only $\hat{F}_3$ and $\hat{F}_8$ commute and the colour states $\chi^c = r, g, b$ are eigenstates of both $\hat{F}_3$ and $\hat{F}_8$. The corresponding eigenvalues for single quark states follow directly from Eqs. (11.10) and (11.11) and are listed in Table 11.2(a). The remaining operators change the colour state and one easily shows, for example, that

$$\hat{F}_1 r = \frac{1}{2} g \quad \hat{F}_1 g = \frac{1}{2} r \quad \hat{F}_1 b = 0. \tag{11.13}$$

Multiquark states h are described by product wavefunctions

$$\chi_h^c \equiv \chi_1^c \chi_2^c \cdots \chi_n^c,$$

where $\chi_1^c = r_1, g_1, b_1$ etc. and the actions of the charges are determined by the distributive law

$$\hat{F}_i \chi_1^c \chi_2^c = \left(\hat{F}_i \chi_1^c\right) \chi_2^c + \chi_1^c \left(\hat{F}_i \chi_2^c\right) \text{ etc.} \tag{11.14}$$

In particular, the values of F_3 and F_8 are obtained by simply adding the values for the constituent quarks using Table 11.2(a). For the corresponding antiquark states $\bar{r}$, $\bar{g}$, $\bar{b}$, the values of the colour charges are reversed[4] in sign, as summarized in Table 11.2(b).

Table 11.2 *Values of the colour charges F_3, F_8 for the colour states of quarks and antiquarks*

(a) Quarks			(b) Antiquarks	
	F_3	F_8	F_3	F_8
r	$1/2$	$1/2\sqrt{3}$	$\bar{r}$ $-1/2$	$-1/2\sqrt{3}$
g	$-1/2$	$1/2\sqrt{3}$	$\bar{g}$ $1/2$	$-1/2\sqrt{3}$
b	0	$-1/\sqrt{3}$	$\bar{b}$ 0	$1/\sqrt{3}$

---

[4] This result, which is like that for electric charge, will be derived in the next section. For the moment it can be regarded as an assumption.

The apparently paradoxical features of the simple quark model can now be shown to follow directly from the hypothesis of colour confinement. This is the requirement that free hadrons exist only in colour singlet states χ_h^c satisfying

$$\hat{F}_i \chi_h^c = 0 \quad (i = 1, 2, \ldots, 8) \tag{11.15}$$

and in particular

$$F_3 = F_8 = 0 \tag{11.16}$$

for any hadron. Using the values given in Table 11.2, one easily sees that Eqs. (11.16) exclude free quarks q, diquarks qq and other fractionally charged states, while the observed combinations $q\bar{q}$ and qqq are allowed. Confinement also resolves the spin-statistics problem for baryons, assumed to be composed of three quarks. From Table 11.2(a) it is clear that Eqs. (11.16) can only be satisfied if baryons are composed of one r, one g and one b quark. This restricts the baryon wavefunction to the form

$$\chi_B^c = \alpha_1 r_1 g_2 b_3 + \alpha_2 g_1 r_2 b_3 + \alpha_3 b_1 r_2 g_3 + \alpha_4 b_1 g_2 r_3 + \alpha_5 g_1 b_2 r_3 + \alpha_6 r_1 b_2 g_3,$$

where, for example, r_3 means that the third quark is in an r state etc. The coefficients $\alpha_i (i = 1, 2, \ldots, 6)$ are constants whose values are restricted by the remaining confinement conditions (11.15). For example, imposing $F_1 \chi_B^c = 0$ and using Eqs. (11.13) and (11.14) gives

$$\alpha_1 = -\alpha_2 \quad \alpha_3 = -\alpha_4 \quad \alpha_5 = -\alpha_6.$$

Similar results follow from the other conditions (11.15). Together they uniquely determine the wavefunction to be

$$\chi_B^c = \varepsilon_{ijk} r_i g_j b_k \tag{11.17}$$

up to an arbitrary normalization, where a sum over repeated indices $i = 1, 2, 3$ etc. is understood as usual. The antisymmetric alternating symbol ε_{ijk} is equal to $+1$ (-1) for (i, j, k) an even (odd) permutation of $(1, 2, 3)$ and vanishes if two or more indices are the same. Because it is totally antisymmetric, applying the Pauli principle to the full wavefunction $\Psi = \psi \chi_B^c$ leads directly to the result that the combined space/spin wavefunction ψ must be symmetric under the interchange of identical quarks, as assumed in the quark model.

11.2.2 Global phase invariance and colour conservation

We next find a set of symmetry transformations which lead to the conservation of the colour charges, rather than electric charge. In order to do this, we start from the free quark Lagrangian density

$$\begin{aligned} \mathscr{L} = & [\overline{\psi}_r^f(x)(i\partial\!\!\!/ - m_f)\psi_r^f(x) \\ & + \overline{\psi}_g^f(x)(i\partial\!\!\!/ - m_f)\psi_g^f(x) + \overline{\psi}_b^f(x)(i\partial\!\!\!/ - m_f)\psi_b^f(x)], \end{aligned} \tag{11.18}$$

where a sum over the repeated flavour index $f = u, d, s, c, b, t$ is understood here and in what follows. The Dirac fields $\psi^f_{r,g,b}$ correspond to quarks of flavour f and colour r, g, b respectively, and are conveniently combined into three component fields

$$\Psi^f(x) = \begin{pmatrix} \psi^f_r(x) \\ \psi^f_g(x) \\ \psi^f_b(x) \end{pmatrix} \tag{11.19a}$$

and correspondingly

$$\overline{\Psi}^f(x) = (\overline{\psi}^f_r(x),\ \overline{\psi}^f_g(x),\ \overline{\psi}^f_b(x)). \tag{11.19b}$$

The Lagrangian density (11.18) then becomes

$$\mathcal{L}(x) = \overline{\Psi}^f(x)(i\,\partial\!\!\!/ - m_f)\Psi^f(x) \tag{11.20}$$

and is clearly invariant under the global transformations

$$\left.\begin{aligned} \Psi^f(x) &\rightarrow \Psi^{f\prime}(x) = U(\alpha)\Psi^f(x) \equiv e^{i\alpha_i\lambda_i/2}\Psi^f(x) \\ \overline{\Psi}^f(x) &\rightarrow \overline{\Psi}^{f\prime}(x) = \overline{\Psi}^f(x)U^\dagger(\alpha) \equiv \overline{\Psi}^f(x)e^{-i\alpha_i\lambda_i/2} \end{aligned}\right\} \tag{11.21}$$

where λ_i are the matrices (11.11b) and the α_i are eight arbitrary real numbers with $\alpha \equiv (\alpha_1, \alpha_2, \ldots, \alpha_8)$. The operators $U(\alpha)$ comprise the group SU(3) of 3×3 unitary matrices with the 'special' property that $\det U(\alpha) = +1$. For this reason the transformations are called SU(3) transformations. They can be regarded as three-dimensional global phase transformations, i.e. as a generalization of the one-dimensional global phase transformations (11.3).

Conservation of the colour charges follows from invariance under the SU(3) transformations (11.21). For infinitesimal α_i, the transformations reduce to

$$\left.\begin{aligned} \Psi^f(x) &\rightarrow \Psi^{f\prime}(x) = (1 + i\alpha_i\lambda_i/2)\Psi^f(x) \\ \overline{\Psi}^f(x) &\rightarrow \overline{\Psi}^{f\prime}(x) = \overline{\Psi}^f(x)(1 - i\alpha_i\lambda_i/2) \end{aligned}\right\} \tag{11.22}$$

and an argument analogous to that following (2.39) leads to the eight conserved currents

$$S^\mu_i(x) = \frac{1}{2}\overline{\Psi}^f(x)\gamma^\mu\lambda_i\Psi^f(x) \quad (i = 1, 2, \ldots, 8). \tag{11.23}$$

The corresponding conserved charges

$$\hat{F}_i \equiv \int d^3x S^0_i(x) = \frac{1}{2}\int d^3x \Psi^{f\,\dagger}(x)\lambda_i\Psi^f(x) \quad (i = 1, 2, \ldots, 8) \tag{11.24}$$

are just the field-theoretic realization of the colour charges introduced earlier. To confirm this, we explicitly consider

$$\hat{F}_3 = \frac{1}{2}\int d^3x N\left[\psi^{f\,\dagger}_r(x)\psi_r(x) - \psi^{f\,\dagger}_g(x)\psi^f_g(x)\right]$$

where we have used Eqs. (11.11) and (11.19) and have emphasized that, on quantization, products of field operators are to be interpreted as normal products as usual. On expanding the fields according to Eq. (4.38), this eventually gives

$$\hat{F}_3 = \frac{1}{2}\left(N_r - \overline{N}_r\right) - \frac{1}{2}\left(N_g - \overline{N}_g\right) \tag{11.25}$$

where $N_c\left(\overline{N}_c\right)$ are the number operators for quarks (antiquarks) of colour index c, summed over all flavour, momentum and spin states. This is consistent with our earlier results (cf. Table 11.2), as are the results obtained in a similar way for the other colour charges (cf. Problem 11.1).

11.2.3 SU(3) gauge invariance

We next generalize the SU(3) transformations (11.22) from global to local phase transformations. The development will be very similar to that for QED in Section 11.1. In order to retain invariance under local phase transformations, we shall have to introduce gauge fields, and this will automatically generate the interactions.

The first step is to replace the global SU(3) transformations (11.21) by the local phase transformations

$$\left. \begin{array}{l} \Psi^f(x) \rightarrow \Psi^{f\prime}(x) = \exp\left[i\, g_s\lambda_j\omega_j(x)/2\right]\Psi^f(x) \\ \overline{\Psi}^f(x) \rightarrow \overline{\Psi}^{f\prime}(x) = \overline{\Psi}^f(x)\exp\left[-i\, g_s\lambda_j\omega_j(x)/2\right] \end{array} \right\}. \tag{11.26a}$$

Here $\omega_j(x)$ ($j = 1, 2, \ldots, 8$) are arbitrary real differentiable functions and g_s is a real constant which will later be identified with a coupling constant.

The free-field Lagrangian density (11.20) is not invariant under the transformation (11.26), but transforms as

$$\mathscr{L}_0 \rightarrow \mathscr{L}_0' = \mathscr{L}_0 - \frac{1}{2}g_s\overline{\Psi}^f(x)\lambda_j\partial\!\!\!/\omega_j(x)\Psi^f(x)$$

for infinitesimal $\omega_j(x)$. As we shall see, an invariant Lagrangian density can be obtained if, in analogy with Eqs. (11.7) and (11.8) in QED, we replace $\partial^\mu\Psi^f(x)$ in Eq. (11.20) by the covariant derivative

$$D^\mu\Psi^f(x) = \left[\partial^\mu + ig_s\lambda_jA_j^\mu(x)/2\right]\Psi^f(x) \tag{11.27}$$

so that $\mathscr{L}_0$ goes over to

$$\mathscr{L}_q(x) = \overline{\Psi}^f(x)\left[i\,D\!\!\!\!/ - m_f\right]\Psi^f(x) = \mathscr{L}_0 + \mathscr{L}_1 \tag{11.28a}$$

where

$$\mathscr{L}_1 = -\frac{1}{2}g_s\overline{\Psi}^f(x)\gamma_\mu\lambda_j\Psi^f(x)A_j^\mu(x). \tag{11.28b}$$

Here we have introduced eight real gauge fields $A_j^\mu(x)$, called gluon fields, as there are eight conserved charges and eight arbitrary real functions $\omega_j(x)$ in the gauge transformations (11.26a). Like the corresponding QED interaction (11.8b), the interaction (11.28b) generates a three-point quark-gluon vertex in perturbation theory, as illustrated in Fig. 11.1(a).

Figure 11.1   *The quark–gluon vertex generated by Eq. (11.28b) and the gluon–gluon vertices generated by the second and third terms in Eq. (11.37)*

Because the matrices λ_j are not all diagonal, this interaction can annihilate quarks of one colour r, g, b and create quarks of a different colour. Hence by colour conservation, the quanta of the gauge fields $A_j^\mu(x)$ – called gluons – must also have non-zero colour charges. This is in contrast to QED, where the interactions are transmitted by photons which have zero electric charge.

For the modified Lagrangian density $\mathcal{L}_q$ to be invariant under the transformations (11.26a), these transformations must be coupled to transformations of the gluon fields $A_j^\mu(x)$ which are chosen so that the covariant derivatives $D^\mu \Psi^f(x)$ transform in the same way as the fields $\Psi^f(x)$ themselves, i.e.

$$D^\mu \Psi^f(x) \rightarrow \exp\left[ig_s \lambda_j \omega_j(x)/2\right] D^\mu \Psi^f(x). \tag{11.29}$$

For finite $\omega_j(x)$, the resulting transformation law is quite complicated. However, it suffices to consider the transformations for infinitesimal functions $\omega_j(x)$. In the appendix to this chapter, we shall show that the required infinitesimal transformations are given by

$$A_i^\mu(x) \rightarrow A_i^{\mu\prime}(x) \equiv A_i^\mu(x) - \partial^\mu \omega_i(x) - g_s f_{ijk} \omega_j(x) A_k^\mu(x) \tag{11.26b}$$

for small $\omega_j(x)$, where f_{ijk} are the structure constants defined in Eqs. (11.12). The coupled transformations (11.26 a, b) are called SU(3) gauge transformations and any theory which is invariant under them is SU(3) gauge invariant.

The Lagrangian density (11.28) described the quark fields and their interactions with the gluon fields. The complete QCD Lagrangian density also contains terms which describe the gluons when no quarks are present. These terms are also required to be SU(3) gauge invariant.

In QED, the Lagrangian density (5.8) contains a term

$$-\frac{1}{4} F_{\mu\nu} F^{\mu\nu} \tag{11.30}$$

which describes free photons, where $F_{\mu\nu}(x)$ is the electromagnetic field tensor (5.5). This free-photon term (11.30) is gauge invariant, as it depends only on the electromagnetic field tensor $F^{\mu\nu}$. An analogous expression

$$-\frac{1}{4} F_{i\mu\nu}(x)\, F_i^{\mu\nu}(x), \tag{11.31}$$

where

$$F_i^{\mu\nu}(x) \equiv \partial^\nu A_i^\mu(x) - \partial^\mu A_i^\nu(x), \tag{11.32}$$

would similarly describe free massless gluons. However, this expression is *not* invariant under gauge transformations, on account of the term $-g_s f_{ijk}\omega_j(k)A_j^\mu(x)$ in Eq. (11.26b). To restore gauge invariance, additional interaction terms must again be introduced. As is shown in the appendix to this chapter, the expression obtained by replacing $F_i^{\mu\nu}(x)$ by

$$G_i^{\mu\nu}(x) \equiv F_i^{\mu\nu}(x) + g_s f_{ijk}A_j^\mu(x)A_k^\nu(x) \tag{11.33}$$

in Eq. (11.31), i.e. the expression

$$\mathscr{L}_G = -\frac{1}{4}G_{i\mu\nu}(x)G_i^{\mu\nu}(x), \tag{11.34}$$

is gauge invariant. Its physical interpretion is discussed below.

11.2.4   Quantum chromodynamics

Quantum chromodynamics is the theory defined by the Lagrangian density

$$\mathscr{L} = \overline{\Psi}^f(x)\left[i\slashed{D} - m_f\right]\Psi^f(x) - \frac{1}{4}G_{i\mu\nu}(x)G_i^{\mu\nu}(x) \tag{11.35}$$

obtained by combining Eqs. (11.28) and (11.34) for the quark and gluon terms, respectively. It incorporates the minimal interactions needed to make the free quark Lagrangian (11.20) gauge invariant. As in QED, it is easy to add more complicated gauge-invariant terms to the Lagrangian, for example,

$$ig_s'\overline{\Psi}^f(x)\Psi^f(x)G_{i\mu\nu}(x)G_i^{\mu\nu}(x). \tag{11.36}$$

However, such terms are excluded by the criterion of renormalizability, as we shall show in Section 11.3.

In contrast to QED, QCD describes a strong interaction and lowest-order perturbation theory is not, in general, a good approximation. We will defer further discussion of this until later.

For the moment, we will summarize some important physical properties of QCD which can be inferred directly from the Lagrangian density (11.35). We start by considering the gluons. On substituting Eq. (11.33) in Eq. (11.34) one obtains

$$\begin{aligned}
\mathscr{L}_G = &-\frac{1}{4}F_{i\mu\nu}(x)F_i^{\mu\nu}(x) + g_s f_{ijk}A_{i\mu}(x)A_{j\nu}(x)\partial^\mu A_k^\nu(x) \\
&-\frac{1}{4}g_s^2 f_{ijk}f_{ilm}A_j^\mu(x)A_k^\nu(x)A_{l\mu}(x)A_{m\nu}(x)
\end{aligned} \tag{11.37}$$

for the purely gluonic contribution to the Lagrangian density.

The first term in Eq. (11.37) is the Lagrangian density (11.31) for eight non-interacting massless spin 1 gluon fields. As already noted, these particles have non-zero colour charges and hence cannot be observed as isolated free particles because of colour confinement. Nonetheless there is ample experimental evidence for their existence.[5]

In contrast, the second and third terms represent interactions of the gluon fields amongst themselves. In perturbation theory, they generate three- and four-point vertices,[6] as illustrated in Figs. 11.1 (b) and (c). These gluon self-interactions are one of the most striking features of the theory. They arise because the gluons, which transmit the interaction between colour charges, themselves have non-zero colour charges. This is in contrast to QED, where the photons have zero electric charge and there are no fundamental photon self-interactions.

Another fundamental property of QCD is *flavour independence*: that is, the strength and form of the quark–gluon interactions in Eq. (11.28b) are independent of the quark flavour $f = u, d, s \ldots$. This has its most striking consequences for u and d quarks, which have almost equal masses. If the masses were exactly equal, the u and d quarks would have identical properties and states like the proton $p \equiv uud$ and the neutron $n \equiv udd$, which differ only by the replacement of u by d quarks, would have exactly equal masses. This symmetry – called *isospin symmetry*[7] – is a very good approximation in nature, although there are small corrections due to quark mass differences and to electromagnetic interactions.

Finally, QCD conserves the six quark numbers

$$N_f \equiv N(f) - N(\overline{f}) \quad f = u, \ d, \ c, \ \ldots \tag{11.38}$$

defined as the number of quarks minus the number of antiquarks of a given flavour f. This follows from the form of the Lagrangian density (11.28) in the same way that the conservation of the lepton numbers (7.53) follows from the QED Lagrangian density (7.52). Some of the quark numbers (11.38) have names, like strangeness $S \equiv -N_s$ and charm $C \equiv N_c$, and experiment confirms that they are indeed conserved in strong interactions.

11.3   Alternative Interactions?

In the previous sections we have formulated QED and QCD as minimal gauge theories. The resulting theories are both renormalizable, although in the case of QCD that remains to be shown. Here we consider possible non-minimal terms and show that they are excluded if a renormalizable theory is required.

11.3.1   Non-minimal interactions

Non-minimal interactions are gauge-invariant terms $\mathscr{L}_1'$ which can be added to the Lagrangian densities (11.8) or (11.35) without spoiling the gauge invariance of the theory.

---

[5] See, for example, B. R. Martin and G. Shaw, *Particle Physics*, 3rd edn., John Wiley & Sons, Ltd, Chichester (2008), in particular Section 7.2.2.

[6] The detailed form of these vertices will be derived in Section 14.3.3.

[7] For a simple discussion of isospin symmetry, see, for example, Section 6.1 and Appendix C of Martin and Shaw, cited earlier in this section.

Figure 11.2   *The three-point vertex of QED, generated by Eq. (11.8b)*

For QED, such terms are easily constructed by noting that the Dirac covariants (A.53) and the electromagnetic field tensor

$$F^{\mu\nu}(x) = \partial^\nu A^\mu(x) - \partial^\mu A^\nu(x)$$

are individually invariant under the gauge transformation (11.2). Hence a Lorentz scalar product of any number of such factors is a candidate for a non-minimal interaction. The simplest and most interesting possibility is

$$\mathscr{L}'_1 = \frac{g_1}{2}\overline{\psi}(x)\sigma^{\mu\nu}\psi(x)F_{\mu\nu}(x) = g_1\overline{\psi}(x)\sigma^{\mu\nu}\psi(x)\partial_\nu A_\mu \tag{11.39}$$

where $\sigma^{\mu\nu}$ is the antisymmetric tensor (A.9). This contains the same fields, $\overline{\psi}(x)$, $\psi(x)$, $A^\mu(x)$, as the QED interaction (11.8b) and generates the same three-point vertex Fig. 11.2, but with a different vertex factor. The derivation of this vertex factor is very similar to the corresponding calculation for QED in Section 7.2.1. We shall therefore only remind the reader of the latter calculation and comment on the new features which arise from the presence of the derivative of the electromagnetic field in Eq. (11.39).

In Section 6.2, we found that, in the interaction picture, the interacting fields satisfy the same equations of motion and the same commutation relations as the free fields, provided the interaction Lagrangian density $\mathscr{L}_1(x)$ does not involve derivatives of the field operators. This result enabled us to obtain the S-matrix expansion (6.23) and hence to derive the Feynman rules for QED in Chapter 7. However, irrespective of whether $\mathscr{L}_1(x)$ involves derivatives or not, a generally valid form of S-matrix expansion is

$$S = \sum_{n=0}^{\infty} \frac{i^n}{n!} \int \cdots \int d^4 x_1 \ldots \, d^4 x_n T\{\mathscr{L}_1(x_1) \ldots \mathscr{L}_1(x_n)\}, \tag{11.40}$$

provided one uses the free-field commutation relations in evaluating the matrix elements of S.[8] For QED we have

$$\mathscr{L}_1(x) = -\mathscr{H}_1(x) = e\left(\overline{\psi}\gamma^\mu\psi A_\mu\right)_x \tag{11.41}$$

---

[8] For the derivation of this S-matrix expansion, see N. N. Bogoliubov and D. V. Shirkov, *Introduction to the Theory of Quantized Fields*, 3rd edn, John Wiley & Sons, Inc., New York, 1979, Section 21.

and Eq. (11.40) reduces to our earlier expansion (6.23). From these results, it follows that the procedures which were used in Chapter 7 for QED can still be applied, even when the non-minimal interaction (11.39) is added to the QED interaction (11.41).

In Section 7.2.1 we obtained the QED vertex factor $ie\gamma^\alpha$ by calculating the matrix elements of the first-order term

$$S^{(1)} = i \int d^4x \mathcal{L}_1(x) \tag{11.42}$$

in the S-matrix expansion (11.40) between free-particle states. This led to the first-order Feynman amplitudes and, omitting the external line factors, to the corresponding vertex factor. In particular, we considered the basic process $e^- \rightarrow e^- + \gamma$ and obtained the first-order Feynman amplitude

$$\mathcal{M} = ie\bar{u}(\mathbf{p}')\gamma^\mu u(\mathbf{p})\varepsilon_\mu(\mathbf{k}' = \mathbf{p} - \mathbf{p}') \tag{7.32}$$

and hence the vertex factor $ie\gamma^\mu$. The same vertex factor is, of course, obtained from any of the other basic processes shown in Fig. 7.1.

When the non-minimal interaction (11.39) is included, the QED interaction (11.41) is replaced by

$$\mathcal{L}_1(x) = e\left(\bar{\psi}\gamma^\mu\psi A_\mu\right)_x + g_1\left(\bar{\psi}\sigma^{\mu\nu}\psi\partial_\nu A_\mu\right)_x. \tag{11.43}$$

Calculating the same basic process $e^- \rightarrow e^- + \gamma$ now yields

$$\mathcal{M} = i\,\bar{u}(\mathbf{p}')\left[e\gamma^\mu + ig_1\sigma^{\mu\nu}k'_\nu\right]u(\mathbf{p})\varepsilon_\mu(\mathbf{k}') \tag{11.44}$$

instead of Eq. (7.32), corresponding to the vertex factor

$$ie\gamma^\mu - g_1\sigma^{\mu\nu}k'_\nu, \tag{11.45}$$

where the presence of the momentum factor k'_ν in the second term is a direct consequence of the derivative of the electromagnetic field in Eq. (11.43). The interpretation of the coupling constant g_1, is obtained by using the Gordon identity (A.80) to rewrite Eq. (11.44) in the form

$$\mathcal{M} = i\,\bar{u}(\mathbf{p}')\left[\frac{e(p+p')^\mu}{2m} - i\left(\frac{e}{2m} - g_1\right)\sigma^{\mu\nu}k'_\nu\right]u(\mathbf{p})\varepsilon_\mu(\mathbf{k}'). \tag{11.46}$$

Hence, g_1 represents an anomalous magnetic moment in addition to that generated by radiative corrections.[9]

The magnetic interaction (11.39) is only the simplest example of a non-minimal interaction. Other forms are easily constructed, for example

$$\mathcal{L}''_1 = g_2\bar{\psi}(x)\psi(x)F_{\mu\nu}(x)F^{\mu\nu}(x), \tag{11.47}$$

which generates a four-point vertex Fig. 11.3. In general, they are characterized by the n.u. dimensions of their coupling constants g_i, which are inferred from the corresponding interaction Lagrangian density using the field dimensions listed in Table 6.1. In this

---

[9] cf. Section 9.6.1.

Figure 11.3   *The four-point vertex generated by the non-minimal electromagnetic interaction*
(11.47)

way, one easily sees that the minimal QED interaction (11.8b) has a dimensionless coupling; i.e. $[q] = [e] = 0$, where from now on the n.u. dimension of any quantity α is denoted $[\alpha]$. Non-minimal interactions necessarily contain more derivatives and/or more field factors than the minimal interaction, leading directly to dimensions $[g_i] < 0$ for the corresponding couplings. For example, the interactions (11.39) and (11.47) have $[g_1]=-1$ and $[g_2]=-3$ respectively. Similar results hold in QCD. On neglecting interactions, the form of the Lagragian density (11.35) implies that the quark and gluon fields have the same natural dimensions $[\Psi^f] = 3/2$, $[A_i^\mu] = 1$ as the electron and electromagnetic fields, respectively, so that the minimal interaction (11.28b) has a dimensionless coupling $[g_s]=0$. Non-minimal interactions again contain more derivatives and/or field factors, leading, as in QED, to coupling constants with dimensions $[g_i] < 0$. In particular, the coupling constant of the non-minimal interaction (11.36) has $[g'_s]=-3$. These results provide simple criteria for the renormalizability of the associated theories. In the next sub-section we shall derive the general theorem that a theory is non-renormalizable if it contains any interaction whose coupling constant has negative dimension.

11.3.2   Renormalizability

In this sub-section we shall investigate the renormalizability of theories which incorporate non-minimal interactions. The discussion is similar to that used to establish the renormalizability of QED in Section 9.9. After reminding the reader of the argument used in QED, we shall use the same argument to show that including the magnetic interaction (11.39) leads to a non-renormalizable theory. We shall then demonstrate that this result generalizes to all non-minimal interactions using a simple dimensional argument.

In Section 9.9 we saw that it was sufficient to consider only primitively divergent graphs, since all other divergences can be built up from these. Furthermore, we saw that the degree of divergence K of such graphs can be found by simple power counting, i.e. by counting powers of momentum variables of integration in the numerators and denominators. The condition $K \geq 0$ which we obtained for a graph to be primitively divergent is a *necessary* condition, i.e. some graphs with $K \geq 0$ may be convergent as a consequence of special symmetries. Examples of this in QED are the three-photon triangle graphs of Figs. 9.6(c) and (d) and the light-light scattering graph of Fig. 9.24.

In Section 9.9 we found that for a theory to be renormalizable, the degree of divergence K of any primitively divergent graph must not increase with the number of vertices; and there must be only a finite number of types of primitively divergent graphs, specified by the number and nature of external lines.

For QED, the necessary results were obtained in two steps. Firstly, we noted that the degree of divergence of a graph with f_i (b_i) internal fermion (photon) lines is

$$K = 4d - f_i - 2b_i, \qquad (9.109)$$

where d is the number of internal momenta not fixed by energy–momentum conservation at the vertices. This was then rewritten using Eqs. (9.110) and (9.111), which follow from energy–momentum conservation at the vertices, together with the fact that each vertex Fig. 11.2 involves two fermion lines and one photon line. In this way, we obtained

$$K = 4 - \frac{3}{2}f_e - b_e \geq 0 \qquad (9.112)$$

as the necessary condition for a graph with $f_e(b_e)$ external fermion (photon) lines to be primitively divergent. This result is independent of the order n of the graph.

Eq. (9.112) is the key result for QED. It shows that only a finite number of types of vertices (f_e, b_e) are primitively divergent and that their degree of divergence does not increase as higher-order graphs are included. In other words, the primitive divergences of the theory can be described by a finite number of divergent constants. These can be absorbed into a finite number of physical constants to define a renormalized theory, as discussed in Section 9.9.

We next repeat this argument for the case when the magnetic interaction (11.39) is added to the theory. The key change from minimal QED is that the new vertex factor (11.45) grows linearly with the momentum k^μ of the photon attached to the vertex. Since each internal photon line is associated with two vertices, the convergence factor k^{-2} arising from each photon propagator is cancelled, and the superficial degree of divergence becomes

$$K = 4d - f_i \qquad (11.48)$$

instead of (9.109). However the vertex Fig. 11.2 still involves two fermion lines and one photon line, so that Eqs. (9.110) and Eq. (9.111) remain unchanged. Together with Eq. (11.48) they yield

$$K = 4 - \frac{3}{2}f_e - 2b_e + n \geq 0 \qquad (11.49)$$

as the condition for a graph of order n with f_e (b_e) external fermion (photon) lines to be primitively divergent. In contrast to the corresponding minimal QED result (9.112), Eq. (11.49) increases with the order n. For sufficiently high order, all vertices (f_e, b_e) are primitively divergent and their degree of divergence increases as n increases. These divergences cannot be described by a finite number of divergent constants and so cannot be absorbed into a finite number of physical constants. The theory incorporating the magnetic interaction (11.39) is not renormalizable.

It remains to generalize the above arguments so that they apply to any non-minimal interaction. The key point is that the superficial degree of divergence of any primitively divergent graph must be of the form

$$K = K_0(b_e, f_e) - n[g] \qquad (11.50a)$$

where $K_0(b_e, f_e)$ is independent of the order n and $[g]$ is the dimension of the coupling constant. This follows because for any vertex (b_e, f_e), increasing the order by Δn implies a compensating change $-\Delta n[g]$ in the number of momentum factors in order to keep the *n.u.* dimension of the vertex itself unchanged. Since the numbers of external lines (b_e, f_e) are fixed, these can only be internal momenta, implying Eq. (11.50a) for the superficial degree of divergence. If several different coupling constants are involved, the result is modified to

$$K = K_0(b_e, f_e) - \sum_i n_i[g_i] \tag{11.50b}$$

for a graph of order $n_1, n_2, \ldots$ in the couplings $g_1, g_2 \ldots$ respectively. For example, both Eqs. (9.112) and (11.49) are seen to be of this form, since the dimensions of the relevant couplings are $[e] = 0$ and $[g_1] = -1$ respectively.

From Eqs. (11.50) we see that if any of the couplings g_i in a theory has $[g_i] < 0$, the theory is non-renormalizable, since every vertex (b_e, f_e) becomes increasingly divergent for large enough values of the corresponding order n_i. For QED and QCD, this implies that adding non-minimal interactions leads to non-renormalizable theories, since all such interactions have couplings with negative dimensions. The only candidates for non-renormalizable theories are QED and QCD themselves, with minimal interactions only.[10] The renormalizability of QED has already been discussed in Section 9.9. QCD is also renormalizable, but to establish this we must first derive the Feynman rules for QCD. This is more complicated than the corresponding case for QED, and will be discussed in Chapter 14.

We conclude by noting that the methods introduced above – i.e. the use of power counting to obtain relations like (9.112) and (11.49) – can be used to assess the renormalizability of any theory for which the Feynman rules are known.

In addition, the derivation of Eqs. (11.50) holds for any theory whose convergence properties are unchanged by setting the particle masses to zero. For theories with spin $\frac{1}{2}$ fermions with non-zero masses, like QED and QCD, this follows from the form of the propagator (7.48); while for massive spin 0 bosons it follows from the corresponding boson propagator (3.59). Eqs. (11.50) then imply the following general theorem.[11]

A theory is non-renormalizable if it contains any interaction whose coupling g_i has negative dimension $[g_i] < 0$.

This theorem is simple to apply and is very useful, as illustrated in Problems 11.4–6.

11.4 Appendix: Two Gauge Transformation Results

In this appendix we shall derive two results which were only quoted earlier in this chapter: (i) the transformation law (11.26b) of the gluon fields $A_i^\mu(x)$, and (ii) the SU(3) gauge invariance of the Lagrangian density of the gluon fields, Eq. (11.34).

---

[10] This statement implicitly assumes CP or equivalently T invariance. If this requirement is relaxed, an additional QCD interaction is allowed, called the Θ-term. This term is excluded by experiment to very high precision, and we shall not consider if further. See, for example, W.Rolinick, *The Fundamental Particles and their Interactions*, Addison-Wesley, 1994, Section 16.1.

[11] For massive spin 1 particles, the propagator contains a term which becomes singular as the mass tends to zero, as we shall see in Section 16.3. However, the presence of this term can only make diagrams more divergent, so that the theorem still holds.

The derivations will gain in clarity if we introduce the 3×3 matrices[12]

$$\omega(x) \equiv \lambda_i \omega_i(x), \quad A^\mu(x) \equiv \lambda_i A^\mu_i(x). \tag{11.51}$$

For simplicity, we shall also in this appendix omit the argument x of all quantities, i.e. $\omega \equiv \omega(x)$, $\Psi^f \equiv \Psi^f(x)$, etc.

It follows from Eqs. (11.11) and (11.12) that the matrices (11.51) satisfy the commutation relations

$$[\omega, A^\mu] = 2i f_{ijk} \lambda_i \omega_j A^\mu_k. \tag{11.52}$$

Expressed in terms of the matrices (11.51), the covariant derivative (11.27) takes the form

$$D^\mu \Psi^f = [\partial^\mu + ig_s A^\mu/2]\Psi^f, \tag{11.53}$$

and the local phase transformation (11.26a) of Ψ^f becomes, for infinitesimal ω,

$$\Psi^f \rightarrow \Psi^{f'} = [1 + ig_s \omega/2]\Psi^f \quad \text{(small } \omega\text{)}. \tag{11.54}$$

11.4.1   The transformation law (11.26b)

As we saw in Section 11.2.3, we require the gauge fields A^μ to undergo an infinitesmal transformation

$$A^\mu \rightarrow A^{\mu\prime} = A^\mu + \delta A^\mu \tag{11.55}$$

(where $\delta A^\mu \equiv \lambda_i \delta A^\mu_i$) such that the covariant derivative $D^\mu \Psi^f$ transforms in the same way as the field Ψ^f itself, i.e.

$$\begin{aligned} D^\mu \Psi^f &\rightarrow [1 + ig_s \omega/2] D^\mu \Psi^f \\ &= [1 + ig_s \omega/2][\partial^\mu + ig_s A/2]\Psi^f, \end{aligned} \tag{11.56}$$

where the $\omega_j(x)$ are infinitesmal functions.

By applying the transformations (11.54) and (11.55) in Eq. (11.53), we obtain the transformation

$$D^\mu \Psi^f \rightarrow [\partial^\mu + ig_s A^\mu/2 + ig_s \delta A^\mu/2][1 + ig_s \omega/2]\Psi^f. \tag{11.57}$$

Comparing the right-hand sides of Eqs. (11.56) and (11.57), neglecting second-order terms, leads to

$$\delta A^\mu = -\partial^\mu \omega + \frac{1}{2} ig_s [\omega, A^\mu]. \tag{11.58}$$

Substituting the commutation relations (11.52) and the definitions (11.51) in Eq. (11.58) we at once obtain the transformation law

$$\delta A^\mu_i(x) = -\partial^\mu \omega_i(x) - g_s f_{ijk} \omega_j(x) A^\mu_k(x) \tag{11.59}$$

quoted in Eq. (11.26b).

---

[12] This notation is only used in this Appendix. Elsewhere $A^\mu(x)$ is the electromagnetic field!

11.4.2   The SU(3) gauge invariance of Eq. (11.34)

We shall now show that Eq. (11.34), i.e. the Lagrangian density of the $A_i^\mu(x)$ fields

$$\mathscr{L}_G = -\frac{1}{4} G_{i\mu\nu} G_i^{\mu\nu}, \tag{11.60}$$

is invariant under SU(3) transformations. It will be convenient to define

$$G^{\mu\nu} \equiv \lambda_i G_i^{\mu\nu}. \tag{11.61}$$

and use

$$\mathrm{Tr}(\lambda_i \lambda_j) = 2\delta_{ij} \tag{11.62}$$

to write $\mathscr{L}_G$ as

$$\mathscr{L}_G = -\frac{1}{8} \mathrm{Tr}(G_{\mu\nu} G^{\mu\nu}) \tag{11.63}$$

In order to prove the invariance of $\mathscr{L}_G$, we shall first derive the transformation properties of $G^{\mu\nu}$. It follows from Eqs. (11.11) and (11.12) that

$$2i\lambda_i f_{ijk} A_j^\mu A_k^\nu = [A^\mu, A^\nu]. \tag{11.64}$$

Substituting Eqs. (11.32) and (11.33) in Eq. (11.61) and using Eq. (11.64), we obtain

$$G^{\mu\nu} = \partial^\nu A^\mu - \partial^\mu A^\nu - \frac{1}{2} i g_s [A^\mu, A^\nu]. \tag{11.65}$$

Under an SU(3) transformation, A^μ experiences a change δA^μ which is given by Eq. (11.58). Consequently $G^{\mu\nu}$ will change by

$$\delta G^{\mu\nu} = \partial^\nu(\delta A^\mu) - \partial^\mu(\delta A^\nu) - \frac{1}{2} i g_s [\delta A^\mu, A^\nu] - \frac{1}{2} i g_s [A^\mu, \delta A^\nu]. \tag{11.66}$$

On substituting Eq. (11.58) in Eq. (11.66) and rearranging, we eventually obtain:

$$\delta G^{\mu\nu} = \frac{1}{2} i g_s [\omega, \partial^\nu A^\mu - \partial^\mu A^\nu] + \frac{1}{4} g_s^2 [[\omega, A^\mu], A^\nu] + \frac{1}{4} g_s^2 [[A^\nu, \omega], A^\mu]. \tag{11.67}$$

By means of the Jacobi identity

$$[[A, \ B], \ C] + [[B, \ C], \ A] + [[C, \ A], \ B] \equiv 0,$$

we can rewrite Eq. (11.67) as

$$\delta G^{\mu\nu} = \frac{1}{2} i g_s [\omega, \partial^\nu A^\mu - \partial^\mu A^\nu] - \frac{1}{4} g_s^2 [[A^\mu, A^\nu], \omega].$$

On account of Eq. (11.65), the last equation reduces to

$$\delta G^{\mu\nu} = \frac{1}{2} i g_s [\omega, G^{\mu\nu}]. \tag{11.68}$$

This change $\delta G^{\mu\nu}$ induces a change in $\mathscr{L}_G$ which, from Eq. (11.63), is given by

$$
\begin{aligned}
\delta\mathscr{L}_G &= -\frac{1}{8}\mathrm{Tr}(\delta G_{\mu\nu}G^{\mu\nu} + G_{\mu\nu}\delta G^{\mu\nu}) \\
&= -\frac{1}{16}ig_s\mathrm{Tr}\{[\omega, G_{\mu\nu}]G^{\mu\nu} + G_{\mu\nu}[\omega, G^{\mu\nu}]\} \qquad (11.69) \\
&= -\frac{1}{16}ig_s\mathrm{Tr}[\omega, G_{\mu\nu}G^{\mu\nu}] = 0,
\end{aligned}
$$

where the last step follows from the identity $\mathrm{Tr}[A, B] \equiv 0$. Eq. (11.69) establishes the SU(3) invariance of $\mathscr{L}_G$.

Problems

11.1. Verify Eq. (11.25) for the free quark theory (11.20) and obtain the corresponding expression for the colour charge $\hat{F}_8$. Use Eq. (11.24) to show that

$$
\hat{F}_1\big|r, \mathbf{p}, s\big\rangle = \frac{1}{2}|g, \mathbf{p}, s\rangle,
$$

where $|r, \mathbf{p}, s\rangle$ represents a state with a free quark with colour r, momentum $\mathbf{p}$ and spin s, etc. [cf. Eq. (11.13)].

11.2. Derive the vertex factor (11.45) corresponding to the interaction (11.43).

11.3. Derive the vertex factor corresponding to the non-minimal interaction (11.47).

11.4. Consider a real scalar field $\phi(x)$ described by a Lagrangian density

$$
\mathscr{L} = \mathscr{L}_0 + g\frac{\phi(x)^p}{p!}
$$

where $\mathscr{L}_0$ is the free-field Lagrangian density (3.4), g is a coupling constant, $p \geq 3$ is an integer and the product of fields is to be interpreted as a normal ordered product as usual.

Deduce the natural dimension $[g]$ of the coupling constant and hence show that an interaction with $p \geq 5$ is non-renormalizable.

Deduce the form of the interaction vertex and the corresponding vertex factor for 'ϕ^3 theory' with p 3. Show that the superficial degree of divergence of a graph of order n with $b_e \geq 2$ external boson lines is

$$
K = 4 - b_e - n
$$

and, using this, show that there is only one primitively divergent graph in ϕ^3 theory. Comment: *theories with only a finite number of divergent graphs are sometimes called 'super-renormalizable'.*

11.5. Consider the interaction of a real scalar field $\phi(x)$ with a spinor field $\psi(x)$, assuming a local Lorentz-invariant interaction $\mathscr{L}_I$ formed using the scalar field ϕ and/or its derivatives and the five basic bi-linear covariants (A.53). By considering the

dimension of the coupling constant, show that the only candidate for a renormalizable interaction is

$$\mathscr{L}_1 = g\overline{\psi}(x)\psi(x)\phi(x) \tag{11.70}$$

if we assume that parity is conserved, and $\phi(x)$ is a proper scalar field

$$\phi(t, \mathbf{x}) = \phi(t, -\mathbf{x}).$$

What is the corresponding renormalizable parity-conserving interaction for a pseudoscalar field

$$\phi(t, \mathbf{x}) = -\phi(t, -\mathbf{x})?$$

*The second of these interactions, called pseudoscalar meson theory, can be used to
describe the long-range nuclear force due to neutral pion exchange.*

11.6. Consider 'scalar meson theory'

$$\mathscr{L} = \mathscr{L}_0(\phi) + \mathscr{L}_0(\psi) + g\overline{\psi}(x)\psi(x)\phi(x), \tag{11.71}$$

where ϕ and ψ are real scalar and spinor fields described in the absence of interactions by the free-field Lagrangian densities (3.4) and (4.20) respectively.

Find the superficial degree of divergence of a graph with b_e external boson and f_e
external fermion lines; and hence show that the only possible primitively divergent
vertices are

$$(b_e, \; f_e) = (2, \; 0), \; (0, \; 2), \; (1, \; 2), \; (3, \; 0) \text{ and } (4, \; 0).$$

*Comment: the first three divergences are similar to those of QED, and contribute to
the renormalization of the bare masses and couplings in (11.71). The last two yield a
renormalization of the bare couplings g_s and λ in the additional interaction terms*

$$g_s \frac{\phi(x)^3}{3!} + \lambda \frac{\phi(x)^4}{4!},$$

*which must be added to (11.71) to yield a well-behaved renormalizable theory. In
general, for each divergent constant generated by the interaction in any renormalizable theory, there must be a corresponding term in the Lagrangian with which it can
be combined to yield a finite, measurable (i.e. renormalized) mass or coupling.*

12

Field Theory Methods

The next four chapters are devoted to QCD and some of the techniques needed to understand it. The ideas are not required in our discussion of the unified theory of electromagnetic and weak interactions in Chapters 16–19, and the reader who wishes to do so may proceed directly to Chapter 16.

QCD is similar to QED in many ways, but there are important differences, which have their origin in the more complicated form of the gauge transformation of the gluon field, Eq. (11.26b). Because of this, the quantization of the gluon field is a more challenging problem than the quantization of the electromagnetic field, and can be accomplished only with great difficulty by the canonical methods we have used hitherto.[1]

We shall instead use the path integral formalism of Feynman, which is particularly well-suited to the discussion of non-Abelian gauge theories. This formalism will be introduced in the next chapter, before applying it to QCD in Chapter 13. Firstly, however, we will introduce some new ideas—namely Green functions and the generating functional – which will play a central role in this development.

12.1 Green Functions

The physical predictions of any quantum field theory are contained in the S-matrix elements whose squared moduli give the probabilities for the physical processes to occur. This is reflected in our applications of perturbation theory, where we have evaluated the S-matrix elements directly. Alternatively one can focus on closely related

---

[1] A systematic presentation of the canonical approach to non-Abelian gauge theories, like QCD, is given by T. Kugo and I. Ojima, *Prog. Theor. Phy.* **60** (1978) 1869; *Suppl. Prog. Theor. Phys.* **66** (1979) 1.

functions, called Green functions. This has several important advantages. Although Green functions are less directly related to observables than S-matrix elements, they are easier to calculate and S-matrix elements are easily obtained from them. Moreover, S-matrix elements for several different processes can be derived from a single Green function, allowing relations between these processes to be inferred. For this reason, more advanced topics in quantum field theory, going beyond low-order perturbation theory, are almost invariably formulated in terms of Green functions. In this section, we will introduce the various types of Green function in the context of QED, before studying their properties in Sections 12.2 and 12.3.

S-matrix elements of QED contain external lines, corresponding to particles (electrons, positrons, photons) present in initial or final states. In the corresponding Green functions, external lines are replaced by propagators. These will be referred to as legs. Four of the simplest Green functions in QED are illustrated diagrammatically in Fig. 12.1, together with some of the leading terms in their perturbative expansion. As can be seen, the first two are just dressed propagators, i.e. the bare boson and fermion propagators, together with all their self-energy modifications.

Mathematically, Green functions are defined as vacuum expectation values of time-ordered products of field operators in the Heisenberg picture.[2] These are of the general form

$$G^{\mu\cdots}(x,y,z,\ \ldots) \equiv {}_H\Big\langle 0\Big|T\Big\{A^{\mu H}(x)\ldots\psi^H(y)\ldots\overline{\psi}^{\,H}(z)\ldots\Big\}\Big|0\Big\rangle_H \qquad (12.1)$$

where there is an electromagnetic field operator $A^{\mu H}$ for each photon leg, an electron field operator ψ^H for each outgoing electron leg and an electron field operator $\overline{\psi}^{\,H}$ for each incoming electron leg respectively. All quantities in Eq. (12.1) are expressed in the Heisenberg picture (H.P.), as indicated by the labels H. In particular, $|0\rangle_H$ is the true vacuum state in the H.P., i.e. it is the eigenstate of lowest energy of the complete

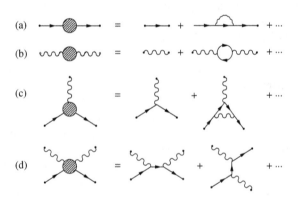

Figure 12.1 *Examples of Green functions in QED, together with some of the leading terms in their perturbative expansion*

---

[2] The Heisenberg picture (H.P.), the interaction picture (I.P.) and the Schrödinger picture (S.P.) and the relations between them, are discussed in Section 1.5.

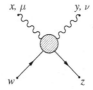

Figure 12.2 *Diagram representing the Green function (12.2)*

Hamiltonian H of the interacting fields. For example, the Green function corresponding to Fig. 12.1(d) is

$$G^{\mu\nu}(x,y,z,\omega) = {}_{\rm H}\Big\langle 0\Big|T\Big\{A^{\mu\rm H}(x)A^{\nu\rm H}(y)\psi^{\rm H}(z)\overline{\psi}^{\,\rm H}(\omega)\Big\}\Big|0\Big\rangle_{\rm H} \qquad (12.2)$$

and is shown again, with the Lorentz indices and space-time points appropriately labelled, in Fig. 12.2.

As previously, we shall work mainly in the interaction picture (I.P.). We therefore, first of all, express the Green function (12.1) in terms of states and operators in the I.P. To simplify the writing, we shall consider, instead of Eq.(12.1), the Green function for two fields,

$$G(x,y) = {}_{\rm H}\langle 0|T\{A^{\rm H}(x)B^{\rm H}(y)\}|0\rangle_{\rm H}, \qquad (12.3)$$

where $A^{\rm H}$ and $B^{\rm H}$ are any two field operators, expressed in the H.P. The generalization to more than two operators will be obvious from our result below.

In Section 1.5 we derived the connection between H.P. and I.P. Labelling quantities in the I.P. by I, we obtain, from Eqs. (1.90) to (1.92)

$$A^{\rm H}(x) = U^{\dagger}(x_0)A^{\rm I}(x)U(x_0), \qquad (12.4a)$$

with a similar equation for $B^{\rm H}(y)$, and

$$|0\rangle_{\rm H} = |0,t_0\rangle_{\rm I}, \qquad (12.4b)$$

where

$$U(t) = e^{iH_0(t-t_0)}e^{-iH(t-t_0)} \qquad (12.4c)$$

with t_0 an arbitrary time. For definiteness, we shall take $x_0 > y_0$. Substituting Eqs. (12.4a) and (12.4b) into Eq. (12.3), we obtain the Green function expressed in the I.P.:

$$G(x,y) = {}_{\rm I}\langle 0,t_0|U^{\dagger}(x_0)A^{\rm I}(x)U(x_0)U^{\dagger}(y_0)B^{\rm I}(y)U(y_0)|0,t_0\rangle_{\rm I}. \qquad (12.5)$$

We rewrite this equation. In Section 1.5, Eqs. (1.93)–(1.95), we obtained the equation of motion of a state vector in the I.P.:

$$|A,t_1\rangle_{\rm I} = U(t_1,t_2)|A,t_2\rangle_{\rm I}, \qquad (12.4d)$$

where the unitary operator $U(t_1,t_2)$ is defined by

$$U(t_1,t_2) = U(t_1)U^{\dagger}(t_2) \qquad (12.4e)$$

and has the property

$$U(t_1, t_2)U(t_2, t_3) = U(t_1, t_3). \tag{12.4f}$$

Using Eq. (12.4e) to replace $U(x_0)U^\dagger(y_0)$ by $U(x_0, y_0)$, Eq. (12.5) becomes

$$G(x, y) = {}_1\langle 0, t_0 | U^\dagger(x_0)A^I(x)U(x_0, y_0)B^I(y)U(y_0)|0, t_0\rangle_I. \tag{12.6a}$$

From the unitarity of $U(t)$, it follows that

$$|0, t_0\rangle_I = U^\dagger(-\infty)U(-\infty)|0, t_0\rangle_I, \quad {}_1\langle 0, t_0| = {}_1\langle 0, t_0|U^\dagger(\infty)U(\infty).$$

Substituting these equations in Eq. (12.6a) and again using Eq. (12.4e), we obtain

$$G(x, y) = {}_1\langle 0, t_0 | U^\dagger(\infty)\{U(\infty, x_0)A^I(x)U(x_0, y_0)B^I(y)$$
$$\times U(y_0, -\infty)\}U(-\infty)|0, t_0\rangle_I. \tag{12.6b}$$

We can write this equation

$$G(x, y) = {}_1\langle 0, t_0 | U^\dagger(\infty)\mathrm{T}\{U(\infty, -\infty)A^I(x)B^I(y)\}U(-\infty)|0, t_0\rangle_I \tag{12.6c}$$

where, from Eq. (12.4f), $U(\infty, -\infty)$ is to be factored into

$$U(\infty, -\infty) = U(\infty, x_0)U(x_0, y_0)U(y_0, -\infty)$$

and the time-ordered product implies the insertion of the various factors in this product in the chronologically correct places, i.e. as in Eq. (12.6b). With the use of the chronological product, Eq. (12.6c) is valid for any time ordering of the operators A and B.

Eq. (12.4d) gives the time development of a state vector $|A, t\rangle_I$ in the I.P. In particular

$$|A, \infty\rangle_I = U(\infty, -\infty)|A, -\infty\rangle_I. \tag{12.4g}$$

Comparison of this equation with Eq. (6.16), which defined the S-matrix operator, at once leads to the identification

$$S = U(\infty, -\infty) \tag{12.4h}$$

with S given by Eqs. (6.22). We can therefore write Eq. (12.6c) as

$$G(x, y) = {}_1\langle 0, t_0 | U^\dagger(\infty)\mathrm{T}\{SA^I(x)B^I(y)\}U(-\infty)|0, t_0\rangle_I \tag{12.6d}$$

Eq. (12.6d) is the interaction picture equivalent of the Green function (12.3). In the next section we shall interpret it in terms of Feynman diagrams, analogous to our analysis of S-matrix elements in Chapter 7. Before we can do this, one further modification of Eq. (12.6d) is required. From Eqs. (12.4d), (12.4e) and $U(t_0) = 1$, it follows that

$$|0, \pm\infty\rangle_I = U(\pm\infty, t_0)|0, t_0\rangle_I = U(\pm\infty)|0, t_0\rangle_I,$$

leading to

$$G(x, y) = {}_1\langle 0, \infty | \mathrm{T}\{SA^I(x)B^I(y)\}|0, -\infty\rangle_I. \tag{12.6e}$$

According to the adiabatic hypothesis, discussed at the end of Section 6.2, the state vectors $|0, \pm\infty\rangle_I$ correspond to the vacuum state of the non-interacting theory. Assuming this state is unique, they can differ only by a phase factor, i.e.

$$|0, \infty\rangle_I = e^{i\phi}|0, -\infty\rangle_I, \tag{12.4i}$$

and hence, from Eqs. (12.4g) and (12.4h), implying

$$_I\langle 0, -\infty|S|0, -\infty\rangle_I = e^{i\phi}. \tag{12.4j}$$

Using the last two equations, Eq. (12.6e) can be rewritten

$$G(x,y) = e^{-i\phi}\,_I\langle 0, -\infty|T\{SA^I(x)B^I(y)\}|0, -\infty\rangle_I$$

$$= \frac{_I\langle 0, -\infty|T\{SA^I(x)B^I(y)\}|0, -\infty\rangle_I}{_I\langle 0, -\infty|S|0, -\infty\rangle_I}. \tag{12.6f}$$

In what follows, we shall generally be working in the interaction picture. To simplify the notation, we shall omit the label I from operators and states and shall write $|0\rangle$ for $|0, -\infty\rangle_I$:

$$|0\rangle \equiv |0, -\infty\rangle_I \tag{12.7}$$

When using the Heisenberg picture, this will be explicitly indicated.

From the derivation of Eq. (12.6f) for two field operators, the generalization to more than two field operators is obvious. In particular, the interaction picture equivalent of Eq. (12.1) is

$$G^{\mu\cdots}(x, \ldots y, \ldots z, \ldots) = \frac{\langle 0|T\{SA^\mu(x)\ldots\psi(y)\ldots\overline{\psi}(z)\ldots\}|0\rangle}{\langle 0|S|0\rangle}, \tag{12.8}$$

where all quantities are expressed in the interaction picture as explained above.

The momentum-space Green functions are defined by the Fourier transforms of the space–time Green functions which we have considered so far. The momentum-space Green function corresponding to Eq. (12.2) and Fig. 12.2 is given by

$$(2\pi)^4\delta^{(4)}(k_1 + k_2 + p_1 + p_2)G^{\mu\nu}(k_1, k_2, p_1, p_2)$$

$$\equiv \int d^4x d^4y d^4z d^4w\, e^{-ik_1x}e^{-ik_2y}e^{-ip_1z}e^{-ip_2w}G^{\mu\nu}(x, y, z, w). \tag{12.9}$$

More generally, momentum-space Green functions are defined by

$$(2\pi)^4\delta^{(4)}(q_1 + q_2 + \ldots + q_n)G^{\mu\cdots}(q_1, q_2, \ldots q_n)$$

$$\equiv \int d^4x_1 d^4x_2 \ldots d^4x_n\left(\prod_{i=1}^n e^{-iq_ix_i}\right)G^{\mu\cdots}(x_1, x_2 \ldots x_n), \tag{12.10}$$

where, by convention, the momenta are directed inwards. When we come to relate a Green function to the S-matrix element of a particular reaction, we shall see that some of the legs correspond to particles present in the initial state, and some to particles present in the final state. A particle corresponding to a leg with inward-directed four-momentum p will have four-momentum p if it is present initially, and it will have four-momentum $-p$ if it is present finally.

The integration over $x_1, \ldots x_n$ in Eq. (12.10) leads to the δ-function $\delta^{(4)}(q_1 + \ldots + q_n)$, as implied by the way the momentum-space Green function is defined in that equation. This δ-function ensures energy–momentum conservation and means that the arguments $q_1, \ldots q_n$ of the Green function satisfy this conservation. (In Section 7.2.1 a similar argument was developed in detail for the case of a very simple S-matrix element.)

The momentum-space definitions (12.9) and (12.10) remain unchanged in the Heisenberg picture.

12.2   Feynman Diagrams and Feynman Rules

Green functions can be evaluated in perturbation theory by substituting the *S*-matrix expansion (11.40) into (12.8) to obtain corresponding expansions for the Green functions themselves. The different terms in these expansions are conveniently represented by Feynman diagrams similar to those for *S*-matrix elements, as illustrated in Fig. 12.1, with identical Feynman rules.

In this section we will verify this for some examples and use them to introduce some important properties of Green functions. Our discussion will be illustrative rather than exhaustive, since the arguments follow closely those given for *S*-matrix elements in Chapters 6 and 7.

12.2.1   The perturbation expansion

We begin by obtaining some general results, which follow on substituting the S- matrix expansion (11.40) into the interaction picture expression for the Green function (12.8). The denominator of the Green function (12.8) then becomes

$$
\begin{aligned}
\langle 0|S|0 \rangle &= \sum_{n=0}^{\infty} \langle 0|S^{(n)}|0 \rangle \\
&\equiv \sum_{n=0}^{\infty} \frac{i^n}{n!} \int \ldots \int d^4x_1 \ldots d^4x_n \langle 0|T\{\mathscr{L}_I(x_1) \ldots \mathscr{L}_I(x_n)\}|0 \rangle
\end{aligned}
\tag{12.11}
$$

where

$$
\mathscr{L}_I(x) = e N\big(\overline{\psi}(x)\slashed{A}(x)\psi(x)\big),
\tag{12.12}
$$

and $\mathscr{L}_I$ has been written as a normal product, as discussed at the beginning of Section 6.2. The corresponding expression for the numerator

$$
\begin{aligned}
\langle 0|T\{SAB \ldots\}|0 \rangle &= \sum_{n=0}^{\infty} \langle 0|T\{AB \ldots S^{(n)}\}|0 \rangle \\
&\equiv \sum_{n=0}^{\infty} \frac{(ie)^n}{n!} \int \ldots \int d^4x_1 \ldots d^4x_n \langle 0|T\{AB \ldots \\
&\quad \ldots N(\overline{\psi}\slashed{A}\psi)_{x_1} \ldots N(\overline{\psi}\slashed{A}\psi)_{x_n}\}|0 \rangle
\end{aligned}
\tag{12.13}
$$

where

$$
AB \ldots \equiv A^\mu(x) \ldots \psi(y) \ldots \overline{\psi}(z) \ldots
\tag{12.14}
$$

is the product of QED fields occurring in Eq. (12.8).

The next step is to evaluate the various terms in the expansions (12.11) and (12.13) using Wick's theorem. This takes a particularly simple form for vacuum expectation values because the vacuum expectation value of any normal ordered product of field operators $AB \ldots$ vanishes:

$$
\langle 0|N(AB \ldots)|0 \rangle = 0.
\tag{12.15}
$$

Hence, for vacuum expectation values, Wick's theorem (6.35) reduces to

$$\langle 0|T(ABCD \ldots WXYZ)|0 \rangle$$
$$= AB\, CD \ldots WX\, YZ + A\, BC\, D \ldots WX\, YZ + \ldots, \qquad (12.16\text{a})$$

where the sum extends over terms in which all the field operators $ABC\ldots$ are paired together in non-vanishing contractions. The same result also applies to mixed T-products of the form

$$\left\langle 0\left|T\left\{AB \ldots N(\overline{\psi}\,A\,\psi)_{x_1} \ldots N(\overline{\psi}\,A\,\psi)_{x_n}\right\}\right|0\right\rangle$$
$$= \left\langle 0\left|T\left\{AB \ldots (\overline{\psi}A\psi)_{x_1} \ldots (\overline{\psi}A\psi)_{x_n}\right\}\right|0\right\rangle_{\text{no e.t.c.,}} \qquad (12.16\text{b})$$

except that terms involving equal-times contractions (e.t.c.) are omitted [cf. Eq. (6.38)].

Before using these results to evaluate particular Green functions, we will obtain some useful results which follow without detailed calculations.

In QED, the only non-vanishing contractions are the propagators (6.32c) and (6.32d):

$$\psi_\alpha(x_1)\overline{\psi}_\beta(x_2) = -\overline{\psi}_\beta(x_2)\psi_\alpha(x_1) = iS_{F\alpha\beta}(x_1 - x_2) \qquad (12.17\text{a})$$

$$A^\mu(x_1)A^\nu(x_2) = A^\nu(x_2)A^\mu(x_1) = iD_F^{\mu\nu}(x_1 - x_2). \qquad (12.17\text{b})$$

Hence, by Eqs. (12.16), all the terms in Eq. (12.13) will vanish, unless the product (12.14) contains equal numbers of fermion fields ψ and antifermion fields $\overline{\psi}$. This in turn implies that the Green function (12.1) and (12.8) itself will vanish, unless it has equal numbers of incoming electron legs (corresponding to factors $\overline{\psi}$) and outgoing electron legs (corresponding to factors ψ). This result embodies the conservation of electron number Eq. (7.53a). As a consequence, Green functions corresponding to graphs like those shown in Figs. 12.3(a) and (b) vanish.

Green functions with an odd number of photon legs and no fermion legs, like those shown in Figs. 12.3 (c, d), also vanish. This is just Furry's theorem, mentioned towards the end of Section 9.1. It follows from the invariance of QED under charge conjugation, which interchanges particles and antiparticles. The unitary charge conjugation operator C leaves the vacuum state invariant and changes the sign of the electromagnetic field; using the Heisenberg picture, we have

$$C|0\rangle_{\text{H}} = |0\rangle_{\text{H}} \quad CA^{\mu\text{H}}(x)C^{-1} = -A^{\mu\text{H}}(x),$$

as shown in Problems 3.5 and 5.4.[3] Hence we obtain from the definition (12.1)

---

[3] A simple account of *C*-invariance is given in, for example, B. R. Martin and G. Shaw, *Particle Physics* 3rd edn., John Wiley & Sons, Ltd, Chichester, 2008, Section 5.4.

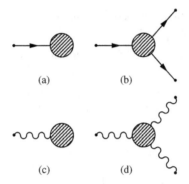

(a)   (b)

(c)   (d)

Figure 12.3 *Examples of vanishing Green functions in QED. The first two violate electron number conservation; the last two violate Furry's theorem*

$$
\begin{aligned}
G^{\mu\nu\cdots}(x_1, x_2 \ldots x_n) &= {}_H\langle 0|T\{A^{\mu H}(x_1)A^{\nu H}(x_2)\ldots A^{\rho H}(x_n)\}|0\rangle_H \\
&= {}_H\langle 0|C^{-1}CT\{A^{\mu H}(x_1)A^{\nu H}(x_2)\ldots A^{\rho H}(x_n)\}C^{-1}C|0\rangle_H \\
&= (-1)^n {}_H\langle 0|T\{A^{\mu H}(x_1)A^{\nu H}(x_2)\ldots A^{\rho H}(x_n)\}|0\rangle_H \\
&= (-1)^n G^{\mu\nu\cdots}(x_1, x_2, \ldots x_n),
\end{aligned}
$$

implying that the Green function vanishes unless n is even.

12.2.2   The vacuum amplitude

Before calculating individual Green functions, we first evaluate the vacuum expectation value $(0|S|0)$ which occurs in the denominator of all Green functions (12.8). Substituting Eq. (12.12) into Eq. (12.11) gives

$$
\langle 0|S|0\rangle = 1 + \langle 0|S^{(2)}|0\rangle + \cdots
\tag{12.18}
$$

since

$$
\langle 0|S^{(1)}|0\rangle = ie \int d^4x_1 \langle 0|N(\bar{\psi}A\!\!\!/\psi)_{x_1}|0\rangle = 0
$$

by Eq. (12.15). The first non-trivial term is

$$
\begin{aligned}
\langle 0|S^{(2)}|0\rangle &= \frac{(ie)^2}{2} \int d^4x_1 d^4x_2 \left\langle 0\Big|T\left\{N(\bar{\psi}A\!\!\!/\psi)_{x_1} N(\bar{\psi}A\!\!\!/\psi)_{x_2}\right\}\Big|0\right\rangle \\
&= -\frac{e^2}{2} \int d^4x_1 d^4x_2\, \bar{\psi}(x_1)\gamma^\mu A_\mu(x_1)\,\psi(x_1)\bar{\psi}(x_2)\gamma^\nu A_\nu(x_2)\psi(x_2)
\end{aligned}
\tag{12.19}
$$

by Wick's theorem, Eq. (12.16). This term is identical to Eq. (7.5f), which is represented by the vacuum diagram shown in Fig. 12.4(a). Higher-order terms in the expansion (12.18) correspond to higher-order vacuum diagrams, two of which are shown in Fig. 12.4(b, c). We do not pursue this further since these vacuum contributions cancel out in the Green functions themselves, as we shall now illustrate.

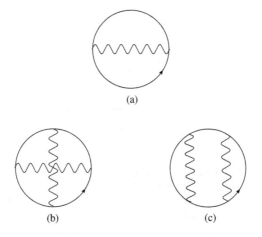

Figure 12.4   (a) *The simplest vacuum diagram, (b, c) some higher-order vacuum diagrams*

12.2.3   The photon propagator

We next consider the photon propagator, that is the two-point Green function

$$G^{\mu\nu}(x, y) = \frac{\langle 0|T(A^{\mu}(x)A^{\nu}(y)S)|0\rangle}{\langle 0|S|0\rangle .} \tag{12.20}$$

To zeroeth order in the S-matrix expansion, $S = S^{(0)} = 1$, and Eq. (12.20) reduces to the free-photon propagator (5.26). Higher-order terms generate the self-energy insertions present in the full propagator, as we shall see.

We start with the S-matrix expansion of the numerator factor

$$F \equiv \langle 0|T(A^{\mu}(x)A^{\nu}(y)S)|0\rangle = \sum_{n=0}^{\infty} F^{(n)} \tag{12.21a}$$

where

$$F^{(n)} \equiv \left\langle 0\left|T\left(A^{\mu}(x)A^{\nu}(y)S^{(n)}\right)\right|0\right\rangle. \tag{12.21b}$$

Using Eq. (12.13) gives

$$F^{(0)} = \langle 0|T\{A^{\mu}(x)A^{\nu}(y)\}|0\rangle = iD_{\mathrm{F}}^{\mu\nu}(x - y) \tag{12.22}$$

while the first-order term vanishes,

$$F^{(1)} = ie \int d^{4}x_{1} \langle 0|T\left\{A^{\mu}(x)A^{\nu}(y)N\left(\overline{\psi}A\!\!\!/\psi\right)_{x_{1}}\right\}|0\rangle = 0, \tag{12.23}$$

since it contains an odd number of electromagnetic field operators. Applying Wick's theorem to the second-order term

$$F^{(2)} = \int d^{4}x_{1}d^{4}x_{2} \langle 0|T\left\{A^{\mu}(x)A^{\nu}(y)N(\overline{\psi}A\!\!\!/\psi)_{x_{1}}N(\overline{\psi}A\!\!\!/\psi)_{x_{2}}\right\}|0\rangle$$

gives

$$F^{(2)} = F_a^{(2)} + F_b^{(2)},.\qquad(12.24)$$

where

$$F_a^{(2)} = \frac{(ie)^2}{2}\int d^4x_1 d^4x_2\, A^\mu(x)A^\nu(y)(\overline{\psi}\,A\!\!\!/\,\psi)_{x_1}(\overline{\psi}\,A\!\!\!/\,\psi)_{x_2}\qquad(12.25a)$$

$$F_b^{(2)} = \frac{(ie)^2}{2}\int d^4x_1 d^4x_2\, A^\mu(x)(\overline{\psi}\,A\!\!\!/\,\psi)_{x_1}(\overline{\psi}\,A\!\!\!/\,\psi)_{x_2}A^\nu(y)$$

$$+ \frac{(ie)^2}{2}\int d^4x_1 d^4x_2\, A^\nu(y)(\overline{\psi}\,A\!\!\!/\,\psi)_{x_1}(\overline{\psi}\,A\!\!\!/\,\psi)_{x_2}A^\mu(x).\qquad(12.25b)$$

On comparing (12.25a) with (12.19) we immediately obtain

$$F_a^{(2)} = iD_F^{\mu\nu}(x - y)\langle 0|S^{(2)}|0\rangle,\qquad(12.26)$$

corresponding to the disconnected Feynman diagram Fig. 12.5. The two terms in (12.25b) are identical since they differ only in the interchange of the integration variables x_1, x_2. Their evaluation is almost identical to that given for the photon self-energy (7.21 and 7.22), giving

$$F_b^{(2)} = \int d^4x_1 d^4x_2 iD_F^{\mu\alpha}(x - x_1)(-1)\mathrm{Tr}\{(ie\gamma_\alpha)iS_F(x_1 - x_2)$$
$$\times (ie\gamma_\beta)iS_F(x_2 - x_1)\}iD_F^{\beta\nu}(x_2 - y).\qquad(12.27)$$

We now have all the ingredients required to calculate the Green function (12.20) to second order. On substituting in Eq. (12.21a) for $F^{(0)}$, $F^{(1)}$ and $F^{(2)}$ from the above equations, we obtain

$$\langle 0|T\{A^\mu(x)A^\nu(y)S\}|0\rangle = iD_F^{\mu\nu}(x - y) + iD_F^{\mu\nu}(x - y)\langle 0|S^{(2)}|0\rangle + F_b^{(2)} + \cdots$$
$$= \left\{iD_F^{\mu\nu}(x - y) + F_b^{(2)} + \cdots\right\}\left[1 + \langle 0|S^{(2)}|0\rangle + \cdots\right]$$
$$= \left\{iD_F^{\mu\nu}(x - y) + F_b^{(2)} + \cdots\right\}\langle 0|S|0\rangle\qquad(12.28)$$

to second order, where we used Eq. (12.18). Eqs. (12.20) and (12.28), then give

Figure 12.5   *The disconnected Feynman diagram corresponding to Eq. (12.26)*

Figure 12.6   *The two lowest-order contributions to the two-point Green function (12.29)*

$$G^{\mu\nu}(x, y) = iD_F^{\mu\nu}(x - y)$$
$$+ \int d^4x_1 d^4x_2 iD_F^{\mu\alpha}(x - x_1)(-1)\mathrm{Tr}\{(ie\gamma_\alpha)iS_F(x_1 - x_2)$$
$$\times (ie\gamma_\beta)iS_F(x_2 - x_1)\}iD_F^{\beta\nu}(x_2 - y) \tag{12.29}$$

corresponding to the Feynman graphs of Fig. 12.6.

This calculation shows that diagrams involving vacuum contributions, like Fig. 12.5 make no contribution to the photon propagator (12.20) in second order. This is a general result for all Green functions to all orders. Essentially, any 'vacuum-bubble-free' diagram that occurs in the perturbative expansion of the numerator of a Green function also occurs accompanied by all vacuum diagrams generated by the expansion (12.11). The resulting multiplicative factor is exactly cancelled by the corresponding factor $\langle 0|S|0\rangle$, which occurs as the denominator of all Green functions; see Eq.(12.8).

We conclude this sub-section by explicitly verifying our assertion that Green functions satisfy the same momentum-space Feynman rules as S-matrix elements. To do this we must calculate the corresponding momentum-space Green function [cf. Eq. (12.10)]

$$(2\pi)^4 \delta^{(4)}(q_1 + q_2)G^{\mu\nu}(q_1, q_2) = \int d^4x_1 d^4x_2 e^{-iq_1 x_1} e^{-iq_2 x_2} G^{\mu\nu}(x_1, x_2). \tag{12.30}$$

As implied by the δ-function in this equation, integration over x_1 and x_2 leads to the equality $q_1 = -q_2 \equiv k$, which ensures energy-momentum conservation. On substituting (12.29) into (12.30) and integrating over x_1 and x_2, using (7.23), we obtain

$$G^{\mu\nu}(k, -k) = iD_F^{\mu\nu}(k)$$
$$+ iD_F^{\mu\alpha}(k)(2\pi)^{-4} \int d^4p(-1)\mathrm{Tr}\{(ie\gamma_\alpha)iS_F(p)$$
$$\times (ie\gamma_\beta)iS_F(p + k)\}iD_F^{\beta\nu}(k) + \ldots. \tag{12.31}$$

The momentum-space Feynman graph corresponding to Eq. (12.31) is shown in Fig. 12.7. That the arguments of the Green function are k and $-k$ results from defining the momenta of both photon propagators in Fig. 12.7 as directed inward, whereas it

Figure 12.7   *The two lowest-order contributions to the momentum-space Green function (12.31)*

follows from energy-momentum conservation that if k denotes the inward directed four-momentum of one of the photon propagators, then k must be the outward directed four-momentum of the other photon propagator. One easily verifies that the relation between Fig. (12.7) and Eq. (12.31) is precisely that specified by Feynman rules 1 to 8 of Appendix B.

12.2.4   Connected Green functions

In QED, Green functions with more than three legs have both connected and disconnected graphs in their perturbative expansion. To illustrate this we consider the four-point Green function

$$G^{\mu\nu}(x, y, z, w) \equiv \frac{\langle 0|T\{A^{\mu}(x)A^{\nu}(y)\psi(z)\overline{\psi}(w)S\}|0\rangle}{\langle 0|S|0\rangle}, \tag{12.32}$$

corresponding to Fig. 12.2. For free fields $S = S^{(0)} = 1$ and Wick's theorem gives

$$G^{(0)\mu\nu}(x, y, z, w) \equiv A^{\mu}(x)A^{\nu}(y)\,\psi(z)\overline{\psi}(w) = iD_F^{\mu\nu}(x - y)iS_F(z - w), \tag{12.33}$$

corresponding to the disconnected Feynman diagram, Fig. 12.8. More generally, the Green function (12.32) can be decomposed into connected and disconnected diagrams, as shown in Fig. 12.9, where we have used the vanishing of single-particle Green functions like Figs. 12.3(a) and (c) to eliminate other possibilities.

Disconnected diagrams correspond to two or more independent processes and contain no information that is not already contained in their component connected diagrams. Consequently it is usual to focus on the connected *Green function*

$$G_c^{\mu\cdots}(x_1 \ldots) \equiv G^{\mu\cdots}(x_1, \ldots) - \sum \text{disconnected diagrams} \tag{12.34}$$

x, μ 〰〰〰 y, ν

w •——▸——• z

Figure 12.8   The Green function (12.32) in lowest order

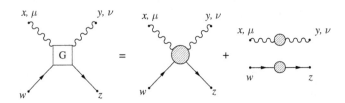

Figure 12.9   Decomposition of the Green function (12.32) into connected and disconnected parts, where shaded 'blobs' indicate sums of connected diagrams

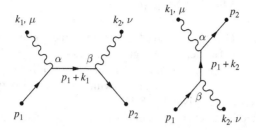

Figure 12.10 *The connected Green function (12.35) to second order. The four momenta are all directed inward, as usual for Green functions*

from which disconnected processes are excluded.[4] We leave it as an exercise for the reader to calculate the connected Green function $G_c^{\mu\nu}(x, y, z, \omega)$ corresponding to Eq. (12.32) to second order, and to show that the related momentum-space Green function is, to second order, given by

$$G_c^{\mu\nu}(k_1, k_2, p_1, p_2)$$
$$= iD_F^{\nu\beta}(k_2) iS_F(-p_2) (ie\gamma_\beta) iS_F(p_1 + k_1)(ie\gamma_\alpha) iS_F(p_1) iD_F^{\alpha\mu}(k_1) \qquad (12.35)$$
$$+ iD_F^{\mu\alpha}(k_1) iS_F(-p_2)(ie\gamma_\alpha) iS_F(p_1 + k_2)(ie\gamma_\beta) iS_F(p_1) iD_F^{\beta\nu}(k_2)$$

corresponding to the Feynman diagrams of Fig. 12.10. The minus signs in the propagators $iS_F(-p_2)$ occurs, since the argument of a fermion propagator is always in the sense of the fermion line [see Eq. (7.48)], whereas, in this case, the inward momentum is in the opposite direction. Alternatively Eq. (12.35) may be obtained directly from the Feynman diagrams using the Feynman rules of Appendix B.

Eq. (12.35) gives the connected momentum-space Green function to second order. In general, this Green function can be written in the form

$$G_c^{\mu\nu}(k_1, k_2, p_1, p_2) = iD_F^{\mu\alpha}(k_1) iS_F(-p_2)\Gamma_{\alpha\beta}(k_1, k_2, p_1, p_2) iS_F(p_1) iD_F^{\beta\nu}(k_2), \qquad (12.36)$$

where the propagator factors correspond to the four legs of either of the two Feynman diagrams in Fig. 12.10, and $\Gamma_{\alpha\beta}(k_1, k_2, p_1, p_2)$ is an example of a *vertex function*. We see from Eq. (12.35) that, to second order, this vertex function is given by

$$\Gamma_{\alpha\beta}(k_1, k_2, p_1, p_2) = (ie\gamma_\beta) iS_F(p_1 + k_1)(ie\gamma_\alpha) + (ie\gamma_\alpha) iS_F(p_1 + k_2)(ie\gamma_\beta). \qquad (12.37)$$

---

[4] More formally, they can be defined by relations like

$$G_c^{\mu\nu}(x, y, z, \omega) \equiv G^{\mu\nu}(x, y, z, \omega) - G^{\mu\nu}(x, y)G(z, \omega)$$

corresponding to Fig. 12.9. For the two-point Green functions corresponding to Figs.12.1(b) and (a), the connected Green functions $G_c^{\mu\nu}(x, y)$, $G_c(z, w)$ and the full Green functions $G^{\mu\nu}(x, y)$, $G(z, \omega)$ are identical.

12.3   Relation to *S*-Matrix Elements

The momentum-space Green functions (12.10) obey the same Feynman rules as the Feynman amplitudes $\mathcal{M}$. Hence, for any given process, the Feynman amplitude can be obtained from the appropriate connected Green function by replacing the external propagators with the appropriate factors associated with ingoing and outgoing particles. The *S*-matrix element itself is then given by (8.1).

This is best clarified by an example. Consider the Compton scattering process

$$\gamma(k,r) + e^{-}(p,s) \rightarrow \gamma((k',r')) + e^{-}(p',s'). \tag{12.38}$$

The Feynman amplitude for Compton scattering can be obtained from the connected Green function $G_c^{\mu\nu}(k_1,k_2,p_1,p_2)$, as illustrated in Fig. 12.11, with the momentum identifications

$$k_1 = k \quad p_1 = p \quad k_2 = -k' \quad p_2 = -p'. \tag{12.39}$$

We replace the four propagator factors corresponding to the legs of the Green function (12.36) by the ingoing and outgoing particle factors which correspond to the Feynman graph 12.11(b). We see from Eq. (12.39) and Eqs. (7.49) that in lowest order perturbation theory, the relevant replacements are

$$\left. \begin{aligned} iD_F^{\mu\alpha}(k_1) &\rightarrow \varepsilon_r^\alpha(\mathbf{k}) \quad & iS_F(p_1) &\rightarrow u_s(\mathbf{p}) \\ iD_F^{\beta\nu}(k_2) &\rightarrow \varepsilon_{r'}^\beta(\mathbf{k}') \quad & iS_F(-p_2) &\rightarrow \bar{u}_{s'}(\mathbf{p}'). \end{aligned} \right\} \tag{12.40a}$$

Making these replacements in Eq. (12.36), we obtain

$$\mathcal{M}(k,k',p,p') = \varepsilon_r^\alpha(\mathbf{k})\bar{u}_{s'}(\mathbf{p}')\Gamma_{\alpha\beta}(k,-k',p,-p')u_s(\mathbf{p})\varepsilon_{r'}^\beta(\mathbf{k}') \tag{12.41}$$

as the expression for the Feynman amplitude for Compton scattering. Replacing the vertex function $\Gamma_{\alpha\beta}$ in the last equation by the second-order expression (12.37) and substituting Eqs. (12.39), the Feynman amplitude (12.41) reduces to our earlier result, Eqs. (7.38a and b).

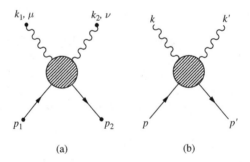

Figure 12.11   *Relation between (a) the Green function $G_c^{\mu\nu}(k_1,k_2,p_1,p_2)$ [Eq. (12.36)] with all momenta directed inward; and (b) the Feynman amplitude for the Compton scattering process (12.38), where the initial momenta p,k are directed inward and the final momenta; p',k' outward*

When one goes beyond lowest non-vanishing order in perturbation theory, Eqs. (12.40a) must be modified to take account of loop diagrams. Specifically, self-energy insertions in the on-shell fermion propagator lead to a multiplicative factor Z_2,[5] whereas the same insertion into an external fermion line only results in a factor $Z_2^{1/2}$ [cf. Eqs. (9.45)]. A similar effect applies to photon lines, so that Eq. (12.40a) becomes

$$
\left.
\begin{aligned}
iD_F^{\mu\alpha}(k_1) &\to Z_3^{-1/2}\varepsilon_r^\alpha(\mathbf{k}) && iS_F(p_1) \to Z_2^{-1/2}u_s(\mathbf{p}) \\
iD_F^{\beta\nu}(k_2) &\to Z_3^{-1/2}\varepsilon_{r'}^\beta(\mathbf{k}') && iS_F(-p_2) \to Z_2^{-1/2}\bar{u}_{s'}(\mathbf{p}'),
\end{aligned}
\right\}
\tag{12.40b}
$$

where $Z_3 = Z_2 = 1$ to lowest order. We shall discuss the renormalization of Green functions further in Chapter 15.

12.3.1 Crossing

Replacing the propagator factors corresponding to the legs of the Green function by the ingoing and outgoing particle factors (12.40) led to the Feynman amplitude (12.41) for Compton scattering. Other replacements of these propagators are possible. Consequently, the Feynman amplitudes for several different physical reactions can be obtained from a single Green function. For example, the Feynman amplitudes for the reactions

$$
\gamma + e^- \to \gamma + e^-
\tag{12.42a}
$$

$$
\gamma + e^+ \to \gamma + e^+
\tag{12.42b}
$$

$$
\gamma + \gamma \to e^+ + e^-
\tag{12.42c}
$$

$$
e^+ + e^- \to \gamma + \gamma
\tag{12.42d}
$$

can all be inferred from the Green function with two photon and two fermion legs. [cf. Fig. 12.11(a)]; or equivalently from the same four-point vertex $\Gamma_{\alpha\beta}(k_1, k_2, p_1, p_2)$ defined by Eq. (12.36). The reactions (12.42) differ from each other in that a particle in the initial state has been replaced by the corresponding antiparticle in the final state, and vice versa. Reactions related in this way are called *crossed reactions*, and the relations between them *crossing relations*. We shall illustrate them for reactions (12.42a and d), leaving reactions (12.42b and c) as exercises for the reader.

Specifically we compare Compton scattering

$$
\gamma(k, r) + e^-(p, s) \to \gamma(k', r') + e^-(p', s')
\tag{12.38}
$$

to the electron–positron annihilation process

$$
e^+(p', s') + e^-(p, s) \to \gamma(k', r') + \gamma(k, r).
\tag{12.43}
$$

The Feynman amplitude for the reaction (12.38) is given by Eq. (12.41). The corresponding result for reaction (12.43) can again be obtained from the same Green function $G_c^{\mu\nu}(k_1, k_2, p_1, p_2)$, as illustrated in Fig. 12.12, with the momentum identifications

---

[5] Cf. Eqs. (9.104), where the factor Z_2 has been used to convert $e_0 \to e$ using Eq. (9.103), and where $\Sigma_c(p) = 0$ on–shell, ie for $\not{p} = m$.

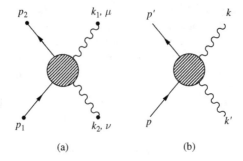

Figure 12.12   *Relation between (a) the Green function $G_c^{\mu\nu}(k_1,k_2,p_1,p_2)$ [Eq. (12.36)] with all momenta directed inward; and (b) the Feynman amplitude for the electron–positron pair annihilation process (12.43), where the initial momenta p, p' are directed inward and the final momenta k, k' outward. The Green function shown here is identical with that shown in Figure 12.11 (a) since the order and position of the legs is of no significance. In each case they are drawn so as to emphasize the relation to the corresponding Feynman amplitudes, where initial and final particles are by convention drawn on the left and right of the graph, respectively*

$$k_1 = -k \quad k_2 = -k' \quad p_1 = p \quad p_2 = p'. \tag{12.44}$$

Replacing the propagator factors with the appropriate ingoing and outgoing particle factors from Eqs. (7.49) corresponding to Fig. 12.12(b), we now obtain

$$\mathcal{M}\left(p, p', k, k'\right) = \varepsilon_r^\alpha(\mathbf{k})\bar{v}_{s'}\left(\mathbf{p}'\right)\Gamma_{\alpha\beta}\left(-k, -k', p, p'\right)u_s(\mathbf{p})\varepsilon_{r'}^\beta\left(k'\right) \tag{12.45}$$

as the appropriate expression for the Feynman amplitude for reaction (12.43). On comparing the Feynman amplitudes (12.41) and (12.45) for reactions (12.38) and (12.43), we see that on replacing a particle in the initial (or final) state with its antiparticle with the same momentum in the final (initial) state, we must reverse the sign of this momentum in the vertex factor $\Gamma_{\alpha\beta}$. This is the crossing relation mentioned above.

Finally we obtain the Feynman amplitude for electron–positron annihilation in second order by substituting (12.37) into (12.45). Suppressing the spin and polarization labels, we obtain

$$\mathcal{M}\left(p, p', k, k'\right) = -e^2\bar{v}(\mathbf{p}')\slashed{\varepsilon}\left(\mathbf{k}'\right)iS_F(p-k)\slashed{\varepsilon}(\mathbf{k})u(\mathbf{p})$$
$$- e^2\bar{v}\left(\mathbf{p}'\right)\slashed{\varepsilon}(\mathbf{k})iS_F\left(p-k'\right)\slashed{\varepsilon}\left(\mathbf{k}'\right)u(\mathbf{p}) \tag{12.46}$$

corresponding to the Feynman diagrams of Fig. 12.13.

12.4   Functionals and Grassmann Fields

All the Green functions of a particular field theory can be simply derived from a single quantity, called the generating functional. Since the S-matrix element for any process can be simply obtained from the corresponding Green function, the generating functional can

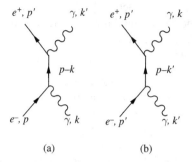

Figure 12.13 *The Feynman diagrams for electron–positron annihilation, corresponding to the second-order Feynman amplitude (12.46)*

be regarded as a complete summary of the physical predictions of the theory. As such it is very useful for establishing general results, as we shall see in the next chapter.

The generating function formations require a knowledge of functional differentiation and of Grassmann fields. A simple introduction to the relevant mathematics is given in the present section and the generating functional itself is then considered in Section 12.5.

12.4.1   Functionals

A functional can be thought of as a generalization of a function. For a function of n variables $x_1, \ldots x_n$, the index $i = 1, \ldots n$ labels the variables x_i, and to a given set of values of these variables corresponds a definite value of the function $f(x_1, \ldots x_n)$.

We go from the discrete set of variables $x_1, \ldots x_n$ to a continuum of infinitely many variables. The index i is now replaced by a continuously varying index x, and the variables $x_1, \ldots x_n$ are replaced by a function $\phi(x)$; we can think of $\phi(x)$ as specifying a variable $\phi(x)$ for each value of the index x. A functional, denoted by $F[\phi]$, will assume a definite value $F = F[\phi]$ for a given function $\phi(x)$. A very simple example of a functional is [6]

$$F[\phi] = \int d^4x \phi(x). \tag{12.47}$$

The integral in this equation specifies a definite value $F[\phi]$ for each function $\phi(x)$. More generally, the integrand in Eq. (12.47) can be a function of $\phi(x)$

$$F[\phi] = \int d^4x f\{\phi(x)\}. \tag{12.48}$$

In quantum field theory, frequently occurring functionals are of the form

$$[AKB] \equiv \int d^4x \int d^4y A(x) K(x,y) B(y), \tag{12.49}$$

---

[6] We shall throughout be considering functions of space–time variables and integrals over all space–time, but it should be obvious to the reader that these considerations apply generally to other cases.

where the kernel $K(x, y)$ is a known function. $[AKB]$ is now a functional with respect to two functions $A(x)$ and $B(x)$. The 'short-hand' notation $[AKB]$ is often convenient, particularly when writing more complicated functionals, such as $e^{\{i[AKB]\}}$, which is an example of a function of a functional $f\{[AKB]\}$.

A functional, in addition to being a functional of a function ϕ, may also depend on some parameter x which is kept fixed in defining the functional. An example is

$$G_x[\phi] = \int d^4y K(x, y)\phi(y). \tag{12.50}$$

$G_x[\phi]$ defines a *different* functional of ϕ for *each* value of x, i.e. $G_x[\phi]$ is a function of x, indicated by the subscript x.

A particular case of interest arises if, for the kernel $K(x, y)$ in Eq. (12.50), we take the Dirac delta function $\delta^{(4)}(x - y)$, resulting in the functional

$$E_x[\phi] = \int d^4y \delta^{(4)}(x - y)\phi(y) = \phi(x). \tag{12.51}$$

Thus, we can think of the function $\phi(x)$, for a fixed value of x, as a functional, i.e. we vary the *function* $\phi(x)$, keeping x fixed. This is in contrast to the usual way of looking on $\phi(x)$ as a definite function of a variable x.

More general forms of functionals are possible. These may involve derivatives of the functions. The action integral, Eq. (2.11), illustrates this case.

We come on now to consider functional differentiation. The derivative of a functional is defined in analogy to that of a function. For a function $f(x_1, \ldots x_n)$, infinitesimal increments $dx_1, \ldots dx_n$ of the variables lead to a change df in the value of the function, given by

$$df = f(x_1 + dx_1, \ldots x_n + dx_n) - f(x_1, \ldots x_n) = \sum_{i=1}^{n} \frac{\partial f}{\partial x_i} dx_i, \tag{12.52}$$

where we have retained only first-order small quantities.

In the same way, we consider the effect on a functional $F[\phi]$, when $\phi(x)$ undergoes an infinitesimal variation to $\phi(x) + \delta\phi(x)$. $\delta\phi(x)$ is not an increment in $\phi(x)$ resulting from a small change in x, but for each value of x, an infinitesimal variation of the function. The first-order change in the functional is then given by

$$\delta F[\phi] = F[\phi + \delta\phi] - F[\phi] = \int d^4x \frac{\delta F[\phi]}{\delta\phi[x]} \delta\phi(x). \tag{12.53}$$

This equation defines the functional derivative $\delta F[\phi]/\delta\phi(x)$. We have again only retained first-order small quantities $\delta\phi(x)$, as we shall ultimately take the limit $\delta\phi \to 0$, as in ordinary differential calculus. Eq. (12.53) is to be compared with Eq. (12.52). Just as df is the sum of contributions due to changes of the discrete set of variables $x_i, i = 1, \ldots n$, so $\delta F[\phi]$ is the integral of contributions from variations of the continuum of variables $\phi(x)$, labelled by x.

Dividing Eq. (12.53) by $\delta\phi(y)$ and going to the limit $\delta(\phi) = 0$, we obtain

$$\frac{\delta F[\phi]}{\delta\phi(y)} = \int d^4x \frac{\delta F[\phi]}{\delta\phi(x)} \frac{\delta\phi(x)}{\delta\phi(y)}. \tag{12.54}$$

Since $F[\phi]$ is an arbitrary functional, it follows from this equation and the definition of the Dirac delta function that

$$\frac{\delta\phi(x)}{\delta\phi(y)} = \delta^{(4)}(x-y). \tag{12.55}$$

This relation merely states that the variations $\delta\phi(x)$ and $\delta\phi(y)$, for $x \neq y$, are independent of each other. It is analogous to $\partial x_i / \partial x_j = \delta_{ij}$, implied by Eq. (12.52) for independent variables.

Because of their close correspondence, functional derivatives can be handled very much-in the same way as ordinary derivatives. For example, for the functional (12.48)

$$F[\phi + \delta\phi] = \int d^4x f\{\phi(x) + \delta\phi(x)\} = \int d^4x \left[f\{\phi(x)\} + \frac{\partial f\{\phi(x)\}}{\partial\phi(x)} \delta\phi(x) \right],$$

and hence from Eqs. (12.53) and (12.55), we obtain

$$\frac{\delta F[\phi]}{\delta\phi(x)} = \frac{\partial f\{\phi(x)\}}{\partial\phi(x)}. \tag{12.56a}$$

It is left as an exercise for the reader to obtain the following general results:

$$\frac{\delta}{\delta\phi(x)}\{aF[\phi] + bG[\phi]\} = a\frac{\delta F[\phi]}{\delta\phi(x)} + b\frac{\delta G[\phi]}{\delta\phi(x)}, \tag{12.56b}$$

$$\frac{\delta}{\delta\phi(x)}F[\phi]G[\phi] = \frac{\delta F[\phi]}{\delta\phi(x)}G[\phi] + F[\phi]\frac{\delta G[\phi]}{\delta\phi(x)}, \tag{12.56c}$$

$$\frac{\delta}{\delta\phi(x)}\{f(F[\phi])\} = \frac{\partial f}{\partial F}\frac{\delta F}{\delta\phi(x)}. \tag{12.56d}$$

The fact that we are able to define functionals which are also functions of another variable, such as $G_x[\phi]$ in Eq. (12.50), allows us to define functionals of functionals:$F = F[G_x[\phi]]$. For these, the basic definition (12.53) leads to the *chain rule*

$$\frac{\delta F[G_x[\phi]]}{\delta\phi(z)} = \int d^4y \frac{\delta F[G_x[\phi]]}{\delta G_y[\phi]}\frac{\delta G_y[\phi]}{\delta\phi(z)}. \tag{12.56e}$$

These general results, Eqs. (12.55) and (12.56a) to (12.56e), enable us to evaluate many functional derivatives easily.

12.4.2 Grassmann algebras and Grassmann fields

Grassmann algebras were invented by the mathematician H. Grassmann in the nineteenth century[7] and were introduced into physics in the 1960s as a means to extend functional methods to fermion systems. Here we will give a simple description of those properties which we shall need, referring to Berezin for a fuller account.[8]

---

[7] H. Grassmann, *Die Ausdehnungslehre von 1844*, Otto Wigand, Leipzig, 1878.
[8] F.A. Berezin, *The Method of Second Quantization*, Academic Press, New York and London, 1966.

A Grassmann algebra is an abstract algebra whose generators $\theta_1, \theta_2, \ldots, \theta_n$ satisfy the anticommutation relations

$$[\theta_i, \theta_j]_+ = 0, \tag{12.57}$$

where the indices i and j each can take on any value in the range $1, \ldots n$. Eq. (12.57) at once implies that

$$\theta_i^2 = \theta_i \theta_i = 0 \quad i = 1, \ldots n. \tag{12.58}$$

This relation is a basic property of a Grassmann algebra which greatly simplifies its use. Consider a product of generators

$$\theta_{i_1} \theta_{i_2} \ldots \theta_{i_{k-1}} \theta_{i_k} \theta_{i_{k+1}} \ldots \theta_{i_w} \tag{12.59}$$

where each of the indices $i_1, \ldots i_w$ can take any value in the range $1, \ldots n$. If any two of these indices are the same, we can use the anticommutation relations (12.57) to bring two generators with the same index into adjacent positions. It follows from Eq. (12.58) that the product of generators (12.59) vanishes if any two indices are the same. Consequently, the most general element of a Grassmann algebra with n generators is of the form

$$f(p) = p_0 + \sum p_i \theta_i + \sum p_{ij} \theta_i \theta_j + \ldots + p_{12 \ldots n} \theta_1 \theta_2 \ldots \theta_n \tag{12.60}$$

where the indices $i, j, \ldots$ run over all values 1 to n, with the restrictions $i \leq j \leq \ldots$. The coefficients $p_0, p_i, p_{ij}, \ldots$ are ordinary numbers, in contrast to the generators θ_i which are abstract entities defined purely by their rules of operation. p on the left-hand side of Eq. (12.60) is shorthand for $p_0, p_i, p_{ij}, \ldots$.

Consider, for simplicity, an algebra with two generators ($n = 2$). The most general element is of the form

$$f(p) = p_0 + p_1 \theta_1 + p_2 \theta_2 + p_{12} \theta_1 \theta_2 \tag{12.61a}$$

which on account of the anticommutation relations (12.57) can also be written

$$f(p) = p_0 + p_1 \theta_1 + p_2 \theta_2 - p_{12} \theta_2 \theta_1. \tag{12.61b}$$

The set of all elements $f(p)$ is called an algebra because, if we consider two different elements $f(p)$ and

$$f(q) = q_0 + q_1 \theta_1 + q_2 \theta_2 + q_{12} \theta_1 \theta_2,$$

then we can *define* a product rule

$$f(p)f(q) = p_0 q_0 + (p_0 q_1 + p_1 q_0)\theta_1 + (p_0 q_2 + p_2 q_0)\theta_2$$
$$+ (p_0 q_{12} + p_1 q_2 - p_2 q_1 + p_{12} q_0)\theta_1 \theta_2$$

which yields another element of the algebra, and is associative.[9] Note that different elements of the algebra, $f(p)$ and $f(q)$, do not, in general, either commute or anticommute with each other. These results obviously generalize to an algebra of n generators.

---

[9] i.e. $f(p)[f(q) + f(r)] = f(p)f(q) + f(p)f(r)$.

The Grassmann generators are abstract entities, defined purely by their operational properties. They are not ordinary numbers nor continuous variables. We define a linear operator $\partial/\partial\theta_i$ with the properties

$$\frac{\partial\theta_i}{\partial\theta_j} = \delta_{ij}, \quad i,j = 1, \ldots n, \tag{12.62}$$

and $\partial/\partial\theta_i$, operating on an ordinary number or ordinary function vanishes. Eq. (12.62) is reminiscent of

$$\frac{\partial x_i}{\partial x_j} = \delta_{ij}$$

for ordinary variables, and for this reason $\partial/\partial\theta_i$ is referred to as differentiation with respect to θ_i. However, we must emphasize that $\partial/\partial\theta_i$ is a purely symbolic operation, defined above, and, unlike ordinary differentiation, it must not be thought of as the result of a limiting process involving a ratio of small quantities.

To extend differentiation to products of generators requires care because of their anticommutation properties (12.57). For the case $n = 2$, we have, from Eqs. (12.61a,b)

$$\frac{\partial f(p)}{\partial\theta_1} = p_1 + p_{12}\frac{\partial}{\partial\theta_1}(\theta_1\theta_2), \quad \frac{\partial f(p)}{\partial\theta_1} = p_1 - p_{12}\frac{\partial}{\partial\theta_1}(\theta_2\theta_1).$$

These two expressions are equal since from the anticommutation relations

$$\frac{\partial}{\partial\theta_1}(\theta_1\theta_2) = \frac{\partial}{\partial\theta_1}(-\theta_2\theta_1) = -\frac{\partial}{\partial\theta_1}(\theta_2\theta_1).$$

We can use this result to *define* differentiation for a product of generators; using Eq. (12.62), we have

$$\frac{\partial}{\partial\theta_1}(\theta_1\theta_2) = \frac{\partial\theta_1}{\partial\theta_1}\theta_2 = \theta_2, \quad \frac{\partial}{\partial\theta_1}(\theta_2\theta_1) = -\frac{\partial}{\partial\theta_1}(\theta_1\theta_2) = -\theta_2,$$

i.e. the generator with respect to which the product is differentiated is moved to the left-hand end of the product, using the anticommutation relations, and then removed on differentiation.

This result easily generalizes. From the anticommutation relations (12.57)

$$\theta_{i_1}\theta_{i_2}\ldots\theta_{i_{k-1}}\theta_{i_k}\theta_{i_{k+1}}\ldots\theta_{i_\omega} = (-1)^{k-1}\theta_{i_k}\{\theta_{i_1}\theta_{i_2}\ldots\theta_{i_{k-1}}\theta_{i_{k+1}}\ldots\theta_{i_\omega}\} \tag{12.63}$$

and hence the general definition of the linear operator $\partial/\partial\theta_i$ follows:[10]

$$\frac{\partial}{\partial\theta_i}\{\theta_{i_1}\theta_{i_2}\ldots\theta_{i_{k-1}}\theta_{i_k}\theta_{i_{k+1}}\ldots\theta_{i_\omega}\}$$

$$= \sum_{k=1}^{\omega}(-1)^{k-1}\delta_{ii_k}\{\theta_{i_1}\theta_{i_2}\ldots\theta_{i_{k-1}}\theta_{i_{k+1}}\ldots\theta_{i_\omega}\}$$

$$= \delta_{ii_1}\{\theta_{i_2}\theta_{i_3}\ldots\theta_{i_\omega}\} - \delta_{ii_2}\{\theta_{i_1}\theta_{i_3}\ldots\theta_{i_\omega}\} + \ldots. \tag{12.64}$$

---

[10] Strictly, this defines a 'left derivative' because the 'differential operator' stands to the left of the operand. Right derivatives can also be defined, but will not be required.

The changes in signs of successive terms in Eq. (12.64) are, as we have seen, a consequence of the anticommutation relations of the generators. The result (12.64) is simpler than it looks. Since for a non-zero product all the indices $i_1 \ldots i_\omega$ in Eq. (12.64) must be different, at most one term survives on the right-hand side of Eq. (12.64). Finally, we note that the definition (12.64), together with the fact that $\partial/\partial\theta_i$ is by definition a linear operator, implies the anticommutation relations

$$\left[\frac{\partial}{\partial\theta_i}, \frac{\partial}{\partial\theta_j}\right]_+ = 0 \quad \left[\theta_i, \frac{\partial}{\partial\theta_j}\right]_+ = \delta_{ij}, \tag{12.65}$$

as the reader can easily verify, for example for $n = 2$.

We next extend the above results to a continuous infinity of Grassmann generators by associating a generator $\theta(x)$ with each space–time point x, so that

$$[\theta(x), \theta(y)]_+ = 0. \tag{12.66}$$

The infinite set of generators $\theta(x)$ is called a Grassmann field. 'Functionals' of this field can be defined by expressions like

$$F[\theta] = f_0 + \int d^4x f_1(x)\theta(x) + \int\int d^4x d^4y f_2(x,y)\theta(x)\theta(y) + \ldots \tag{12.67}$$

where the f_i are ordinary functions. These Grassmann functional are analogous to the 'ordinary' functionals, considered in the last sub-section, such as the $F[\phi]$ defined in Eqs. (12.47) and (12.48). However, in the latter case we could choose specific examples for the function $\phi(x)$ and explicitly evaluate the space–time integrals in Eqs. (12.47) or (12.48). This is not possible for Grassman functionals, such as Eq. (12.67): $\theta(x)$ is not an ordinary function, but a Grassmann field, and equations such as Eq. (12.67) are purely formal expressions, with the properties of Grassmann fields being defined analogously to those of Grassmann generators θ_i for the finite-dimensional case.

In analogy to the definition (12.64) of a linear operator $\partial/\partial\theta_i$, we *define* functional differentiation with respect to the Grassmann field $\theta(x)$ as the linear operator $\delta/\delta\theta(x)$ with the property:

$$\frac{\delta}{\delta\theta(x)}\{\theta(x_1)\theta(x_2)\theta(x_3)\ldots\theta(x_n)\}$$
$$= \delta^{(4)}(x-x_1)\{\theta(x_2)\ldots\theta(x_n)\} - \delta^{(4)}(x-x_2)\{\theta(x_1)\theta(x_3)\ldots\theta(x_n)\}$$
$$+\ldots(-1)^{n-1}\delta^{(4)}(x-x_n)\{\theta(x_1)\theta(x_2)\ldots\theta(x_{n-1})\}. \tag{12.68}$$

From this definition, relations similar to Eqs. (12.62) and (12.65) for the finite dimensional case follow:

$$\frac{\delta\theta(x)}{\delta\theta(y)} = \delta^{(4)}(x-y), \tag{12.69}$$

analogous to Eq. (12.55) for ordinary functionals, and the anticommutation relations

$$\left[\frac{\delta}{\delta\theta(x)}, \frac{\delta}{\delta\theta(y)}\right]_+ = 0 \quad \left[\theta(x), \frac{\delta}{\delta\theta(y)}\right]_+ = \delta^4(x-y). \tag{12.70}$$

In practice, we will usually be concerned with relatively simple functionals, in which two associated, but *independent*, Grassmann fields, which we denote by $\theta(x)$ and $\widetilde{\theta}(x)$, occur in pairs. The field $\widetilde{\theta}(x)$ satisfies the relations analogous to Eqs. (12.66) and (12.68) to (12.70) for $\theta(x)$. The independence of $\theta(x)$ and $\widetilde{\theta}(x)$ is expressed by the relations

$$\frac{\delta\theta(x)}{\delta\widetilde{\theta}^{(y)}} = 0 \qquad \frac{\delta\widetilde{\theta}(x)}{\delta\theta(y)} = 0. \tag{12.71}$$

Anticommutators involving two independent fields or their functional derivatives all vanish by definition. For example

$$\left[\theta(x),\widetilde{\theta}(y)\right]_{+} = 0, \qquad \left[\frac{\delta}{\delta\theta(x)},\frac{\delta}{\delta\widetilde{\theta}(y)}\right]_{+} = 0 \tag{12.72a}$$

$$\left[\theta(x),\frac{\delta}{\delta\widetilde{\theta}(y)}\right]_{+} = 0, \qquad \left[\widetilde{\theta}(x),\frac{\delta}{\delta\theta(y)}\right]_{+} = 0. \tag{12.72b}$$

A typical functional is

$$\left[\theta K\widetilde{\theta}\right] = \int d^4x d^4y\,\theta(x)K(x,y)\widetilde{\theta}(y), \tag{12.73}$$

where $K(x, y)$ is a known kernel and $\left[\theta K\widetilde{\theta}\right]$ is the short-hand notation introduced in Eq. (12.49). Using Eqs. (12.71) and (12.72) we then obtain, for example,

$$\frac{\delta\left[\theta K\widetilde{\theta}\right]}{\delta\widetilde{\theta}(z)} = -\int d^4x\,\theta(x)K(x,z). \tag{12.74}$$

A more complicated example is

$$\exp\,i\left[\theta K\widetilde{\theta}\right] = \sum_{n=0}^{\infty}\frac{i^n}{n!}\int\ldots\int d^4x_1 d^4y_1 \ldots d^4x_n d^4y_n$$
$$\times\left[\theta(x_1)K(x_1,y_1)\widetilde{\theta}(y_1)\right]\ldots\left[\theta(x_n)K(x_n,y_n)\widetilde{\theta}(y_n)\right]. \tag{12.75}$$

Note that the order of the factors $\left[\theta(x_i)K(x_i,y_i)\widetilde{\theta}(y_i)\right]$ on the right-hand side of this equation is unimportant since, being bilinear in the Grassmann fields, they commute with each other. For the same reason, they also commute with the functional differential operators $\delta/\delta\theta(z)$ and $\delta/\delta\widetilde{\theta}(z)$. We leave it as an exercise to the reader to show, using these results, that

$$\frac{\delta}{\delta\theta(z)}\exp\,i\left[\theta K\widetilde{\theta}\right] = i\frac{\delta\left[\theta K\widetilde{\theta}\right]}{\delta\theta(z)}\exp\,i\left[\theta K\widetilde{\theta}\right]. \tag{12.76}$$

12.5   The Generating Functional

After these mathematical preliminaries, we turn to the generating functional itself. As discussed at the beginning of Section 12.3, it can be regarded as a complete summary of the theory, and as such it will play a central role in our discussion of path integrals and QCD in the next two chapters. Here, we shall introduce it in the context of QED and show how it is related to Green functions. We shall then show how it can be evaluated in perturbation theory, starting from the exact solution for non-interacting fields.

We shall start by constructing the generating functional for QED. For this purpose, we introduce a set of fictitious *classical* sources, one for each quantized field, and augment the interaction Lagrangian density $\mathscr{L}_I(x)$ by a source term $\mathscr{L}_S(x)$. For QED, the usual interaction Lagrangian density (12.12) is replaced by

$$\mathscr{L}_I'(x) = \mathscr{L}_I(x) + \mathscr{L}_S(x) \tag{12.77}$$

with the Lagrangian source density $\mathscr{L}_S(x)$ given by

$$\mathscr{L}_S(x) = J_\kappa(x)A^\kappa(x) + \overline{\sigma}_\alpha(x)\psi_\alpha(x) + \overline{\psi}_\alpha(x)\sigma_\alpha(x). \tag{12.78a}$$

Here $J_\kappa(x)$ is a classical four-vector source, and $\sigma_\alpha(x)$ and $\overline{\sigma}_\alpha(x)$, with $\alpha = 1, \ldots 4$, are independent spinor Grassmann sources which anticommute both with themselves and with the fermion fields $\psi_\alpha(x)$ and $\overline{\psi}_\alpha(x)$. As in previous chapters, the spinor index α will usually be suppressed so that Eq. (12.78a) becomes

$$\mathscr{L}_S(x) = J_\kappa(x)A^\kappa(x) + \overline{\sigma}(x)\psi(x) + \overline{\psi}(x)\sigma(x). \tag{12.78b}$$

In what follows, we shall require the equations of motion for the case when the physical interaction is neglected, but the source term is retained, so that

$$\mathscr{L}_I'(x) = \mathscr{L}_S(x). \tag{12.79}$$

As in Chapters 4 and 5, we start by considering the corresponding classical field theory, in which all the fields and sources commute with each other. For the case of the electromagnetic field, we note that the fictitious source $J_\kappa(x)$ in Eq.(12.78) is defined with the opposite sign to that of the physical electromagnetic current $s^\mu(x)$ in the Lagrangian density (5.10). Because of this, the equation of motion (5.12) is replaced by

$$\Box A^\mu(x) = -J^\mu(x). \tag{12.80}$$

In the same approximation (12.79), the equations of motion for the Dirac fields are (see Problem 12.3)

$$i\gamma^\mu \frac{\partial \psi(x)}{\partial x^\mu} - m\psi(x) = -\sigma(x) \tag{12.81a}$$

$$i\frac{\partial \overline{\psi}(x)}{\partial x^\mu}\gamma^\mu + m\overline{\psi}(x) = -\overline{\sigma}(x). \tag{12.81b}$$

These classical equations of motion are taken over into the quantum field theory by replacing the classical fields $\psi(x)$ and $\overline{\psi}(x)$ by quantized fermion operators. It is also necessary to treat the classical sources $\sigma(x)$ and $\overline{\sigma}(x)$ as independent spinor Grassmann sources, in order to retain compatibility with the Pauli Principle, as we shall see below.[11]

---

[11] Because the complex conjugate of a Grassmann field is not defined, we cannot write $\sigma^\dagger = (\sigma^*)^T$. However, we can still choose to *define* new fields $\sigma^\dagger \equiv \overline{\sigma}\gamma^0$ and $\overline{\sigma}^\dagger \equiv \gamma^0\sigma$ if we so wish, so that relations like $\overline{\sigma} = \sigma^\dagger\gamma^0$ still hold.

Returning to the full quantum Lagrangian (12.77, 12.78), the S-matrix corresponding to the augmented interaction density $\mathscr{L}'_1(x)$ is given by

$$S' = \sum_{n=0}^{\infty} \frac{i^n}{n!} \int \cdots \int d^4x_1 \ldots d^4x_n T\{\mathscr{L}'_1(x_1) \ldots \mathscr{L}'_1(x_n)\}. \tag{12.82}$$

In the absence of sources (i.e. $J_\kappa = \sigma = \overline{\sigma} = 0$), we have $\mathscr{L}_S(x) = 0$ and $\mathscr{L}'_1(x) = \mathscr{L}_1(x)$, and S' reduces to the usual S-matrix of QED, Eq. (11.40). The generating functional $Z[J_\kappa, \sigma, \overline{\sigma}]$ is then defined by

$$Z[J_\kappa, \sigma, \overline{\sigma}] = \frac{\langle 0|S'|0\rangle}{\langle 0|S|0\rangle}. \tag{12.83}$$

In the absence of sources $S' \to S$, and

$$Z[0,0,0] = 1. \tag{12.84}$$

We next establish the relation between Green functions and the generating functional Z. For the three-point Green function $G^\mu(x, y, z)$, which from Eq. (12.32) is defined as

$$G^\mu(x, y, z) = \frac{\langle 0|T\{SA^\mu(x)\psi(y)\overline{\psi}(z)\}|0\rangle}{\langle 0|S|0\rangle}, \tag{12.85}$$

we shall show that

$$G^\mu(x, y, z) = \left(\frac{1}{i}\frac{\delta}{\delta J_\mu(x)}\right)\left(\frac{1}{i}\frac{\delta}{\delta\overline{\sigma}(y)}\right)\left(-\frac{1}{i}\frac{\delta}{\delta\sigma(z)}\right)Z[J_\kappa, \sigma, \overline{\sigma}]\Big|_0 \tag{12.86}$$

where the symbol $|_0$ on the right-hand side indicates that the source terms are set equal to zero after the functional differentiations. We note that to each field factor in the Green function (12.85) there corresponds a functional differentiation with respect to the associated source term in Eq. (12.86).

We consider the functional derivative of S' with respect to $\eta(x)$, where $\eta(x)$ represents any of the sources $J_\kappa(x), \sigma(x)$ or $\overline{\sigma}(x)$. From Eq.(12.82)

$$\frac{\delta S'}{\delta\eta(x)} = \frac{i}{1!}\int d^4x_1 \frac{\delta\mathscr{L}'_1(x_1)}{\delta\eta(x)}$$

$$+ \frac{i^2}{2!}\int\int d^4x_1 d^4x_2 T\left\{\mathscr{L}'_1(x_1)\frac{\delta\mathscr{L}'_1(x_2)}{\delta\eta(x)} + \frac{\delta\mathscr{L}'_1(x_1)}{\delta\eta(x)}\mathscr{L}'_1(x_2)\right\} + \cdots$$

$$= i\int d^4x_1 \frac{\delta\mathscr{L}'_1(x_1)}{\delta\eta(x)} + 2\times\frac{i^2}{2!}\int\int d^4x_1 d^4x_2 T\left\{\mathscr{L}'_1(x_1)\frac{\delta\mathscr{L}'_1(x_2)}{\delta\eta(x)}\right\} + \cdots$$

$$= i\int d^4z\, T\left\{S'\frac{\delta\mathscr{L}'_1(z)}{\delta\eta(x)}\right\}. \tag{12.87}$$

From Eqs. (12.77) and (12.78), together with the properties of functional derivatives discussed in the previous section, it follows that

$$\frac{\delta\mathscr{L}'_1(z)}{\delta J_\mu(x)} = A^\mu(x)\delta^{(4)}(x-z) \tag{12.88a}$$

and

$$\frac{\delta \mathscr{L}'_1(z)}{\delta \sigma(x)} = -\bar{\psi}(x)\delta^{(4)}(x-z) \qquad \frac{\delta \mathscr{L}'_1(z)}{\delta \bar{\sigma}(x)} = \psi(x)\delta^{(4)}(x-z). \qquad (12.88b)$$

Hence from Eq. (12.87), we have

$$\frac{\delta S'(z)}{\delta J_\mu(x)} = iT\{S'A^\mu(x)\}, \qquad \frac{\delta S'}{\delta \bar{\sigma}}(x) = iT\{S'\psi(x)\}, \qquad \frac{\delta S'}{\delta \sigma(x)} = -iT\{S'\bar{\psi}(x)\}. \qquad (12.89)$$

Using the definition (12.83) of $Z[J_\kappa, \sigma, \bar{\sigma}]$, Eqs.(12.89) can be expressed as functional derivatives of Z. Since $\langle 0|S|0 \rangle$ does not depend on the source terms, we obtain from the first of Eqs. (12.89).

$$\frac{1}{i}\frac{\delta Z[J_\kappa, \sigma, \bar{\sigma}]}{\delta J_\mu(x)} = \frac{\left\langle 0 \left| \dfrac{1}{i}\dfrac{\delta S'}{\delta J_\mu(x)} \right| 0 \right\rangle}{\langle 0|S|0 \rangle} = \frac{\left\langle 0|T\{S'A^\mu(x)\}|0 \right\rangle}{\langle 0|S|0 \rangle}. \qquad (12.90)$$

Successive functional differentiations of Eq.(12.90) with respect to $\sigma(z)$ and $\bar{\sigma}(y)$, and finally going to the limit of no source terms, so that $S' \to S$, gives the desired result (12.86) for the three-point Green function $G^\mu(x, y, z)$.

This result at once generalizes. For Green functions of the general form (12.8), one obtains

$$G^{\mu\cdots}(x_1, \ldots y_1, \ldots z_1, \ldots)$$

$$= (-1)^{\bar{n}} \left(\frac{1}{i}\right)^n \frac{\delta^n Z[J_\kappa, \sigma, \bar{\sigma}]}{\delta J_\mu(x_1) \ldots \delta \bar{\sigma}(y_1) \ldots \delta \sigma(z_1) \ldots}\Bigg|_0 \qquad (12.91)$$

where n and $\bar{n}$ are respectively the total number of field operators and the number of adjoint field operators $\bar{\psi}(x_1), \ldots, \bar{\psi}(x_\pi)$ in Eq. (12.8). Equivalently, n and $\bar{n}$ are the total number of functional derivatives and the number of functional derivatives with respect to Grassmann sources $\sigma(z_1), \ldots \sigma(z_\pi)$ in Eq. (12.91). Eq. (12.91) is antisymmetric in the variables $y_1, y_2, \ldots$ and $z_1, z_2, \ldots$, because functional derivatives with respect to Grassmann sources also anticommute [see Eq. (12.70)]:

$$\frac{\delta}{\delta \bar{\sigma}(y_1)} \frac{\delta}{\delta \bar{\sigma}(y_2)} = -\frac{\delta}{\delta \bar{\sigma}(y_2)} \frac{\delta}{\delta \bar{\sigma}(y_1)}, \text{etc.}$$

In other words, the choice of Grassmann sources $\sigma, \bar{\sigma}$ for the spinor fields $\bar{\psi}, \psi$ is required for compatibility with the Pauli principle.

From Eq. (12.91), it is clear that an exact evaluation of the generating functional (12.83) would constitute an exact solution of the theory for all physical processes. Unfortunately, this has not been achieved for QED and one must still have recourse to approximation schemes. Here, we will evaluate the generating functional in perturbation theory, starting from the exact solution for non-interacting fields.

12.5.1   The free-field case

For the free, i.e. non-interacting, fields, we set the interaction term $\mathscr{L}_I(x)$ equal to zero but retain the source term $\mathscr{L}_S(x)$, Eq. (12.78). The QED S-matrix (11.40) then reduces to $S_0 = 1$, and the augmented interaction Lagrangian density $\mathscr{L}'_I(x)$, Eq. (12.77), reduces to

$$\mathscr{L}'_I = \mathscr{L}_S(x), \tag{12.92}$$

where the subscript zero indicates that we have set $\mathscr{L}_I(x) = 0$. The expansion (12.82) of S' now becomes

$$S'_0 = \sum_{n=0}^{\infty} \frac{i^n}{n!} \int \cdots \int d^4x_1 \ldots d^4x_n T\{\mathscr{L}_S(x_1) \ldots \mathscr{L}_S(x_n)\}, \tag{12.93}$$

and the free-field generating functional is, from Eq. (12.83),

$$Z_0[J_\kappa, \sigma, \bar{\sigma}] = \langle 0 | S'_0 | 0 \rangle \tag{12.94}$$

since

$$\langle 0 | S_0 | 0 \rangle = 1. \tag{12.95}$$

Here we will show how an explicit expression for the free-field generating functional can be obtained by solving separately for the dependences on the electromagnetic source $J_\mu(x)$ and on the fermion sources $\sigma(x)$ and $\bar{\sigma}(x)$.

The explicit dependence on the source $J_\mu(x)$ is found by obtaining and then solving a functional differential equation for $Z_0[J_\kappa, \sigma, \bar{\sigma}]$.

To obtain this differential equation, we take as starting point, the equation

$$\frac{\delta Z_0[J_\kappa, \sigma, \bar{\sigma}]}{\delta J_\mu(x)} = i\langle 0 | T\{S'_0 A^\mu(x)\} | 0 \rangle \tag{12.96}$$

which is derived from S'_0, in complete analogy to the derivation of equation (12.90) from S', taking into account Eq. (12.95). We transform the right-hand side of Eq. (12.96) into the H.P. From Eqs. (12.1) and (12.8)

$$\langle 0 | T\{S'_0 A^\mu(x)\} | 0 \rangle = {}_H\langle 0 | A^{\mu H}(x) | 0 \rangle_H \langle 0 | S'_0 | 0 \rangle, \tag{12.97}$$

and using this equation and the definition (12.94) of Z_0, Eq. (12.96) becomes

$$\frac{\delta Z_0[J_\kappa, \sigma, \bar{\sigma}]}{\delta J_\mu(x)} = i\,{}_H\langle 0 | A^{\mu H}(x) | 0 \rangle_H Z_0[J_\kappa, \sigma, \bar{\sigma}]. \tag{12.98}$$

The desired differential equation for Z_0 is obtained by operating on Eq. (12.98) with the four-dimensional Laplacion operator $\Box_x$. Since $A^{\mu H}(x)$ satisfies the equation of motion (12.80), i.e.

$$\Box_x A^{\mu H}(x) = -J^\mu(x), \tag{12.99}$$

and $Z_0[J_\kappa, \sigma, \bar{\sigma}]$ is not a function of x and $J_\mu(x)$ is a classical source, we obtain from Eq. (12.98) the functional differential equation

$$\Box_x \frac{\delta Z_0[J_\kappa, \sigma, \bar{\sigma}]}{\delta J_\mu(x)} = -iJ^\mu(x) Z_0[J_\kappa, \sigma, \bar{\sigma}]. \tag{12.100}$$

We solve this equation by the standard method. Let $\Delta^{\mu\nu}(x - x')$ be a solution of

$$\Box_x \Delta^{\mu\nu}(x - x') = g^{\mu\nu}\delta^{(4)}(x - x'), \tag{12.101}$$

together with boundary conditions, to be determined later. Multiplying this equation with $\{-iJ_\nu(x')Z_0[J_\kappa, \sigma, \bar{\sigma}]\}$, and summing over ν and integrating over x', we obtain

$$\Box_x \int d^4x' \Delta^{\mu\nu}(x - x')\{-iJ_\nu(x')Z_0[J_\kappa, \sigma, \bar{\sigma}]\} = -iJ^\mu(x)Z_0[J_\kappa\sigma, \bar{\sigma}] \tag{12.102}$$

and comparing this equation with Eq. (12.100) gives

$$\frac{\delta Z_0[J_\kappa, \sigma, \bar{\sigma}]}{\delta J_\mu(x)} = \left\{-i\int d^4x' \Delta^{\mu\nu}(x - x')J_\nu(x')\right\}Z_0[J_\kappa\sigma, \bar{\sigma}]. \tag{12.103}$$

Just as the ordinary differential equation

$$\frac{df(x)}{dx} = \Phi(x)f(x)$$

has the solution

$$f(x) = \text{const. } e^{F(x)},$$

where $F(x)$ is found by substituting this expression in the differential equation, so Eq. (12.103) has a solution of the form

$$Z_0[J_\kappa, \sigma, \bar{\sigma}] = N[\sigma, \bar{\sigma}]e^{F[J]}, \tag{12.104}$$

where $N[\sigma, \bar{\sigma}]$ is an arbitrary functional of the spinor sources σ and $\bar{\sigma}$. The functional F[J] is found by substituting Eq. (12.104) into Eq. (12.103). Since, from Eq. (12.56d)

$$\frac{\delta}{\delta J_\mu(x)}e^{F[J]} = \frac{\delta F[J]}{\delta J_\mu(x)}e^{F[J]},$$

we obtain in this way

$$\frac{\delta F[J]}{\delta J_\mu(x)} = \Phi, \tag{12.105a}$$

with

$$\Phi \equiv -i\int d^4x' \Delta^{\mu\nu}(x - x')J_\nu(x').$$

To solve this equation for the functional F[J], we rewrite Φ. Since

$$\delta J_\sigma(x'')/\delta J_\mu(x) = \delta^\mu_\sigma \delta^{(4)}(x'' - x'),$$

we obtain

$$\Phi = -i\int d^4x' d^4x'' \frac{\delta J_\sigma(x'')}{\delta J_\mu(x)} \Delta^{\sigma\nu}(x'' - x')J_\nu(x').$$

From Eq. (12.101), we have $\Delta^{\sigma v}(x''-x') = \Delta^{v\sigma}(x''-x')$, hence

$$\Phi = \frac{-i}{2} \int \int d^4x' d^4x'' \left\{ \frac{\delta J_\sigma(x'')}{\delta J_\mu(x)} \Delta^{\sigma v}(x''-x') J_v(x') + J_\sigma(x'') \Delta^{\sigma v}(x''-x') \frac{\delta J_v(x')}{\delta J_\mu(x)} \right\}$$

$$= \frac{\delta}{\delta J_\mu(x)} \left\{ \frac{-i}{2} \int \int d^4x' d^4x'' J_\sigma(x'') \Delta^{\sigma v}(x''-x') J_v(x') \right\} \equiv \frac{\delta}{\delta J_\mu(x)} \left\{ \frac{-i}{2} [J_\sigma \Delta^{\sigma v} J_v] \right\},$$

(12.105b)

where the last equation defines the functional $[J_\sigma \Delta^{\sigma v} J_v]$ in agreement with the notation (12.49). From Eqs. (12.105a) and (12.105b) we have

$$F[J] = \left\{ \frac{-i}{2} [J_\sigma \Delta^{\sigma v} J_v] \right\}$$

and hence from Eq. (12.104) we obtain the solution

$$Z_0[J_\kappa, \sigma, \bar\sigma] = N[\sigma, \bar\sigma] \exp \left\{ \frac{-i}{2} [J_\sigma \Delta^{\sigma v} J_v] \right\}. \tag{12.106}$$

It remains to identify the appropriate solution $\Delta^{\mu v}(x - x')$ of Eq. (12.101). To do this, we remember that for free fields $S \to S_0 = 1$ and the two-point Green function (12.20) reduces to

$$G_0^{\mu v}(x - y) = \langle 0|T\{A^\mu(x)A^v(y)\}|0\rangle = iD_F^{\mu v}(x - y) \tag{12.107a}$$

where $D_F^{\mu v}(x - y)$ is the Feynman free-photon propagator (5.26)–(5.27). On the other hand, Eq. (12.91) gives, for the free-field case,

$$G_0^{\mu v}(x - y) = \left(\frac{1}{i}\right)^2 \frac{\delta^2 Z_0[J_\kappa \sigma, \bar\sigma]}{\delta J_\mu(x)\delta J_v(y)} \bigg|_0 \tag{12.108}$$

and evaluating this from Eq. (12.106) leads to

$$G_0^{\mu v}(x - y) = i\Delta^{\mu v}(x - y). \tag{12.107b}$$

Comparing Eqs. (12.107a) and (12.107b), we see we must identify $\Delta^{\mu v}(x - y)$ with $D_F^{\mu v}(x - y)$, so that our final expression for the dependence of the free-field generating functional on the four-vector source $J_\mu(x)$ becomes

$$Z_0[J_\kappa, \sigma, \bar\sigma] = N[\sigma, \bar\sigma] \exp \left\{ -\frac{i}{2} [J_\mu D_F^{\mu v} J_v] \right\}. \tag{12.109}$$

The dependence of the free-field generating functional on the spinor sources $\sigma(x)$ and $\bar\sigma(x)$ is obtained in a similar manner to the dependence on the current $J_\mu(x)$, but starting from the second of Eqs. (12.89) instead of from the first (see Problem 12.6). In this way one obtains the final expression

$$Z_0[J_\kappa, \sigma, \bar\sigma] = Z_0[J_\kappa]Z_0[\sigma, \bar\sigma] \tag{12.110a}$$

for the free-field generating functional, where

$$Z_0[J_\kappa] = \exp \left\{ -\frac{i}{2} [J_\mu D_F^{\mu v} J_v] \right\} \tag{12.110b}$$

$$Z_0[\sigma, \overline{\sigma}] = \exp\{-i[\overline{\sigma} S_F \sigma]\}. \tag{12.110c}$$

Here $S_F(x - x')$ is the free-electron propagator, Eqs. (4.61) and (4.63), and we have again used the short-hand notation

$$[\overline{\sigma} S_F \sigma] = \int dx' dx'' \overline{\sigma}(x') S_F(x' - x'') \sigma(x''). \tag{12.111}$$

12.5.2   The perturbation expansion

The generating functional (12.83) for QED has not been evaluated in closed form. However, a useful expression valid to all orders of perturbation theory can be derived by replacing the QED interaction $\mathcal{L}_I(x)$ by

$$\mathcal{L}_{I\lambda}(x) = \lambda \mathcal{L}_I(x) = \lambda e \overline{\psi}(x) A\!\!\!/(x) \psi(x) \tag{12.112}$$

where λ is a real parameter and, for the moment, we ignore the normal ordering of the fields in the QED interaction $\mathcal{L}_I(x)$, Eq. (12.12). Correspondingly, the QED S-matrix (11.40) becomes

$$S_\lambda = \sum_{n=0}^{\infty} \frac{i^n}{n!} \int \ldots \int d^4x_1 \ldots d^4x_n T\{\mathcal{L}_{I\lambda}(x_1) \ldots \mathcal{L}_{I\lambda}(x_n)\}. \tag{12.113}$$

The augmented interaction (12.77) is now replaced by

$$\mathcal{L}'_{I\lambda}(x) = \mathcal{L}_{I\lambda}(x) + \mathcal{L}_s(x) = \lambda \mathcal{L}_I(x) + \mathcal{L}_s(x) \tag{12.114}$$

and S', Eq. (12.82), by

$$S'_\lambda = \sum_{n=0}^{\infty} \frac{i^n}{n!} \int \ldots \int d^4x_1 \ldots d^4x_n \, T\left\{\mathcal{L}'_{I\lambda}(x_1) \ldots \mathcal{L}'_{I\lambda}(x_n)\right\}. \tag{12.115}$$

The generating functional (12.83) now becomes a function of λ:

$$Z_\lambda[J_\kappa, \sigma, \overline{\sigma}] = \frac{\langle 0|S'_\lambda|0 \rangle}{\langle 0|S_\lambda|0 \rangle}. \tag{12.116}$$

For $\lambda = 0$, the above expressions go over into the free-field quantities; for $\lambda = 1$ into the QED quantities of the interacting fields.

To obtain from Eqs. (12.115) and (12.116) an expansion of $Z_\lambda[J_\kappa, \sigma, \overline{\sigma}]$ in powers of λ, we differentiate Eq. (12.115) with respect to λ. From Eq. (12.114) we have $\partial \mathcal{L}'_{I\lambda}(x)/\partial \lambda = \mathcal{L}_I(x)$ and hence from Eq. (12.115)

$$\frac{\partial S'_\lambda}{\partial \lambda} = i \int d^4x T\left\{S'_\lambda \mathcal{L}_I(x)\right\}. \tag{12.117}$$

In the same manner in which Eqs. (12.89) were derived by functional differentiation of S', we obtain from S'_λ, Eq. (12.115), that

$$e I_\delta(x) S'_\lambda = T\left\{S'_\lambda \mathcal{L}_I(x)\right\}, \tag{12.118}$$

where we have introduced the interaction operator $I_\delta(x)$, defined by

$$I_\delta(x) = \left(-\frac{1}{i}\frac{\delta}{\delta\sigma(x)}\right)\gamma_\mu\left(\frac{1}{i}\frac{\delta}{\delta\overline{\sigma}(x)}\right)\left(\frac{1}{i}\frac{\delta}{\delta J_\mu(x)}\right). \tag{12.119}$$

Combining Eqs. (12.117) and (12.118) gives the differential equation in λ:

$$\frac{\partial S'_\lambda}{\partial\lambda} = ie\int d^4x I_\delta(x) S'_\lambda,$$

and taking the expectation value of this equation leads to

$$\frac{\partial}{\partial\lambda}\langle 0|S'_\lambda|0\rangle = ie\int d^4x I_\delta(x)\langle 0|S'_\lambda|0\rangle.$$

The solution of this equation is

$$\langle 0|S'_\lambda|0\rangle = \exp\left\{ie\lambda\int d^4x I_\delta(x)\right\}\langle 0|S'_0|0\rangle, \tag{12.120}$$

as is easily verified by differentiating the last equation with respect to λ. Eq. (12.120) satisfies the correct boundary condition that $\langle 0|S'_\lambda|0\rangle \to \langle 0|S'_0|0\rangle$ as $\lambda \to 0$. Using Eqs. (12.116) and (12.94), we can rewrite Eq. (12.120) as

$$\langle 0|S_\lambda|0\rangle Z_\lambda[J_\kappa,\sigma,\overline{\sigma}] = \exp\left\{ie\lambda\int d^4x I_\delta(x)\right\}Z_0[J_\kappa,\sigma,\overline{\sigma}].$$

Setting $\lambda = 1$ gives the result

$$\langle 0|S|0\rangle Z[J_\kappa,\sigma,\overline{\sigma}] = \exp\left\{ie\int d^4x I_\delta(x)\right\}Z_0[J_\kappa,\sigma,\overline{\sigma}] \tag{12.121}$$

and expanding the exponential gives the perturbation expansion

$$\langle 0|S|0\rangle Z[J_\kappa,\sigma,\overline{\sigma}] = \sum_{n=0}^{\infty}\frac{(ie)^n}{n!}\left[\int d^4x I_\delta(x)\right]^n Z_0[J_\kappa,\sigma,\overline{\sigma}] \tag{12.122}$$

where the free-field generating functional Z_0 and the interaction operator $I_\delta(x)$ are given by Eqs. (12.110) and (12.119), respectively. On substituting Eqs. (12.121) or (12.122) into Eq. (12.91) to calculate Green functions, one sees that the factor $\langle 0|S|0\rangle$ in these equations gives rise to the same factor in the denominator of the Green functions (12.91). As shown in Section 12.2.3, this factor cancels the disconnected vacuum diagrams that would otherwise occur in the perturbative expansion of Green functions. When calculating connected Green functions, it can be ignored.

In deriving these results, we have omitted to normal order the fields in the interaction Lagrangian (12.112). As we saw in sub-section 12.2.1, the effect of normal ordering the interaction in calculating Green functions was to omit equal-times contractions from the Wick expansion of vacuum expectation values of the form (12.16b). The only equal-times contractions which occur on applying Wick's theorem to the S-matrix expansion of the Green functions (12.113), and which do not obviously vanish, are of the form

$$F^\mu A_\mu(x_i) \equiv \overline{\psi}(x_i)\mathcal{A}(x_i)\psi(x_i) \tag{12.123}$$

Figure 12.14   *The vanishing sub-graph arising from the equal-times contraction (12.123) in QED*

corresponding to the sub-graph shown in Fig. 12.14. In fact, this sub-graph and hence all terms in the *S*-matrix expansion containing it vanish identically. To see this, we first display the spinor indices to obtain

$$F^\mu \equiv \overline{\psi}_\alpha(x_i)\gamma^\mu_{\alpha\beta}\psi_\beta(x_i) = (-1)\mathrm{Tr}[\gamma^\mu S_F(0)]$$

in the usual way. Substituting from Eq. (4.63), and using dimensional regularization to define the integral,[12] gives

$$F^\mu = \frac{-1}{(2\pi)^4}\int d^D p \frac{4p^\mu}{p^2 - m^2 + i\varepsilon} = 0,$$

where we have evaluated the trace and used Eq. (10.32). Because of this, the normal ordering of the interaction may be ignored in calculating Green functions and, by implication, the generating functional, as we have assumed.

Problems

12.1. Calculate the Green function (12.32) to second order, retaining the connected part only; and show that it leads to the momentum-space Green function (12.35).

12.2. Express the Feynman amplitudes for the reactions

$$\gamma(k,r) + e^+(p,s) \to \gamma(k',r') + e^+(p',s')$$
$$\gamma(k,r) + \gamma(k',r') \to e^-(p,s) + e^+(p',s')$$

in terms of the vertex function $\Gamma_{\alpha\beta}(k_1, k_2, p_1, p_2,)$ defined in Eq. (12.36), and use Eq. (12.37) to obtain explicit expressions for these amplitudes in second order.

12.3. In the absence of interactions, the appropriate classical Lagrangian density for the Dirac field in the presence of sources is given by

$$\mathscr{L}(x) = \mathscr{L}_0(x) + \overline{\sigma}(x)\psi(x) + \overline{\psi}(x)\sigma(x),$$

where $\mathscr{L}_0(x)$ is given by Eq. (4.20). Derive Eqs. (12.81a) and (12.81b) from the Euler–Lagrange equations obtained from $\mathscr{L}(x)$ by varying with respect to $\overline{\psi}(x)$ and $\psi(x)$, respectively.

12.4. This problem will establish that Eq. (12.81a), which was obtained as a classical equation in the previous problem, also holds in the quantum theory.

---

[12] Throughout this chapter, we have implicitly assumed some form of regularization, for example dimensional regularization, which defines the loop integrals without spoiling the symmetries of the theory.

In the approximation (12.79), where the physical interaction terms are neglected but the source terms are retained, the Hamiltonian corresponding to Eq. (12.77) is

$$H = H_0 - \int d^3 z \, \overline{\sigma}(z)\psi(z) - \int d^3 z \, \overline{\psi}(z)\sigma(z),$$

where H_0 is given by Eq. (4.22) and we have ignored terms which do not contain fermion fields.

Use the equal-times anticommutation relations derived in Problem 4.1 to show that

$$[H, \psi(x)] = i\gamma^0 \gamma^j \frac{\partial \psi(x)}{\partial x^j} - m\gamma^0 \psi(x) + \gamma^0 \sigma(x),$$

where the index j is summed over 1, 2, 3 as usual and the Grassmann sources anticommute with the fermion fields:

$$[\sigma, \psi]_+ = [\sigma, \overline{\psi}]_+ = [\overline{\sigma}, \psi]_+ = [\overline{\sigma}, \overline{\psi}]_+ = 0.$$

Hence, show that the Heisenberg equations of motion for the field operator $\psi(x)$,

$$i\frac{\partial \psi(x)}{\partial t} = [\psi(x), H],$$

coincides with Eq. (12.81a), obtained as a classical equation in the previous problem.

12.5. Derive Eq. (12.107b) by explicitly performing the functional derivatives in Eq. (12.108) and taking the limit that the sources go to zero.

12.6. In the text, we explicitly derived Eq. (12.110b) for the dependence of the free-field generating functional on the four-vector source $J_\mu(x)$. Derive the corresponding result Eq. (12.110c) for its dependence on the spinor sources $\sigma(x), \overline{\sigma}(x)$ by an analogous argument, but starting from the second of Eqs. (12.89) rather than the first.

12.7. Consider scalar meson theory, Eq. (11.71). If normal ordering is not imposed on the interaction, then a sub-graph analogous to Fig. 12.14 in QED is generated, as shown in Fig. 12.15(a). Show that, in contrast to QED, this sub-graph does not vanish.

In second order, this leads to an extra fermion self-energy term, represented by the Feynman graph Fig. 12.15(b). Show that this additional term contributes solely to the mass renormalization constant δm^2, analogous to Eqs. (9.26)–(9.28) in QED, and obtain a value for δm^2 using the dimensional regularization scheme.

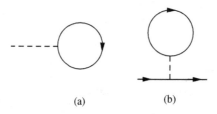

(a)   (b)

Figure 12.15 (a) The non-vanishing sub-graph arising from the equal-times contraction in scalar meson theory, Eq. (11.71), and (b) its contribution to the fermion self-energy in second order

13

Path Integrals

So far we have used the canonical formulation of quantum field theory, in which the fields are represented by non-commuting operators. This approach is familiar from non-relativistic quantum mechanics and brings out the physical meaning of the formalism in a simple and direct way. However, it is difficult to apply to non-Abelian gauge theories; that is, to gauge theories in which the conserved charges which act as the sources of the interaction, like the colour charges $\hat{F}_i$ in QCD, do not commute with each other. Here we will introduce the alternative 'path integral' formalism, which is particularly well suited to such theories.

In this chapter, we begin by introducing the properties of functional integrals, which will be required later in the chapter. We will then introduce the path integral formalism itself in the context of QED, before applying it to QCD in Chapter 14.

13.1 Functional Integration

Functional integrals were first introduced by Wiener in the study of Brownian motion and later, in a slightly different form, by Dirac and Feynman in the context of quantum mechanics.[1] In these early applications, the functions integrated over often represented particle trajectories or paths, and for that reason the terms 'path integral' and 'functional integral' are often used interchangeably. In the present section, we will briefly introduce functional integration, and obtain some of the more important mathematical results required later.

---

[1] An interesting account and comparison of the two approaches is given in C. Nash, *Relativistic Quantum Fields*, Academic Press, 1978.

13.1.1   Classical fields

We begin by defining the functional integral

$$\int \mathcal{D}\phi F[\phi].$$
(13.1)

Here $F[\phi]$ is a functional of a real scalar field $\phi(x)$. We can think of this integral as a summation of functionals $F[\phi]$ over a whole class of functions, as specified below. One way to define this integral is to introduce a complete set of orthonormal functions $u_i(x)$, $(i=1, 2, \ldots)$, such that

$$\int d^4 x u_i(x) u_j(x) = \delta_{ij},$$
(13.2)

and expand $\phi(x)$ in the infinite series

$$\phi(x) = \sum_{i=1}^{\infty} \alpha_i u_i(x).$$
(13.3)

The class of functions over which summation is carried out corresponds to the expansion coefficients $\alpha_1, \alpha_2, \ldots$ taking on all possible values: $-\infty < \alpha_i < \infty$, $(i = 1, 2, \ldots)$. If we cut off the series (13.3) after n terms, the functional $F[\phi]$ in Eq. (13.1) is replaced by a function of n parameters $\alpha_1, \ldots, \alpha_n$:

$$F[\phi] \rightarrow F(\alpha_1, \ldots, \alpha_n).$$

We then define the functional integral (13.1) by

$$\int \mathcal{D}\phi F[\phi] \equiv \lim_{n\to\infty} \left(\int_{-\infty}^{\infty} \frac{d\alpha_1}{\sqrt{2\pi}} \cdots \int_{-\infty}^{\infty} \frac{d\alpha_n}{\sqrt{2\pi}} F(\alpha_1, \ldots, \alpha_n) \right)$$
(13.4)

provided that a finite non-zero limit exists. The 'weight factors' $1/\sqrt{2\pi}$ are chosen so that this is the case for the 'Gaussian' type of functional integrals which occur in field theory. These are typically of the form

$$I \equiv \int \mathcal{D}\phi \exp\left\{-\frac{1}{2}[\phi K \phi]\right\}$$
(13.5)

where we have employed the 'short-hand' notation (12.49), i.e.

$$[AKB] \equiv \int d^4 x \int d^4 y A(x) K(x, y) B(y),$$
(13.6)

and $K(x, y)$ is a real symmetric operator, $K(x, y) = K(y, x)$.

To illustrate these ideas, we will explicitly evaluate the standard integral (13.5). Using the definition (13.4), together with Eqs. (13.2) and (13.3), this is given by

$$I = \lim_{n\to\infty} I_n = \lim_{n\to\infty} \int_{-\infty}^{\infty} \left(\prod_{i=1}^{n} \frac{d\alpha_i}{\sqrt{2\pi}} \right) \exp\left\{ -\frac{1}{2} \sum_{i,j=1}^{n} \alpha_i K_{ij} \alpha_j \right\},$$
(13.7)

where the coefficients

$$K_{ij} = \int d^4x \int d^4y\, u_i(x) K(x,y) u_j(y). \tag{13.8}$$

For $i, j = 1, 2, \ldots n$, these coefficients are the elements of a real, symmetric n-dimensional square matrix K_n, which can be diagonalized by an n-dimensional orthogonal matrix O_n, with elements O_{ij}. Introducing new variables

$$\beta_i = \sum_{j=1}^{n} O_{ij}\alpha_j, \tag{13.9}$$

we can write

$$\sum_{i,j=1}^{n} \alpha_i K_{ij} \alpha_j = \sum_{i,j=1}^{n} \beta_i \lambda_i \delta_{ij} \beta_j = \sum_{i=1}^{n} \lambda_i \beta_i^2 \tag{13.10}$$

where λ_i, $i = 1, 2, \ldots n$, are the eigenvalues of K_n. For the product of differentials in Eq. (13.7) we have

$$\prod_{i=1}^{n} d\beta_i = \det O_n \prod_{i=1}^{n} d\alpha_i = \prod_{i=1}^{n} d\alpha_i, \tag{13.11}$$

since the determinant of an orthogonal matrix is unity. Substituting Eqs. (13.10) and (13.11) into Eq. (13.7) and using

$$\exp\left\{ -\frac{1}{2} \sum_{i=1}^{n} \lambda_i \beta_i^2 \right\} = \prod_{i=1}^{n} \exp\left\{ -\frac{1}{2} \lambda_i \beta_i^2 \right\}$$

gives

$$I_n = \prod_{i=1}^{n} \left(\int_{-\infty}^{\infty} \frac{d\beta_i}{\sqrt{2\pi}} \exp\left\{ -\frac{1}{2} \lambda_i \beta_i^2 \right\} \right) = \prod_{i=1}^{n} \lambda_i^{-1/2} = (\det K_n)^{-1/2},$$

since $\det K_n = \prod_{i=1}^{n} \lambda_i$. Hence, using Eq. (13.7), we finally obtain

$$\int \mathcal{D}\phi \exp\left\{ -\frac{1}{2} [\phi K \phi] \right\} = (\det K)^{-1/2}, \tag{13.12}$$

where

$$\det K = \lim_{n\to\infty} \det K_n. \tag{13.13}$$

In particular, one finds

$$\int \mathcal{D}\phi\, e^{-[\phi^2]/2} = \lim_{n\to\infty} \left(\prod_{i=1}^{n} \int_{-\infty}^{\infty} \frac{d\alpha_i}{\sqrt{2\pi}} e^{-\alpha_i^2/2} \right) = 1 \tag{13.14}$$

where we use the notation

$$[\phi\psi] \equiv \int d^4x \phi(x)\psi(x).$$ (13.15)

A common alternative definition of a functional integral is obtained by dividing space–time into tiny cells of space–time volume Δ centered at lattice points X_i. Within each cell, ϕ is approximated by the constant value $\phi_i = \phi(x_i)$ so that, for example,

$$[\phi^2] \equiv \int d^4x \{\phi(x)\}^2 = \sum_i \Delta\phi_i^2$$

and any functional

$$F[\phi] \to F(\phi_1, \phi_2, ..).$$

The functional integral is defined by

$$\int \mathcal{D}\phi F[\phi] \equiv \lim_{\Delta\to 0}\left[\left(\frac{\Delta}{2\pi}\right)^{1/2}\int_{-\infty}^{\infty} d\phi_1 \left(\frac{\Delta}{2\pi}\right)^{1/2}\int_{-\infty}^{\infty} d\phi_2 \cdots F(\phi_1, \phi_2, \cdots)\right]$$ (13.16)

provided that a finite non-zero limit exists. The weight factors are again chosen so that this is the case for the type of functional integrals which occur in field theory, and one easily obtains our previous result Eq. (13.14) for the case of a simple Gaussian integral. The other integrals we require can be obtained from Eq. (13.14) using general properties of functional integrals which follow from either definition. Hence either definition is equally satisfactory. In what follows we shall adopt Eq. (13.4).

The most important general properties of functional integrals follow from the above definitions. They are:

  (i) *linearity*:

$$\int \mathcal{D}\phi\,(aF[\phi] + bG[\phi]) = a\int \mathcal{D}\phi F[\phi] + b\int \mathcal{D}\phi G[\phi]$$ (13.17)

where a and b are arbitrary constants; and
  (ii) *'translation invariance'*:

$$\int \mathcal{D}\phi F[\phi + f] = \int \mathcal{D}\phi F[\phi]$$ (13.18)

where $f(x)$ is any function, independent of $\phi(x)$.

To see this, we expand $f(x)$ in terms of the $u_i(x)$:

$$f(x) = \sum_{i=1}^{\infty} f_i u_i(x).$$

From the definition (13.4), it follows that

$$\int \mathcal{D}\phi F[\phi] = \lim_{n\to\infty}\left\{\left(\prod_{i=1}^{n}\int_{-\infty}^{\infty}\frac{d\alpha_i}{\sqrt{2\pi}}\right)F(\alpha_1 + f_1, \ldots, \alpha_n + f_n)\right\}.$$

Introducing new variables of integration $\beta_i = \alpha_i + f_i$, Eq. (13.18) follows at once.

It will also be useful to consider a change of integration variables from ϕ to $\overline{\phi}$, defined by

$$\overline{\phi}(x) = \int d^4 y K(x, y) \phi(y). \tag{13.19}$$

Expanding both ϕ and $\overline{\phi}$ in the form (13.3) yields coefficients α_i and $\overline{\alpha}_i$, which are related by

$$\overline{\alpha}_i = \sum_{j=1}^{\infty} K_{ij} \alpha_j, \tag{13.20}$$

where

$$K_{ij} = \int d^4 x \int d^4 y u_i(x) K(x, y) u_j(y).$$

Curtailing the series after n terms, we obtain

$$\prod_{i=1}^{n} d\overline{\alpha}_i = \det K_n \prod_{i=1}^{n} d\alpha_i, \tag{13.21}$$

where K_n is the n-dimensional square matrix with coefficients K_{ij}. Together with the definition (13.4), this gives

$$\int \mathcal{D}\overline{\phi} F[\phi] = \det K \int \mathcal{D}\phi F[\phi] \tag{13.22}$$

for the change of variables (13.19), where

$$\det K \equiv \lim_{n \to \infty} \det K_n.$$

In applications in field theory, one deals with functionals $I[J]$ of a real scalar source $J(x)$, defined by the functional integral

$$I[J] = N \int \mathcal{D}\phi \exp\left(A[J, \phi]\right), \tag{13.23}$$

where

$$A[J, \phi] = -\frac{1}{2} \int d^4 x \int d^4 y \phi(x) K(x, y) \phi(y) + \int d^4 x J(x) \phi(x). \tag{13.24}$$

Here $K(x, y)$ is a real symmetric operator $K(x, y) = K(y, x)$ and N is a normalization constant, to be fixed by the subsidiary condition

$$I[J = 0] = 1. \tag{13.25}$$

Eq. (13.23) for $I[J]$ can be greatly simplified. As we shall now show, it can be transformed so as not to involve any functional integrals. Because of the source term, Eq. (13.23) is not of the Gaussian form (13.5). However, it can be brought into this form by 'completing the square.' Defining an 'inverse operator' K^{-1} by

$$\int d^4x' K(x, x')K^{-1}(x', x'') = \delta^{(4)}(x - x''), \tag{13.26}$$

we introduce a shifted scalar field

$$\phi'(x) = \phi(x) - \int d^4x' K^{-1}(x, x')J(x'). \tag{13.27}$$

Substituting this shifted field in Eq. (13.24) leads to

$$
\begin{aligned}
A[J, \phi] = &-\frac{1}{2} \int d^4x \int d^4y \left\{ \phi'(x) + \int d^4x' K^{-1}(x, x')J(x') \right\} \\
&\times K(x, y) \left\{ \phi'(y) + \int d^4y' K^{-1}(y, y')J(y') \right\} \\
&+ \int d^4x J(x) \left\{ \phi'(x) + \int d^4x' K^{-1}(x, x')J(x') \right\}.
\end{aligned}
\tag{13.28}
$$

Multiplying this equation out, it can by repeated use of Eq. (13.26) be simplified to

$$A[J, \phi] = \frac{1}{2} [JK^{-1}J] - \frac{1}{2} [\phi' K \phi'], \tag{13.29}$$

where we again use the notation (13.6). Substituting Eq. (13.29) into Eq. (13.23) gives

$$I[J] = N \int \mathcal{D}\phi \exp \left\{ \frac{1}{2} [JK^{-1}J] - \frac{1}{2} [\phi' K \phi'] \right\}. \tag{13.30}$$

Since all the factors in the exponent of this exponential commute with each other, it factorizes into

$$\exp \left\{ \frac{1}{2} [JK^{-1}J] \right\} \times \exp \left\{ -\frac{1}{2} [\phi' K \phi'] \right\}. \tag{13.31}$$

Furthermore, the first of these exponentials does not depend on ϕ, so that it can be taken outside the functional integral of Eq. (13.30), which becomes

$$I[J] = N \exp \left\{ \frac{1}{2} [JK^{-1}J] \right\} \int \mathcal{D}\phi \exp \left\{ -\frac{1}{2} [\phi' K \phi'] \right\}. \tag{13.32}$$

Since from Eq. (13.27) ϕ' is of the form

$$\phi'(x) = \phi(x) + f(x),$$

where $f(x)$ is independent of $\phi(x)$, it follows from translation invariance, Eq. (13.18), that

$$\int \mathcal{D}\phi \exp \left\{ -\frac{1}{2} [\phi' K \phi'] \right\} = \int \mathcal{D}\phi \exp \left\{ -\frac{1}{2} [\phi K \phi] \right\},$$

which is a constant. It can therefore be absorbed into the normalization constant N in Eq. (13.32), giving as our final result

$$I[J] = \exp \left\{ \frac{1}{2} [JK^{-1}J] \right\}, \tag{13.33}$$

where we have imposed the normalization condition (13.25).

13.1.2 Grassmann generators

In order to discuss integrals over Grassmann fields, we must first define integration with respect to discrete Grassmann generators $\theta_1, \theta_2, \ldots, \theta_n$. Differentiation with respect to such variables was defined in Section 12.4.2, but integration cannot be defined as the inverse of differentiation, since there is no element (12.60) of the Grassmann algebra which gives the result θ_i when differentiated with respect to θ_i. Instead we define operators $\int d\theta_i$ as follows:

(i) $\int d\theta_i 1 = 0, \quad \int d\theta_i \theta_i = 1.$ (13.34)

(ii) Writing the general element (12.60) of the Grassmann algebra in the form

$$f(p) = A + \theta_i B,$$ (13.35)

where A and B are independent of θ_i, we define

$$\int d\theta_i f(p) = \int d\theta_i (A + \theta_i B) = \left(\int d\theta_i 1\right)A + \left(\int d\theta_i \theta_i\right)B = B.$$ (13.36)

It follows from these definitions that integration is essentially the same as differentiation. For example, for the most general element (12.61a) of a two-dimensional Grassmannn algebra, one easily verifies that

$$\int d\theta_1 f(p) = p_1 + p_{12}\theta_2 = \frac{\partial f}{\partial \theta_1}.$$

As for differentiation, care must be taken with the order of factors, so that, for example,

$$\int d\theta_1 \theta_1 \theta_2 = \left(\int d\theta_1 \theta_1\right)\theta_2 = \theta_2, \quad \int d\theta_1(\theta_2\theta_1) = \int d\theta_1(-\theta_1\theta_2) = -\theta_2.$$ (13.37)

From the definition of $\int d\theta_i$ linearity easily follows, i.e.

$$\int d\theta_i \{f(p) + f(q)\} = \int d\theta_i f(p) + \int d\theta_i f(q)$$

where $f(p)$ and $f(q)$ are two elements of a Grassmann algebra. Translation invariance also holds, i.e. if α is another Grassmann variable, independendent of θ_i, then

$$\int d\theta_i F(\theta_1, \ldots \theta_i \ldots \theta_n) = \int d\theta_i F(\theta_1, \ldots \theta_i + \alpha, \ldots \theta_n).$$ (13.38)

Eq. (13.38) is easily derived by writing $F(\theta_1, \ldots \theta_i, \ldots \theta_n)$ in the form (13.35).
The generalization to multiple integrals is straightforward. We define

$$\int d\theta_1 d\theta_2 F(\theta_1, \theta_2) = \int d\theta_1 \left\{\int d\theta_2 F(\theta_1, \theta_2)\right\},$$ (13.39)

with obvious extensions to more than two generators; for example

$$\int d\theta_1 d\theta_2 \theta_1 \theta_2 = \int d\theta_1 \left\{\int d\theta_2 \theta_1 \theta_2\right\} = \int d\theta_1(-\theta_1) = -1$$ (13.40)

and

$$\int d\theta_1 d\theta_2 \ldots d\theta_n \theta_1 \theta_2 \ldots \theta_n = \int d\theta_1 d\theta_2 \ldots d\theta_n \theta_n \theta_{n-1} \ldots \theta_1 (-1)^\kappa = (-1)^\kappa, \quad (13.41)$$

where

$$\kappa = (n-1) + (n-2) + \cdots + 1 = n(n-1)/2. \quad (13.42)$$

Note that, for multiple integrals, the order of the differentials matters; for example

$$\int d\theta_2 \, d\theta_1 \theta_1 \theta_2 = \int d\theta_2 \left\{ \int d\theta_1 \theta_1 \theta_2 \right\} = 1,$$

which should be compared with Eq. (13.40).

Linearity and translation invariance also hold for multiple integrals; for example:

$$\int d\theta_i d\theta_j F\left(\theta_1, \ldots \theta_i, \ldots \theta_j, \ldots \theta_n\right) = \int d\theta_i d\theta_j F\left(\theta_1, \ldots \theta_i + \alpha_i, \ldots \theta_j + \alpha_j, \ldots \theta_n\right),$$

$$(13.43)$$

where α_i and α_j are two other Grassmann variables, independent of θ_i and θ_j.

In what follows, we shall require integrals of the standard form

$$I_n \equiv \int \left(\prod_{i=1}^n d\theta_i d\tilde{\theta}_i \right) \exp \left(-\sum_{i,j=1}^n \theta_i A_{ij} \tilde{\theta}_j \right), \quad (13.44)$$

where A_{ij} are arbitrary numbers, and θ_i and $\tilde{\theta}_i$ are two independent sets of Grassmann generators, i.e.

$$\left[\theta_i, \theta_j\right]_+ = \left[\tilde{\theta}_i, \tilde{\theta}_j\right]_+ = \left[\theta_i, \tilde{\theta}_j\right]_+ = 0, \quad (13.45)$$

for $i, j = 1, 2, \ldots n$. The order of the differentials in Eq. (13.44) will always be taken as

$$\left(\prod_{i=1}^n d\theta_i d\tilde{\theta}_i \right) = d\theta_1 d\tilde{\theta}_1 d\theta_2 d\tilde{\theta}_2 \ldots d\theta_n d\tilde{\theta}_n. \quad (13.46)$$

To evaluate the integral (13.44), we introduce new variables

$$\tilde{\eta}_i = \sum_{j=1}^n A_{ij} \tilde{\theta}_j, \quad i = 1, \ldots n, \quad (13.47)$$

so that the integrand in Eq. (13.44) becomes

$$\exp \left(-\sum_{i,j=1}^n \theta_i A_{ij} \tilde{\theta}_j \right) = \exp \left(-\sum_{i=1}^n \theta_i \tilde{\eta}_i \right). \quad (13.48)$$

Since the new variables $\tilde{\eta}_i$ also satisfy anticommutation relations

$$\left[\tilde{\eta}_i, \tilde{\eta}_j\right]_+ = \left[\tilde{\eta}_i, \theta_j\right]_+ = 0,$$

it follows that $\left(\theta_i \tilde{\eta}_i\right)$ and $\left(\theta_j \tilde{\eta}_j\right)$ commute and the exponential on the right-hand side of Eq. (13.48) can be factorized:

$$\exp\left(-\sum_{i=1}^{n} \theta_i \tilde{\eta}_i\right) = \prod_{i=1}^{n} e^{-\theta_i \tilde{\eta}_i} = \prod_{i=1}^{n}\left(1 - \theta_i \tilde{\eta}_i\right), \qquad (13.49)$$

where the last step follows since $\theta_i^2 = \tilde{\eta}_i^2 = 0$.

Since $\int d\theta_i = 0$, it follows that the only term from the integrand (13.49) which contributes to the integral (13.44) is that which contains all the variables $\theta_1, \theta_2, \ldots \theta_n$, so that

$$I_n = \int d\theta_1 d\tilde{\theta}_1 \ldots d\theta_n d\tilde{\theta}_n (-1)^n \left(\theta_1 \tilde{\eta}_1\right) \ldots \left(\theta_n \tilde{\eta}_n\right). \qquad (13.50)$$

From Eq. (13.47) we obtain

$$\begin{aligned}
\tilde{\eta}_1 \tilde{\eta}_2 \ldots \tilde{\eta}_n &= \sum_{\alpha,\beta,\ldots} A_{1\alpha} A_{2\beta} \ldots A_{n\nu} \tilde{\theta}_\alpha \tilde{\theta}_\beta \ldots \tilde{\theta}_\nu \\
&= \tilde{\theta}_1 \tilde{\theta}_2 \ldots \tilde{\theta}_n \sum_{\alpha,\beta,\ldots} \varepsilon_{\alpha\beta\ldots\nu} A_{1\alpha} A_{2\beta} \ldots A_{n\nu} \\
&= \tilde{\theta}_1 \tilde{\theta}_2 \ldots \tilde{\theta}_n \det A_n, \qquad (13.51)
\end{aligned}$$

where A_n is the n-dimensional square matrix with elements A_{ij}. Hence Eq. (13.50) becomes

$$I_n = \det A_n \int d\theta_1 d\tilde{\theta}_1 \ldots d\theta_n d\tilde{\theta}_n (-1)^n \left(\theta_1 \tilde{\theta}_1\right) \ldots \left(\theta_n \tilde{\theta}_n\right), \qquad (13.52)$$

since the order of the variables $\tilde{\theta}_1, \ldots \tilde{\theta}_n$ in this equation is the same as that of $\tilde{\eta}_1 \ldots \tilde{\eta}_n$ in Eq. (13.50). From the anticommutation relations (13.45), we have

$$(-1)^n \left(\theta_1 \tilde{\theta}_1\right) \ldots \left(\theta_n \tilde{\theta}_n\right) = \left(\tilde{\theta}_n \theta_n\right) \ldots \left(\tilde{\theta}_1 \theta_1\right).$$

Substituting this relation into Eq. (13.52), we see from Eq. (13.41) that the integral on the right-hand side of Eq. (13.52) has the value 1, and we obtain our final result

$$I_n \equiv \int \left(\prod_{i=1}^{n} d\theta_i d\tilde{\theta}_i\right) \exp\left(-\sum_{i,j=1}^{n} \theta_i A_{ij} \tilde{\theta}_j\right) = \det A_n. \qquad (13.53)$$

13.1.3   Grassmann fields

In quantum field theory, we are interested in double functional integrals

$$\int \mathcal{D}\theta \mathcal{D}\tilde{\theta} F[\theta, \tilde{\theta}] \qquad (13.54)$$

of functionals $F[\theta, \tilde{\theta}]$ of two independent Grassmann fields $\theta(x)$, $\tilde{\theta}(x)$. These are defined in close analogy to our treatment of integrals over classical fields in Sub-section 13.1.1. To do this, we formally write

$$\theta(x) = \sum_{i=1}^{\infty} \theta_i u_i(x) \qquad \tilde{\theta}(x) = \sum_{i=1}^{\infty} \tilde{\theta}_i u_i(x) \tag{13.55}$$

where $u_i(x)$ are the orthonormal functions introduced in Eq. (13.2) and the coefficients $\theta_1, \theta_2, \ldots$ and $\tilde{\theta}_1, \tilde{\theta}_2, \ldots$, are two independent, infinite sets of discrete Grassmann generators. If we cut off the series after n terms, the functional $F\left[\theta, \tilde{\theta}\right]$ is replaced by a function of $2n$ generators

$$F\left[\theta, \tilde{\theta}\right] \rightarrow F\left(\theta_1, \theta_2, \ldots \theta_n; \tilde{\theta}_1, \tilde{\theta}_2, \ldots \tilde{\theta}_n\right)$$

and we can define

$$\int \mathcal{D}\theta \mathcal{D}\tilde{\theta} F\left[\theta, \tilde{\theta}\right] \equiv \lim_{n \to \infty} \left(\int \left[\prod_{i=1}^{n} d\theta_i d\tilde{\theta}_i \right] F\left(\theta_1, \theta_2, \ldots \theta_n; \tilde{\theta}_1, \tilde{\theta}_2, \ldots \tilde{\theta}_n\right) \right). \tag{13.56}$$

General properties which follow are:

(i) *linearity*:

$$\int \mathcal{D}\theta \mathcal{D}\tilde{\theta} \left(aF\left[\theta, \tilde{\theta}\right] + bG\left[\theta, \tilde{\theta}\right] \right) = a\int \mathcal{D}\theta \mathcal{D}\tilde{\theta} F\left[\theta, \tilde{\theta}\right] + b\int \mathcal{D}\theta \mathcal{D}\tilde{\theta} G\left[\theta, \tilde{\theta}\right] \tag{13.57}$$

where a and b are arbitrary constants; and

(ii) *'translation invariance'*:

$$\int \mathcal{D}\theta \mathcal{D}\tilde{\theta} F\left[\theta + \xi, \tilde{\theta} + \tilde{\xi}\right] = \int \mathcal{D}\theta \mathcal{D}\tilde{\theta} F\left[\theta, \tilde{\theta}\right] \tag{13.58}$$

where $\xi(x)$ and $\tilde{\xi}(x)$ are Grassmann fields, independent of $\theta(x)$ and $\tilde{\theta}(x)$. Using the definition (13.56) of the functional integral, we see that Eq. (13.58) is an obvious generalization for functional integrals of Eq. (13.43) for multiple integrals over discrete Grassmann variables.

In practice, we are concerned solely with 'Gaussian' integrals, and in particular the standard integral

$$\int \mathcal{D}\theta \mathcal{D}\tilde{\theta} e^{-[\theta A \tilde{\theta}]}$$

where we again use the standard notation (13.6). Using the definition (13.56) and substituting the expansions (13.55) gives

$$\int \mathcal{D}\theta \mathcal{D}\tilde{\theta} e^{-[\theta A \tilde{\theta}]} = \lim_{n \to \infty} \int \left(\prod_{i=1}^{n} d\theta_i d\tilde{\theta}_i \right) \exp\left(-\sum_{i,j=1}^{n} \theta_i A_{ij} \tilde{\theta}_j \right) \tag{13.59}$$

where the matrix elements A_{ij} are given by

$$A_{ij} = \int d^4x \int d^4y \, u_i(x) A(x, y) u_j(y). \tag{13.60}$$

Comparing Eq. (13.59) with Eq. (13.53), we see that

$$\int \mathcal{D}\theta \mathcal{D}\tilde{\theta} e^{-[\theta A \tilde{\theta}]} = \det A \tag{13.61}$$

where $\det A$ is the limit as $n \to \infty$ of the determinant of the n-dimensional square matrix A_n whose elements are A_{ij}.

13.2   Path Integrals

In Chapter 12 we employed the canonical formalism to define the Green functions of QED and to relate them to the QED generating functional $Z[J_\kappa, \sigma, \overline{\sigma}]$ through Eq. (12.91):

$$G^{\mu\cdots}\,(x_1, \ldots y_1, \ldots z_1, \ldots)$$

$$= (-1)^{\overline{n}}\,\left(\frac{1}{i}\right)^n \frac{\delta^n Z[J_\kappa, \sigma, \overline{\sigma}]}{\delta J_\mu(x_1) \ldots \delta\overline{\sigma}(y_1) \ldots \delta\sigma(z_1)\ldots}\bigg|_0, \tag{13.62a}$$

where $Z[J_\kappa, \sigma, \overline{\sigma}]$ is given in terms of vacuum expectation values of non-commuting field operators, Eq. (12.83):

$$Z[J_\kappa, \sigma, \overline{\sigma}] = \frac{\langle 0|S'|0\rangle}{\langle 0|S|0\rangle}, \tag{13.62b}$$

where S and S' are the S-matrix (11.40) and the augmented S-matrix (12.82). Eqs. (13.62a–b) express the essential results of the theory, since Green functions, or, more accurately, the S-matrix elements derived from them, embody all predictions of the theory.

The same results can be obtained using the path integral formulation of quantum field theory. In this approach, one usually first derives Feynman's path integral formulation of quantum mechanics from the familiar canonical formalism and then generalizes it to quantum fields to obtain expressions for the Green functions.[2]

We shall content ourselves with quoting the path integral expression for $Z[J_\kappa, \sigma, \overline{\sigma}]$ obtained in the Feynman formalism, and with showing that it is equivalent, to all orders of perturbation theory, to the corresponding expression derived in Chapter 12 using the canonical formalism.[3] This has the advantage of brevity and, more importantly, of bringing out the relationship between the two formulations. Substituting the path integral expression for $Z[J_\kappa, \sigma, \overline{\sigma}]$ in Eq. (13.62a), one obtains path integral expression for Green functions which are the starting point for applications of the path integral methods.

In this section, we shall define the path integral form of the generating functional, and show it is equivalent to our previous canonical form. The latter will be accomplished in two steps. In the first, in Section 13.2.2, we shall show that the two forms are equivalent in the presence of interactions if they are equivalent in the free-field case. Finally, in

---

[2] See, for example, L.H. Ryder, *Quantum Field Theory*, 2nd edn, Cambridge University Press, 1996, Chapters 5–9.
[3] This equivalence can be established independently of perturbation theory by using the axial gauge. See, for example, S. Coleman, *Aspects of Symmetry*, Cambridge University Press, 1985, Section 5.4.

Sections 13.2.3 and 13.2.4, we shall complete the proof by showing they are indeed equivalent in the free-field case.

13.2.1 The generating functional

The path integral expression for the generating functional of QED is given by

$$Z[J_\kappa, \sigma, \overline{\sigma}] = \frac{1}{N} \int \mathcal{D}A \mathcal{D}\overline{\psi} \mathcal{D}\psi e^{iX'}, \tag{13.63a}$$

where

$$X' = \int d^4x (\mathcal{L}_0 + \mathcal{L}_I + \mathcal{L}_S), \tag{13.63b}$$

with $\mathcal{L}_0$ and $\mathcal{L}_I$ the free–field and interaction Lagrangian densities

$$\mathcal{L}_0(x) = \overline{\psi}(x)(i\slashed{\partial} - m)\psi(x) - \frac{1}{2}(\partial_\nu A_\mu(x))(\partial^\nu A^\mu(x)), \tag{13.64}$$

$$\mathcal{L}_I(x) = e\overline{\psi}(x)\slashed{A}(x)\psi(x), \tag{13.65}$$

and $\mathcal{L}_S$ the Lagrangian density of the source term, Eq. (12.78),

$$\mathcal{L}_S(x) = J_\kappa(x)A^\kappa(x) + \overline{\sigma}(x)\psi(x) + \overline{\psi}(x)\sigma(x). \tag{13.66}$$

X' is seen to be the action

$$X = \int d^4x (\mathcal{L}_0 + \mathcal{L}_I) \tag{13.67}$$

of QED, augmented by the action of the source term $\mathcal{L}_S$. In Eq. (13.63a) and subsequent equations, functional integration over a multicomponent field implies integration over all its components; for example

$$\int \mathcal{D}A = \prod_{\mu=0}^{3} \mathcal{D}A^\mu. \tag{13.68}$$

In contrast to the canonical formalism where $A^\mu(x)$, $\psi(x)$ and $\overline{\psi}(x)$ are non-commuting operators, in the path integral formalism $A^\mu(x)$ are classical commuting fields, and $\psi(x)$ and $\overline{\psi}(x)$ are classical anticommuting Grassmann fields.

The factor N^{-1} in Eq. (13.63a) is chosen so that the source-free generating functional is normalized to unity:

$$Z[0, 0, 0] = 1 \tag{13.63c}$$

From Eq. (13.63a) this implies

$$N = \int \mathcal{D}A \mathcal{D}\overline{\psi} \mathcal{D}\psi e^{iX}. \tag{13.69}$$

Substituting Eq. (13.63a) for $Z[J_\kappa, \sigma, \overline{\sigma}]$ in Eq. (13.62a), we at once obtain the path integral expression for Green functions:

$$G^{\mu\cdots}(x_1, \ldots y_1, \ldots z_1, \ldots)$$
$$= \frac{1}{N} \int \mathcal{D}A\mathcal{D}\overline{\psi}\mathcal{D}\psi \left\{ e^{iX}A^{\mu}(x_1) \ldots \psi(y_1) \ldots \overline{\psi}(z_1) \ldots \right\}. \tag{13.70}$$

Eqs. (13.63a) and (13.70) for the generating functional and Green functions state the basic results of QED, expressed in the path integral formalism, Analogous results hold for other field theories. In particular, the generating functional is given by the appropriately normalized functional integral, over all fields, of $\exp(iX')$ where X' is the action of the interacting fields augmented by a term due to the sources. In Chapter 14, these methods will be applied to derive the Feynman rules for QCD.

13.2.2 Free and interacting fields

It remains to establish the equivalence of the two formalisms. In the canonical formalism of Chapter 12, we started from Eq. (13.62b) for the generating functional $Z[J_\kappa, \sigma, \overline{\sigma}]$ and transformed it into Eq. (12.121), which expresses $Z[J_\kappa, \sigma, \overline{\sigma}]$ in terms of the free-field generating functional $Z_0[J_\kappa, \sigma, \overline{\sigma}]$, Eq. (12.110a–c). In this section we shall show that the same expression (12.121) holds in the path integral formalism. This reduces the problem to demonstrating the equivalence of the free-field generating functions, which will be done in Section 13.2.3 and 13.2.4.

We start by rewriting the augmented action X', Eq. (13.63b). One verifies directly that

$$eI_\delta(x)\exp\left\{i\int d^4x(\mathscr{L}_0 + \mathscr{L}_S)\right\} = \mathscr{L}_I(x)\exp\left\{i\int d^4x(\mathscr{L}_0 + \mathscr{L}_S)\right\} \tag{13.71}$$

where $I_\delta(x)$ is the 'interaction operator' (12.119). From this relation one clearly shows that[4]

$$\exp\left\{ie\int d^4x I_\delta(x)\right\}\exp\left\{i\int d^4x(\mathscr{L}_0 + \mathscr{L}_S)\right\}$$
$$= \exp\left\{i\int d^4x \mathscr{L}_I(x)\right\}\exp\left\{i\int d^4x(\mathscr{L}_0 + \mathscr{L}_S)\right\}$$
$$= \exp\left\{i\int d^4x(\mathscr{L}_0 + \mathscr{L}_I + \mathscr{L}_S)\right\} = e^{iX'}. \tag{13.72}$$

Substituting the first line of this equation for $e^{iX'}$ is Eq. (13.63a), we obtain

$$Z[J_\kappa, \sigma, \overline{\sigma}] = \frac{1}{N}\int \mathcal{D}A\mathcal{D}\overline{\psi}\mathcal{D}\psi \exp\left\{ie\int d^4x I_\delta(x)\right\}\exp\left\{i\int d^4x(\mathscr{L}_0 + \mathscr{L}_S)\right\}. \tag{13.73}$$

Since the first exponential in this equation is independent of the fields A^μ, ψ and $\overline{\psi}$, it can be taken outside the functional integral, and we can write Eq. (13.73) as

$$Z[J_\kappa, \sigma, \overline{\sigma}] = \frac{1}{N_I}\exp\left\{ie\int d^4x I_\delta(x)\right\}Z_0[J_\kappa, \sigma, \overline{\sigma}], \tag{13.74}$$

---

[4] In obtaining this result, we have used the fact that $\mathscr{L}_0$, $\mathscr{L}_I$ and $\mathscr{L}_S$ commute with each other, since they contain classical fields and bilinear combinations of Grassmann fields only. This enables us to use the identity e^A, $e^B = e^{(A+B)}$ where A and B commute with each other.

where

$$Z_0[J_\kappa, \sigma, \bar{\sigma}] = \frac{1}{N_0} \int \mathcal{D}A \mathcal{D}\bar{\psi} \mathcal{D}\psi \exp\left(iX_0'\right) \tag{13.75a}$$

with

$$X_0' = \int d^4x (\mathcal{L}_0 + \mathcal{L}_S). \tag{13.75b}$$

The normalizing constant N_0 is chosen so that

$$Z_0[0, 0, 0] = 1 \tag{13.75c}$$

i.e.

$$N_0 = \int \mathcal{D}A \mathcal{D}\bar{\psi} \mathcal{D}\psi \exp\left\{i \int d^4x \mathcal{L}_0\right\}. \tag{13.76}$$

It follows from Eqs. (13.74), (13.75a) and (13.73) that

$$N_1 = \frac{N}{N_0}. \tag{13.77}$$

The exponential in Eq. (13.74) operating on $Z_0[J_\kappa, \sigma, \bar{\sigma}]$ produces the generating functional $Z[J_\kappa, \sigma, \bar{\sigma}]$ of the interacting fields from the free-field generating functional $Z_0[J_\kappa, \sigma, \bar{\sigma}]$. Expanding the exponential, one obtains a perturbation series for $Z[J_\kappa, \sigma, \bar{\sigma}]$ in powers of the coupling constant e.

We can now show that the expressions for the generating functional $Z[J_\kappa, \sigma, \bar{\sigma}]$ using the canonical and the path integral formalisms, i.e. Eqs. (12.121) and (13.74), are identical. In both cases, $Z[J_\kappa, \sigma, \bar{\sigma}]$ satisfies the same normalization condition $Z[0,0,0,]=1$ [see Eqs. (12.84) and (13.63c)]. It follows that the canonical and path integral forms of $Z[J_\kappa, \sigma, \bar{\sigma}]$ are identical, provided the corresponding canonical and path integral forms for the free-field generating functional $Z_0[J_\kappa, \sigma, \bar{\sigma}]$ are identical. We see from Eqs. (13.75) that $Z_0[J_\kappa, \sigma, \bar{\sigma}]$ factorizes into

$$Z_0[J_\kappa, \sigma, \bar{\sigma}] = Z_0[J_\kappa] Z_0[\sigma, \bar{\sigma}] \tag{13.78}$$

where $Z_0[J_\kappa]$ and $Z_0[\sigma, \bar{\sigma}]$ are the generating functionals for the free electromagnetic and free spinor fields respectively. From Eqs. (13.75a–c) one obtains:
for the free electromagnetic field:

$$Z_0[J_\kappa] = \frac{1}{N_1} \int \mathcal{D}A \exp\{iX[J_\kappa]\}, \tag{13.79a}$$

with the action

$$X[J_\kappa] = \int d^4x \left\{ -\frac{1}{2} (\partial_\nu A_\mu(x))(\partial^\nu A^\mu(x)) + J_\kappa(x) A^\kappa(x) \right\} \tag{13.79b}$$

and the constant N_1 determined by the normalization condition

$$Z_0[J_\kappa = 0] = 1; \tag{13.79c}$$

for the free spinor fields:

$$Z_0[\sigma, \bar{\sigma}] = \frac{1}{N_2} \int \mathcal{D}\bar{\psi}\mathcal{D}\psi \exp\{iX[\sigma, \bar{\sigma}]\}, \tag{13.80a}$$

with the action

$$X[\sigma, \bar{\sigma}] = \int d^4x\{\bar{\psi}(x)(i\partial\!\!\!/ - m)\psi(x) + \bar{\sigma}(x)\psi(x) + \bar{\psi}(x)\sigma(x)\} \tag{13.80b}$$

and the constant N_2 determined by

$$Z_0[\sigma = 0, \bar{\sigma} = 0] = 1. \tag{13.80c}$$

In Sections 13.2.3 and 13.2.4 we shall complete our proof of the equivalence of the formalisms by establishing the identity of the free-field generating functionals $Z_0[J_\kappa]$ and $Z_0[\sigma, \bar{\sigma}]$ in the canonical and path integral formalisms.

13.2.3 The free electromagnetic field

In order to evaluate the generating functional (13.79a) for the electromagnetic field, we first transform the action (13.79b). Substituting

$$\int d^4x \left(\partial_\nu A_\mu(x)\right)\left(\partial^\nu A^\mu(x)\right) = -\int d^4x A_\mu(x) \Box_x A^\mu(x) \tag{13.81}$$

in Eq. (13.79b), we obtain

$$X[J_\kappa] = \frac{1}{2}\int d^4x A_\mu(x)\Box_x A^\mu(x) + \int d^4x J_\kappa(x)A^\kappa(x). \tag{13.82}$$

Defining the symmetric wave operator

$$K^{\mu\nu}(x, x') = \left\{-g^{\mu\nu}\delta^{(4)}(x - x')\Box_{x'}\right\}, \tag{13.83}$$

we have

$$\int d^4x A_\mu(x)\int d^4x' \left\{-g^{\mu\nu}\delta^{(4)}(x - x')\Box_{x'}\right\}A_\nu(x') = -\int d^4x A_\mu(x)\Box_x A^\mu(x) \tag{13.84}$$

and substituting this equation into Eq. (13.82) we obtain

$$X[J_\kappa] = -\frac{1}{2}\int d^4x d^4x' A_\mu(x)K^{\mu\nu}(x,x')A_\nu(x') + \int d^4x J_\kappa(x)A^\kappa(x). \tag{13.85}$$

Eq. (13.85) is of the some form as Eq. (13.24) in Section (13.1.1). Hence we can simplify Eq. (13.85) in exactly the same way which was used in Section 13.1.1 to go from Eq. (13.24) to Eq. (13.33). Because of the close similarity, we shall just indicate the main steps of this derivation, leaving the details to the reader.

We define an inverse wave operator $K_{\nu\tau}^{-1}(x', x'')$ and a shifted field $A'_\mu(x)$ through the equations

$$\int d^4x' K^{\mu\nu}(x, x')K_{\nu\tau}^{-1}(x', x'') = g_\tau^\mu \delta^{(4)}(x - x''), \tag{13.86}$$

$$A'_\mu(x) = A_\mu(x) - \int d^4x' \, K^{-1}_{\mu\nu}(x, x') \, J^\nu(x'). \tag{13.87}$$

Utilizing these equations, we can rewrite Eq. (13.85) for $X[J_\kappa]$ as

$$X[J_\kappa] = \frac{1}{2} \left[J^\kappa K^{-1}_{\kappa\lambda} J^\lambda \right] - \frac{1}{2} \left[A'_\mu K^{\mu\nu} A'_\nu \right], \tag{13.88}$$

in analogy to the derivation of Eq. (13.29) from Eq. (13.24) in Section 13.1.1. The generating functional (13.79a) now becomes

$$Z_0[J_\kappa] = \frac{1}{N_1} \int \mathcal{D}A \, \exp\left\{ \frac{i}{2} \left[J^\kappa K^{-1}_{\kappa\lambda} J^\lambda \right] - \frac{i}{2} \left[A'_\mu K^{\mu\nu} A'_\nu \right] \right\}, \tag{13.89}$$

which can be transformed into

$$Z_0[J_\kappa] = \exp\left\{ \frac{i}{2} \left[J^\kappa K^{-1}_{\kappa\lambda} J^\lambda \right] \right\}, \tag{13.90}$$

[where the normalization condition (13.79c) has been imposed], in analogy to the derivation of Eq. (13.33) from Eq. (13.30).

It remains to identify the inverse wave operator $K^{-1}_{\nu\tau}(x', x'')$ which we write as a Fourier transform

$$K^{-1}_{\nu\tau}(x', x'') = \frac{1}{(2\pi)^4} \int d^4k \, D_{\nu\tau}(k) e^{-ik(x' - x'')}. \tag{13.91}$$

Substituting Eqs. (13.91), (13.83) and the Fourier representation of the δ-function into Eq. (13.86) immediately gives

$$D^{\mu\nu}(k) = \frac{g^{\mu\nu}}{k^2}. \tag{13.92}$$

Unfortunately, this has poles on the real k^0 axis at $k^0 = \omega_k \equiv |k|$, and the integral over k^0 in Eq. (13.91), and hence the generating functional (13.90), is ill-defined without a prescription for dealing with the poles.

At this point, we remind the reader that there was a similar ambiguity in our solution for the canonical generating functional (12.106), corresponding to the choice of the appropriate solution $\Delta^{\mu\nu}(x - x')$ of Eq. (12.101). As discussed following Eq. (12.106), this ambiguity was resolved by the requirement that the generating functional should lead to the correct free photon propagator (5.26-27). The same requirement here leads us to interpret the k^0-integral as a contour integral, and to displace the poles from the real axis by modifying Eq. (13.92) to

$$D^{\mu\nu}(k) = \frac{g^{\mu\nu}}{k^2 + i\varepsilon}. \tag{13.93}$$

Substituting this expression in Eq. (13.91) and comparing the resulting equation with Eq. (5.27) for the Feynman photon propagator $D_{F\mu\nu}(x - x')$, we see that

$$K^{-1}_{\mu\nu}(x, x') = -D_{F\mu\nu}(x - x'), \tag{13.94}$$

and that the generating functional (13.90) is given by

$$Z_0[J_\kappa] = \exp\left\{\frac{-i}{2}\left[J^\kappa D_{F\kappa\lambda} J^\lambda\right]\right\}, \tag{13.95}$$

which is identical to the canonical expression (12.110b). The same result can also be obtained more rigorously by first defining the generating functional in 'imaginary time' $x^0 = ix^4$ (x^4 real), when it can be evaluated without ambiguity, and then analytically continuing the result to real time x^0, when the result (13.95) is again obtained.[5]

13.2.4   The free spinor fields

The evaluation of the generating functional $Z_0[\sigma, \bar{\sigma}]$, Eqs. (13.80a–c), follows closely the same pattern as that of $Z_0[J_\kappa]$. We shall therefore only indicate the main steps and leave the rest to the reader. We transform the action $X[\sigma, \bar{\sigma}]$, Eq. (13.80b), into

$$X[\sigma, \bar{\sigma}] = \int d^4x d^4x' \,\bar{\psi}(x) K(x, x') \psi(x) + \int d^4x \{\bar{\psi}(x)\sigma(x) + \bar{\sigma}(x)\psi(x)\}, \tag{13.96}$$

where the symmetric wave operator $K(x, x')$ is defined by

$$K(x, x') = \delta^{(4)}(x - x')(i\partial_{x'} - m). \tag{13.97}$$

Defining the inverse wave operator $K^{-1}(x', x'')$ by

$$\int d^4x' \, K(x, x') \, K^{-1}(x', x'') = -\delta^{(4)}(x - x''), \tag{13.98}$$

and the shifted fields by

$$\psi'(x) = \psi(x) - \int d^4x' \, K^{-1}(x, x')\sigma(x') \tag{13.99a}$$

$$\bar{\psi}'(x) = \bar{\psi}(x) - \int d^4x' \,\bar{\sigma}(x') \, K^{-1}(x', x) \tag{13.99b}$$

one expresses $X[\sigma, \bar{\sigma}]$ in terms of the shifted fields. Using this expression in Eq. (13.80a), one obtains

$$Z_0[\sigma, \bar{\sigma}] = \exp\left\{i\left[\bar{\sigma} K^{-1}\sigma\right]\right\}. \tag{13.100}$$

To identify the inverse wave operator $K^{-1}(x, x')$, we write it as

$$K^{-1}(x', x'') = \frac{1}{(2\pi)^4}\int d^4p S(p) e^{-ip(x' - x'')}, \tag{13.101}$$

and substituting this and Eq. (13.97) in Eq. (13.98) leads to

$$S(p) = -\frac{\not{p} + m}{p^2 - m^2}. \tag{13.102}$$

---

[5] See, for example, D. Bailin and A. Love, *Introduction to Gauge Field Theories*, Rev. Edn., Institute of Physics Publishing Ltd. (1993), Section 4.2.

This expression in singular at $p^0 = \pm\sqrt{(\mathbf{p}^2 + m^2)}$. The ambiguities to which this gives rise are resolved by ensuring that the generating functional (13.80a) leads to the correct free-fermion propagator. This is achieved by interpreting the p^0-integral as a contour integral and displacing the poles an infinitesimal amount from the real axis by modifying Eq. (13.102) to

$$S(p) = -\frac{\not{p} + m}{p^2 - m^2 + i\varepsilon} \tag{13.103}$$

From the defining equations (7.23a) and (7.24a) of the fermion propagator and Eq. (13.101) it then follows that

$$K^{-1}(x, x') = -S_F(x - x'), \tag{13.104}$$

and from Eq. (13.100) we obtain our final result for the generating functional for the free spinor fields

$$Z_0[\sigma, \bar{\sigma}] = \exp\{-i[\bar{\sigma} S_F \sigma]\}. \tag{13.105}$$

This result is identical with the corresponding canonical expression, Eq. (12.110c).

This completes the demonstration that the path integral form of the generating functionals $Z_0[J_\kappa]$, $Z_0[\sigma, \bar{\sigma}]$ and hence of the free-field generating functional $Z_0[J_\kappa, \sigma, \bar{\sigma}]$ are identical to those obtained in the canonical formalism, Eqs. (12.110a–c). As discussed in Section 13.2.1, it follows from this that the path integral and canonical forms of the generating functional $Z[J_\kappa, \sigma, \bar{\sigma}]$ of the interacting fields are identical.

13.3   Perturbation Theory

In Section 13.2, we expressed the generating functional of QED in the form (13.74)–(13.75) to demonstrate the equivalence of the path integral and canonical formalisms to all orders in perturbation theory. However, for applications, it is usually much simpler to apply perturbation theory directly to the Green functions themselves, by expanding the Green functions of the interacting fields in a series in terms of free-field Green functions. In this section we will explain how to do this within the path integral framework. We start by showing how the free-field Green functions can be expressed in a form identical to Wick's theorem, Eq. (12.16a), in the canonical formalism.

13.3.1   Wick's theorem

It will be convenient to denote Green functions of the interacting fields by

$$\langle A^\mu(x_1) \ldots \psi(y_1) \ldots \bar{\psi}(z_1) \ldots \rangle \equiv G^{\mu \cdots}(x_1, \ldots y_1, \ldots z_1 \ldots), \tag{13.106}$$

where $G^{\mu \cdots}(x_1, \ldots y_1, \ldots z_1, \ldots)$ are given either by Eq. (13.62a) in the canonical formalism or by Eq. (13.70) in the path integral formalism. The corresponding free-field Green functions, denoted by $\langle A^\mu(x_1) \ldots \psi(y_1) \ldots \bar{\psi}(z_1) \ldots \rangle_0$ are most easily obtained by substituting the free-field generating functional $Z_0[J_\kappa, \sigma, \bar{\sigma}]$, given by Eqs. (13.78), (13.95) and (13.105), into the general expression (13.62a):

$$\langle A^\mu(x_1) \dots \psi(y_1) \dots \overline{\psi}(z_1) \dots \rangle_0$$
$$= (-1)^{\bar{n}} \left(\frac{1}{i}\right)^n \frac{\delta^{(n)} Z_0[J_\kappa, \sigma, \overline{\sigma}]}{\delta J_\mu(x_1) \dots \delta \overline{\sigma}(y_1) \dots \delta \sigma(z_1) \dots}\Bigg|_0.$$

Note that the subscript 0 in $\langle \ \rangle_0$ on the left-hand side of this equation denotes a free-field Green function, whereas the symbol $|_0$ on the right hand side indicates that the source terms are set equal to zero after the functional differentiation. It follows that the free-field Green function $\langle A^\mu(x_1) \dots \psi(y_1) \dots \overline{\psi}(z_1) \dots \rangle_0$ vanishes unless the functional differentiations remove all source terms. For example,

$$\langle A^\nu(y) \rangle_0 = \frac{1}{i} Z_0[\sigma, \overline{\sigma}] \frac{\delta Z_0[J_\kappa]}{\delta J_\nu(y)}\Bigg|_0 = 0,$$

since

$$Z_0[J_\kappa] = \exp\left\{-\frac{i}{2}\left[J^\kappa D_{F\kappa\lambda} J^\lambda\right]\right\}$$

and

$$\frac{\delta Z_0[J_\kappa]}{\delta J_\nu(y)} = \exp\left\{\frac{-i}{2}\left[J^\kappa D_{F\kappa\lambda} J^\lambda\right]\right\} \left(\frac{-i}{2}\right) \frac{\delta}{\delta J_\nu(y)}\left[J^\kappa D_{F\kappa\lambda} J^\lambda\right],$$

which vanishes when the source terms are set to zero after differentiation. From the last equation one obtains the two-point Green function

$$\langle A^\mu(x) A^\nu(y) \rangle_0 = \left(\frac{1}{i}\right)^2 \frac{\delta^2 Z_0[J_\kappa]}{\delta J_\mu(x) \delta J_\nu(y)}\Bigg|_0$$
$$= \left(\frac{1}{i}\right)^2 \left(\frac{-i}{2}\right) \frac{\delta^2}{\delta J_\mu(x) \delta J_\nu(y)}\left[J^\kappa D_{F\kappa\lambda} J^\lambda\right]\Bigg|_0$$
$$= i D_F^{\mu\nu}(x-y) = \overline{A^\mu(x) A^\nu(y)}.$$

Since other results are obtained in a similar manner, we shall summarize them very briefly. One finds at once that

$$\langle \psi(x) \rangle_0 = \langle \overline{\psi}(x) \rangle_0 = 0,$$

and for the two-point Green functions one obtains

$$\langle AB \rangle_0 = \overline{A B} \tag{13.107}$$

where A, B are any field operators A^μ, ψ, $\overline{\psi}$ and where, for QED, the only non-vanishing contractions are the photon and fermion propagators $iD_F^{\mu\nu}(x)$ and $iS_{F\alpha\beta}(x)$, Eqs. (12.17).

More complicated free-field Green functions $\langle ABCD \dots WXYZ \rangle_0$ vanish if they contain an odd number of field operators $AB \dots YZ$; for an even number one obtains

$$\langle ABCD \cdots WXYZ \rangle_0$$
$$= \overline{AB}\,\overline{CD} \cdots \overline{WX}\,\overline{YZ} + \overline{A}\,\overline{B}\,\overline{C}\,\overline{D} \cdots \overline{WX}\,\overline{YZ} + \cdots, \tag{13.108}$$

where the sum extends over all terms in which all the field operators $ABC\ldots$ are paired together in non-vanishing contractions. This is identical to Wick's theorem (12.16a) for the T-products $\langle 0|T(ABCD\cdots WXYZ)|0\rangle$. As in the canonical case, we shall state it without proof, which is again by induction. If we write the free-field Green functions as functional integrals by setting $\mathscr{L}_1=0$ in Eq. (13.70), we obtain

$$\langle ABC\ldots\rangle_0 = \frac{1}{N_0}\int \mathcal{D}A\mathcal{D}\bar{\psi}\mathcal{D}\psi e^{iX_0}(ABC\ldots)$$ (13.109)

where

$$X_0 = \int d^4x\,\mathscr{L}_0(x)$$ (13.110)

and N_0 is given by Eq. (13.76). Eq. (13.109) expresses the free-field Green function as a functional integral. With this interpretation of the free-field Green function, Eq. (13.108) is a statement of Wick's theorem for functional integrals. It is our final result for free-field Green functions, expressing them in terms of the photon and fermion propagators $iD_F^{\mu\nu}(x)$ and $iS_{F\alpha\beta}(x)$.

13.3.2   Interactions

Green functions for interacting fields are most conveniently evaluated perturbatively by relating them to free-field Green functions and evaluating the latter using Wick's theorem. Using the notation of Eq. (13.106), we can write Eq. (13.70) as

$$\langle ABC\ldots\rangle = \frac{1}{N}\int \mathcal{D}A\mathcal{D}\bar{\psi}\mathcal{D}\psi e^{iX}(ABC\ldots).$$ (13.111)

From Eqs. (13.67) and (13.110)

$$X = X_0 + \int d^4x\,\mathscr{L}_1(x).$$ (13.112)

Since X_0 and $\mathscr{L}_1(x)$ commute, we have

$$e^{iX} = \exp\left\{iX_0 + i\int d^4x\,\mathscr{L}_1(x)\right\} = e^{iX_0}\exp\left\{i\int d^4x\,\mathscr{L}_1(x)\right\},$$

so that Eq. (13.111) can be written

$$\langle ABC\ldots\rangle = \frac{1}{N}\int \mathcal{D}A\mathcal{D}\bar{\psi}\mathcal{D}\psi\, e^{iX_0}\exp\left\{i\int d^4x\,\mathscr{L}_1(x)\right\}(ABC\ldots).$$

Comparing this equation with Eq. (13.109), we obtain

$$\langle ABC\ldots\rangle = \frac{N_0}{N}\left\langle \exp\left\{i\int d^4x\,\mathscr{L}_1(x)\right\}(ABC\ldots)\right\rangle_0.$$

From Eq. (13.69) one similarly obtains

$$N = N_0 \left\langle \exp\left\{ i \int d^4x \, \mathscr{L}_I(x) \right\} \right\rangle_0,$$

and combining the last two equations

$$\langle ABC \ldots \rangle = \frac{\left\langle \exp\left\{ i \int d^4x \, \mathscr{L}_I(x) \right\} (ABC \ldots) \right\rangle_0}{\left\langle \exp\left\{ i \int d^4x \, \mathscr{L}_I(x) \right\} \right\rangle_0}. \tag{13.113}$$

The perturbation expansion for this Green function is then obtained by expanding the exponential factors in Eq. (13.113):

$$\exp\left\{ i \int d^4x \, \mathscr{L}_I(x) \right\} = \sum_{n=0}^{\infty} \frac{i^n}{n!} \int d^4x_1 \, \ldots \, d^4x_n \{ \mathscr{L}_I(x_1) \ldots \mathscr{L}_I(x_n) \}. \tag{13.114}$$

The expressions (13.113) and (13.114) are closely analogous to the corresponding expressions (12.8) and (12.11) obtained in the canonical formalism. In the canonical case, we saw, in Sections 12.2.2 and 12.2.3, that the denominator in Eq. (12.8) merely serves to cancel vacuum diagrams from the perturbation expansion of the numerator, and can be ignored in calculating *connected* Green functions. The same result can be shown to hold for the denominator factor in Eq. (13.113), which can similarly be ignored in calculating connected Green functions.

As an example of the path integral form of perturbation theory, we consider the Green function $\langle A^\mu(x_1)\psi(x_2)\overline{\psi}(x_3)\rangle$. Using Eq. (13.113) and neglecting the denominator factor, we obtain

$$\langle A^\mu(x_1)\psi(x_2)\overline{\psi}(x_3)\rangle = \left\langle \exp\left\{ i \int d^4x \, \mathscr{L}_I(x) \right\} A^\mu(x_1)\psi(x_2)\overline{\psi}(x_3) \right\rangle_0$$

$$= i \int d^4x \langle \mathscr{L}_I(x) \, A^\mu(x_1)\psi(x_2)\overline{\psi}(x_3)\rangle_0,$$

where we have used

$$\langle A^\mu(x_1)\psi(x_2)\overline{\psi}(x_3)\rangle_0 = \langle A^\mu(x_1)\rangle_0 \langle \psi(x_2)\overline{\psi}(x_3)\rangle_0 = 0,$$

which follows from $\langle A^\mu(x_1)\rangle_0 = 0$, and neglected terms beyond the first order in $\mathscr{L}_I(x)$. Substituting the explicit form (13.65) for $\mathscr{L}_I(x)$ then gives

$$\langle A^\mu(x_1)\psi(x_2)\overline{\psi}(x_3)\rangle = ie \int d^4x F, \tag{13.115}$$

where

$$F = \langle [\overline{\psi}(x)\gamma_\nu \psi(x)] A^\nu(x) A^\mu(x_1) \psi(x_2) \overline{\psi}(x_3) \rangle_0$$
$$= \langle A^\mu(x_1) A^\nu(x) \psi(x_2) [\overline{\psi}(x)\gamma_\nu \psi(x)] \overline{\psi}(x_3) \rangle_0. \tag{13.116}$$

The last line follows, since there is an even number of fermion interchanges in going from the first to the second line of Eq. (13.116), so that no sign change is required, and since we have kept the factors in the square bracket together, so that there is no need to keep track of the fermion indices, which are suppressed as usual. Wick's theorem (13.108) then gives

$$F = A^\mu(x_1)A^\nu(x)\,\underline{\psi(x_2)\overline{\psi}(x)}\,\gamma_\nu\,\underline{\psi(x)\overline{\psi}(x_3)}$$
$$+A^\mu(x_1)A^\nu(x)\,\underline{\psi(x_2)}\,\underline{\overline{\psi}(x)\gamma_\nu\psi(x)}\,\underline{\overline{\psi}(x_3)}.$$

(13.117)

Substituting the second term of Eq. (13.117) into Eq. (13.115) leads to an amplitude corresponding to the disconnected Feynman diagram of Fig. 13.1, which can be neglected in calculating the connected Green function. On substituting the first term into Eq. (13.115), and replacing the contractions with the appropriate propagator factors (12.17), one obtains, to first order in the interaction $\mathscr{L}_1(x)$,

$$\langle A^\mu(x_1)\,\psi(x_2)\overline{\psi}(x_3)\rangle$$
$$= ie \int d^4x\,i D_F^{\mu\nu}(x_1 - x)\,iS_F(x_2 - x)\,\gamma_\nu\,iS_F(x - x_3),$$

(13.118)

which corresponds to the Feynman diagram of Fig. 13.2 and is identical to the result which would be obtained in the canonical approach. In the next chapter we shall use the path integral form of perturbation theory, which has here been illustrated, to obtain the Feynman rules for QCD.

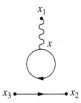

Figure 13.1   *The disconnected contribution to the Green function $\langle A^\mu(x_1)\,\psi(x_2)\overline{\psi}(x_3)\rangle$ which arises from substituting the second term of Eq. (13.117) into Eq. (13.115)*

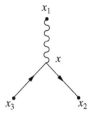

Figure 13.2   *The connected Green function $\langle A^\mu(x_1)\,\psi(x_2)\overline{\psi}(x_3)\rangle$ to first order, Eq. (13.118)*

13.4   Gauge Independent Quantization?

The free electromagnetic field can be described by the gauge invariant Lagrangian density

$$\mathscr{L} = -\frac{1}{4} F_{\mu\nu}(x) F^{\mu\nu}(x) \tag{13.119}$$

given by Eq. (5.8) with $s_\mu(x)=0$. We were unable to quantize this by canonical methods, since there is no field conjugate to the scalar potential $A^0(x)$, as discussed in Section 5.1. For this reason, we have hitherto worked in the Lorentz gauge, defined by Eq. (5.13). However, before considering the more complicated case of QCD, it will be instructive to determine whether the gauge invariant Lagrangian density (13.119) can be quantized using path integral methods. To do this, we simply try to evaluate the path integral form of the generating functional corresponding to the Lagrangian density (13.119):

$$Z_0[J_\kappa] = \frac{1}{N_Z} \int \mathcal{D}A \, \exp\{iX[J_\kappa]\}, \tag{13.120a}$$

where the action

$$X[J_\kappa] = \int d^4 x \left\{ -\frac{1}{4} F_{\mu\nu}(x) \, F^{\mu\nu}(x) + J_\lambda(x) A^\lambda(x) \right\} \tag{13.120b}$$

and the normalization is, as usual, fixed by the condition

$$Z_0[J_\kappa = 0] = 1. \tag{13.120c}$$

The argument which follows adheres closely to that used to evaluate the generating functional of the electromagnetic field in the Lorentz gauge in Section 13.2.3. Using

$$-\frac{1}{4} \int d^4 x F_{\mu\nu}(x) \, F^{\mu\nu}(x) = \frac{1}{2} \int d^4 x A^\mu(x) \left[g_{\mu\nu} \Box_x - \partial_\mu \partial_\nu \right] A^\nu(x),$$

the action (13.120b) is conveniently rewritten is the form

$$X[J_\kappa] = -\frac{1}{2} \int d^4 x \int d^4 x' \, A_\mu(x) K^{\mu\nu}(x, \, x') A_\nu(x') + \int d^4 x J_\kappa(x) A^\kappa(x),$$

analogous to Eq. (13.85), where the kernel $K^{\mu\nu}(x, \, x')$ is now given by

$$K^{\mu\nu}(x, \, x') = -\delta^4(x - x') \left[g_{\mu\nu} \Box_{x'} - \partial'_\mu \partial'_\nu \right], \tag{13.121}$$

and $\partial'_\mu = \partial/\partial x'_\mu$ etc. The argument following Eq. (13.85) again leads to the expression (13.90) for the generating functional, i.e.

$$Z_0[J_\kappa] = \exp\left\{ \frac{i}{2} [J^\tau K_{\tau\lambda}^{-1} J^\lambda] \right\}, \tag{13.122}$$

where $K_{\tau\lambda}^{-1}(x, \, x')$ is now the inverse of the kernel (13.121), defined by Eq. (13.86).

It remains to identify the inverse wave operator $K_{\nu\lambda}^{-1}(x, \, x')$. To do this, we substitute Eq. (13.91) into Eq. (13.86) and use Eq. (13.121) to obtain

$$\left(g^{\mu\nu} k^2 - k^\mu k^\nu \right) D_{\nu\tau}(k) = g_\tau^\mu, \tag{13.123}$$

where $D_{v\tau}(k)$ is the Fourier transform of $K_{v\tau}^{-1}(x, x')$ defined by Eq. (13.91). On the other hand, $D_{v\tau}(k)$ is a Lorentz tensor dependent only on the four-vector k^μ. Since the only scalar which can be formed from k^μ is k^2, the most general possible form for any such tensor is

$$D_{v\tau}(k) = a(k^2) g_{v\tau} + b(k^2) k_v k_\tau, \tag{13.124}$$

where $a(k^2)$ and $b(k^2)$ are arbitrary scalar functions. On substituting Eq. (13.124) into Eq. (13.123), one easily sees that Eq. (13.124) has no solution, and the generating functional (13.120) is undefined. As in the canonical formulation, the electromagnetic field cannot be consistently quantized using path integrals without 'fixing a gauge.'

Problems

13.1. Evaluate the second-order contributions to the connected Green function $\langle A^\mu(x_1) A^v(x_2) \psi(x_3) \overline{\psi}(x_4) \rangle$ using the methods of Section 13.3.2, and draw the corresponding configuration space Feynman diagrams.

13.2. A massive, neutral spin-1 vector boson is described by the Lagrangian density

$$\mathscr{L}(x) = -\frac{1}{4} F_{\mu v}(x) F^{\mu v}(x) + \frac{1}{2} m^2 A_\mu(x) A^\mu(x), \tag{13.125}$$

where m is the mass, $A^\mu(x)$ is a real four-vector field and the field tensor $F^{\mu v}(x)$ is given by Eq. (5.5) as usual.
Show that the corresponding equation of motion is the Proca equation:

$$\Box A^\mu(x) - \partial^\mu(\partial_v A^v(x)) + m^2 A^\mu(x) = 0.$$

The corresponding generating functional $Z_0[J_\kappa]$ is given by Eqs. (13.120), except that now

$$-\frac{1}{4} F_{\mu v}(x) F^{\mu v}(x) \rightarrow -\frac{1}{4} F_{\mu v}(x) F^{\mu v}(x) + \frac{1}{2} m^2 A_\mu(x) A^\mu(x)$$

Show that

$$Z_0[J_\kappa] = \exp\left\{ -\frac{i}{2} \left[J^\kappa D_{F\kappa\lambda} J^\lambda \right] \right\},$$

where the propagator is given by

$$D_{F\kappa\lambda}(x - x') = \frac{1}{(2\pi)^4} \int d^4 k \, D_{F\kappa\lambda}(k) e^{-ik(x' - x'')},$$

where

$$D_F^{\kappa\lambda}(k) = \frac{-g^{\kappa\lambda} + k^\kappa k^\lambda / m^2}{k^2 - m^2 + i\varepsilon}$$

14

Quantum Chromodynamics

Quantum chromodynamics is the standard theory of strong interactions, and was introduced in Section 11.2. However, as we shall shortly see, the gauge-invariant Lagrangian density (11.35) given in Section 11.2 is not suitable for quantization, and an equivalent Lagrangian, which can be quantized and which is suitable for perturbation theory, must be found. The precise form of this Lagrangian was first inferred by de Witt, and then re-derived by Faddeev and Popov, using a method based on path integral quantization, which we shall outline here.[1]

Since the problems which arise in quantizing QCD are associated entirely with gluon fields, it is convenient to first discuss the dynamics of gluon fields in the absence of quarks. The Faddeev–Popov procedure is introduced in this context in Section 14.1, and the theory generalized to include the quark fields in Section 14.2. The chapter continues with a discussion of perturbation theory and the Feynman rules for QCD in Sections 14.3 and 14.4, and concludes with a discussion of the renormalizability of the theory in Section 14.5.

14.1 Gluon Fields

In the absence of quarks, the gluon fields are described by the gauge-invariant Lagrangian density (11.33, 34), i.e.

$$\mathscr{L}_G = -\frac{1}{4} G_{i\mu\nu}(x) G_i^{\mu\nu}(x), \qquad (14.1)$$

---

[1] A systematic presentation of the canonical approach to gauge field theory is given by T. Kugo and I. Ojima, Prog. Theor. Phys. 60 (1978) 1869; Suppl. Prog. Theor. Phys. 66 (1979)1.

Quantum Field Theory, Second Edition   Franz Mandl and Graham Shaw
© 2010 John Wiley & Sons, Ltd

where

$$G_i^{\mu\nu}(x) = F_i^{\mu\nu}(x) + g_s f_{ijk} A_j^{\mu}(x) A_k^{\nu}(x). \qquad (14.2)$$

Like the gauge-invariant Lagrangian density for the free electromagnetic field, this Lagrangian density is not suitable for quantization. In this section, we shall first, in Section 14.1.1, outline the problems encountered in finding an appropriate Lagrangian density which can be quantized; and then show how they can be surmounted in the path integral framework in Sections 14.1.2–5.

14.1.1   The generating functional

We start from the generating functional corresponding to the gauge-invariant Lagrangian (14.1). This is of the standard form

$$Z[J_{i\kappa}] = \frac{1}{N} \int \mathcal{D}A e^{iX[J_{i\kappa}]} \qquad (14.3a)$$

where there is a classical source $J_{i\kappa}(x)$, $i=1, 2, \ldots 8$, associated with each gluon field $A_i^{\kappa}(x)$, and the augmented action

$$X[J_{i\kappa}] = \int d^4x \left(\mathcal{L}_G + J_{i\kappa}(x) A_i^{\kappa}(x) \right), \qquad (14.3b)$$

where the Lagrangian density $\mathcal{L}_G$ is given by Eq. (14.1). The integral extends over all the components, of all eight gluon fields, i.e.

$$\mathcal{D}A \equiv \prod_{i=1}^{8} \prod_{\mu=0}^{3} \mathcal{D}A_i^{\mu}(x) \qquad (14.4)$$

and the normalization constant is fixed by the condition

$$Z[J_{i\kappa}=0]=1. \qquad (14.3c)$$

In the absence of interactions, $G_i^{\mu\nu}(x)$ reduces to $F_i^{\mu\nu}(x)$ and the Lagrangian density (14.1) reduces to a sum of terms

$$-\frac{1}{4} F_{i\mu\nu}(x) F_i^{\mu\nu}(x), \qquad (14.5)$$

of the same form as the Lagrangian density (13.119) of the electromagnetic field. In this case, the generating functional (14.3) reduces to a straightforward generalization of Eq. (13.120) for the free electromagnetic field discussed in Section 13.4, and is similarly ill defined.

At this point, it is tempting to impose Lorentz conditions

$$\partial_\mu A_i^{\mu}(x) = 0 \quad i=1, 2, \ldots 8, \qquad (14.6)$$

on each of the gluon fields $A_i^{\mu}(x)$, and to proceed in analogy to our treatment of the electromagnetic field in Section 13.2.3. Unfortunately, this argument is not correct. In the case of QED, we were able to show that, for an arbitrary set of classical fields $A^{\mu}(x)$, one can always find a gauge transformation such that the transformed potentials obey the

Lorentz condition. However the proof [following Eq. (5.13)] that this is possible relies on the simple form (5.7) of the electromagnetic gauge transformation, and it cannot be generalized to the more complicated gauge transformations (11.26b) of QCD. We therefore cannot, in general, impose the Lorentz conditions (14.6), and doing so regardless leads to inconsistencies when interactions are considered.

Instead, we note that the path integral formalism suggests a natural reason for the generating functional (14.3) to be ill defined. This is that only the two transverse field components are independent dynamical variables, as emphasized in our discussion of Coulomb gauge quantization of the electromagnetic field in Chapter 1;[2] whereas, in defining the generating functional (14.3) we have integrated over all four components $\mu = 0, 1, 2, 3$ for each of the eight gluon fields $A_i^\mu(x)$, $i = 1, 2, \ldots 8$. The implication is that we should eliminate the integrals over the additional degrees of freedom if we are to obtain a well-defined generating functional.

One way of doing this is to restrict the integration to fields which satisfy a given gauge condition. For example, for the electromagnetic field in the absence of charges and currents, the scalar field $\phi(x) = A^0(x)$ and the longitudinal field[3] $\mathbf{A}_L(x)$ vanish in the Coulomb gauge, as we saw in Section 1.2.1. On the other hand, if we choose a Lorentz gauge, they do not vanish. The values of the scalar and longitudinal fields corresponding to given physical fields $\mathbf{E}$, $\mathbf{B}$ depend on the choice of gauge condition; and, conversely, their values can be fixed by making a specific gauge choice.

The first essential step in defining a well-behaved generating functional for QCD is, then, to restrict the functional integral to those field configurations satisfying a chosen set of gauge conditions. We specify these by defining

$$f_i(A_i^\mu(x)) = \partial_\mu A_i^\mu(x) - h_i(x) \qquad i = 1, 2, \ldots 8, \tag{14.7a}$$

where the $h_i(x)$, $i = 1, 2, \ldots 8$ are arbitrary functions; and restricting ourselves to gauge fields $A_i^\mu(x)$ which satisfy the subsidiary conditions

$$f_i(A_i^\mu(x)) = 0, \qquad i = 1, 2, \ldots 8. \tag{14.7b}$$

The second step is to show that the result of this is equivalent to a generating functional of the original form (14.3), but with the gauge-invariant Lagrangian density (14.1) replaced by a suitable modified Lagrangian density. This procedure is by no means simple and we will make no attempt to give a full and rigorous derivation here.[4] Our aim is rather to make the main features of the final result plausible; and then to use this result to obtain the Feynman rules for QCD in Section 14.3, where quarks will also be included.

14.1.2 A mathematical analogy

The first step in obtaining a well-defined generating functional is to restrict the functional integration in Eq. (14.3) to field configurations satisfying subsidiary conditions of the

---

[2] See in particular Sections 1.2 and 1.4.1

[3] Any three-vector field $\mathbf{A}(x)$ can be decomposed into transverse and longitudinal fields $\mathbf{A} = \mathbf{A}_T + \mathbf{A}_L$ such that $\mathbf{\nabla} \cdot \mathbf{A}_T = 0$, $\mathbf{\nabla} \times \mathbf{A}_L = 0$. (Cf. the discussion of the electromagnetic field in Section 1.4.1.)

[4] For a more rigorous account, see L.D. Faddeev and A.A. Slavnov, *Gauge Fields: An Introduction to Quantum Theory*, 2nd edn., Addison-Wesley, 1991; or S. Weinberg, *The Quantum Theory of Fields*, Volume II, Cambridge University Press, 1996.

form (14.7b). Some useful insights can be obtained by first considering an analogous problem in ordinary integration, following Coleman.[5]

Consider a well-defined, non-zero n-dimensional integral

$$Z = \int_{-\infty}^{\infty} \cdots \int_{-\infty}^{\infty} dz_{k+1} \ldots dz_{k+n} e^{-X} \tag{14.8}$$

where the 'action' X depends on a subset $z_{k+1}, \cdot \cdot z_{k+n}$ of $k + n$ real variables $z_1, z_2, \ldots z_{k+n}$. The $(k + n)$-dimensional integral

$$Z = \int_{-\infty}^{\infty} \cdots \int_{-\infty}^{\infty} dz_1 \ldots dz_{k+n} e^{-X} \tag{14.9}$$

is then obviously divergent, and our aim is to find a well-defined $(k + n)$-dimensional integral, which is equivalent to Eq. (14.8) To do this, for any given values of $z_{k+1}, \cdot \cdot z_{k+n}$ we assign values to $z_1, z_2, \ldots z_k$ such that

$$f_i(z_1, z_2, \ldots z_{k+n}) = 0, \quad i = 1, 2, \ldots k, \tag{14.10}$$

where f_i are suitable given functions. Since X is independent of $z_1, z_2, \cdots z_k$, and hence of the choice of functions f_i, Eq. (14.8) can be rewritten in the form

$$Z = \int_{-\infty}^{\infty} \cdots \int_{-\infty}^{\infty} df_1 df_2 \ldots df_k dz_{k+1} \ldots dz_{k+n} e^{-X} \prod_{i=1}^{k} \delta(f_i)$$

where the $\delta(f_i)$ are Dirac δ-functions. Changing variables from $f_1, f_2, \ldots f_k$ to $z_1, z_2, \cdots z_k$ then gives

$$Z = \int_{-\infty}^{\infty} \cdots \int_{-\infty}^{\infty} dz_1 dz_2 \ldots dz_{k+n} e^{-X} \det\left(\frac{\partial f}{\partial z}\right) \prod_{i=1}^{k} \delta(f_i) \tag{14.11a}$$

where the Jacobian $\det\left(\frac{\partial f}{\partial z}\right)$ can be recast in the form

$$\det\left(\frac{\partial f}{\partial z}\right) = \int \left(\prod_{i=1}^{k} d\theta_i d\widetilde{\theta}_i\right) \exp\left(-\sum_{i,j=1}^{k} \theta_i \frac{\partial f_i}{\partial z_j} \widetilde{\theta}_j\right) \tag{14.11b}$$

using Eq. (13.53), where $\theta_i, \widetilde{\theta}_i$, are again two independent sets of Grassmann generators. In contrast to the $(k + n)$-dimensional integral (14.9), the integral (14.11a) is well-defined and equivalent to Eq. (14.8). Substituting Eq. (14.11b) into Eq. (14.11a) then gives our final result

$$Z = \int_{-\infty}^{\infty} \left(\prod_{i=1}^{k+n} dz_i\right) \int \left(\prod_{i=1}^{k} d\theta_i d\widetilde{\theta}_i\right) e^{-\widetilde{X}} \prod_{i=1}^{k} \delta(f_i), \tag{14.12a}$$

---

where

$$\widetilde{X} = X + \sum_{i,j=1}^{k} \theta_i \frac{\partial f_i}{\partial z_j} \widetilde{\theta}_j. \tag{14.12b}$$

14.1.3   The Faddeev–Popov Method

In the last section we showed how the ill-defined integral (14.9) could be replaced by an appropriate well-defined integral (14.11) over the same variables by using δ-functions to impose the conditions (14.10). The result (14.11a) involves a determinant arising from a change of variables, which can be written in the form (14.11b) by introducing an integration over Grassmann variables $\theta_i, \widetilde{\theta}_i, i=1, 2, \ldots k$.

In the field theory case, we can achieve a similar result by using functional δ-functions to impose the gauge conditions (14.7). This was done by Faddeev and Popov, who introduced integrals

$$\int \mathcal{D}\eta \mathcal{D}\widetilde{\eta} \equiv \prod_{i=1}^{8} \left(\int \mathcal{D}\eta_i \mathcal{D}\widetilde{\eta}_i \right) \tag{14.13}$$

over independent Grassmann fields $\eta_i(x), \widetilde{\eta}_i(x), i=1, 2, \cdots 8$, rather than discreet Grassmann variables. Specifically they showed that, at least for the calculation of S-matrix elements, the ill-defined generating functional (14.3a), analogous to Eq. (14.9), can be replaced by

$$Z[J_{i\mu}] \propto \int \mathcal{D}A\, e^{iX} \det\left(\frac{\delta f_i}{\delta \omega_j} \right) \left(\prod_{i=1}^{8} \delta[f_i] \right) \tag{14.14a}$$

with the functional determinant given by[6]

$$\det\left(\frac{\delta f_i}{\delta \omega_j} \right) \propto \int \mathcal{D}\eta \mathcal{D}\widetilde{\eta} \exp\left\{ i \int d^4x d^4x'\, \eta_i(x) \frac{\delta f_i}{\delta \omega_j} \widetilde{\eta}_j(x') \right\}. \tag{14.14b}$$

Here the f_i are given by Eq. (14.7a) and the $\delta[f_i]$ are functional δ-functions, defined to satisfy

$$\int \mathcal{D}f_i F[f_i] \delta[f_i] = F[f_i = 0]$$

for any functional $F[f_i]$. The functional derivatives in Eqs. (14.14) also require more careful definition. To do this, we start with a gauge field which satisfies the gauge condition (14.7b) and which we shall call $A_{i\mu}^{(0)}(x)$. From the latter, any gauge field $A_{i\mu}(x)$ can be obtained by a gauge transformation with suitably chosen gauge functions

---

[6] Eq. (14.14b) is a generalization of Eq. (13.61). See, for example, S. Podorski, *Gauge Field Theories*, 2nd edn., CambridgeUniversityPress (2000), Section 5.2.

$\omega(x) = (\omega_1(x), \omega_2(x), \cdots \omega_8(x))$. In other words, we can regard $A_{i\mu}(x)$ as a function of $A_{i\mu}^{(0)}(x)$ and of the $\omega_j(x), j = 1, 2, \ldots 8$, which takes the form

$$A_{i\mu}(x) = A_{i\mu}^{(0)}(x) - \partial_\mu \omega_i(x) - g_s f_{ijk} \omega_j(x) A_{k\mu}^{(0)}(x) \tag{14.15}$$

for infinitesimal $\omega_j(x)$ [cf. Eq. (11.26b)]. Because of the δ-functionals in Eq. (14.14a), the functional derivatives in Eq. (14.14) need only be evaluated in the limit when the gauge conditions (14.7) are satisfied, i.e. when $A_{i\mu}(x) = A_{i\mu}^{(0)}(x)$, corresponding to $\underset{\sim}{\omega}(x) = 0$, as follows from Eq. (14.15). The functional derivatives in Eqs. (14.14) are then given by

$$\frac{\delta f_i}{\delta \omega_j} \equiv \frac{\delta f_i(A_{i\mu}(x))}{\delta \omega_j(x')}\Bigg|_{\underset{\sim}{\omega} = 0}. \tag{14.16}$$

Eqs. (14.14a,b) are the foundation of the path integral quantization of QCD, and are a functional analogue of the results (14.11a, b), the example discussed in the previous section. In particular the subsidiary conditions (14.7b) are imposed in Eq. (14.14a) by the presence of δ-functionals, and the determinant, which arises from a change of integration variables, is replaced in Eq. (14.14b) by an integral over Grassmann fields. Because these fields $\eta_i(x)$, $\tilde{\eta}_i(x)$ are scalar Grassmann fields, their quanta are spin-zero fermions. For consistency with the spin-statistics theorem,[7] such particles must be absent from physical states, as we will discuss in Section 14.1.4.

However, despite these similarities, the analogy is not exact. In Eq. (14.8) we assumed that the 'action' X was independent of a sub-set of integration variables $z_1, z_2, \ldots z_k$. Gauge invariance is a more complicated symmetry than that, and the corresponding action (14.3b) is not even gauge invariant, because of the source terms. Because of this, the derivation of Eqs. (14.14), though similar is some ways to the derivation of Eq. (14.11) from Eq. (14.8), is more complicated. For this reason we will not reproduce it here,[8] but will content ourselves with using Eqs. (14.14a, b) to infer a more convenient form of the generating functional for perturbation theory. This will be done in the following section, and used to obtain the Feynman rules for QCD in Section 14.3.

14.1.4   Gauge fixing and ghosts

In this section we will manipulate the generating functional (14.14) into a form suitable for perturbation theory. To do this we need to evaluate the functional determinant (14.14b), and then eliminate the functional δ-functions from Eq. (14.14a).

We start by evaluating the functional determinant (14.14b). From Eq. (14.15), we obtain

$$\frac{\delta A_i^\mu(x)}{\delta \omega_j(x')}\Bigg|_{\underset{\sim}{\omega} = 0} = -\left[\delta_{ij} \partial_{x'}^\mu + g_s f_{ijk} A_k^\mu(x')\right] \delta^{(4)}(x - x')$$

---

[7] See the discussion at the end of Section 4.3

[8] Many authors reproduce this derivation on the false assumption that the action (14.3) is gauge invariant. For a clear account, together with a reference to where this shortcoming is dealt with, see D. Bailin and A. Love, *Introduction to Gauge Field Theories*, Rev. Edn., Institute of Physics Publishing (1993), Section 10.4.

and hence, from Eqs. (14.7a) and (14.16),

$$\frac{\delta f_i}{\delta \omega_j} = -\partial_{\mu x'} \left[\delta_{ij} \partial^\mu_{x'} + g_s f_{ijk} A^\mu_k(x') \right] \delta^{(4)}(x - x').$$

Substituting this expression in Eq. (14.14b) and integrating by parts enables Eq. (14.14b) to be written in the form

$$\det\left(\frac{\delta f_i}{\delta \omega_j}\right) \propto \int \mathcal{D}\eta \mathcal{D}\tilde{\eta} \exp\left\{ i \int d^4 x \mathcal{L}_g \right\},$$

where the 'ghost term' $\mathcal{L}_g$ is given by

$$\mathcal{L}_g = \partial_\mu \eta_i(x) \left[\partial^\mu \tilde{\eta}_i(x) + g_s f_{ijk} \tilde{\eta}_j(x) A^\mu_k(x) \right]. \tag{14.17}$$

Hence Eq. (14.14a) can be written in the form

$$Z[J_{i\mu}] \propto \int \mathcal{D}A \mathcal{D}\eta \mathcal{D}\tilde{\eta} \exp\left\{ i \int d^4 x (\mathcal{L}_G + \mathcal{L}_g + J_{i\kappa} A^\kappa_i) \right\} \prod_i \delta[f_i] \tag{14.18}$$

where $\mathcal{L}_G$ is given by Eq. (14.1).

The last step is to remove the awkward δ-functionals from Eq. (14.18). To do this, we multiply Z by the constant

$$C \equiv \int \left(\prod_i \mathcal{D}h_i \right) \exp\left\{ -\frac{i}{2} \sum_i (h_i(x))^2 \right\}$$

and note that since $Z[J_{i\mu}]$ is gauge invariant, it must be independent of the particular choice of the functions $h_i(x)$. Hence we obtain

$$Z[J_{i\mu}] \propto C Z[J_{i\mu}] \propto \int \left(\prod_i \mathcal{D}h_i \right) Z \exp\left\{ -\frac{i}{2} \sum_i (h_i(x))^2 \right\}. \tag{14.19}$$

We then substitute Eq. (14.18) into the right-hand side of Eq. (14.19) and eliminate the δ-functions $\delta[f_i]$ by using the relation

$$\int \left(\prod_i \mathcal{D}h_i \right) \exp\left\{ -\frac{i}{2} \int d^4 x \sum_i (h_i(x))^2 \right\}$$
$$\times \left(\prod_i \delta[\partial_\mu A^\mu_i(x) - h_i(x)] \right) = \exp\left\{ -\frac{i}{2} \int d^4 x (\partial_\mu A^\mu_i(x))^2 \right\},$$

where, on the right-hand side, summation over the index i is implied, as for a repeated index. In this way, we obtain our final result

$$Z[J_{i\mu}] = \frac{1}{N} \int \mathcal{D}A \, \mathcal{D}\eta \mathcal{D}\tilde{\eta} \, \exp\left\{ i \int d^4 x (\mathcal{L} + J_{i\kappa} A^\kappa_i) \right\}, \tag{14.20}$$

where the Lagrangian density $\mathcal{L}$ is given by

$$\mathcal{L} = \mathcal{L}_{G'} + \mathcal{L}_g, \tag{14.21}$$

$$\mathscr{L}_{G'} = -\frac{1}{4} G_{i\mu\nu}(x) G_i^{\mu\nu}(x) - \frac{1}{2} \left(\partial_\mu A_i^\mu(x) \right)^2 \tag{14.22}$$

and the ghost term $\mathscr{L}_g$ is given by Eq. (14.17). The constant N is fixed by the normalization condition Eq. (14.3c).

Eq. (14.20) is our final result for the generating functional for the gluon fields in the absence of quarks. Its most striking feature is the occurrence of the ghost term $\mathscr{L}_g$, which describes 'ghost fields' $\eta_i(x)$, $\widetilde{\eta}_i(x)$, $i = 1, 2 \ldots 8$, interacting with the gauge fields $A_i^\mu(x)$. These ghost fields are scalar Grassmann fields, so their quanta are spin-zero fermions. Such particles are not allowed in physical states, since this would violate the spin-statistics theorem, but they nonetheless contribute as virtual intermediate particles via their propagator, as we shall see in Section 14.3. That they are a convenient mathematical artefact, rather than real particles, is further emphasized by the fact that they can be eliminated completely from the theory by choosing a different gauge, for example the axial gauge, in which the gauge conditions (14.7) are replaced by

$$f_i(A_i^\mu(x)) = n_\mu A_i^\mu(x) - h_i(x) = 0, \qquad i = 1, 2, \ldots 8, \tag{14.23}$$

where $n_\mu = (0, 0, 0, 1)$ or any other fixed space-like four-vector. The axial gauge is important because it can be used to establish the equivalence of the canonical and path integral formations for gauge theories, independent of perturbation theory, as noted in Section 13.2.1. It is also often used in non-perturbative discussions of QCD. However, it is not always well suited to perturbation theory, as will be illustrated in Problem 14.2. For this reason we will not pursue axial gauge quantization further.

14.1.5   The electromagnetic field revisited

Finally, before introducing quarks into the theory and deducing the Feynman rules for QCD, it will be instructive to comment briefly on the result of applying the Faddeev–Popov procedure to the electromagnetic field. The argument follows closely that given for the gluon fields, except that the terms proportional to f_{ijk} are absent; and there is only a single gauge field $A^\mu(x)$ and, correspondingly, a single pair of ghost fields $\eta(x)$, $\widetilde{\eta}(x)$. In this way, one obtains the generating functional

$$Z[J_\mu] = \frac{1}{N} \int \mathcal{D}A \mathcal{D}\eta \mathcal{D}\widetilde{\eta} \, \exp\left\{ i \int d^4x \left(\mathscr{L} + \mathscr{L}_g + J_\kappa A^\kappa \right) \right\} \tag{14.24}$$

where $\mathscr{L}$ is given by

$$\mathscr{L} = -\frac{1}{4} F_{\mu\nu}(x) F^{\mu\nu}(x) - \frac{1}{2} \left[\partial_\mu A^\mu(x) \right]^2 \tag{14.25}$$

and the ghost term $\mathscr{L}_g$ by

$$\mathscr{L}_g = \partial_\mu \eta(x) \partial^\mu \widetilde{\eta}(x).$$

Since this term is independent of $A^\mu(x)$, the integral in Eq. (14.24) factorizes and the factor

$$\int \mathcal{D}\eta\,\mathcal{D}\tilde{\eta}\,\exp\left\{i\int d^4x\,\mathcal{L}_g\right\}$$

can be absorbed into the normalization constant to give

$$Z[J_\mu] = \frac{1}{N}\int \mathcal{D}A\exp\left\{i\int d^4x\,(\mathcal{L} + J_\kappa A^\kappa)\right\}. \tag{14.26}$$

As noted in problem 5.1, the Lagrangian density (14.25) is equivalent to our previous Lagrangian density (5.48), i.e.

$$\mathcal{L} = -\frac{1}{2}\left(\partial_\nu A_\mu(x)\right)\left(\partial^\nu A^\mu(x)\right), \tag{14.25a}$$

since the two differ by a total four-divergence which does not contribute to the action. Hence the generating functional (14.26) is identical to the generating functional (13.79) obtained earlier from conventional Lorentz gauge quantization. Finally, we note that the second term in Eq. (14.25) and the second term in Eq. (14.22) are called 'gauge-fixing terms'. The corresponding gauge-fixing terms for other gauges are explored in problems 14.1 and 14.2.

14.2   Including Quarks

In the previous section we ignored quarks in order to focus on the dynamics of gluon fields. In this section we will restore the quark fields, before exploring perturbation theory and the Feynman rules in Section 14.4. For simplicity we shall restrict ourselves to a single quark flavor, dropping the flavour index $f = u, d, s \dots$ throughout. This parallels our treatment of QED, where we have generally restricted ourselves to a single charged lepton (the electron), before commenting briefly on the inclusion of other charged leptons in Section 7.4.

14.2.1   The QCD Lagrangian

Quarks and their interactions with gluons are described by the Lagrangian density (11.28). Explicitly indicating the quark colour indices $a, b, c, \dots = r, g, b$, which are suppressed in Eq. (11.28), this becomes

$$\mathcal{L}_q = \bar{\psi}_a(x)[i\slashed{D}_{ab} - m\delta_{ab}]\psi_b(x), \tag{14.27}$$

where

$$D^\mu_{ab} = \delta_{ab}\partial^\mu + ig_s\left(\lambda_j\right)_{ab}A^\mu_j(x)/2, \tag{14.28}$$

and repeated indices are summed over as usual.[9]

---

[9] Here and throughout we use the beginning of the Latin alphabet $a, b, c, \dots$ to denote quark colour indices, which run over the three values r, g, b; and the middle of the alphabet $i, j, k, \dots$ to denote gluon colour indices, which run over the values $1, 2, \dots 8$.

This Lagrangian density $\mathscr{L}_q$ is then added to the Lagrangian density for gluons Eq. (14.21), obtained from the Faddeev–Popov approach, to yield the corresponding Lagrangian density for QCD:

$$\mathscr{L} = \mathscr{L}_{G'} + \mathscr{L}_q + \mathscr{L}_g, \tag{14.29}$$

where the purely gluonic term $\mathscr{L}_{G'}$ and the terms $\mathscr{L}_q$ and $\mathscr{L}_g$, containing the quark and ghost fields respectively, are given by Eqs. (14.22), (14.27) and (14.17).

As usual, we can decompose the Lagrangian density into free-field terms, corresponding to a vanishing coupling constant $g_s = 0$, and interaction terms. The former are given by

$$\begin{aligned} \mathscr{L}_0 = &-\frac{1}{4}F_{i\mu\nu}(x)F_i^{\mu\nu}(x) - \frac{1}{2}\left(\partial_\mu A_i^\mu(x)\right)^2 \\ &+ \overline{\psi}_a(x)[i\partial\!\!\!/ - m]\psi_a(x) + \partial_\mu\eta_i(x)\partial^\mu\widetilde{\eta}_i(x), \end{aligned}$$

or, more conveniently, by

$$\begin{aligned} \mathscr{L}_0 = &-\frac{1}{2}\left(\partial_\nu A_{i\mu}(x)\right)\left(\partial^\nu A_i^\mu(x)\right) \\ &+ \overline{\psi}_a(x)[i\partial\!\!\!/ - m]\psi_a(x) + \partial_\mu\eta_i(x)\partial^\mu\widetilde{\eta}_i(x), \end{aligned} \tag{14.30}$$

where the gluon field terms have been replaced by a physically equivalent form, analogous to that used in our discussion of QED. [cf. Eqs (14.25) and (14.25a).] The full Lagrangian density can then be written in the form

$$\mathscr{L} = \mathscr{L}_0 + \mathscr{L}_I, \tag{14.31}$$

with the interaction term $\mathscr{L}_I$ given by

$$\begin{aligned} \mathscr{L}_I = &-\frac{1}{2}g_s\overline{\psi}_a(x)\gamma_\mu\left(\lambda_j\right)_{ab}\psi_b(x)A_j^\mu(x) \\ &+ g_s f_{ijk}A_{i\mu}(x)A_{j\nu}(x)\partial^\mu A_k^\nu(x) \\ &-\frac{1}{4}g_s^2 f_{ijk}f_{ilm}A_j^\mu(x)A_k^\nu(x)A_{l\mu}(x)A_{m\nu}(x) \\ &+ g_s f_{ijk}\left(\partial_\mu\eta_i(x)\right)\widetilde{\eta}_j(x)A_k^\mu(x). \end{aligned} \tag{14.32}$$

We have already met the first three terms of Eq. (14.32), in Eqs. (11.28b) and (11.37). They give rise to the three interactions of Fig. 11.1. The additional fourth term corresponds to an interaction between the gluon field and the ghost field, symbolized by Figure 14.1.

$\widetilde{\eta}(j)$

$g(k, \mu)$

$\eta(i)$

Figure 14.1   *The ghost-gluon vertex generated by the last term of Eq. (14.32), where from now on we use a dashed line with arrows to represent the particles associated with the ghost fields $\eta_i, \widetilde{\eta}_i$*

14.2.2 The generating functional

The generating functional corresponding to the Lagrangian density (14.29) is given by

$$Z\left[J_{i\kappa}, \sigma_a, \bar{\sigma}_a, S_i, \tilde{S}_i\right] = \frac{1}{N} \int \mathcal{D}A \, \mathcal{D}\bar{\psi} \, \mathcal{D}\psi \, \mathcal{D}\eta \, \mathcal{D}\tilde{\eta} \, e^{iX'}. \tag{14.33a}$$

Here

$$X' = \int d^4x (\mathcal{L}_0 + \mathcal{L}_1 + \mathcal{L}_S) \tag{14.33b}$$

is the action of QCD

$$X = \int d^4x (\mathcal{L}_0 + \mathcal{L}_1) \tag{14.34}$$

augmented by the action of the source terms,

$$\mathcal{L}_S(x) = J_{i\kappa}(x)A_i^\kappa(x) + \bar{\sigma}_a(x)\psi_a(x) + \bar{\psi}_a(x)\sigma_a(x)$$
$$+ S_i(x)\eta_i(x) + \tilde{S}_i(x)\tilde{\eta}_i(x), \tag{14.35}$$

where $S_i(x)$, $\tilde{S}_i(x)$, $i=1,2,\ldots 8$, and $\sigma_a(x)$, $\bar{\sigma}_a(x)$, $a=r,g,b$, are scalar and spinor Grassmann sources corresponding to the ghost and quark fields respectively. As usual, the integrals extend over all the components of the fields, as indicated by Eqs. (14.4) and (14.13) for the gluon and ghost fields respectively, and by

$$\int \mathcal{D}\bar{\psi} \mathcal{D}\psi = \prod_a \prod_\alpha (\mathcal{D}\bar{\psi}_{a,\alpha} \mathcal{D}\psi_{a,\alpha}) \tag{14.36}$$

for the quark fields, where a, α are the colour and spinor indices respectively. Finally, the normalization constant N in Eq (14.33a) is fixed by the usual condition

$$Z[0,0,0,0,0] = 1 \tag{14.33c}$$

implying

$$N = \int \mathcal{D}A \, \mathcal{D}\bar{\psi} \, \mathcal{D}\psi \, \mathcal{D}\eta \, \mathcal{D}\tilde{\eta} \, e^{iX}. \tag{14.37}$$

Green functions are defined in analogy to Eq. (13.111) for QED, i.e.

$$\langle A_1(x_1)A_2(x_2)\ldots A_n(x_n)\rangle$$
$$= \frac{1}{N} \int \mathcal{D}A \, \mathcal{D}\bar{\psi} \, \mathcal{D}\psi \, \mathcal{D}\eta \, \mathcal{D}\tilde{\eta} \, \{e^{iX} A_1(x_1)\ldots A_n(x_n)\} \tag{14.38}$$

where X is the action (14.34) and $A_1(x_1)\ldots A_n(x_n)$ is any combination of quark and gluon fields. These can, as usual, be obtained from appropriate derivatives of the generating functional (14.33). For example, for Green functions involving quark and gluon fields only, they are given by [cf. Eq. (13.62a) for QED]:

$$\langle A_i^\mu(x_1)\ldots \psi_a(y_1)\ldots \bar{\psi}_b(z_1)\ldots \rangle$$
$$= (-1)^{\bar{n}} \left(\frac{1}{i}\right)^n \frac{\delta^n Z}{\delta J_{i\mu}(x_1)\ldots \delta\bar{\sigma}_a(y_1)\ldots \delta\sigma_b(z_1)\ldots}\bigg|_0 , \tag{14.39a}$$

where n is the total number of fields and $\bar{n}$ is the number of adjoint quark fields $\bar{\psi}_b$. Strictly speaking, these are the only Green functions required to calculate physical processes, since ghost particles cannot occur in physical states. Nonetheless, in formulating perturbation theory it is convenient to also consider Green functions involving ghost fields when, for example,

$$\langle A_i^\mu(x) \ldots \tilde{\eta}_j(x) \ldots \eta_k(x) \ldots \rangle$$
$$= \left(\frac{1}{i}\right)^n \frac{\delta^n Z}{\delta J_{i\mu}(x) \ldots \delta\tilde{S}_j(x) \ldots \delta S_k(x) \ldots}\bigg|_0. \tag{14.39b}$$

The corresponding momentum-space Green functions are given by [cf. Eq. (12.10)]

$$(2\pi)^4 \delta^{(4)}(q_1 + q_2 + \cdots + q_n) \langle A_1(q_1)A_2(q_2) \cdots A_n(q_n)\rangle$$
$$= \int d^4x_1 d^4x_2 \cdots \cdot d^4x_n \left(\prod_{i=1}^n e^{-iq_ix_i}\right) \langle A_1(x_1)A_2(x_2)\cdots A_n(x_n)\rangle, \tag{14.40}$$

where again, by convention, the momenta are directed inwards.

14.2.3   Free fields

In the absence of interactions, we have $\mathscr{L}_1 = 0$, and the generating functional (14.33) reduces to

$$Z_0\left[J_{i\kappa}, \sigma, \bar{\sigma}, S_i, \tilde{S}_i\right] = \frac{1}{N_0} \int \mathcal{D}A\,\mathcal{D}\bar{\psi}\,\mathcal{D}\psi\,\mathcal{D}\eta\,\mathcal{D}\tilde{\eta}\,e^{iX_0'}, \tag{14.41a}$$

where

$$X_0' = \int d^4x(\mathscr{L}_0 + \mathscr{L}_S) \tag{14.41b}$$

and the constant N_0 is again fixed by the normalization condition

$$Z_0[0,0,0,0,0] = 1. \tag{14.41c}$$

On substituting the explicit formulae (14.30) and (14.35) for $\mathscr{L}_0$ and $\mathscr{L}_S$ in Eq. (14.41), we see that

$$Z_0[J_{i\kappa}, \sigma_a, \bar{\sigma}_a, S_i, \tilde{S}_i] = Z_0[J_{i\kappa}]Z_0[\sigma_a, \bar{\sigma}_a]Z_0[S_i, \tilde{S}_i] \tag{14.42}$$

where $Z_0[J_{i\kappa}]$, $Z_0[\sigma_a, \bar{\sigma}_a]$ and $Z_0\left[S_i, \tilde{S}_i\right]$ are the generating functionals of the free gluon, quark and ghost fields, respectively. Explicitly, one obtains:

For the free gluon fields:

$$Z_0[J_{i\kappa}] = \frac{1}{N_1} \int \mathcal{D}A\exp\{iX[J_{i\kappa}]\} \tag{14.43a}$$

with the action

$$X[J_{i\kappa}] = \int d^4x \left\{-\frac{1}{2}(\partial_\nu A_{i\mu}(x))(\partial^\nu A_i^\mu(x)) + J_{i\kappa}(x)A_i^\kappa(x)\right\} \tag{14.43b}$$

and the constant N_1 determined by the normalization condition

$$Z_0[J_{i\kappa} = 0] = 1;$$
(14.43c)

for the free quark fields:

$$Z_0[\sigma_a, \overline{\sigma}_a] = \frac{1}{N_2} \int \mathcal{D}\overline{\psi}\,\mathcal{D}\psi\ \exp\{iX[\sigma_a, \overline{\sigma}_a]\}$$
(14.44a)

with the action

$$X[\sigma_a, \overline{\sigma}_a] = \int d^4x \{\overline{\psi}_a(x)(i\slashed{\partial} - m)\psi_a(x) + \overline{\sigma}_a(x)\psi_a(x) + \overline{\psi}_a(x)\sigma_a(x)\}$$
(14.44b)

and the constant N_2 determined by the normalization condition

$$Z_0[\sigma_a = 0, \overline{\sigma}_a = 0] = 1;$$
(14.44c)

for the free ghost fields:

$$Z_0\left[S_i, \widetilde{S}_i\right] = \frac{1}{N_3} \int \mathcal{D}\eta\,\mathcal{D}\widetilde{\eta}\ \exp\left\{iX\left[S_i, \widetilde{S}_i\right]\right\}$$
(14.45a)

with the action

$$X\left[S_i, \widetilde{S}_i\right] = \int d^4x \left\{\partial_\mu \eta_i(x)\partial^\mu \widetilde{\eta}_i(x) + S_i(x)\eta_i(x) + \widetilde{S}_i(x)\widetilde{\eta}_i(x)\right\}$$
(14.45b)

and the constant N_3 determined by the normalization condition

$$Z_0\left[S_i = 0, \widetilde{S}_i = 0\right] = 1.$$
(14.45c)

The free-field generating functionals (14.43), (14.44) and (14.45) are similar in form to Eqs. (13.79) and (13.80) for the free electromagnetic and free electron fields in QED. The latter were evaluated in detail in Sections 13.2.3 and 13.2.4 respectively, and the same techniques can be used to evaluate the free-field generating functionals for QCD. Here, we will simply state the results, which are:

for the gluon fields [cf. Eqs. (13.95)]

$$Z_0[J_{i\kappa}] = \exp\left\{-\frac{i}{2}\left[J_{i\kappa}D_{\text{F}ij}^{\kappa\lambda}J_{j\lambda}\right]\right\}.$$
(14.46)

The gluon propagator is given by

$$D_{\text{F}ij}^{\kappa\lambda}(x) = \delta_{ij}D_{\text{F}}^{\kappa\lambda}(x)$$
(14.47a)

and $D_{\text{F}}^{\kappa\lambda}(x)$ is identical to the photon propagator (5.27), i.e.

$$D_{\text{F}}^{\kappa\lambda}(x) = \frac{1}{(2\pi)^4}\int d^4k\, e^{-ikx}D_{\text{F}}^{\kappa\lambda}(k), \quad D_{\text{F}}^{\kappa\lambda}(k) = \frac{-g^{\kappa\lambda}}{k^2 + i\varepsilon};$$
(14.47b)

for the quark fields [cf. Eq. (13.105)]

$$Z_0[\sigma_a, \overline{\sigma}_a] = \exp\{-i[\overline{\sigma}_a S_{\text{F}ab}\sigma_b]\},$$
(14.48)

where the quark propagator is given by

$$S_{Fab}(x) = \delta_{ab}S_F(x) \tag{14.49a}$$

and $S_F(x)$ is identical to the fermion propagator (4.63), i.e.

$$S_F(x) = \frac{1}{(2\pi)^4}\int d^4p\, e^{-ipx}S_F(p), \quad S_F(p) = \frac{1}{\not{p}-m+i\varepsilon}; \tag{14.49b}$$

for the ghost fields

$$Z_0\left[S_i, \tilde{S}_i\right] = \exp\left\{-i\left[S_i\Delta_{Fij}\tilde{S}_j\right]\right\}, \tag{14.50}$$

where the ghost propagator is given by

$$\Delta_{Fij}(x) = \delta_{ij}\Delta_F(x) \tag{14.51a}$$

and $\Delta_F(x)$ is identical to the meson propagator (3.59) with zero mass, i.e.

$$\Delta_F(x) = \frac{1}{(2\pi)^4}\int d^4k\, e^{-ikx}\Delta_F(k), \quad \Delta_F(k) = \frac{1}{k^2+i\varepsilon}. \tag{14.51b}$$

14.3   Perturbation Theory

In this and the following section, we shall extend the perturbation theory methods introduced in Section 13.3 for QED to the case of QCD, and then use them to evaluate suitably chosen Green functions in order to infer a set of Feynman rules for QCD. Where this follows closely the analogous developments in QED, we will be content to merely outline the procedure, leaving the details to the reader. However there are a number of novel features, to which we will wish to draw attention.

14.3.1   Wick's theorem and propagators

The first step is to evaluate the free-field Green functions $<ABC\ldots>_0$, where $ABC\ldots$ is any combination of quark, gluon and ghost fields. These can be evaluated from the free-field generating functionals (14.46), (14.48) and (14.50), in close analogy to the treatment given in Section 13.3.1 for QED. Again, one finds that for an odd number of fields $ABC\ldots$, the corresponding free-field Green functions vanish. For an even number of fields, they are again given by Wick's theorem (13.108), where the only non-vanishing contractions are:

$$A_i^\mu(x)A_j^\nu(y) = A_j^\nu(y)A_i^\mu(x) = iD_{Fij}^{\mu\nu}(x-y), \tag{14.52a}$$

$$\psi_a(x)\overline{\psi}_b(y) = -\overline{\psi}_b(y)\psi_a(x) = iS_{Fab}(x-y), \tag{14.52b}$$

$$\tilde{\eta}_i(x)\eta_j(y) = -\eta_j(y)\tilde{\eta}_i(x) = i\Delta_{Fij}(x-y), \tag{14.52c}$$

where the gluon, quark and ghost propagators, $\left(D^{\mu\nu}_{Fij},\, S_{Fab}\, \text{and}\, \Delta_{Fij} \right)$, are given by Eqs. (14.47), (14.49) and (14.51), respectively.

14.3.2   The perturbation expansion

The perturbation expansion for QCD is a direct generalization of that discussed for QED in Section 13.3.2. That discussion was based on Eq. (13.113), which followed from Eq. (13.111) for the generating functional on decomposing the action into free-field and interaction terms. The same equation, i.e.

$$\langle ABC \ldots \rangle = \frac{\langle \exp\left\{ i \int d^4x\, \mathscr{L}_{\mathrm{I}}(x) \right\} (ABC \ldots) \rangle_0}{\langle \exp\left\{ i \int d^4x\, \mathscr{L}_{\mathrm{I}}(x) \right\} \rangle_0}, \tag{14.53}$$

where the interaction Lagrangian $\mathscr{L}_I$ is now given by Eq. (14.32), follows in the same way from the generating functional for QCD, Eq. (14.33). The perturbation series is then obtained by expanding the exponentials in Eq. (14.53) in powers of $\mathscr{L}_I$, and evaluating the terms in the resulting expansion using Wick's theorem, as we shall illustrate in the next sections. In addition, since we are solely interested in connected Green functions, the denominator factor in Eq. (14.53) can again be ignored, as discussed for the case of QED following Eq. (13.114).

14.3.3   The vertex factors

We start by considering first-order perturbation theory, when the expansion of Eq. (14.53) reduces to

$$\langle ABC \ldots \rangle = i \int d^4x \langle \mathscr{L}_{\mathrm{I}}(x)(ABC \ldots) \rangle_0. \tag{14.54}$$

Here $\mathscr{L}_I(x)$, Eq. (14.32), contains a sum of terms, and, by Wick's theorem, the contribution from each term will vanish unless the fields within it match the fields $ABC\ldots$, so that they can be combined together in non-vanishing contractions. This can be exploited to infer the 'vertex factors' associated with each term of the interaction Lagrangian density (14.32) by making an appropriate choice of the fields $ABC\ldots$, as we shall immediately illustrate.

The quark-gluon vertex

If we consider the choice

$$ABC \ldots = A^\nu_i(x_1)\psi_c(x_2)\overline{\psi}_d(x_3),$$

then only the first term in Eq. (14.32) contributes to Eq. (14.54), giving

$$\langle A^\nu_i(x)\psi_c(x_2)\overline{\psi}_d(x_3) \rangle = -\frac{i}{2} g_s \int d^4xF, \tag{14.55a}$$

where

$$F = \langle [\overline{\psi}_a(x)\gamma_\mu(\lambda_j)_{ab}\psi_b(x)]A_j^\mu(x)A_i^\nu(x_1)\psi_c(x_2)\overline{\psi}_d(x_3)\rangle_0$$
$$= \langle A_j^\mu(x)A_i^\nu(x_1)\psi_c(x_2)[\overline{\psi}_a(x)\gamma_\mu(\lambda_j)_{ab}\psi_b(x)]\overline{\psi}_d(x_3)\rangle_0. \tag{14.55b}$$

This is closely analogous to Eq. (13.115), which was studied in the case of QED to obtain Eq. (13.118) for the connected Green function, corresponding to the Feynman diagram of Fig. 13.2. An exactly analogous calculation in this case gives

$$\langle A_i^\nu(x_1)\psi_c(x_2)\overline{\psi}_d(x_3)\rangle$$
$$= -\frac{i}{2}g_s \int d^4x i D_F^{\mu\nu}(x_1 - x) i S_F(x_2 - x)\gamma_\mu(\lambda_i)_{cd} i S_F(x - x_3) \tag{14.56}$$

for the connected Green function in first order. The corresponding momentum-space Green function is defined by

$$(2\pi)^4 \delta^{(4)}(k + p + p')\langle A_i^\nu(k)\psi_c(p')\overline{\psi}_d(p)\rangle$$
$$= \int d^4x_1 d^4x_2 d^4x_3 e^{-ikx_1} e^{-ip'x_2} e^{-ipx_3} \langle A_i^\nu(x_1)\psi_c(x_2)\overline{\psi}_d(x_3)\rangle,$$

where all the moments are directed inwards [cf. Eq. (14.40)]. On substituting Eq. (14.56) into this expression, and evaluating the integrals using the explicit expressions for the propagators, one finally obtains

$$\langle A_i^\nu(k)\psi_c(p')\overline{\psi}_d(p)\rangle$$
$$= i D_F^{\mu\nu}(k) i S_F(p + k)\left[-\frac{ig_s}{2}(\lambda_i)_{cd}\gamma_\mu\right] i S_F(p), \tag{14.57}$$

where we have imposed momentum conservation $p' + p + k = 0$. This is represented by the Feynman graph of Fig. 14.2, where the lines are associated with the propagator factors, as indicated in Fig. 14.3, and where in applying propagator factors like $i S_F(p)_{ab} = i S_F(p)\delta_{ab}$ and $i D_{Fij}^{\mu\nu}(k)\delta_{ij}$, we routinely set $a = b$ and $i = j$, and drop the Kronecker deltas. The remaining factor in Eq. (14.57) is the desired vertex factor

$$-ig_s\left(\frac{\lambda_i}{2}\right)_{cd}\gamma_\mu = -ig_s(T_i)_{ab}\gamma_\mu \tag{14.58a}$$

associated with the quark–gluon vertex, where we have introduced the common notation

$$T_i \equiv \frac{\lambda_i}{2}, \quad i = 1, 2, \ldots 8. \tag{14.58b}$$

The quark–gluon vertex, and its associated factor, are illustrated in Fig. 14.4(a)

The three-gluon vertex

This vertex arises from the second term in Eq. (14.32), and is most easily obtained by evaluating the first-order Green function (14.54) for the field choice

$$ABC = A_l^\alpha(x_1)A_m^\beta(x_2)A_n^\gamma(x_3). \tag{14.59}$$

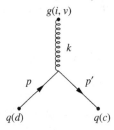

Figure 14.2 The first-order Green function (14.57), where all momenta are directed inwards and $p'=-(p+k)$

Figure 14.3 The propagator factors arising from the contractions (14.52), where i, j are ghost or gluon colour indices and a, b are quark color indices

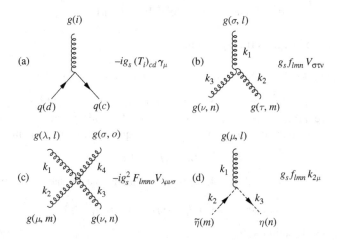

Figure 14.4 The vertex factors of QCD, corresponding to Eqs. (14.58), (14.68), (14.70) and (14.72) for the vertices (a), (b), (c) and (d) respectively. The four-momenta are defined to be directed towards the vertices, as usual

Because of Wick's theorem, only the second term in Eq. (14.32) gives a non-vanishing contribution, so that

$$\langle A_l^\alpha(x_1)A_m^\beta(x_2)A_n^\gamma(x_3)\rangle = \mathrm{i}g_s f_{ijk}g_{\mu\sigma}g_{\nu\tau}\int \mathrm{d}^4 xF \qquad (14.60\mathrm{a})$$

where

$$F = \langle A_i^\sigma(x)A_j^\tau(x)\partial_x^\mu A_k^\nu(x)A_l^\alpha(x_1)A_m^\beta(x_2)A_n^\gamma(x_3)\rangle_0. \qquad (14.60\mathrm{b})$$

On applying Wick's theorem (13.108), there are six non-vanishing contributions, corresponding to the 3! ways of contracting the three fields (14.59) with the three fields in the interaction. However, these terms differ from each other by interchanges of the set, of variables

$$(l,\alpha,x_1) \quad (m,\beta,x_2) \quad (n,\gamma,x_3) \qquad (14.61)$$

corresponding to the fields (14.59), and it will be sufficient to evaluate just one term and then use symmetry to infer the others. We therefore initially choose the term

$$F_1 = A_i^\sigma(x)A_l^\alpha(x_1) \, A_j^\tau(x)A_m^\beta(x_2)\partial_x^\mu \, A_k^\nu(x)A_n^\gamma(x_3), \qquad (14.62)$$

which correspondingly gives

$$\langle A_l^\alpha(x_1)A_m^\beta(x_2)A_n^\gamma(x_3)\rangle_1 = \mathrm{i}g_s g_{\mu\sigma}g_{\nu\tau}f_{lmn}\int \mathrm{d}^4 x\mathrm{i}D_F^{\sigma\alpha}(x-x_1)\mathrm{i}D_F^{\tau\beta}(x-x_2)\partial_x^\mu\mathrm{i}D_F^{\nu\gamma}(x-x_3)$$

on substituting into Eq. (14.60a) and using the explicit expressions (14.52a) for the propagator factors. Converting to momentum-space using

$$(2\pi)^4\delta^{(4)}(k_1+k_2+k_3)\langle A_l^\alpha(k_1)A_m^\beta(k_2)A_n^\gamma(k_3)\rangle_1$$
$$= \int \mathrm{d}^4 x_1\mathrm{d}^4 x_2\mathrm{d}^4 x_3 \mathrm{e}^{-\mathrm{i}k_1 x_1}\mathrm{e}^{-\mathrm{i}k_2 x_2}\mathrm{e}^{-\mathrm{i}k_3 x_3}\langle A_l^\alpha(x_1)A_m^\beta(x_2)A_n^\gamma(x_3)\rangle_1 \qquad (14.63)$$

and evaluating the integrals finally yields

$$\langle A_l^\alpha(k_1)A_m^\beta(k_2)A_n^\gamma(k_3)\rangle_1 = g_s \, g_{\mu\sigma}g_{\nu\tau}f_{lmn}k_3^\mu \mathrm{i}D_F^{\sigma\alpha}(k_1) \, \mathrm{i}D_F^{\tau\beta}(k_2) \, \mathrm{i}D_F^{\nu\delta}(k_3) \qquad (14.64)$$

as the contribution to the momentum-space Green function $\langle A_l^\alpha(k_1)A_m^\beta(k_2)A_n^\gamma(k_3)\rangle$ arising from the choice (14.62), where $k_1 + k_2 + k_3 = 0$. The other contributions arise from interchanges between the combinations (14.61) or equivalently, in momentum space

$$(l,\alpha,k_1) \quad (m,\beta,k_2) \quad (n,\gamma,k_3). \qquad (14.65)$$

Interchanging the last pair in Eq. (14.64) therefore leads to a second contribution

$$g_s \, g_{\mu\sigma} \, g_{\nu\tau}f_{lnm}k_2^\mu \mathrm{i}D_F^{\sigma\alpha}(k_1)\mathrm{i}D_F^{\tau\gamma}(k_3)\mathrm{i}D_F^{\nu\beta}(k_2),$$

which can be rewritten in the form

$$-g_s g_{\mu\sigma}g_{\nu\tau}f_{lmn}k_2^\mu \mathrm{i}D_F^{\sigma\alpha}(k_1)\mathrm{i}D_F^{\tau\beta}(k_2)\mathrm{i}D_F^{\nu\gamma}(k_3) \qquad (14.66)$$

Figure 14.5 *The first-order Green function (14.67), where l, m, n are colour indices.*

by interchanging the dummy indices v and τ and taking account of the antisymmetry of the structure constants f_{lmn}. The contributions arising from the other four permutations of the variables in Eq. (14.66) are deduced in the same way, and involve the same propagator factors as Eqs. (14.64) and (14.66). Collecting them all together, we finally obtain

$$\langle A_l^\alpha(k_1)A_m^\beta(k_2)A_n^\gamma(k_3)\rangle = g_s f_{lmn} \, V_{\sigma\tau v} \, \mathrm{i} D_F^{\sigma\alpha}(k_1) \, \mathrm{i} D_F^{\tau\beta}(k_2) \, \mathrm{i} D_F^{\tau\gamma}(k_3) \tag{14.67}$$

corresponding to the Feynman Figure 14.5, where the vertex factor is given by

$$g_s f_{lmn} V_{\sigma\tau v} = g_s f_{lmn} \left[g_{v\tau}(k_3 - k_2)_\sigma + g_{\tau\sigma}(k_2 - k_1)_v + g_{\sigma v}(k_1 - k_3)_\tau \right] \tag{14.68}$$

This three-gluon vertex factor is illustrated in Figure 14.4(b).

The four-gluon vertex

This vertex arises from the third term in Eq. (14.32) and can be obtained by evaluating the first-order Green function (14.54) for the field choice

$$ABCD = A_l^\alpha(x_1)A_m^\beta(x_2)A_n^\gamma(x_3)A_o^\delta(x_4).$$

The derivation is similar to that given for the three-gluon vertex. It leads to the momentum-space Green function

$$\begin{aligned}
&\langle A_l^\alpha(k_1)A_m^\beta(k_2)A_n^\gamma(k_3)A_o^\delta(k_4)\rangle \\
&= -\mathrm{i}g_s^2 \, F_{lmno} \, V_{\lambda\mu v\sigma} \, \mathrm{i}D_F^{\lambda\sigma}(k_1)\mathrm{i}D_F^{\mu\beta}(k_2)\mathrm{i}D_F^{v\gamma}(k_3)\,\mathrm{i}D_F^{\sigma\delta}(k_4)
\end{aligned} \tag{14.69}$$

corresponding to the Feynman diagram of Fig. 14.6, where

$$\begin{aligned}
F_{lmno}V_{\lambda\mu v\sigma} = &f_{ilm}f_{ino}\left(g_{\lambda v}g_{\mu\sigma} - g_{\mu v}g_{\lambda\sigma}\right) \\
&+ f_{inm}f_{ilo}\left(g_{v\lambda}g_{\mu\sigma} - g_{\mu\lambda}g_{v\sigma}\right) \\
&+ f_{iln}f_{imo}\left(g_{\lambda\mu}g_{v\sigma} - g_{v\mu}g_{\lambda\sigma}\right)
\end{aligned} \tag{14.70a}$$

The four-gluon vertex factor is given by

$$-\mathrm{i}g_s^2 F_{lmno} V_{\lambda\mu v\sigma} \tag{14.70b}$$

and is illustrated in Fig. 14.4(c).

Figure 14.6   *The first-order Green function (14.69), where l, m, n, o are colour indices.*

Figure 14.7   *The first-order Green function (14.71)*

The ghost-gluon vertex

The final vertex factor in QCD arises from the last term in the interaction Lagrangian density (14.32). It can be obtained by evaluating the first-order Green function (14.54) for the choice

$$ABC = A_l^\alpha(x_1)\,\widetilde{\eta}_m(x_2)\,\eta_n(x_3).$$

The derivation is similar to that given for the quark–gluon vertex and leads to the momentum-space Green function

$$\langle A_l^\alpha(k_1)\,\widetilde{\eta}_m(k_2)\,\eta_n(k_3)\rangle = g_s f_{lmn}\,k_{2\mu}\,iD_F^{\mu\alpha}(k_1)\,i\Delta_F(k_2)\,i\Delta_F(k_3),\tag{14.71}$$

corresponding to the Feynman graph of Fig. 14.7. The vertex factor is given by

$$g_s f_{lmn}\,k_{2\mu}\tag{14.72}$$

and is illustrated in Fig. 14.4(d)

14.4   Feynman Rules for QCD

On calculating Green functions to higher orders in perturbation theory, one finds terms in the expansion corresponding to all the topologically distinct Feynman graphs which can be drawn using the propagators and vertices of Figs. 14.3 and 14.4 respectively. Furthermore, on evaluating such contributions one can identify individual factors and features in the resulting formulae with different aspects of a Feynman graph. As in QED, one can derive a set of Feynman rules that enable one to write down, directly from a Feynman graph, the contribution it makes to a Green function or Feynman amplitude.

The vertex factors and propagators for QCD have been derived in the previous section. The remaining rules have much in common with those obtained for QED in Section 7.3, as we shall see.

The Feynman rules for QCD are:

1. For each vertex, write the corresponding vertex factor specified in Fig. 14.4.
2. For each propagator line, write the corresponding propagator factor specified in Fig. 14.3.
3. The Feynman rules for external lines are closely analogous to those for QED [cf. Eqs. (7.49a–f)]. For each external line, write one of the factors

    (a) for each initial quark:

    $$u_{ar}(\mathbf{p}) \qquad p \longrightarrow\!\!\!\!\!\bullet$$

    (b) for each final quark:

    $$\bar{u}_{ar}(\mathbf{p}) \qquad \bullet\!\!\!\!\!\longrightarrow\!\!\!- p$$

    (c) for each initial antiquark:

    $$\bar{v}_{ar}(\mathbf{p}) \qquad p \longrightarrow\!\!\!\longleftarrow\!\!\!\bullet$$

    (d) for each final antiquark:

    $$v_{ar}(\mathbf{p}) \qquad \bullet\!\!\!\longleftarrow\!\!\!- p$$

    (e) for each initial gluon:

    $$\varepsilon_{ir\alpha}(\mathbf{k}) \qquad k \sim\!\!\sim\!\!\sim\!\!\sim\!\!\sim\!\!\bullet^{(\alpha)}$$

    (f) for each final gluon:

    $$\varepsilon^*_{ir\alpha}(\mathbf{k}) \qquad {}^{(\alpha)}\!\bullet\sim\!\!\sim\!\!\sim\!\!\sim\!\!\sim k$$

Here $\mathbf{p}$ and $\mathbf{k}$ denote the three-momenta of the external particles, $r(=1,2)$ labels the quark spin states and the gluon polarization states. Only transverse gluons are allowed because, in the absence of interactions, the free gluon field obeys exactly the same equations as the free electromagnetic field, $a(=r, g, b)$ and $i (= 1,2, \ldots 8)$ specify the colour states of the quarks and gluons [see Section 14.2.1, the footnote following Eq. (14.28)]. α is a Lorentz index.

4. The spinor factors for each quark line are ordered so that, reading from right to left, they occur in the same sequence as following the quark line in the direction of its arrows.
5. For each closed quark loop, take the trace and multiply by a factor (-1).
6. The four-momenta associated with the lines meeting at any vertex satisfy energy–momentum conservation. For each four-momentum q, which is not fixed by energy–momentum conservation, carry out the integration $(2\pi)^{-4} \int d^4q$. One such integration with respect to an internal momentum variable q occurs for each closed loop.
7. Multiply the expression by a phase factor δ_P which is equal to $+1(-1)$ if an even (odd) number of interchanges of neighboring Grassmann fields is required to write the

Grassmann fields in the correct normal order. (Compare the corresponding Feynman rule 8 for QED, section 7.3.)

These Feynman rules 1 to 7 are analogous to the Feynman rules 1 to 8 of QED (cf. section 7.3). QCD has some new features which lead to the following two additional Feynman rules, to be discussed below.

8. For each closed ghost loop, multiply by a factor (-1).
9. For each closed gluon loop, multiply by the appropriate 'symmetry factor' S.

Rule 8 associates a factor (-1) with ghost loops, like that shown in Figure 14.8, and arises because ghosts, like quarks and electrons, are described by anti-commuting fields. It has exactly the same origin as the factor (-1) occurring in the photon self-energy diagram, Fig. 7.9. [cf. the discussion following Eq. (7.22.).]

Fig. 14.9 illustrates two Feynman diagrams involving closed gluon loops, leading to additional symmetry factors S. These factors arise because the second and third terms in the interaction Lagrangian density (14.32) contain the gluon fields, not linearly, but to a higher power, resulting in extra combinational factors in the Green functions.

Consider, for example, the second term in Eq. (14.32):

$$\mathscr{L}_3(x) \equiv g_s f_{ijk} A_{i\mu}(x) A_{j\nu}(x) \partial_\nu^\mu A_k^\nu(x). \tag{14.73}$$

The vertex factor (14.68) arising from this term was obtained by considering the first-order Green function (14.60), corresponding to Fig. 14.5, which contains the vacuum expectation value

$$\langle \mathscr{L}_3(x) A_l^\alpha(x_1) A_m^\beta(x_2) A_n^\gamma(x_3) \rangle_0. \tag{14.74}$$

Clearly there are 3! ways of contracting the three fields in $\mathscr{L}_3(x)$ with fields at x_1, x_2, x_3, and all 3! combinations were incorporated into the symmetric vertex factor (14.68). More generally, for higher-order diagrams without loops (tree diagrams) that contain n three-gluon vertices, the corresponding combinational factor is $(3!)^n$, and this is taken into account in the corresponding n vertex factors.

For graphs in which the same vertex occurs more than once in a gluon loop, additional factors can occur. Consider, for example, the gluon self-energy graph Fig. 14.9(a). The above argument would suggest a combinational factor $(3!)$ for each vertex, i.e. a factor $(3!)^2$ in all. In fact, the correct combinational factor is $(3!)^2/2$.

To see how the factor $(3!)^2/2$ comes about for the gluon self-energy loop, we shall look at the gluon Green function $\langle A_l^\alpha(x) A_m^\beta(y) \rangle$, In second order, this contains the term

$$\langle \mathscr{L}_3(x_1) \mathscr{L}_3(x_2) A_l^\alpha(x) A_m^\beta(y) \rangle_0$$

corresponding to Fig. 14.10 which is, of course, just the gluon self-energy loop. In this case, we can choose any one of the three fields in $\mathscr{L}_3(x_1)$ to contract with $A_l^\alpha(x)$ and any one of the three fields in $\mathscr{L}_3(x_2)$ to contract with $A_m^\beta(y)$, leaving two pairs of fields in

Figure 14.8 *A closed ghost loop, with which an extra factor (–1) is associated [cf. rule 8]*

Figure 14.9 *Examples of closed gluon loops, for which the associated symmetry factor S has the value (a) S = 1/2 and (b) S = 1/6, respectively [cf. rule 9]*

Figure 14.10 *A second-order contribution to the gluon self-energy arising from the gluon loop of Figure 14.9(a)*

$\mathcal{L}_3(x_1)$ and $\mathcal{L}_3(x_2)$ which can only be contracted in two ways, giving $3.3.2 = (3!)^2/2$ combinations overall rather than the $(3!)^2$ ways allowed for in the vertex factors. The extra factor of $1/2$ is just the symmetry factor $S = 1/2$ specified in rule 9 and Fig. 14.9(a). Such arguments can easily be extended to other cases, involving both three-gluon and four-gluon vertices, leading to, for example, the symmetry factor $S = 1/6$ associated with the gluon loops of Fig. 14.9(b).

14.5   Renormalizability of QCD

In this section, we shall discuss the renormalizability of QCD, using the method that has already been discussed for minimal QED in Sections 9.9 and, more generally, in Section 11.3.2. We saw there that renormalizability is assured if the number of types of graph which are primitively divergent is finite; and that the maximum degree of divergence K of a primitively divergent graph can be determined by simple power counting of momentum factors in the numerator and denominator of the Feynman amplitude of that graph.

Consider a primitively divergent graph G with $f_i(f_e)$ internal (external) quark lines, $b_i(b_e)$ internal (external) gluon lines and $\eta_i(\eta_e)$ internal (external) ghost lines.[10] Let G contain n_a, n_b, n_c and n_d quark-gluon, three-gluon, four-gluon and ghost-gluon vertices respectively. These basic QCD vertices are illustrated in Figs. 14.4(a)–(d) and correspond to the vertex factors of Eqs. (14.58), (14.68), (14.70) and (14.72).

Counting the number of momentum factors in the numerator and the denominator of the Feynman amplitude of the graph G, we obtain the degree of divergence K of this graph. If d

---

[10] The graph G we are considering is a sub-graph of a larger graph. If it represented a physical process, there would be no external ghost lines, $\eta_e = 0$.

is the number of internal momenta not fixed by energy–momentum conservation, then K is given by

$$K = 4d - f_{\mathrm{i}} - 2b_{\mathrm{i}} - 2\eta_{\mathrm{i}} + n_{\mathrm{b}} + (n_{\mathrm{d}} - \eta_{\mathrm{e}}/2). \tag{14.75}$$

The second, third and fourth terms on the right-hand side of this equation result from the momentum dependence of the basic quark, gluon and ghost propagators. [Figs. 14.3 and Eqs. (14.52)]. The last two terms stem from the momentum dependence of the three-gluon and ghost-gluon vertices. Since the vertex factor (14.68) of the three-gluon vertex is linear in the gluon momenta, each such vertex contributes one momentum factor to the amplitude, leading to the term n_b in Eq. (14.75) for K.

The ghost-gluon vertices require careful treatment. We see from Fig. 14.4(d) and Eq. (14.72) that, at each ghost-gluon vertex, it is only the ghost line *entering* at it (i.e. with arrow pointing towards it) which contributes a momentum factor to the Feynman amplitude. Of the n_{d} ghost-gluon vertices, $\eta_{\mathrm{e}}/2$ are connected to external ghost lines which *start* at them and therefore do not contribute momentum factors. Hence only $(n_{\mathrm{d}} - \eta_{\mathrm{e}}/2)$ of the ghost-gluon vertices contribute momentum factors to the amplitude, resulting in the last term in Eq. (14.75).

The number of vertices n of the graph G is given by

$$n = n_{\mathrm{a}} + n_{\mathrm{b}} + n_{\mathrm{c}} + n_{\mathrm{d}}. \tag{14.76}$$

Each of these vertices is associated with a δ-function. One of these δ-functions ensures overall energy–momentum conservation and only involves external momenta. Hence of the $(f_{\mathrm{i}} + b_{\mathrm{i}} + \eta_{\mathrm{i}})$ internal momenta, only

$$d = f_{\mathrm{i}} + b_{\mathrm{i}} + \eta_{\mathrm{i}} - (n - 1) \tag{14.77}$$

are independent variables. We also have the relations

$$2f_{\mathrm{i}} + f_{\mathrm{e}} = 2n_{\mathrm{a}}, \tag{14.78a}$$

$$2b_{\mathrm{i}} + b_{\mathrm{e}} = n_{\mathrm{a}} + 3n_{\mathrm{b}} + 4n_{\mathrm{c}} + n_{\mathrm{d}}, \tag{14.78b}$$

$$2\eta_{\mathrm{i}} + \eta_{\mathrm{e}} = 2n_{\mathrm{d}}, \tag{14.78c}$$

which are easily derived. For example, Eq. (14.78a) follows since there are n_a quark–gluon vertices, each connected to two quark lines, and each internal quark line is shared between two vertices. The other two relations (14.78) follow in a similar manner. Using Eqs. (14.76) to (14.78) to eliminate $d, n, f_{\mathrm{i}}, b_{\mathrm{i}}$ and η_{i} from Eq. (14.75) for the degree of divergence K, we obtain as our final result

$$K = 4 - b_{\mathrm{e}} - \frac{3}{2}(f_{\mathrm{e}} + \eta_{\mathrm{e}}). \tag{14.79}$$

$K \geq 0$ is a *necessary* condition for the graph G to be primitively divergent. $K = 0, 1, \ldots$ means at most a logarithmic, linear...divergence.

Eq. (14.79) has the property that the degree of divergence K of a graph depends only on the number of external quark, gluon and ghost lines, but is independent of the number of vertices

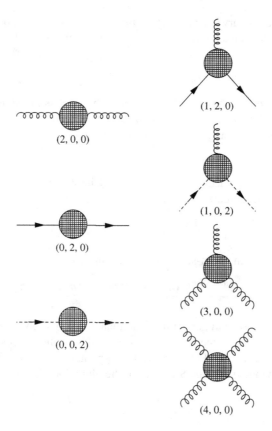

Figure 14.11 *The seven types of graphs which may be primitively divergent, specified by the external line numbers* (b_e, f_e, η_e)

n_a, n_b, n_c and n_d of the graph. From Eq. (14.79) follows the important result that there exist only a finite number of types of graph which may be primitively divergent. They are specified by $(b_e, f_e, \eta_e) = (2, 0, 0), (0, 2, 0), (0, 0, 2), (1, 2, 0), (1, 0, 2), (3, 0, 0), (4, 0, 0)$ and are represented by the seven diagrams in Fig. 14.11. The three graphs on the left-hand side of this figure are seen to represent the basic gluon, quark and ghost propagators of Fig. 14.3 or self-energy corrections to them. The four graphs on the right are the four basic interactions of QCD, i.e. the basic vertices of Fig. 14.4, or higher-order corrections to them. Since there are only a finite number of types of primitively divergent graphs, QCD is a renormalizable theory.

Problems

14.1. In a Lorentz gauge, the free-field Lagrangian density (14.25) for the electromagnetic field can be replaced by the more general form

$$\mathcal{L} = -\frac{1}{4} F_{\mu\nu}(x) F^{\mu\nu}(x) - \frac{1}{2\alpha} \left(\partial_\mu A^\mu(x) \right)^2,$$

where α is an arbitrary real parameter. Use the path integral methods of Section 13.2.3 to show that the momentum space photon propagator is given by

$$D_F^{\mu\nu}(k) = \frac{-g^{\mu\nu} + \delta k^\mu k^\nu / k^2}{k^2 + i\varepsilon}$$

where $\delta = 1 - \alpha^{-1}$. (The choice $\alpha = 1$ used in this book is called the Fermi–Feynman gauge, while the choices $\alpha \to 0$ is called the Landan or transverse gauge.)

14.2. Consider the electromagnetic field in the axial gauge defined by

$$n_\mu A^\mu(x) = 0,$$

where n_μ is a fixed space-like four-vector. A suitable Lagrangian density, analogous to (14.25) in the Lorentz gauge, is

$$\mathscr{L} = -\frac{1}{4} F_{\mu\nu}(x) F^{\mu\nu}(x) - \frac{1}{2} \left(n_\mu A^\mu(x) \right)^2.$$

Show that the momentum-space propagator in this gauge is

$$D_F^{\mu\nu}(k) = \frac{-g^{\mu\nu} - k^\mu k^\nu (n^2 + k^2)/(kn)^2 + (n^\mu k^\nu + n^\nu k^\mu)/(kn)}{k^2 + i\varepsilon}$$

14.3. Evaluate the ghost field generating functional (14.45) to obtain Eq. (14.50).

14.4. Verify the existence of the factor (-1) associated with ghost loops by considering the self-energy contribution Fig. 14.8 to the two-point Green function $\langle A_l^\mu(x_1) A_n^\nu(x_2) \rangle$.

14.5. Derive the symmetry factor $S = 1/6$ for the gluon loop Fig. 14.9(b).

15

Asymptotic Freedom

At distances of the order of 1fm, the strong interaction is indeed strong, as its name implies. Otherwise, its most striking feature is that it becomes progressively weaker at shorter and shorter distances. This property is now known as 'asymptotic freedom.' It first and most obviously manifested itself in the success of the quark–parton model of the electromagnetic interactions of hadrons at high energies. This model preceded QCD, and rested on two main assumptions: that the electromagnetic interactions of quarks are those given by QED, but with the electron charge e replaced by the quark charge ee_f, where $f = u, d, s. \ldots$; and that the strong interactions between quarks can be neglected at distances much less than 1 fm. The latter assumption was a great puzzle, since interaction strengths were expected to increase at short distances, as we shall show explicitly for the case of QED. However, in 1972, 't Hooft suggested that non-Abelian gauge theories might be asymptotically free. This result was explicitly demonstrated by Gross and Wilczek and, independently, by Politzer in 1973, using the so-called 'renormalization group,' and it was this discovery which led directly to the rapid development of QCD as the theory of strong interactions.

Perhaps the most direct experimental evidence for asymptotic freedom is obtained from the observation of jets of hadrons in many high-energy processes. We therefore begin this chapter with a brief account of this for the particularly straightforward case of electron–positron annihilation to hadrons at high energies. Sections 15.2 and 15.3 introduce the renormalization group and show how it can be used to study the short-distance or, equivalently, high-momentum limit of the theory in the familiar context of QED. The arguments are extended to QCD, which is shown to be asymptotically free, in Section 15.4, before returning to applications of QCD in the final Section 15.5.

15.1 Electron–Positron Annihilation

In this section we shall consider reactions of the type

$$e^+ + e^- \rightarrow \text{hadrons}, \tag{15.1}$$

Quantum Field Theory, Second Edition   Franz Mandl and Graham Shaw
© 2010 John Wiley & Sons, Ltd

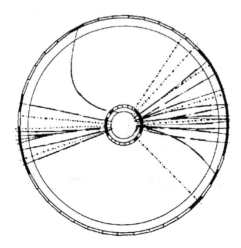

Figure 15.1 *Computer reconstruction of a typical 'two-jet' event observed by the JADE collaboration at DESY, where the election and positron have annihilated at the centre of the inner circle. The solid lines indicate the trajectories of the charged particles and the dotted lines those of neutral particles which are unaffected by the applied magnetic field. (After S. Orito, Proceedings of the 1979 International Conference on Lepton and Photon Interactions at High Energies, FNAL (1979)*

which have been extensively studied in electron–positron colliding beam experiments. Such experiments were historically important in that they gave clear evidence for the existence of color and gluons. Their study will also introduce us to jets of hadrons, like those shown in Figure 15.1. Such jets are produced in a wide variety of high-energy processes, and are closely related to the underlying quark and gluon interactions. In all this, the relative weakness of the strong interaction at short distances plays a crucial role.

15.1.1   Two-jet events

In the centre-of-mass energy range 15–40 GeV, electron–positron annihilation into hadrons (15.1) is dominated by processes like that shown in Fig. 15.2.[1] These can be regarded as two-step processes: a primary electromagnetic process

$$e^+ + e^- \to q + \bar{q}, \tag{15.2}$$

in which the quark and antiquark are emitted in opposite directions in the centre-of-mass frame to conserve momentum; followed by a strong interaction process called *fragmentation*, which converts the high energy $q\bar{q}$ pair into two jets of hadrons. In the process of fragmentation, colour must be transferred by a strong interaction between the quark and its associated jet on the one hand, and the antiquark and its associated jet on the other. This is

---

[1] At lower energies, $q\bar{q}$ bound states (mesons) play a role, while at higher energies weak interaction processes, like those discussed in Section 19.4, must be taken into account.

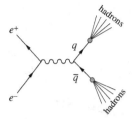

Figure 15.2   *Basic mechanism of two-jet production in electron–positron annihilation*

necessary because, among other reasons, quarks have non-zero colour quantum numbers while hadrons, and hence, the resultant jets, are colourless.

The fragmentation process which converts the quarks and antiquarks into hadrons is very complicated and the composition of the jets – i.e. the number and types of particles in each jet and their individual momenta – varies from event to event. However the direction of a jet, defined by its total momentum

$$\mathbf{p} = \sum_i \mathbf{p}_i, \tag{15.3}$$

where the sum extends over all particles i within the jet, reflects closely the parent quark or antiquark direction. This is only possible if the quark and antiquark produced in the initial electromagnetic reaction (15.2) do not interact strongly until they are separated by distances of order $r \approx 1$ fm, giving rise to momentum transfers of order $\hbar/r$ between them. This is very small compared to the momenta of the quark and antiquark produced in the initial reaction (15.2), so that the total momenta (15.3) of the jets, which subsequently develop, point almost exactly in the initial quark and antiquark directions. In other words, over many events, the observed angular distribution of the jets, relative to the incoming beam direction, should coincide with the calculated angular distribution of the quark and antiquark produced in reaction (15.2).

To confirm this, we need to calculate the cross-section for the basic reaction (15.2), which is most easily done by exploiting its similarity to the reaction

$$e^+ + e^- \rightarrow \mu^+ + \mu^-. \tag{15.4}$$

This proceeds by the mechanism of Fig. 7.17, which is the same as that of reaction (15.2). The cross-section for reaction (15.4) was calculated in Section 8.4, and is given by Eq. (8.46) in the approximation that the energy is sufficiently high for the muon mass to be neglected. The corresponding cross-section for producing quark–antiquark pairs of a given flavour $f = u, \ d, \ s, \ldots$ is then obtained by replacing the muon charge e by the quark charge ee_f, and since this charge is the same for all three colour states r, g and b, we must multiply by three to account for the three colour states $r\bar{r}, \ g\bar{g}$ and $b\bar{b}$. We thus obtain the centre-of-mass cross-section

$$\frac{d\sigma}{d\cos\theta} = 3e_f^2 \left[\frac{\pi\alpha^2}{8E^2} \ (1 + \cos^2\theta) \right], \tag{15.5}$$

where θ is the angle between the electron and quark directions, and the factor in square brackets is the $e^+ + e^- \rightarrow \mu^+ + \mu^-$ cross-section (8.46). In particular, the angular

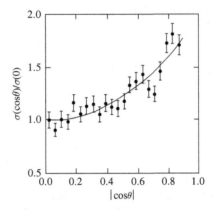

Figure 15.3 *Comparison between the angular distribution of two-jet events observed in an electron–positron colliding beam experiment by the CELLO collaboration at DESY and the theoretically predicted $(1 + \cos^2\theta)$ behaviour (15.5). (After H. J. Behrend et. al., Physics Letters **183B** (1987) 400.)*

distribution of the jets should be proportional to $(1 + \cos^2\theta)$ if the jet directions reflect the initial quark and antiquark directions. This is indeed the case, as is shown in Fig. 15.3

15.1.2   Three-jet events

The dominant process in e^+e^- annihilation to hadrons is the formation of 'back-to-back' jets in the centre-of-mass frame, as illustrated in Fig. 15.4(a). However, sometimes we might expect a high-momentum gluon to be emitted at a wide angle by the quark or antiquark before fragmentation occurs, leading to the formation of a three-jet event, as illustrated in Fig. 15.4(b). Three-jet events are indeed observed, as shown in Fig. 15.5.

The first step in analysing such events is to calculate the differential cross-section for the initial reaction

$$e^+ + e^- \rightarrow q + \bar{q} + g, \tag{15.6}$$

which is given in lowest order by the Feynman graphs of Fig. 15.6, using the Feynman rules for QED and QCD given in Sections 7.3 and 14.4 respectively. The angular

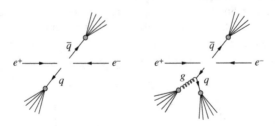

Figure 15.4 *Schematic diagrams representing (a) two-jet and (b) three-jet formation in electron–positron annihilation in the centre-of-mass frame*

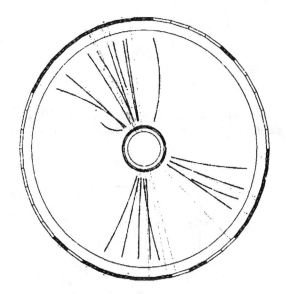

Figure 15.5 *Computer reconstruction of a three-jet event observed by the JADE collaboration at DESY. (After S.-L. Wu,* Physics Reports **107** *(1984) 150.)*

distribution of the resulting jets is then obtained by assuming that particles produced in reaction (15.6) do not interact strongly again until they are separated by large distances of order 1 fm, so the resulting jet directions coincide closely with the initial quark, antiquark and gluon directions. It is, however, not obvious which of the jets in events like that shown in Fig. 15.5 is due to the gluon. Instead, one focusses on the total energy of each jet, and the three jets are classified in the order $E_1 > E_2 > E_3$ in the overall centre-of-mass frame, before Lorentz transforming to the centre-of-mass frame of jets 2 and 3 alone. In this frame, jets 2 and 3 are emitted back-to-back, and the angular distribution of jet 1 is plotted with respect to the angle ϕ between its direction and the mutual line of flight of jets 2 and 3. The latter is usually coincident with the line of flight of the gluon, since we would expect jet 1, which has the highest energy in the centre-of-mass frame, to be associated with the quark or antiquark which does not emit the gluon; and detailed calculations show that the distribution described is sensitive to the gluon spin. The measured distribution is compared to the theoretical expectation for spin-1 gluons and hypothetical spin-0 gluons in Fig. 15.7 and is in excellent agreement with QCD. The magnitude of the cross-section is directly proportional to the interaction strength

$$\alpha_s \equiv g_s^2/4\pi$$

of the quark–gluon vertex Fig. 14.4(a), and the measured rate corresponds to $\alpha_s \approx 0.14$.[2] To understand how this relatively weak coupling strength can be reconciled with the much stronger forces at ranges of order 1 fm requires discussion of renormalization and the renormalization group, to which we now turn.

---

[2] A more precise determination requires higher-order corrections to be taken into account. We will return to this point in Section 15.5.

Figure 15.6   *One of the two Feynman graphs which contribute to the reaction (15.6) in lowest order. In the other graph, the gluon is emitted by the antiquark rather than the quark*

Figure 15.7   *Angular distribution of three-jet events measured by the TASSO collaboration, where ϕ is the angle defined in the text. The dashed and solid lines show the theoretical expectations for spin-0 and spin-1 gluons respectively. (After R. Brandelik et al.,* Physics Letters **B97** (1980) 453; and S.-L. Wu, Physics Reports **107** (1984) 59.)

15.2   The Renormalization Scheme

In our discussion of renormalization in Chapters 9 and 10 we adopted a method, called the 'on-shell scheme,' which uses the static properties of a free electron as the basic parameters of the theory. This is not very convenient for the case of QCD, since quarks are confined and the properties of 'free quarks' cannot be measured. There are, however, alternative renormalization schemes which lead to identical physical results. The most popular of such schemes is the 'modified minimal subtraction scheme,' denoted $\overline{\text{MS}}$. This introduces an arbitrary mass scale into the theory, as we shall shortly see, and the freedom to vary this mass scale without changing the physical predictions leads to the so-called renormalization group equation. This equation can then be exploited to derive the behavior of the theory at high momenta, corresponding to short-range interactions, leading, in QCD, to the derivation of asymptotic freedom.

In this and the following section we will introduce the $\overline{\text{MS}}$ renormalization scheme and the renormalization group equations in the familiar context of QED, where we stress they do not lead to asymptotic freedom. We will then go on to apply the same methods to QCD in Section 15.4 in order to understand the differences which do give rise to asymptotic freedom in QCD.

15.2.1 The electron propagator

In this section we will introduce the $\overline{\text{MS}}$ scheme by considering the momentum-space electron propagator in second-order perturbation theory. As we saw in Section 9.3,[3] this is given by [cf. Eq. (9.24)]:

$$G(p) = \frac{i}{\not p - m_0 + e_0^2 \sum(p) + i\varepsilon}, \tag{15.7a}$$

where m_0 and e_0 are the bare mass and bare charge respectively. The divergent self-energy term $ie_0^2 \sum(p)$ arises from the loop correction of Fig. 9.9, and is given by [cf. Eq. (9.4)]:

$$ie_0^2 \sum(p) = \frac{(ie_0)^2}{(2\pi)^4} \int d^4 k i D_{F\alpha\beta}(k) \gamma^\alpha i S_F(p-k) \gamma^\beta \tag{15.7b}$$

Eqs. (15.7a) and (15.7b) are independent of the regularization and renormalization schemes. However the $\overline{\text{MS}}$ scheme is defined in the context of dimensional regularization, where Eq. (15.7b) is generalized to

$$D = 4 - \eta$$

dimensions, when it becomes

$$ie_0^2 \sum(p) = \frac{(i\tilde e_0)^2 \mu^2}{(2\pi)^D} \int d^4 k i D_{F\alpha\beta}(k) \gamma^\alpha i S_F(p-k) \gamma^\beta. \tag{15.8}$$

Here μ is the mass scale associated with dimensional regularization, as discussed in Section 10.3.1, and

$$\tilde e_0 = \mu^{-\eta/2} e_0 \tag{15.9}$$

is the dimensionless bare coupling [cf. Eq. (10.31)]. The integral in Eq. (15.8) can then be evaluated using the techniques discussed in Sections 10.3–5. In this way, one eventually obtains [cf. Eqs. (10.73a,b)]

$$e_0^2 \sum(p) = \frac{\tilde e_0^2}{16\pi^2} (\not p - 4m_0) \left[\frac{2}{\eta} - \gamma + \ln 4\pi \right] + \tilde e_0^2 \sum_r(p), \tag{15.10a}$$

where

$$16\pi^2 \sum_r(p) = 2m_0 - \not p$$
$$- 2 \int_0^1 dz [\not p(1-z) - 2m_0] \ln \left(\frac{m_0^2 z - p^2 z(1-z)}{\mu^2} \right) \tag{15.10b}$$

is finite and we have omitted terms which vanish when $\eta \to 0$, and the original theory is restored.

Within dimensional regularization, different renormalization schemes correspond to different ways of separating Eq. (15.10) into a finite part and divergent constants, before

---

[3] The arguments which follow are similar to those given for the on-shell scheme in Chapter 9.

absorbing the latter into redefined, or renormalized, physical parameters. In the $\overline{\text{MS}}$ scheme these divergent constants are always chosen to be proportional to the factor[4]

$$\left[\frac{2}{\eta} - \gamma + \ln 4\pi\right], \tag{15.11}$$

i.e. they include the pole term together with the associated factors of the Euler constant γ and $\ln 4\pi$. In second order, this is achieved by defining constants

$$A' = \frac{-m_0}{4\pi^2}\left[\frac{2}{\eta} - \gamma + \ln 4\pi\right] \tag{15.12a}$$

$$B' = \frac{1}{16\pi^2}\left[\frac{2}{\eta} - \gamma + \ln 4\pi\right] \tag{15.12b}$$

so that Eq. (15.10a) becomes

$$\sum(p) = A' + B'\not{p} + \sum_r(p). \tag{15.13}$$

The remaining development is similar to that of the on-shell scheme in Section 9.3. Specifically, on substituting Eq. (15.13) into Eq. (15.7a), one finds that the resulting expression can be written in the form

$$G(p) = \frac{iZ_2}{\not{p} - m_r + e_r^2\sum_r(p) + i\varepsilon} + O(\tilde{e}_0^4), \tag{15.14}$$

where the various constants are defined by[5]

$$Z_2 = 1 - e_r^2 B' \quad m_r = Z_2(m_0 - e_r^2 A') \tag{15.15}$$

and

$$e_r \equiv \tilde{e}_0 Z_2^{1/2}, \tag{15.16a}$$

implying

$$e_r = \tilde{e}_0 + O(e_r^3). \tag{15.16b}$$

The renormalization constant Z_2 in the *bare propagator* Eq. (15.14) is then absorbed into a partial renormalization (15.16a) of the charges in the two adjoining vertices, yielding

$$G_r(p) = \frac{i}{\not{p} - m_r + e_r^2\sum_r(p) + i\varepsilon} + O(e_r^4) \tag{15.17}$$

for the '*renormalized propagator*' defined by

$$G_r(p) \equiv Z_2^{-1}G(p). \tag{15.18}$$

---

[4] In the alternative minimal subtraction scheme MS, they are instead chosen to be proportional to η^{-1}, i.e. to the pole terms only
[5] We stress that the renormalization constant Z_2 defined by Eq. (15.15) is not the same as that defined by Eq. (9.35) in the on-shell scheme, although it plays a similar role.

The identity sign in Eqs. (15.16a) and (15.18) is used to indicate that these definitions hold to all orders in perturbation theory, whereas the value of Z_2 is modified when higher-order terms are taken into account.

Eq. (15.17) represents our final result for the electron propagator to second order. As can be seen, it depends on two parameters, the renormalized mass and the renormalized charge, and on the function $\Sigma_r(p)$, which is finite and well-defined in the limit $\eta \to 0$, when the original theory is restored. However, even though we started from a bare propagator, Eqs. (15.7), which is obviously independent of the chosen mass scale μ, both $\Sigma_r(p)$ and hence the renormalized propagator $G_r(p)$ do depend on this mass scale, even in the limit $\eta \to 0$, as can be seen from Eq. (15.10b).

In the on-shell renormalization scheme of Chapter 9, the renormalized electron mass is identical to its inertial or 'physical' mass. The renormalized mass m_r in Eq. (15.17) is not equal to the physical mass, but is easily related to it by remembering that the propagator must have a pole at the physical mass $\not{p} = m$. This pole can only arise from the vanishing of the denominator in Eq. (15.17), which implies that

$$m - m_r + e_r^2 \Sigma_r(\not{p} = m) = 0, \qquad (15.19)$$

where $\Sigma_r(\not{p} = m)$ is given to second order by substituting $\not{p} = m$ in Eq. (15.10b) and noting that $m = m_r + O(e_r^2)$. Since the inertial mass is obviously independent of μ and finite, and we know from Eq. (15.10b) that $\Sigma_r(\not{p} = m)$ is finite and μ-dependent, one sees immediately that the renormalized mass $m_r = m_r(\mu)$ is both finite and mass-scale dependent.

15.2.2 The photon propagator

To obtain the corresponding results for the electric charge, we must also consider the contributions to charge renormalization from the radiative corrections to the photon propagator and the vertex factor.

To second order, the radiative correction to the photon propagator arises from the photon self-energy term [cf. Eq. (9.8)]:

$$ie_0^2 \Pi^{\mu\nu}(k) = \frac{(ie_0)^2}{(2\pi)^4} \int d^4p \, \frac{T_r[\gamma^\mu(\not{p} + \not{k} + m_0)\gamma^\nu(\not{p} + m_0)]}{[(p+k)^2 - m_0^2 + i\varepsilon][p^2 - m_0^2 + i\varepsilon]}, \qquad (15.20)$$

which, by Lorentz invariance, is of the form [cf. Eq. (9.11)]:

$$\Pi^{\mu\nu}(k) = -g^{\mu\nu}A(k^2) + k^\mu k^\nu B(k^2). \qquad (15.21)$$

When this is taken into account, as illustrated in Fig. 9.8, the photon propagator becomes [cf. Eq. (9.13)]

$$G^{\alpha\beta}(k) = \frac{-ig^{\alpha\beta}}{k^2 + i\varepsilon + e_0^2 A(k^2)} + O(e_0^4), \qquad (15.22)$$

where we have again, for simplicity, neglected the gauge-dependent term.[6]

We now specialize to the dimensional regularization scheme. The photon self-energy (15.20) has been explicitly evaluated in Section 10.4, and the result for $A(k^2)$ is obtained by combining Eqs. (9.11), (10.48) and (10.52b). Explicitly

$$e_0^2 A(k^2) = \tilde{e}_0^2 \frac{k^2}{12\pi^2} \left[\frac{2}{\eta} - \gamma + \ln 4\pi \right] + \tilde{e}_0^2 k^2 \Pi_r(k^2), \tag{15.23}$$

where

$$\Pi_r(k^2) = \frac{-1}{2\pi^2} \int_0^1 dz \, z(1-z) \ln \left[\frac{m^2 - k^2 z(1-z)}{\mu^2} \right] \tag{15.24}$$

and terms which vanish as $\eta \to 0$ have again been neglected.

As illustrated for the case of the electron propagator, the $\overline{\text{MS}}$ scheme is characterized by absorbing divergent constant, proportional to the factor (15.11) into redefined physical parameters. In this case, we define the partially renormalized charge to be[7]

$$e_r \equiv Z_3^{1/2} \tilde{e}_0, \tag{15.25}$$

where the renormalization constant

$$Z_3 = 1 - \frac{\tilde{e}_0^2}{12\pi^2} \left[\frac{2}{\eta} - \gamma + \ln 4\pi \right]$$

in second order, or more conveniently

$$Z_3 = 1 - \frac{\tilde{e}_r^2}{12\pi^2} \left[\frac{2}{\eta} - \gamma + \ln 4\pi \right] \tag{15.26}$$

since $e_r = \tilde{e}_0 + O(e_r^3)$. Eq. (15.23) then becomes

$$e_0^2 A(k^2) = k^2 [1 - Z_3 + e_r^2 \Pi_r(k^2)] + O(e_r^4)$$

and substituting this into Eq. (15.22) gives

$$G^{\alpha\beta}(k^2) = \frac{-i Z_3 \, g^{\alpha\beta}}{k^2 + i\varepsilon + e_r^2 k^2 \Pi_r(k^2)} \tag{15.27}$$

for the bare propagator. Finally the factor Z_3 in the bare propagator (15.27) is then absorbed into a partial renormalization (15.25) of the bare charges in the two adjoining vertices, yielding

$$G_r^{\alpha\beta}(k^2) = \frac{-i g^{\alpha\beta}}{k^2 + i\varepsilon + e_r^2 \Pi_r(k^2)} + O(e_r^4) \tag{15.28}$$

---

[6] cf. the discussion in Section 9.2 following Eq. (9.11).
[7] In discussing each of the three contributions to charge renormalization, we ignore the other two contributions. The three will be combined in Eq. (15.35) below.

as our final result for the renormalized photon propagator

$$G_r^{\alpha\beta}(k^2) \equiv Z_3^{-1} G^{\alpha\beta}(k^2) \tag{15.29}$$

Like the renormalized electron propagator (15.17), the renormalized photon propagator is finite and μ-dependent in the limit $\eta = 0$, where the original theory is restored.

15.2.3   Charge renormalization

The third contribution to charge renormalization arises in second order from the vertex modification of Fig. 9.13, resulting in the vertex factor [cf Eq. (9.47)]:

$$i\Gamma^{\mu}(p', p) = ie_0 [\gamma^{\mu} + e_0^2 \Lambda^{\mu}(p', p)], \tag{15.30}$$

where $\Lambda^{\mu}(p',p)$ is given by the divergent integral (9.48). On generalizing to $D = 4 - \eta$ dimensions, this becomes

$$i\Gamma^{\mu}(p', p) = ie_0 [\gamma^{\mu} + e_0^2 \Lambda^{\mu}(p', p)], \tag{15.31}$$

where $e_0^2 \Lambda^{\mu}(p',p)$ is now given by Eq. (10.55). The evaluation of this term is discussed in detail in Section 10.5, where it is shown that the only divergent contribution to $\Lambda^{\mu}(p',p)$ occurs in the term (10.65). On carrying out the integrals over the term in round brackets in Eq. (10.65), we arrive at the desired result that $\Lambda^{\mu}(p',p)$ is of the form

$$e_0^2 \Lambda^{\mu}(p', p) = \frac{\widetilde{e}_0^2 \gamma^{\mu}}{16\pi^2} \left[\frac{2}{\eta} - \gamma + \ln 4\pi \right] + \widetilde{e}_0^2 \Lambda_r^{\mu}(p', p), \tag{15.32}$$

where $\Lambda_r^{\mu}(p',p)$ is finite and μ-dependent. Its explicit form is not needed below, but as shown in Section 10.5 it gives rise, among other things, to the anomalous magnetic moment of the electron, which does not depend on μ. Here we simply note that on substituting Eq. (15.32) into Eq. (15.31), and defining

$$Z_1 = 1 - \frac{e_r^2}{16\pi^2} \left[\frac{2}{\eta} - \gamma + \ln 4\pi \right] \tag{15.33}$$

we obtain

$$i\Gamma^{\mu}(p', p) = ie_r \mu^{\eta/2} [\gamma^{\mu} + \widetilde{e}_0^2 \Lambda_r^{\mu}(p', p)] + O(e_0^5), \tag{15.34}$$

where the partially renormalized charge

$$e_r \equiv \frac{\widetilde{e}_0}{Z_1} . \tag{15.35}$$

We finally combine the charge renormalization (15.35) with the charge renormalizations (15.16a) and (15.25) arising from the election and photon self-energies, respectively. In doing this, we must remember that in our discussion of the propagators, we absorbed factors of $Z_2^{1/2}$ and $Z_3^{1/2}$ into each vertex from each adjoining fermion and photon lines respectively. Since there are two fermion lines and one photon line at each vertex, this gives rise to a renormalized vertex

$$\Gamma_r^\mu(p', p) \equiv Z_2 Z_3^{1/2} \Gamma^\mu(p', p) \tag{15.36}$$

and substituting Eq. (15.34) and (15.35) yields

$$i\Gamma_r^\mu(p', p) = ie_r \mu^{\eta/2}\left[\gamma^\mu + e_r^2 \Lambda^\mu(p', p)\right] + O(e_r^5), \tag{15.37}$$

where e_r is now the fully renormalized charge, defined by

$$e_r \equiv \frac{\tilde{e}_0 Z_3^{1/2} Z_2}{Z_1} = e_0 \mu^{-\eta/2} \frac{Z_3^{1/2} Z_2}{Z_1} \tag{15.38}$$

using Eq. (15.10). Furthermore, since the Ward identity

$$Z_1 = Z_2 \tag{15.39}$$

is satisfied [cf Eqs. (15.12b), (15.15) and (15.33)], this result simplifies to

$$e_r = e_0 \mu^{-\eta/2} Z_3^{1/2}, \tag{15.40}$$

where Z_3 is given by Eq. (15.26).

The renormalized charge e_r, defined by Eq. (15.40) is, like the renormalized mass m_r, finite and dependent on the chosen mass scale μ. Because of this dependence $m_r = m_r(\mu)$ and $e_r = e_r(\mu)$ are referred to as the 'running mass' and 'running charge' or 'running coupling,' respectively. For a given choice of mass scale μ, they are the basic parameters of the theory. However, all physical observables are independent of μ and this invariance under the choice of scale leads to the renormalization group equations, from which one can show that, as the mass-scale μ gets bigger, $e_r(\mu)$ measures the strength of the interaction at shorter and shorter distances. This will be covered in Section 15.3. Here, we note that Eqs. (15.40) and (15.26) imply that

$$\mu \frac{\partial e_r}{\partial \mu} = \frac{e_r^3}{24\pi^2} + O(e_r^5), \tag{15.41}$$

where we have taken the limit of four dimensions, i.e. $\eta \to 0$. From this we see that $e_r(\mu)$ increases with μ, although the effect is small in QED, as we discuss in more detail in Section 15.3.3. In QCD, by contrast, the effect is larger and, moreover, the coupling strength decreases with increasing μ, leading to asymptotic freedom, as will be shown in Section 15.4. Firstly, however, we turn to the renormalization group equations to establish the link between the strength of the interaction at short distances, and the behavior of the running coupling constant at large values of the mass scale μ.

15.3   The Renormalization Group

In this section, we will first derive the renormalization group equations, and then use them in Sections 15.3.2 and 15.3.3 to derive the behavior of the theory at large momentum transfers and, correspondingly, short distances.

15.3.1 The renormalization group equations

The renormalization group equations are formulated in terms of the proper vertex functions $\Gamma^{fb}(p_1, p_2, \ldots, p_{b+f})$, where f and b are the number of external fermions and photons, respectively, and p_i $(i = 1, 2, \ldots, b+f)$ are the associated four-momenta, subject to energy–momentum conservation

$$p_1 + p_2 + \ldots + p_{b+f} = 0.$$

Vertex functions were first introduced for the case $b = 2$, $f = 2$ in Section 12.2.4 [cf. Eq. (12.36)] and shown to be closely related to the matrix elements for physical processes in Section 12.3. More generally, for $n = b+f > 2$, vertex functions are just connected Green functions with the propagator factors associated with the external legs removed. The proper vertex function is the sum of all one-particle irreducible graphs which contribute to a given vertex function, where one-particle irreducible means it cannot be split in two by cutting a single line. Proper vertices are convenient for general discussions, because any Green function can be expressed in terms of proper vertex functions and propagators, as illustrated in Fig. 15.8. The two-point vertices

$$\Gamma^{20}(p) = e_0^2 \Sigma(p) \qquad \Gamma^{02}(k) = e_0^2 \Pi^{\mu\nu}(k) \qquad (15.42)$$

are defined in terms of the electron and photon self-energies $e_0^2 \Sigma(p)$ and, $e_0^2 \Pi^{\mu\nu}(k)$, which were discussed in Sections 9.2, 9.3 and 9.8. The electron and photon propagators can, in turn, be expressed in terms of the two-point vertices of Eq. (15.42) by using Eqs. (9.101) and (9.105), respectively. As in Eq. (15.42), we often suppress Lorentz indices when referring to vertex functions, so that the three-point vertex of Eq. (15.30), for example, is denoted by

$$i\Gamma^{21}(p', p) = i\Gamma^\mu(p', p) = ie_0 \left[\gamma^\mu + e_0^2 \, \Lambda^\mu(p', p) \right]. \qquad (15.43)$$

The bare vertex functions, like Eq. (15.43) and the self-energies (15.9) and (15.20), are dependent on the bare mass and charge m_0 and e_0, and are ill defined in the limit $\eta \to 0$, when the original theory is restored. However, on renormalizing, factors of $Z_2^{1/2}$ and $Z_3^{1/2}$ are absorbed from the adjacent fermion and photon propagators, and the resulting renormalized vertex functions are expressed in terms of the μ-dependent renormalized mass and charge m_r and e_r. We then have

$$\Gamma_r^{fb}(p_i, m_r, e_r, \mu) = Z_2^{f/2} Z_3^{b/2} \Gamma^{fb}(p_i, m_0, e_0), \qquad (15.44)$$

where the renormalized vertex function Γ^{fb} remain finite and μ-dependent in the limit when $\eta \to 0$ and the original theory is restored. A differential equation governing this μ-dependence is obtained by substituting

$$\Gamma^{fb}(p_i, m_0, e_0) = Z_2^{-f/2} Z_3^{-b/2} \Gamma_r(p_i, m_r, e_r, \mu)$$

Figure 15.8   *Decomposition of a connected Green function into proper vertex factors (open circles) and full propagators*

into the trivial equation

$$\mu \frac{d\Gamma^{fb}}{d\mu}(p_i, m_0, e_0) = 0$$

to give

$$\left[\mu \frac{\partial}{\partial \mu} + \beta \frac{\partial}{\partial e_r} - f\gamma_2 - b\gamma_3 + m_r \gamma_m \frac{\partial}{\partial m_r}\right]\Gamma_r^{fb}(p_i, m_r e_r, \mu) = 0, \qquad (15.45)$$

where

$$\beta \equiv \mu \frac{\partial e_r}{\partial \mu} \qquad (15.46a)$$

$$\gamma_2 \equiv \frac{\mu}{2} \frac{\partial}{\partial \mu}(\ln Z_2) \qquad (15.46b)$$

$$\gamma_3 \equiv \frac{\mu}{2} \frac{\partial}{\partial \mu}(\ln Z_3) \qquad (15.46c)$$

and

$$m_r \gamma_m \equiv \mu \frac{\partial m_r}{\partial \mu}. \qquad (15.46d)$$

Equation (15.45) is called the 't Hooft–Weinberg equation and is the appropriate renormalization group equation for the minimal subtraction scheme used here. Before exploiting it, however, we need to combine it with a differential equation deduced from the scaling properties of the vertices, as we shall immediately see.

15.3.2   Scale transformations

In order to derive the behavior of the theory at high momenta, we consider the vertex function

$$\Gamma_r^{fb}(tp_i, m_r, e_r, \mu), \tag{15.47}$$

where we have introduced the dimensionless scale parameter t. A differential equation for this quantity is obtained by combining the renormalization group equation (15.45) with another equation obtained by considering a scale transformation

$$p_i, \; t^{-1}m_r, \; t^{-1}\mu \; \rightarrow \; tp_i, \; m_r, \; \mu. \tag{15.48}$$

Under such a transformation, the proper vertices must transform as

$$\Gamma_r^{fb}(tp_i, m_r, e_r, \mu) = t^d \, \Gamma_r^{fb}(p_i, t^{-1}m_r, e_r, t^{-1}\mu), \tag{15.49}$$

where d is the natural dimension of the vertex, since all the dimensioned arguments are scaled by the same parameter t. On differentiating both sides of Eq. (15.49) with respect to t, and using the identity

$$\frac{\partial}{\partial t}\Gamma_r^{fb}(p_i, t^{-1}m_r, e_r, t^{-1}\mu) = -\left(m_r \frac{\partial}{\partial m_r} + \mu \frac{\partial}{\partial \mu}\right)\Gamma_r^{fb}(p_i, t^{-1}m_r, e_r, t^{-1}\mu)$$

together with Eq. (15.49), one finds that $\Gamma_r^{fb}(tp_i, m_r, e_r, \mu)$ satisfies the differential equation

$$\left(t\frac{\partial}{\partial t} + m_r \frac{\partial}{\partial m_r} + \mu \frac{\partial}{\partial \mu} - d\right)\Gamma_r^{fb}(tp_i, m_r, e_r, \mu) = 0. \tag{15.50}$$

On comparing Eq. (15.50) with Eq. (15.45), the term in $\partial \Gamma_r/\partial \mu$ can be eliminated to give

$$\left[t\frac{\partial}{\partial t} - \beta\frac{\partial}{\partial e_r} - m_r(\gamma_m - 1)\frac{\partial}{\partial m_r} + f\gamma_2 + b\gamma_3 - d\right]\Gamma_r^{fb}(tp_i, m_r, e_r, \mu) = 0. \tag{15.51}$$

Eq. (15.51) is the desired differential equation governing the dependence of the vertices on a scale transformation on the momenta alone. Before solving it, however, we need to evaluate the natural dimension d of the vertices Γ_r^{fb}. From the natural dimensions of the fields given in Table 6.1, one easily sees that the natural dimension of the Green function (12.1) is $(b + 3f/2)$, where b and f are the number of photon and fermion legs. The dimension of the associated momentum-space Green function then follows from Eq. (12.10) and is

$$4 - \frac{5}{2}f - 3b.$$

The vertices are then obtained by eliminating the propagators associated with the external lines, giving

$$d = 4 - \frac{3}{2}f - b \qquad (15.52)$$

as the natural dimension of the proper vertices Γ_r^{fb}.

We will now solve Eq. (15.51) for a particular choice of the renormalization scale $\mu = \mu_0$, when it becomes

$$\left[t\frac{\partial}{\partial t} - \beta\frac{\partial}{\partial e_r} - m_r(\gamma_m - 1)\frac{\partial}{\partial m_r} + f\gamma_2 + b\gamma_3 - d \right] \Gamma_r^{fb}(tp_i, m_r, e_r, \mu_0) = 0. \qquad (15.53)$$

This equation implies that, for a fixed renormalization scale, a small change in t can be compensated by small changes in m_r and e_r, together with a multiplicative factor. We therefore look for a solution of the form

$$\Gamma_r^{fb}(tp_i, m_r, e_r, \mu_0) = \Lambda(t)\,\Gamma_r^{fb}(p_i, \tilde{m}(t), \tilde{e}(t), \mu_0), \qquad (15.54)$$

where $\Lambda(t)$, $\tilde{m}(t)$ and $\tilde{e}(t)$ are functions to be determined, subject to the boundary conditions

$$\Lambda(1) = 1, \quad \tilde{m}(1) = m_r(\mu_0), \quad \tilde{e}(1) = e_r(\mu_0). \qquad (15.55)$$

Differentiating Eq. (15.54) with respect to t then gives

$$t\frac{\partial}{\partial t}\Gamma_r^{fb}(tp_i, m_r, e_r, \mu_0)$$
$$= \left(t\frac{d\Lambda}{dt} + t\Lambda\frac{\partial\tilde{m}}{\partial t}\frac{\partial}{\partial\tilde{m}} + t\Lambda\frac{\partial\tilde{e}}{\partial t}\frac{\partial}{\partial\tilde{e}} \right) \Gamma_r^{fb}(p_i, \tilde{m}(t), \tilde{e}(t), \mu_0)$$

from which, using Eq. (15.54),

$$t\frac{\partial}{\partial t}\Gamma_r^{fb}(tp_i, m_r, e_r, \mu_0)$$
$$= \left(\frac{t}{\Lambda}\frac{d\Lambda}{dt} + t\frac{\partial\tilde{m}}{\partial t}\frac{\partial}{\partial\tilde{m}} + t\frac{\partial\tilde{e}}{\partial t}\frac{\partial}{\partial\tilde{e}} \right) \Gamma_r^{fb}(tp_i, m_r, e_r, \mu_0), \qquad (15.56a)$$

where m_r and e_r are evaluated at μ_0, i.e.

$$m_r \equiv m_r(\mu_0) = \tilde{m}(1) \quad e_r \equiv e_r(\mu_0) = \tilde{e}(1).$$

On comparing Eq. (15.56) with Eq. (15.53) we see that they are compatible provided that

$$t\frac{\partial\tilde{m}}{\partial t} = m_r(\gamma_m - 1), \qquad (15.57a)$$

$$t\frac{\partial\tilde{e}}{\partial t} = \beta, \qquad (15.57b)$$

and

$$\frac{t}{\Lambda}\frac{d\Lambda}{dt} = d - f\gamma_2 - b\gamma_3. \tag{15.58}$$

On comparing Eqs. (15.57a, b) with Eqs. (15.46a, b), we see that Eqs. (15.57a, b) are satisfied, subject to the boundary conditions (15.55), if

$$\tilde{m}(t) = t^{-1}m_r(\mu = t\mu_0) \quad \tilde{e}(t) = e_r(\mu = t\mu_0). \tag{15.59}$$

Finally Eq. (15.58) can be explicitly integrated to give

$$\Lambda(t) = t^d \exp\left[-\int_1^t \frac{(f\gamma_2 + b\gamma_3)\,dt}{t}\right], \tag{15.60}$$

and substituting Eqs. (15.59) and (15.60) into Eq. (15.54) gives

$$\begin{aligned}
&\Gamma_r^{fb}(tp_i, m_r, e_r, \mu_0) \\
&= t^d \exp\left[-\int_1^t \frac{(f\gamma_2 + b\gamma_3)}{t}\,dt\right]\Gamma_r^{fb}(p_i, t^{-1}m_r(t\mu_0), e_r(t\mu_0), \mu_0)
\end{aligned} \tag{15.61}$$

as the solution for the momentum-scale dependence of the vertices in terms of the running mass[8] $\tilde{m}(t) = t^{-1}m_r(t\mu_0)$ and running coupling constant $e_r(t\mu_0)$.

An interesting feature of Eq. (15.61) is revealed by considering the behavior in the limit of massless electrons. The only other parameter in QED is the charge, which is dimensionless in natural units, so that the Lagrangian becomes invariant under a change of energy scale in this limit; and one would naively expect the vertex functions to exhibit simple dimensional scaling behavior, as expressed by the factor t^d in Eq. (15.61). That is, one would expect any function Γ of momenta p_i to scale according to its natural dimension d:

$$\Gamma(tp_i, e_r) = t^d \Gamma(p_i, e_r)$$

since there are no dimensioned parameters in the theory. However, renormalization necessarily introduces the mass scale μ into the theory, and Eq. (15.61) deviates from naive dimensional scaling, i.e.

$$\Gamma_r^{fb}(tp_i, m_r = 0, e_r, \mu) \neq t^d \Gamma_r^{fb}(p_i, m_r = 0, e_r, \mu)$$

because of the running mass and the running coupling constant and because of the non-vanishing 'anomalous dimensions' γ_2 and γ_3. The introduction of the mass scale μ has physical consequences, even though its value is completely arbitrary.

15.3.3   The running charge

In Sections 15.3.1 and 15.3.2, we have avoided using explicit expressions for the renormalization constants, and the results have been written in a form which holds to all orders

---

[8] Both $\tilde{m}(t)$ and $m_r(\mu)$ are referred to as 'running masses' in the literature. It is important to remember that they are different and related by Eq. (15.59).

in perturbation theory. We now specialize to lowest-order radiative corrections in order to investigate the behavior of the running charge at high mass-scales $\mu = t\mu_o$, which in turn controls the behavior of the vertex functions at high momenta tp_i. To do this, we evaluate Eq. (15.46a) using Eq. (15.40) together with the second-order result Eq. (15.26) to obtain [cf. Eq. (15.41)]

$$\beta = \mu \frac{\partial e_r}{\partial \mu} = \beta_o e_r^3 + O(e_r^5) \tag{15.62}$$

in the limit $\eta = D - 4 \rightarrow 0$, where

$$\beta_0 = \frac{1}{24\pi^2}. \tag{15.62b}$$

Eq. (15.62a) can be easily solved to give

$$e_r^2(\mu) = \frac{e_r^2(\mu_0)}{1 - \beta_0 e_r^2(\mu_0) \ln (\mu^2/\mu_0^2)} \tag{15.63}$$

to second order. In QED, the running coupling, i.e. the running charge $e_r(\mu)$, increases with increasing μ because $\beta_0 > 0$, although $t = \mu/\mu_0$ has to be very large before the effect is appreciable.

This result can be qualitatively understood by analogy with that observed when an electric charge Q is immersed in a dielectric medium. If the charge is positive, then the molecules of the dielectric will tend to align themselves, as shown in Fig. 15.9; and if a small test charge is introduced at a distance which is large compared to the size of the molecules, it will experience a force which is the same as if the charge Q had been replaced by a smaller 'effective charge' Q_{eff}. However, at distances between charges which are of the same order as the molecular sizes, this screening becomes less effective, so that Q_{eff} increases as the charges get closer together. In QED, there is a similar effect, even in the vacuum, associated with the electron–positron pairs produced by quantum fluctuations. For example, in electron–electron scattering (Møller scattering), the electric charges can be regarded as partially screened from each other by the virtual electron–positron pair in the Feynman diagram Fig. 9.7(b). It is this effect that is described quantitatively by

Figure 15.9 *Schematic diagram representing the polarization of the molecules of dielectric by a positive charge placed within it*

Figure 15.10 *Quark–quark scattering by gluon exchange*

Figure 15.11 *Two lowest-order vacuum polarization corrections to one-gluon exchange in quark–quark scattering*

Eq. (15.63). The same vacuum polarization effect was discussed, albeit in a different approach, in Chapter 9, where it led, in the non-relativistic limit, to the modified Coulomb potential Eq. (9.90). This result also parameterizes the increase in the interaction strength at high momentum transfers, or equivalently short distances, and is essential to a full understanding of the Lamb shift, as discussed at the end of Section 9.6.

Quantum fluctuations also exist in QCD, and, as in QED, they lead to a variation in the interaction strength with distance. Specifically, if we consider quark–quark scattering, there are two lowest-order vacuum polarization corrections to the one-gluon exchange diagram Fig. 15.10. These are shown in Fig. 15.11. The first of these is a direct analogue of Fig. 9.7(b) in electron–electron scattering, and similarly leads to a screening correction. If this were the only contribution, the interaction would grow stronger at short distances, as in QED. However, there is also the second diagram Fig. 15.11(b), involving a gluon–gluon pair produced in a gluon self-interaction of the type shown in Fig. 11.1(b). This diagram has no counterpart in QED, and the nature of its contribution is far from obvious. In fact, as we shall show in the next section, it leads to an *antiscreening* effect; that is, it causes the interaction to grow weaker at short distances. As we shall also show, this effect is bigger than the screening correction from Fig. 15.11(a), and the net result is that the interaction grows weaker at short distances, or, equivalently, large momentum transfers. That is, there is asymptotic freedom.

15.4 The Strong Coupling Constant

In this section we shall extend the arguments of Sections 15.2 and 15.3 from QED to QCD. In particular, we shall obtain a relation

$$g_r = g_0 \mu^{-\eta/2} \frac{Z_3^{1/2} Z_2}{Z_1}, \tag{15.64}$$

where instead of denoting the strong coupling by g_s as before, we now use g_0 and g_r to specify the bare and renormalized couplings, respectively. Here μ is the renormalization scale and Z_3, Z_2 and Z_1 are renormalization constants associated with the gluon and quark propagators and the quark–gluon vertex respectively. Eq. (15.64) is closely analogous to Eq. (15.38) in QED, and like Eq. (15.38) it can be used to determine the scale dependence of the coupling.

The application of the renormalization group to QCD is essentially identical to that in QED, with only trivial modifications. It will be dealt with in Section 15.5, where we shall also discuss applications. In this section we shall obtain Eq. (15.64), together with explicit values of the renormalization constants, and use them to determine the scale dependence of the strong coupling. Firstly, however, we discuss the 'colour factors' which are a feature of all perturbation calculations in QCD, since they will be required in what follows.

15.4.1   Colour factors

The vertex factors of Fig. 14.4 involve the structure constants f_{ijk} and the three by three colour matrices

$$T^i = \frac{\lambda^i}{2}, \quad i = 1, 2, \ldots 8, \tag{15.65}$$

and it is necessary to evaluate combination of these quantities, called 'colour factors,' when calculating Feynman diagrams. In this short section, we will discuss the evaluation of these colour factors to the extent to which we will need them later.[9]

The structure constants f_{ijk} are defined by Table 11.1 and Eq. (11.12b). They are totally antisymmetric

$$f_{ijk} = f_{jki} = f_{kij} = -f_{jik} = -f_{kji} = -f_{ikj} \tag{15.66}$$

and one can also verify that

$$f_{ijk}f_{ljk} = 3\delta_{il}, \tag{15.67}$$

where repeated indices are summed over as usual. The matrices T^i satisfy the commutation relations (11.12a):

$$[T^i, T^j] = \mathrm{i}f_{ijk}T^k. \tag{15.68}$$

Other properties include

$$Tr\,T^i = \frac{1}{2}Tr\,\lambda_i = 0 \tag{15.69}$$

---

[9] A much more extensive discussion of color factors may be found in, for example, R.D. Field, *Applications of Perturbative QCD*, Addison-Wesley (1989), Appendix D.

$$Tr(T^i T^j) = \frac{1}{4} Tr(\lambda_i \lambda_j) = \frac{1}{2} \delta_{ij} \tag{15.70}$$

and

$$T^i T^i = \frac{1}{4} \lambda_i \lambda_i = \frac{4}{3}, \tag{15.71}$$

as may be verified from Eq. (11.11b). Further identities can be derived from these basic relations. For example, two which will be required later are:

$$if_{ijk} T^j T^k = -\frac{3}{2} T^i \tag{15.72}$$

and

$$T^j T^i T^j = -\frac{1}{6} T^i. \tag{15.73}$$

These are derived as follows. To obtain Eq. (15.72) we write

$$if_{ijk} T^j T^k = if_{ijk} [T^j, T^k] + if_{ijk} T^k T^j$$
$$= -f_{ijk} f_{jkl} T^l - if_{ikj} T^k T^j$$
$$= -f_{ijk} f_{ljk} T^l - if_{ijk} T^j T^k$$

using Eq. (15.68) and the antisymmetry of the structure constants f_{ijk}. Eq. (15.72) then follows directly using Eq. (15.67). Eq. (15.73) is then obtained by writing

$$T^j T^i T^j = T^j [T^i, T^j] + T^j T^j T^i$$
$$= if_{ijk} T^j T^k + \frac{4}{3} T^i$$

using Eqs. (15.68) and (15.71). Eq. (15.73) then follows directly using Eq. (15.72).

15.4.2   Null diagrams

In order to calculate the renormalization constants Z_2 and Z_3 in Eq. (15.64), we need to calculate the quark and gluon self-energy diagrams, respectively. These include diagrams containing loops like those shown in Fig. 15.12, which vanish as a direct consequence of the Feynman rules of Section 14.4. For Figs. 15.12(a, b), this follows from the colour factors associated with the loops, which in both cases are

$$f_{ijk} \delta_{jk} = f_{ijj} = 0,$$

where the f_{ijk} stems from the vertex factors of Fig. 14.4 and the δ_{jk} from the propagators of Fig. 14.3. The corresponding colour factor from the quark loop of Fig. 15.12(c) is

$$(T^i)_{dc} \delta_{cd} = Tr \ T^i = 0$$

by Eq. (15.69). Finally, the Lorentz and colour factors associated with the loop in Fig. 15.12(d) are

$$F_{ijkl} V_{\mu\nu\sigma\tau} g^{\mu\nu} \delta_{jk} = 0,$$

as is easily confirmed using Eq. (14.70).

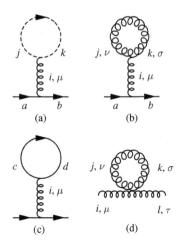

Figure 15.12   *Examples of Feynman graphs which vanish due to their loop contributions*

In what follows, diagrams containing loops like those shown in Fig. 15.12 will be set to zero without further discussion.[10]

15.4.3   Renormalization of the coupling constant

In this section we will derive Eq. (15.64) and the explicit form of the renormalization constants Z_2, Z_3 and Z_1 by considering the renormalization of the quark and gluon propagators and the quark–gluon vertex in turn. We shall only give the discussion to the extent that it is required to derive Eq. (15.64), and we shall work to second-order throughout.

The quark self-energy

The renormalization of the quark propagator to second order is very similar to that of the electron propagator summarized in Fig. 9.9, where the only non-vanishing quark self-energy diagram is shown explicitly in Fig. 15.13. Using the Feynman rules of Section 14.4, the resulting self-energy is given by

$$i g_0^2 \Sigma_{ac}(p) = (i g_0)^2 C_{ac} \int \frac{d^4 k}{(2\pi)^4} i D_{F\alpha\beta}(k) \gamma^\alpha i S_F(p-k)\gamma^\beta \tag{15.74a}$$

where the colour factor is given by

$$C_{ac} = T^i_{ab} T^i_{bc} = \frac{4}{3}\delta_{ac}, \tag{15.74b}$$

using Eq. (15.71). On setting $a = c$, which is just colour conservation, the quark self-energy (15.74) is identical to the electron self-energy (15.8b), except for the replacement

$$e_0^2 \rightarrow 4 g_0^2/3. \tag{15.75}$$

---

[10] Note that in the canonical formalism with normal ordering, such diagrams are eliminated automatically by the 'no e.t.c' condition in Wick's theorem (6.38).

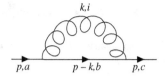

Figure 15.13 *The second-order quark self-energy*

Consequently, the discussion of the electron propagator given in Section 15.2.1 is easily adapted to the case of the quark propagator for any flavor $a = c$ using the substitution (15.75). In particular, the bare and renormalized propagator are related by Eq. (15.18),

$$G_r(p) \equiv Z_2^{-1} G(p), \tag{15.76}$$

where the renormalization constant and the partially renormalized coupling constant are given by [cf. Eqs. (15.12b), (15.15) and (15.16b)]

$$Z_2 = 1 - \frac{g_r^2}{12\pi^2} \left[\frac{2}{\eta} - \gamma + \ln 4\pi \right] \tag{15.77}$$

and

$$g_r \equiv \tilde{g}_0 Z_2^{1/2} \tag{15.78}$$

respectively, and

$$\tilde{g}_0 = g_0 \mu^{-\eta/2} \tag{15.79}$$

in analogy to Eq. (15.9) in QED.

The gluon self-energy

The renormalization of the gluon propagator proceeds in a very similar manner to the photon propagator, and results from the substitution of Fig. 15.14. It differs from QED in that there are now three non-zero self-energy diagrams at one-loop level, as shown in Fig. 15.15. The evaluation of these diagrams using the Feynman rules of Section 14.4 is straightforward, but somewhat lengthy. It is discussed in Appendix 15.6.1. Here we simply quote the results, which are that the contributions of the quark, ghost and gluon loop diagrams of Fig. 15.15 are given by

$$ig_0^2 \Pi_{ij}^{\mu\nu}(a) = \delta_{ij} \left(k^\mu k^\nu - k^2 g^{\mu\nu} \right) \frac{i\tilde{g}_0^2}{24\pi^2} \left[\frac{2}{\eta} - \gamma + \ln 4\pi \right] + \cdots \tag{15.80a}$$

$$ig_0^2 \Pi_{ij}^{\mu\nu}(b) = \delta_{ij} \left(k^\mu k^\nu + \frac{k^2 g^{\mu\nu}}{2} \right) \frac{i\tilde{g}_0^2}{32\pi^2} \left[\frac{2}{\eta} - \gamma + \ln 4\pi \right] + \cdots \tag{15.80b}$$

$$ig_0^2 \Pi_{ij}^{\mu\nu}(c) = \delta_{ij} \left(11 k^\mu k^\nu - \frac{19}{2} k^2 g^{\mu\nu} \right) \frac{-i\tilde{g}_0^2}{32\pi^2} \left[\frac{2}{\eta} - \gamma + \ln 4\pi \right] + \cdots, \tag{15.80c}$$

Figure 15.14   *The modified gluon propagator, where the second-order self-energy is given by the three Feynman diagrams of Fig. 15.15*

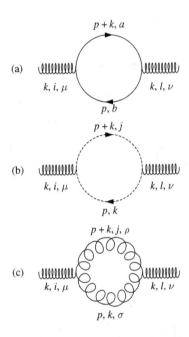

Figure 15.15   *The second-order contributions to the gluon self-energy*

respectively, where the dots represent non-singular terms, which are not required in what follows.

In combining the contributions (15.80a, b, c), we must remember that there are several different quark flavors u, d, s, ..., which can all contribute to the quark loop Fig. 15.15(a). Assuming n_f flavors, where $n_f = 6$ in the standard model, we finally obtain

$$ig_0^2 \, \Pi_{ij}^{\mu\nu}(k) = i\delta_{ij} \, g_0^2 \, \widetilde{\Pi}^{\mu\nu}(k) \tag{15.81a}$$

as our final result for the gluon self-energy, where

$$g_0^2 \, \widetilde{\Pi}^{\mu\nu}(k) = \left(k^\mu k^\nu - k^2 g^{\mu\nu}\right) \frac{\tilde{g}_0^2}{16\pi^2} \left(\frac{2n_f}{3} - 5\right)\left[\frac{2}{\eta} - \gamma + \ln 4\pi\right] + \dots \tag{15.81b}$$

This should be compared to the corresponding result

$$e_0^2 \Pi^{\mu\nu}(k) = \left(k^\mu k^\nu - k^2 g^{\mu\nu}\right) \frac{\tilde{e}_0^2}{12\pi^2}\left[\frac{2}{\eta} - \gamma + \ln 4\pi\right] + \dots \tag{15.82}$$

obtained [cf. Eqs. (10.48) and (10.52b)] for the photon self-energy in QED.

It remains to use this result to renormalize the gluon propagator. From Fig. 15.13 we have

$$G_{ij}^{\mu\nu}(k) = iD_F^{\mu\nu}(k)\delta_{ij} + iD_F^{\mu\nu}(k)\delta_{il}ig_0^2\Pi_{\alpha\beta,\,lm}(k)iD_F^{\beta\nu}(k)\delta_{mj}$$

$$= \left[iD_F^{\mu\nu}(k) + iD_F^{\mu\alpha}(k)ig_0^2\tilde{\Pi}_{\alpha\beta}(k)iD_F^{\beta\gamma}\right]\delta_{ij} \qquad (15.83)$$

for the gluon propagator to second-order. For $i = j$, color conservation, this is identical to the corresponding expression (9.10a) for the photon propagator, except that the gluon self-energy (15.81) differs from the photon self-energy (15.82) by the replacement

$$\frac{\tilde{e}_0^2}{12\pi^2} \rightarrow \frac{\tilde{g}_0^2}{16\pi^2}\left(\frac{2n_f}{3} - 5\right). \qquad (15.84)$$

Consequently the analysis is identical to that given for the photon propagator in Section 15.2.2, and the results (15.25), (15.27), (15.28) and (15.29) can be taken over directly, with a trivial modification. In this way one obtains

$$g_r \equiv Z_3^{1/2}\tilde{g}_0 \qquad (15.85)$$

for the partially renormalized coupling constant, where the renormalization constant is given by

$$Z_3 = 1 - \frac{g_r^2}{16\pi^2}(2n_f - 5)\left[\frac{2}{\eta} - \gamma + \ln 4\pi\right] \qquad (15.86)$$

to second order. The renormalized propagator

$$G_{r,ij}^{\mu\nu}(k) = Z_3^{-1}G_{ij}^{\mu\nu}(k) \qquad (15.87)$$

is given by [cf. Eqs. (15.25) and (15.27)]

$$G_{r,ij}^{\mu\nu}(k) = \frac{-ig^{\mu\nu}\delta_{ij}}{k^2 + i\varepsilon + g_r^2\,\Pi_r(k^2)} + O(g_r^4), \qquad (15.88)$$

where $\Pi_r(k^2)$ remains finite in the limit of four space-time dimensions, $\eta \to 0$.

On comparing the expression (15.86) with the corresponding expression (15.26) in QED, we see that the quark-loop contribution to Z_3 has the same sign as the electron-loop contribution in QED. In QCD the additional gluon and ghost loop contributions reverse this sign. However, before pursuing this, we must also take into account the quark self-energy correction and the vertex correction, to which we now turn.

The quark–gluon vertex correction

When second-order radiative corrections are included, the quark–gluon vertex is given by

$$i\Gamma^{i,\mu}(p',p) = ig_0\left[\gamma^\mu T^i + g_0^2\Lambda^{i,\mu}(p',p)\right] \qquad (15.89)$$

corresponding to Fig. 15.16. Here, we have suppressed the quark color indices, T^i are the three by three color matrices (14.58b), and the vertex correction $g_0^2\Lambda^{i,\mu}(p',p)$ is given by the two Feynman diagrams of Fig. 15.17. Their evaluation is discussed in Appendix 15.6.2, and yields

$$g_0^2 \Lambda^{i,\mu}(p',p) = -\gamma^\mu T^i \frac{\tilde{g}_0^2}{96\pi^2} \left[\frac{2}{\eta} - \gamma + \ln 4\pi \right] + \cdots \tag{15.90a}$$

$$g_0^2 \Lambda^{i,\mu}(p',p) = \gamma^\mu T^i \frac{9\tilde{g}_0^2}{32\pi^2} \left[\frac{2}{\eta} - \gamma + \ln 4\pi \right] + \cdots \tag{15.90b}$$

for the contributions of Figs. 15.17(a,b) respectively, where the dots again indicate finite terms whose form is not required in what follows. On combining these contributions we obtain

$$g_0^2 \Lambda^{i,\mu}(p',p) = \gamma^\mu T^i \frac{13\tilde{g}_0^2}{48\pi^2} \left[\frac{2}{\eta} - \gamma + \ln 4\pi \right] + \cdots \tag{15.91}$$

for the full vertex correction in second order. On substituting Eq. (15.91) into Eq. (15.89), and defining

$$Z_1 = 1 - \frac{13g_r^2}{48\pi^2} \left[\frac{2}{\eta} - \gamma + \ln 4\pi \right] \tag{15.92}$$

to second order, we obtain

$$i\Gamma^{i,\mu}(p',p) = ig_r \mu^{\eta/2} \left[T^i \gamma^\mu + \tilde{g}_0^2 \Lambda^{i,\mu}(p',p) \right], \tag{15.93}$$

where the partially renormalized coupling

$$g_r \equiv \frac{\tilde{g}_0}{Z_1} \tag{15.94}$$

and $\Lambda_r^{i,\mu}(p',p)$ is the finite part of the vertex correction, whose form is not required.

Finally, we must combine Eq. (15.94) with the corresponding partial renormalizations (15.78) and (15.85). As in QED,[11] this is done by absorbing factors of $Z_2^{1/2}$ and $Z_3^{1/2}$ into each vertex from each adjoining fermion and boson line respectively, to compensate for

Figure 15.16   *The modified quark-gluon vertex, where the second-order vertex correction is given by the Feynman graphs of Figure 15.17*

---

[11] cf. the discussion following Eq. (15.35).

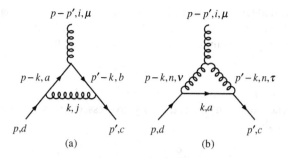

Figure 15.17 *The second-order quark–gluon vertex corrections*

those removed in defining the renormalized propagators (15.76) and (15.87). This gives rise to a renormalized vertex.

$$\Gamma_r^{i,\mu}(p',p) \equiv Z_2 Z_3^{1/2} \Gamma^{i,\mu}(p',p) \tag{15.95}$$

and substituting Eqs. (15.93) and (15.94) yields

$$i\Gamma_r^{i,\mu}(p',p) = ig_r\mu^{\eta/2}\left[T^i\gamma^\mu + g_r^2\Lambda_r^{i,\mu}(p',p) + O(g_r^5)\right], \tag{15.96}$$

where g_r is now the fully renormalized coupling defined by

$$g_r \equiv \tilde{g}_0 \frac{Z_3^{1/2} Z_2}{Z_1} = g_0\mu^{-\eta/2}\frac{Z_3^{1/2} Z_2}{Z_1} \tag{15.97}$$

using Eq. (15.79). We note that in this case $Z_1 \neq Z_2$, and Eq. (15.97) does not reduce to an expression of the form of our final result (15.40) in QED.

15.4.4   The running coupling

In this section, we will use Eq. (15.97) to deduce the scale dependence of the coupling constant, and consider its value as determined by experiment.

The first step is to substitute the explicit values for the renormalization constants given in Eqs. (15.77), (15.86) and (15.94) into Eq. (15.97) to give

$$g_r = g_0\mu^{-\eta/2}\left[1 + \frac{g_r^2}{32\pi^2}\left(11 - \frac{2n_f}{3}\right)\left[\frac{2}{\eta} - \gamma + \ln 4\pi\right] + O(g_r^4)\right] \tag{15.98}$$

as our final result for the relationship between the bare and renormalized coupling. The μ-dependence of the renormalized coupling is determined by solving the equation

$$\beta(g_r) \equiv \mu\frac{\partial g_r}{\partial\mu} = \frac{-\beta_0 g_r^3}{16\pi^2}, \tag{15.99}$$

which follows from Eq. (15.98) in the limit of four space–time dimensions $\eta \to 0$, where

$$\beta_0 = 11 - \frac{2}{3}n_f \tag{15.100}$$

and we have neglected terms of order g_r^5. Provided the number n_f of quark flavors u, d, s, ... is less than 17, β_0 is positive, so that g_r decreases with increasing μ. The form of this decrease is usually expressed in terms of the coupling strength

$$\alpha_s(\mu) \equiv g_r^2/4\pi, \tag{15.101}$$

defined in analogy to the fine-structure constant $\alpha \approx 1/137$.

The μ-dependence of $\alpha_s(\mu)$ is commonly expressed in two different ways. From Eq. (15.99) one obtains

$$\mu\frac{\partial\alpha_s}{\partial\mu} = -\frac{\beta_0}{2\pi}\alpha_s^2,$$

which has the solution

$$\alpha_s(\mu) = \frac{4\pi}{\beta_0\ln\left(\mu^2/\Lambda^2\right)}, \tag{15.102}$$

where Λ is a 'scale parameter,' which must be determined from experiment, and which characterizes the scale $\mu \approx \Lambda$ at which α_s becomes large as μ decreases. Alternatively, Eq. (15.99) also implies

$$\mu\frac{\partial\alpha_s^{-1}}{\partial\mu} = \frac{\beta_0}{2\pi}$$

and hence

$$\alpha_s(\mu) = \frac{\alpha_s(\mu_0)}{1 + (\beta_0/4\pi)\alpha_s(\mu_o)\ln\left(\mu^2/\mu_0^2\right)}, \tag{15.103}$$

where $\alpha_s(\mu_0)$, the value at an arbitrarily chosen reference scale μ_0, is now the parameter which must be determined experimentally. The latter approach is now more commonly used, where the reference scale is conventionally chosen to be the mass of the Z^0 boson, $\mu = m_Z = 91.29$ GeV. Furthermore, for precision work it is necessary to include corrections to Eq. (15.103) arising from terms in Eq. (15.99) which are of higher order in the renormalized coupling g_r. When this is done, the most precise value obtained from the analysis of current experiments is

$$\alpha_s(m_Z) = 0.118 \pm 0.002, \tag{15.104}$$

where the predicted μ-dependence is shown in Fig. 15.18. The origin of this value will be discussed in the following section.

15.5   Applications

In Sections 15.3.1 and 15.3.2, we showed that, in QED, at high energy and momentum transfers, the strength of the interaction was characterized by the running coupling constant at high mass scales. This result was obtained by deriving the renormalization group

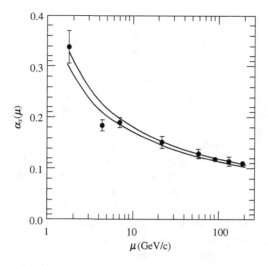

Figure 15.18 *Values of the running coupling constant α_s obtained from the following experiments at increasing values of μ: τ-decay; γ-decay; deep inelastic lepton scattering; e^+e^- annihilation at 22 and 50 GeV; Z^0 decay; and e^+e^- annihilation at 135 and 189 GeV. The solid lines show the evolution of α_s with μ, as predicted by QCD, assuming the value (15.104) for $\alpha_s(m_Z)$. Reproduced by permission of the Institute of Physics. W.M. Yao et al, Journal of Physics, Vol 33, 2006, section 9, p. 112–116*

equations from Eq. (15.44), and then solving them. In QCD, the corresponding relation between the bare and renormalized vertex functions is

$$\Gamma_r^{fb}(p_i, m_r, g_r, \mu) \equiv Z_2^{f/2} Z_3^{b/2} \Gamma^{fb}(p_i, m_0, g_0), \qquad (15.105)$$

where f and b are now the number of external quarks and gluons, rather than electrons and photons. Eq. (15.105) is of the same form as Eq. (15.44), and the argument which led from Eq. (15.44) to Eq. (15.61) goes through unchanged, except for the trivial replacement of the charge by the strong coupling constant. Since in this case the running coupling decreases with increasing scale, as shown in the last section, in QCD we have asymptotic freedom and perturbation theory should be applicable at high energy–momentum transfers, corresponding to short-distance interactions.

This is best illustrated by an example. Let us consider the total cross-section for electron–positron annihilation into hadrons, reaction (15.1), which is conveniently expressed in terms of its ratio to the total cross-section for annihilation into muon pairs:

$$R \equiv \frac{\sigma(e^+e^- \rightarrow \text{hadron})}{\sigma(e^+e^- \rightarrow \mu^+ + \mu^-)}. \qquad (15.106)$$

As discussed in Section 15.1.1, in the centre-of-mass energy range 15–40 GeV, the main contribution arises from the two-step mechanism of Fig. 15.2, in which the rate is determined by the primary electromagnetic process (15.2), corresponding to the Feynman diagram 15.19(a). In the approximation that the energy is large enough for the quark masses to be neglected, this leads to the cross-section formula (15.5), implying

$$R = R_0 \equiv 3 \sum_f e_f^2 = 11/3, \qquad (15.107)$$

on comparing with the $e^+e^- \rightarrow \mu^+\mu^-$ cross-section (8.46). In Eq. (15.107), the factor 3 arises from the existence of three colour states for each quark flavour, and we have summed over the five flavours $f = u, d, s, c, b$ allowed by energy conservation in the centre-of-mass energy range considered.

Eq. (15.107) is a parton model result, and is the biggest single contribution to the cross-section ratio. However, there are other contributions which arise, to first order in α_s, from the Feynman diagrams of Fig. 15.19b, c and d. Here Figs. 15.19b, c correspond to the 'gluon bremsstrahlung' process (15.6), which gives rise to the three-jet events discussed in Section 15.1.2; while interference between Fig. 15.9a and Fig. 15.19d gives rise to an order α_s correction to quark-pair production (15.2). In calculating these contributions to the total cross-section, infrared divergences arise which cancel between the different contributions, in much the same way as in QED [cf. Sections 8.9 and 9.7.] Here we omit the details of the calculation[12] and simply quote the result, which is that Eq. (15.107) is replaced by

$$R = R_0 [1 + \alpha_s(\mu)/\pi] \qquad (15.108)$$

when these corrections are included.

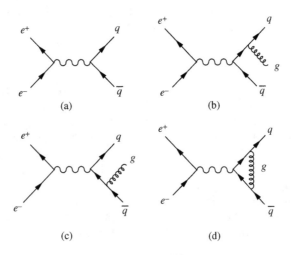

Figure 15.19   *Contributions to the reaction $e^+e^- \rightarrow$ hadrons to second order in α_s, where the final state particles manifest themselves as hadron jets*

---

[12] This calculation is explicitly given by R.D. Field, *Applications of Perturbative QCD*, Addison-Wesley (1989), Sections 2.1–2.7.

At this point we note that the leading order result (15.13) holds for any choice of the renormalization scale, since [cf. Eq. (15.103)]

$$\alpha_s(\mu) = \alpha_s(\mu_0) + O\left[\alpha_s^2(\mu_0)\right],$$

and the value of $\alpha_s(\mu)$ is significantly scale dependent, as shown in Fig. 15.18. The same result (15.108) also holds for all renormalization schemes, since the values of α_s in different schemes differ by terms of order α_s^2. These ambiguities in leading order perturbation theory[13] are greatly reduced by working to higher order, since all formulations of the theory must lead to the same results for physical quantities when calculated exactly. In addition, one normally chooses a value for the renormalization scale μ of the same order as the characteristic momentum scale of the process being studied, for reasons which will be discussed below.

The calculation of such higher-order corrections is long and complicated, and we will again simply quote the results. In doing so, it is important to specify the renormalization scheme used and the chosen renormalization scale μ. Here, we use the $\overline{MS}$ scheme and choose $\mu = Q$, where Q is the invariant mass of the electron–positron pair or, equivalently, the total energy in the centre-of-mass frame. The cross-section ratio (15.106) has been calculated to third order, and is given by

$$R = R_0 \left(1 + \frac{\alpha_s}{\pi} + 1.41 \left(\frac{\alpha_s}{\pi}\right)^2 - 12.8 \left(\frac{\alpha_s}{\pi}\right)^3 + \cdots \right) \qquad (15.109a)$$

in the $\overline{MS}$ scheme, where

$$\alpha_s = \alpha_s \left(\mu^2 = Q^2\right). \qquad (15.109b)$$

From Eqs. (15.109) and the values of α_s shown in Fig. 15.18, one sees that the series converges quite rapidly provided Q^2 is reasonably large. Furthermore, although small, the magnitude and Q^2 dependence of the predicted corrections to the zeroeth order prediction (15.107) are in good agreement with experiment and can be used to extract a value of α_s which is in agreement with Eq. (15.104) to within an uncertainty of order 20%.

Before continuing further, we consider briefly the renormalization scheme and scale dependence of the expansion (15.109).

Firstly, we reiterate that beyond leading order, the numerical coefficients in (15.109a) and the value of the coupling constant α_s at a given scale depend on the renormalization scheme used. For example, in the MS scheme mentioned earlier,[14] the coefficient of α_s^3 (but not in this case α_s^2) differs from that given in Eq. (15.109a). Here we consider only the $\overline{MS}$ scheme.

Secondly, and more interestingly, in Eq. (15.109) we have chosen to 'run the coupling constant.' That is, we have chosen a renormalization scale $\mu = Q$, which varies to reflect

---

[13] Note that such ambiguities are also present in lowest order calculations in QED, but are less serious because the differences between the coupling defined in different renormalization schemes and at different renormalization scales are numerically much smaller.

[14] See footnote 4 on p. 332.

the momentum scale at which the process is observed, rather than a fixed scale $\mu = \mu_0$. From Eq. (15.103) we see that

$$\alpha_s(Q) = \alpha_s(\mu_0) - \frac{\beta_0}{2\pi} \ln\left(\frac{Q}{\mu_0}\right)\alpha_s(\mu_0)^2 + O\left(\alpha_s(\mu_0)^3\right)$$

so that Eq. (15.109a) implies

$$R = R_0 \left\{ 1 + \frac{\alpha_s(\mu_0)}{\pi} + \left[1.41 - \frac{\beta_0}{2}\ln\left(\frac{Q}{\mu_0}\right)\right]\left(\frac{\alpha_s(\mu_0)}{\pi}\right)^2 + O\left(\frac{\alpha_s(\mu_0)}{\pi}\right)^3 \right\}$$

for the expansion in $\alpha_s(\mu_0)$ up to second order. On comparing this with Eq. (15.109), we see that the coefficient of $\alpha_s(\mu_0)^2$ has a log-dependent term which becomes large when Q is large, and which is not present in the expansion (15.109). More generally, it can be shown that if one calculates to all orders in $\alpha_s(\mu_0)$, but retaining only the leading term in $\ln(Q/\mu_0)$ at each order, the result is

$$1 + \frac{\alpha_s(\mu_0)}{\pi}\sum_{n=0}^{\infty}\left[-\frac{\beta_0}{2\pi}\ln\left(\frac{Q}{\mu_0}\right)\alpha_s(\mu_0)\right]^n.$$

This is precisely equal to

$$1 + \frac{\alpha_s(Q)}{\pi},$$

as may be seen by expanding the denominator in Eq. (15.103). In other words, in Eq. (15.103) the first order term in $\alpha_s(Q^2)$ is equal to a sum over all orders in $\alpha_s(\mu_0)$, where each term is evaluated in the 'leading log approximation.' The terms in $\alpha_a^2(Q_0)$ include the next leading logs and so on, so that the 'renormalization group improved' perturbation expansion (15.103) is equivalent to a systematic expansion in leading, next leading logs etc. at a fixed renormalization scale μ_0.

Returning to experiment, one is not merely interested in the total cross-section for a given reaction, but also in the angular distribution of, and the energy correlation between, the resulting hadron jets, as illustrated for one simple case in Fig. 15.3. In such processes, the jets will not always be clearly separated, as shown in Fig. 15.5, but may overlap, so that the assignment of the individual hadrons to the various jets is not trivial. In such cases, a careful definition of a jet, together with a reliable model of the non-perturbative hadronization process is required to fully analyse the event. In addition, most high-energy processes involve initial-state protons, and it is necessary to know the distributions of quarks and gluons within the proton. These cannot be calculated perturbatively, but can be measured in electron and neutrino scattering experiments. We will not pursue such matters here, but refer the reader to the excellent textbooks devoted to this subject.[15] Instead, we shall conclude by merely illustrating the

---

[15] General introductions to applications of QCD, essentially starting from the Feynman rules of Section 14.4, are given by, for example, R.D. Field, *Applications of Perturbative QCD*, Addison Wesley (1989) and R.K. Ellis, W.J. Stirling and B.R. Webber, *QCD and Collider Physics*, Cambridge University Press (1996).

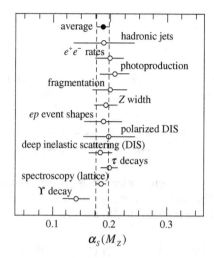

$\alpha_s(M_Z)$

Figure 15.20 *The values of $\alpha_s(m_Z)$ obtained from the analysis of a variety of processes. Reproduced by permission of the Institute of Physics. W.M. Yao et al, Journal of Physics, Vol 33, 2006, section 9, p. 112–116*

wide-ranging success of the theory by noting the high degree of consistency between values of the strong coupling obtained from a variety of different processes,[16] as shown in Fig. 15.20.

15.6   Appendix: Some Loop Diagrams in QCD

In this Appendix, we shall discuss the evaluation of the gluon self-energy graphs of Fig. 15.15, leading to Eqs. (15.80 a–c); and the quark–gluon vertex corrections of Fig. 15.17, leading to Eqs. (15.90).

15.6.1   The gluon self-energy graphs

The gluon self-energy is given in lowest order by the three graphs Fig. 15.15a, b, and c. We discuss the contribution of each in turn.

(a) *The quark-loop contribution*
   On applying the Feynman rules of Section 14.4 to Fig. 15.15a, one obtains

$$\mathrm{i}g_0^2\,\Pi_{il}^{\mu\nu}(a) = C_{il}^{(a)}\,(-1)(-\mathrm{i}g_0)^2 \int \frac{\mathrm{d}^4p}{(2\pi)^4}\,\mathrm{Tr}[\gamma^\mu \mathrm{i}S_\mathrm{F}(p+k)\gamma^\nu \mathrm{i}S_\mathrm{F}(p)], \qquad (15.110)$$

---

[16] For a discussion of the origin of the various data points on Figs. 15.18 and 15.20, see C. Hinchliffe, "Quantum Chromodynamics and its coupling", in "Review of Particle Physics", *J. Phys. G : Nucl. Pait. Phys.* **33** (2006) 1.

where the colour factor

$$C_{il}^{(a)} = T_{ab}^i T_{ba}^l = \frac{1}{2}\delta_{il} \tag{15.111}$$

using Eq. (15.70). On substituting Eq. (15.111) and the explicit forms for the propagators into Eq. (15.110), one obtains

$$ig_0^2\Pi_{il}^{\mu\nu}(a) = -\frac{\delta_{il}}{2}g_0^2 \int \frac{d^4p}{(2\pi)^4} \frac{\mathrm{Tr}[\gamma^\mu(\not{p} + \not{k} + m_0)\gamma^\nu(\not{p} + m_0)]}{\left[(p+k)^2 - m_0^2 + i\varepsilon\right]\left[p^2 - m_0^2 + i\varepsilon\right]}. \tag{15.112}$$

This is identical to the photon self-energy term (9.8), except for the factor δ_{il} and the trivial substitution $e_0^2 \to g_0^2/2$. The photon self-energy was evaluated using dimensional regularization in Section 10.5, and from Eqs. (10.48) and (10.52b) we obtain

$$ig_0^2\Pi_{il}^{\mu\nu}(a) = \delta_{il}(k^\mu k^\nu - k^2 g^{\mu\nu})\frac{i\widetilde{g}_0}{24\pi^2}\left[\frac{2}{\eta} - \gamma + \ln 4\pi\right] + \cdots, \tag{15.80a}$$

where $\widetilde{g}_0$ is given by Eq. (15.79) and the dots represent finite terms which do not contribute to the renormalization constant.

(b) *The ghost loop*

We next turn to the evaluation of the ghost loop Fig. 15.15(b). Using the Feynman rules of Section 14.4, not forgetting the factor (-1) associated with closed ghost loops, we obtain

$$ig_0^2\Pi_{il}^{\mu\nu}(b) = C_{il}^{(b)}(-1)g_0^2 \int \frac{d^4p}{(2\pi)^4}(p+k)^\mu k^\nu i\Delta_F(p+k)i\Delta_F(k), \tag{15.113a}$$

where the colour factor

$$C_{il}^{(b)} = f_{ijk}f_{lkj} = -f_{ijk}f_{ljk} = -3\delta_{il} \tag{15.113b}$$

and where we will set $i = l$ from now on. On generalizing to $D = 4 - \eta$ dimensions, Eqs. (15.113) can be written in the form

$$ig_0^2\Pi_{il}^{\mu\nu}(b) = -3\widetilde{g}_0^2 \, I^{\mu\nu}(k)\delta_{il}, \tag{15.114}$$

where

$$I^{\mu\nu}(k) = \frac{\mu^\eta}{(2\pi)^D} \int d^D p \frac{N^{\mu\nu}(p,k)}{\left[(p+k)^2 + i\varepsilon\right]\left[k^2 + i\varepsilon\right]} \tag{15.115}$$

and

$$N^{\mu\nu}(p+k) = (p+k)^\mu k^\nu. \tag{15.116}$$

To evaluate the integral (15.115), we first Feynman parameterize it using Eq. (10.10) to give

$$I^{\mu\nu}(k) = \frac{\mu^\eta}{(2\pi)^D} \int_0^1 dz \int d^D p \frac{N^{\mu\nu}(p,k)}{[p^2 + (k^2 + 2pk)z + i\varepsilon]^2}.$$

Introducing, the new variable

$$q = p + kz,$$

this becomes

$$I^{\mu\nu}(k) = \frac{\mu^\eta}{(2\pi)^D} \int_0^1 dz \int d^D q \, \frac{N^{\mu\nu}(q - kz, k)}{[q^2 + k^2 z(1-z) + i\varepsilon]^2}, \tag{15.117}$$

where

$$N^{\mu\nu}(q - kz, k) = q^\mu q^\nu - k^\mu k^\nu z(1-z) + \cdots$$

and where the dots indicate that we have omitted terms linear in q, since these terms will vanish by Eq. (10.32). Eq. (15.117) then becomes

$$I^{\mu\nu}(k) = \frac{\mu^\eta}{(2\pi)^D} \int_0^1 dz \left(I_1^{\mu\nu}(k, z) + I_2^{\mu\nu}(k, z) \right), \tag{15.118}$$

where, using Eqs. (10.23) and (10.33), we have

$$I_1^{\mu\nu}(k, z) \equiv \int d^D q \, \frac{q^\mu q^\nu}{[q^2 + k^2 z(1-z) + i\varepsilon]^2}$$

$$= -i \frac{\pi^{D/2} \Gamma(1 - D/2)}{2} \frac{g^{\mu\nu}}{[-k^2 z(1-z)]^{1-D/2}} \tag{15.119a}$$

and

$$I_2^{\mu\nu}(k, z) \equiv -k^\mu k^\nu z(1-z) \int \frac{d^D q}{[q^2 + k^2 z(1-z) + i\varepsilon]^2}$$

$$= -z(1-z) i \pi^{D/2} \Gamma(2 - D/2) \frac{k^\mu k^\nu}{[-k^2 z(1-z)]^{2-D/2}}. \tag{15.119b}$$

It remains to substitute Eqs. (15.119a, b) into Eq. (15.117), and take the limit $\eta \to 0$ using the relations (10.28) and (10.29). In this way, after substituting the result into Eq. (15.114) we finally obtain

$$ig_0^2 \Pi_{il}^{\mu\nu}(b) = \delta_{il} \left(k^\mu k^\nu + \frac{k^2 g^{\mu\nu}}{2} \right) \frac{i\tilde{g}_0^2}{32\pi^2} \left[\frac{2}{\eta} - \gamma + \ln 4\pi \right] + \cdots \tag{15.80b}$$

and the dots indicate finite contributions which we have neglected.

(c) *The gluon loop*

The final contribution arises from the gluon-loop diagram Fig. 15.15(c). Using Eq. (14.68) and Fig. 14.4(b), the two three-point vertex factors are:

$$g_0 f_{ijk} V_{\mu\rho\sigma}(-k, -p - k, p) = g_0 f_{ijk} \left[g_{\rho\sigma}(k + 2p)_\mu + g_{\mu\rho}(-2k - p)_\sigma + g_{\mu\sigma}(k - p)_\rho \right] \tag{15.120a}$$

$$g_0 f_{lkj} V_{v\sigma\rho}(-k, -p, p + k) = g_0 f_{lkj} \left[g_{\rho\sigma}(k + 2p)_v + g_{\sigma v}(k - p)_\rho + g_{v\rho}(-2k - p)_\sigma \right]. \tag{15.120b}$$

Remembering the symmetry factor of 1/2 associated with the gluon loop [cf. Fig. 14.9], the contribution of Fig. 15.15(c) to the gluon self-energy is given by

$$ig_0^2\Pi_{il}^{\mu\nu}(c) = \delta_{il}\frac{3}{2}\frac{g_0^2}{(2\pi)^4}\int d^4p\,\frac{N^{\mu\nu}(p,k)}{\left[(p+k)^2+i\varepsilon\right][k^2+i\varepsilon]}, \tag{15.121a}$$

where we have evaluated the color factor using Eq. (15.66) and (15.67), and where

$$N^{\mu\nu}(p,k) = V_{\rho\sigma}^{\mu}(k,-k-p,p)V^{\nu\sigma\rho}(-k,-p,p+k). \tag{15.121b}$$

On extending to $D = 4 - \eta$ dimensions, substituting Eqs. (15.120) into Eq. (15.121b), and using $g_\rho^\rho = \delta_\rho^\rho = D$, these equations become

$$ig_0^2\Pi_{il}^{\mu\nu}(c) = \frac{3\tilde{g}_0^2}{2}\delta_{il}\frac{\mu^\eta}{(2\pi)^D}\int d^Dp\,\frac{N^{\mu\nu}(p,k)}{\left[(p+k)^2+i\varepsilon\right][k^2+i\varepsilon]}, \tag{15.122a}$$

where

$$\begin{aligned} N^{\mu\nu}(p,k) &= g^{\mu\nu}\left[(2k+p)^2+(p-k)^2\right]+p^\mu p^\nu(4D-6)\\ &\quad +k^\mu k^\nu(D-6)+(p^\mu k^\nu+p^\nu k^\mu)(2D-3). \end{aligned} \tag{15.122b}$$

Eq. (15.122a) in of the same form as Eq. (15.115), and can be evaluated in the same way, except that there are more terms to keep track of in Eq. (15.122b) than in Eq. (15.116). In this way, we finally obtain

$$ig_0^2\Pi_{il}^{\mu\nu}(c) = \delta_{il}\left(11k^\mu k^\nu-\frac{19}{2}k^2g^{\mu\nu}\right)\left(\frac{-i\tilde{g}_0^2}{32\pi^2}\right)\left[\frac{2}{\eta}-\gamma+\ln 4\pi\right]+\cdots \tag{15.80c}$$

as the contribution of the gluon loop to the gluon self-energy, where the dots, as usual, represent non-singular terms whose form is not required.

15.6.2   The quark–gluon vertex corrections

The quark–gluon vertex corrections are given, in lowest order, by the two graphs in Figs. 15.17(a, b). We shall discuss the contribution of each in turn.

The first diagram, Fig. 15.17(a), is closely analogous to the QED vertex correction, Fig. 9.2(d). On applying the Feynman rules of Section 14.4, it gives rise to a contribution

$$ig_0^3\Lambda_{(a)}^{i,\mu}(p',p) = ig_0^3 C_{(a)}\Lambda^\mu(p',p), \tag{15.123}$$

where $\Lambda^\mu(p',p)$ is given by Eq. (9.48) and the colour factor

$$C_{(a)} = T_{cb}^j T_{ba}^i T_{ad}^j = -\frac{1}{6}T_{cd}^i \tag{15.124}$$

using Eq. (15.73). The integral $\Lambda^\mu(p',p)$ was evaluated in Section 10.5 using dimensional regularization, and combining Eqs. (10.63) and (10.65) gives

$$\Lambda^\mu(p',p) = \mu^{-\eta}\frac{\gamma^\mu}{16\pi^2}\left[\frac{2}{\eta}-\gamma+\ln 4\pi\right]+\cdots$$

for the divergent contribution. Hence, from Eqs. (15.123) and (15.124) we obtain

$$g_0^2 \Lambda^{i,\mu}_{(a)}(p',p) = -\tilde{g}_0^2 \frac{\gamma^\mu T^i}{96\pi^2} \left[\frac{2}{\eta} - \gamma + \ln 4\pi \right] + \cdots, \tag{15.90a}$$

where we have suppressed the quark indices and the dots represent finite terms whose form is not required.

We next calculate the corresponding contribution arising from Fig. 15.15(b). Using the Feynman rules of Section 14.4, this is given by

$$ig_0^3 \Lambda^{i,\mu}_{(b)}(p',p) = \int \frac{d^4 k}{(2\pi)^4} g_0 f_{inm} V^{\mu\tau\nu}(p'-p, k-p', p-k)$$

$$\times \frac{-i}{(p-k)^2 + i\varepsilon} \frac{-i}{(p'-k)^2 + i\varepsilon} (-ig_0) T^n_{ca} \gamma_\tau \frac{i(k+m_0)}{k^2 - m^2 + i\varepsilon} (-ig_0) T^m_{ad} \gamma_\nu, \tag{15.125}$$

where

$$V^{\mu\tau\nu}(p'-p, k-p', p-k)$$
$$= g^{\tau\nu}(p+p'-2k)^\mu + g^{\mu\tau}(k+p)^\nu + g^{\mu\nu}(p'-2p+k)^\tau \tag{15.126}$$

and we have explicitly displayed the terms corresponding to each feature of the Feynman graph. On grouping the various factors together in a more convenient way, Eq. (15.125) can be written in the form

$$ig_0^3 \Lambda^{i,\mu}_{(b)}(p',p)$$
$$= ig_0^3 C_{(b)} \int \frac{d^4 k}{(2\pi)^4} \frac{N^\mu(p',p,k)}{\left[(p-k)^2 + i\varepsilon\right]\left[(p'-k)^2 + i\varepsilon\right][k^2 - m^2 + i\varepsilon]}, \tag{15.127}$$

where

$$N^\mu(p',p,k) = V^{\mu\tau\nu}(p-p', k-p', p-k)\gamma_\tau(k+m)\gamma_\nu \tag{15.128}$$

and the colour factor

$$C_{(b)} = -f_{inm} T^n_{ca} T^m_{ad} = -\frac{3}{2} i T^i_{cd}, \tag{15.129}$$

where we have used Eqs. (15.66) and (15.72).

Eq. (15.127) is evaluated using the standard techniques of dimensional regularization. On generalizing to $4 - \eta$ dimensions and using Eq. (15.129), it can be written in the form

$$ig_0^3 \Lambda^{i,\mu}_{(b)}(p',p) = \frac{3}{2} g_0^3 T^i_{cd} I^\mu, \tag{15.130a}$$

where

$$g_0^2 I^\mu = \tilde{g}_0^2 \mu^\eta \int \frac{d^D k}{(2\pi)^D} \frac{N^\mu(p',p,k)}{\left[(p'-k)^2 - \lambda^2 + i\varepsilon\right]\left[(p-k)^2 - \lambda^2 + i\varepsilon\right][k^2 - m^2 + i\varepsilon]} \tag{15.130b}$$

and we have introduced an 'infrared cut-off' λ to guard against divergences arising when the internal gluon four-momenta go to zero. Eq. (15.130) is of the same form as Eq. (10.55), albeit with a different $N^{\mu\nu}(p',p,k)$, and its divergent part can be evaluated using the same steps which led from Eq. (10.55) to Eq. (10.65). When this is done, one obtains

$$g_0^2 \Lambda_{(b)}^{i,\mu}(p',p) = \gamma^\mu T_{cd}^i \frac{9\tilde{g}_0^2}{32\pi^2}\left[\frac{2}{\eta} - \gamma + \ln 4\pi\right] + \dots \tag{15.90b}$$

for the divergent contribution arising from Fig. 15.17(b). This is the result quoted in Section 15.4.3, albeit with the quark color indices c, d suppressed.

Problems

15.1. Find a relation between the on-shell charge e and the running charge $e_r(\mu)$ to second order. For what value of μ does $e_r(\mu) = e$?

15.2. Derive Eq. (15.80c) for the gluon-loop contribution to the gluon self-energy, starting from Eq. (15.121).

15.3. Derive Eq. (15.90b) for the vertex correction arising from Fig. 15.17(b), starting from Eq. (15.130).

16

Weak Interactions

The rest of this book is devoted to the theory of weak interactions. It follows on directly from Chapter 11, and a knowledge of the content of Chapters 12–15, inclusive, is not required in what follows.

16.1   Introduction

Like the strong and electromagnetic interactions, the weak interaction is also associated with the exchange of elementary spin-1 bosons that act as 'force carriers,' between quarks and/or leptons. However, in contrast to photons and gluons, these are very massive particles. There are three such 'intermediate vector bosons': the charged bosons W^+ and W^- and the neutral Z^0 boson, with masses

$$M_W = 80.40 \, \text{GeV} \qquad M_Z = 91.19 \, \text{GeV}. \tag{16.1}$$

These large masses have two obvious consequences.

Firstly, very general arguments show that the range of any force is given by the Compton wavelength of the particles transmitting it.[1] Consequently, the range of the weak interactions between quarks and leptons is of order 10^{-3}fm and at low energies the weak interaction can be treated as a zero-range interaction, as discussed in Section 16.6.1 below.

Secondly, very large energies are required to produce $W^\pm$ and Z^0 bosons in the laboratory, so that they were not discovered until 1983, long after their existence and masses had been theoretically predicted.[2]

---

[1] See e.g. B.R. Martin and G. Shaw, *Particle Physics*, 3rd edn., John Wiley & Sons Ltd, 2008, Section 1.4.
[2] For an account of the discovery of neutral currents and of the $W^\pm$ and Z^0 bosons, see, for example, Martin and Shaw, loc.cit., in particular Sections 8.1 and 4.5.1, respectively.

Quantum Field Theory, Second Edition   Franz Mandl and Graham Shaw
© 2010 John Wiley & Sons, Ltd

Figure 16.1   *Mechanism for neutral current reactions of the type* $v_\mu + N \rightarrow v_\mu + X$, *where X represents any set of hadrons allowed by the conservation laws*

The idea that weak interactions were due to the exchange of massive charged bosons seems to have been first proposed by Klein in 1938,[3] and until 1973 all observed weak interactions were consistent with the hypothesis that they were mediated by the exchange of heavy charged bosons $W^\pm$ only. However, in the 1960s, Glashow, Salam and Weinberg developed a theory that unified electromagnetic and weak interactions in a way that is often compared to the unification of electric and magnetic interactions by Faraday and Maxwell a century earlier. This new theory made several remarkable predictions, including the existence of a neutral vector boson Z^0, and of weak reactions arising from its exchange. These processes are called *neutral-current* reactions to distinguish them from the so-called *charged-current* reactions arising from charged $W^\pm$ boson exchange. In particular, neutral current reactions of the type

$$v_\mu + N \rightarrow v_\mu + X$$

were predicted to occur via the mechanism of Figure 16.1, where N is a nucleon and X is any set of hadrons allowed by the conservation laws. Such reactions were finally observed at CERN in 1973.

The prediction of the existence and properties of neutral currents, prior to their discovery, is only one of many spectacular successes of the unified theory of electromagnetic and weak interactions. Others include the prediction of the existence of the charmed quark, prior to its discovery in 1974; and the prediction of the masses of the $W^\pm$ and Z^0 bosons prior to the long-awaited detection of these particles in 1983. In general, the theory is in complete agreement with all the data on both weak and electromagnetic interactions, which are now usually referred to collectively as the *electroweak interaction*. However the new unification only becomes manifest at very high energies, and at lower energies, weak and electromagnetic interactions can still be clearly separated, as we shall see.

In this chapter, we shall develop and apply the intermediate vector boson (IVB) theory, which takes account of $W^\pm$ exchange only. This theory is the forerunner and basis of the modern theory, and describes many processes successfully. However, we shall see that it also leads to serious difficulties, which will be resolved by the formulation of the unified electroweak theory in Chapters 17–19.

---

[3] O. Klein, Proceedings of the Symposium on Les Nouvelles Theories de la Physique, Warsaw, 1938; Nature, **161** (1948) 897.

Before proceeding, we note that it is conventional to divide weak interaction processes into three categories, depending on whether leptons and/or hadrons are involved. Accordingly, we distinguish: (i) purely leptonic processes like

$$\mu^- \rightarrow e^- + \overline{\nu}_e + \nu_\mu, \tag{16.2}$$

which involve only the charged leptons $e^\pm, \mu^\pm, \tau^\pm$ and their associated neutrinos ν_e, ν_μ, ν_τ and antineutrinos $\overline{\nu}_e, \overline{\nu}_\mu, \overline{\nu}_\tau$; (ii) semileptonic processes involving both hadrons and leptons, like neutron β-decay

$$n \rightarrow p + e^- + \overline{\nu}_e; \tag{16.3}$$

and (iii) purely hadronic processes like the Λ-decay

$$\Lambda \rightarrow p + \pi^-. \tag{16.4}$$

Unfortunately we have only a limited understanding of hadron structure, since the strong forces which bind the quarks act over distances of order 1 fm, where they cannot be treated in perturbation theory. In contrast, we believe that perturbation theory in valid for weak and electromagnetic interactions, and that we understand the latter. Consequently purely leptonic processes afford an unambiguous and far simpler field for studying weak interactions, and we shall restrict ourselves to purely leptonic processes. This is analogous to our treatment of QED, where we also did not consider hadrons.[4]

16.2 Leptonic Weak Interactions

The electron–positron field enters the QED interaction in the bilinear combination of the electromagnetic current, i.e.

$$\mathscr{H}_{\text{QED}}(x) = -e\overline{\psi}(x)\gamma^\alpha\psi(x)A_\alpha(x). \tag{16.5}$$

The weak interaction Hamiltonian density responsible for leptonic processes is similarly constructed from bilinear forms of the lepton field operators. The experimental data on a wide range of leptonic and semileptonic processes are consistent with the assumption that the lepton fields enter the interaction only in the combinations

$$J_\alpha(x) = \sum_l \overline{\psi}_l(x)\gamma_\alpha(1-\gamma_5)\psi_{\nu_l}(x) \tag{16.6a}$$

$$J_\alpha^\dagger(x) = \sum_l \overline{\psi}_{\nu_l}(x)\gamma_\alpha(1-\gamma_5)\psi_l(x). \tag{16.6b}$$

In Eqs. (16.6), l labels the various charged lepton fields, $l = e, \mu, \ldots$, and ν_l the corresponding neutrino fields. ψ_l and ψ_{ν_l} are the corresponding quantized fields. ψ_l is linear in the absorption operators of the l^- leptons and in the creation operators of the l^+ leptons, etc. In analogy to the electromagnetic current in Eq. (16.5), one calls $J_\alpha(x)$ and $J_\alpha^\dagger(x)$

---

[4] For an excellent account of all aspects of weak interactions, see D. Bailin, *Weak Interactions*, 2nd edn., Adam Hilger, Bristol, 1982.

leptonic currents, since they too transform like vectors under continuous Lorentz transformations and imply certain lepton number conservation laws, as we shall see.

There is of course no unique way of constructing the leptonic interaction from the currents (16.6). In analogy with our description of the electromagnetic interaction as being transmitted by photons, we would like to describe the weak interactions as due to the transmission of quanta. These are called W particles. The QED interaction (16.5) then suggests the leptonic interaction of the IVB theory

$$\mathscr{H}_I(x) = g_W J^{\alpha\dagger}(x) W_\alpha(x) + g_W J^\alpha(x) W_\alpha^\dagger(x), \tag{16.7}$$

where g_W is a dimensionless coupling constant and the field $W_\alpha(x)$ describes the W particles. With this interaction, processes such as the muon decay process (16.2) or the neutrino scattering process

$$\nu_e + e^- \rightarrow \nu_e + e^-$$

are described, in lowest-order perturbation theory, by the Feynman graphs in Figs. 16.2 and 16.3. In each case the interaction between the two leptonic currents is brought about through the exchange of one W particle, analogously to the one-photon exchange of electron–electron scattering, Fig. 7.14.

We shall now study the interaction (16.7) in more detail. We note, first of all, that this interaction couples the field $W_\alpha(x)$ to the leptonic vector current. Hence, it must be a vector field, and the W particles are vector bosons with spin 1.[5] Since each term in the leptonic currents (16.6) (i.e. each vertex in a Feynman graph) involves a charged and a neutral lepton, the W particles are electrically charged and the $W(x)$ field is non-Hermitian.

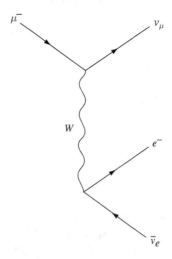

Figure 16.2   *The leading contribution to muon decay, $\mu^- \rightarrow e^- + \bar{\nu}_e + \nu_\mu$, in the IVB theory*

---

[5] We shall see in Section 16.3 that $W_\alpha(x)$ satisfies the Lorentz condition $\partial^\alpha W_\alpha(x) = 0$. Consequently, the field $W_\alpha(x)$ possesses only three independent states of polarization and describes spin-1 particles.

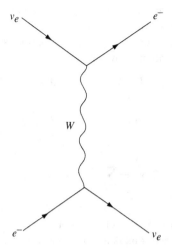

Figure 16.3   *The leading contribution to the scattering process $v_e + e^- \to v_e + e^-$ in the IVB theory*

In Section 7.4 we introduced lepton numbers

$$N(e) = N(e^-) - N(e^+),\tag{16.8}$$

etc., and saw that these are conserved in electromagnetic processes. It is obvious that the interaction (16.7) does not conserve these numbers, nor indeed are they conserved in the reactions (16.2)–(16.4). However, if we modify the definition of lepton numbers to

$$\left.\begin{array}{l} N(e) = N(e^-) - N(e^+) + N(v_e) - N(\bar{v}_e) \\ N(\mu) = N(\mu^-) - N(\mu^+) + N(v_\mu) - N(\bar{v}_\mu) \\ N(\tau) = N(\tau^-) - N(\tau^+) + N(v_\tau) - N(\bar{v}_\tau) \end{array}\right\}\tag{16.9}$$

then the currents $J_\alpha(x)$ and $J_\alpha^\dagger(x)$, Eqs. (16.6), do conserve lepton numbers. For example, the term

$$\bar{\psi}_e(x)\gamma_\alpha(1 - \gamma_5)\psi_{v_e}(x)$$

in Eq. (16.6a) is linear in electron creation and positron absorption operators, and in v_e absorption and $\bar{v}_e$ creation operators. It follows that any interaction built up from these leptonic currents, as the interaction (16.7) is, conserves lepton numbers. This is in agreement with experiment, where lepton number conservation is found to hold for all processes, whatever particles or interactions are involved. For example, in bombarding nuclei with muon neutrinos, the process

$$v_\mu + (Z, A) \to (Z + 1, A) + \mu^-\tag{16.10}$$

is allowed by lepton number conservation, while

$$v_\mu + (Z, A) \to (Z - 1, A) + \mu^+\tag{16.11}$$

is forbidden. No forbidden processes, such as (16.11), have been observed, within very small upper bounds, in experiments especially designed to detect them.

The interaction (16.7) is known as a 'V–A' interaction, since the current $J^\alpha(x)$, from which it is built up, can be written as the difference

$$J^a(x) = J_V^\alpha(x) - J_A^\alpha(x) \tag{16.12a}$$

of the vector current

$$J_V^\alpha(x) = \sum_l \overline{\psi}_l(x)\gamma^\alpha\psi_{\nu_l}(x) \tag{16.12b}$$

and the axial vector current

$$J_A^\alpha(x) = \sum_l \overline{\psi}_l(x)\gamma^\alpha\gamma_5\psi_{\nu_l}(x). \tag{16.12c}$$

Under the parity transformation $(\mathbf{x}, t) \to (-\mathbf{x}, t)$, $J_V^\alpha(x)$ changes sign, while $J_A^\alpha(x)$ does not.[6] Hence, the interaction (16.7) is clearly not invariant under spatial inversion, and parity is not conserved. (Indeed, this is not peculiar to purely leptonic processes, but is a characteristic of all weak interactions.) This has its most striking consequences for the neutrinos. Assume, first of all, that the neutrinos have zero mass. In this case we know from Appendix A, Eq. (A.43), that $(1 - \gamma_5)/2$ is a helicity projection operator. Since $\psi_{\nu_l}(x)$ is linear in neutrino absorption operators and in antineutrino creation operators, it follows from Eqs. (A.40) that the operator

$$\psi_{\nu_l}^L(x) \equiv \frac{1}{2}(1 - \gamma_5)\psi_{\nu_l}(x), \tag{16.13}$$

which occurs in the interaction (16.6)–(16.7) can annihilate only negative helicity neutrinos and create only positive helicity antineutrinos. Hence, in weak interactions, only these states would play a role, and positive helicity neutrinos and negative helicity antineutrinos would not partake in weak interaction processes.

The operators

$$\left.\begin{matrix} P_L \\ P_R \end{matrix}\right\} = \frac{1}{2}(1\pm\gamma_5) \tag{16.14}$$

are of course always projection operators (since $P_L^2 = P_L$ and $P_R = 1 - P_L$), independently of the particle mass m of the fermion field. However, for particles of non-zero mass m, the states projected out by P_L and P_R are helicity eigenstates only in the high-energy limit in which the particle energy is very large compared to the particle mass m. For neutrinos, this will always be a very good approximation in what follows, even though their masses are not precisely zero. However, it may also be a good approximation for high-energy charged leptons. In analogy to Eq. (16.13), we define the 'left-handed' charged lepton fields

$$\psi_l^L(x) \equiv P_L\psi_l(x) = \frac{1}{2}(1 - \gamma_5)\psi_l(x). \tag{16.15}$$

---

[6] See Appendix A, Eqs. (A.53) and Problem A.1.

One easily shows that the leptonic current (16.6a) can be written

$$J_\alpha(x) = 2 \sum_l \overline{\psi}_l^{\text{L}}(x) \gamma_\alpha \psi_{\nu_l}^{\text{L}}(x) \tag{16.16}$$

so that, as for the neutrinos, only the left-handed fields are involved for the charged leptons. Thus, if the electrons (positrons) emitted in a weak-interaction process are highly relativistic, they will have negative (positive) helicity. For example, this will be the case for most of the electrons and positrons in the muon decay processes

$$\mu^- \to e^- + \overline{\nu}_e + \nu_\mu, \quad \mu^+ \to e^+ + \nu_e + \overline{\nu}_\mu. \tag{16.17}$$

16.3   The Free Vector Boson Field

The simplest equation for a vector field $W^\alpha(x)$, describing particles of mass m_W and spin 1, is the Proca equation

$$\Box W^\alpha(x) - \partial^\alpha \left(\partial_\beta W^\beta(x) \right) + m_W^2 W^\alpha(x) = 0. \tag{16.18}$$

On taking the divergence of this equation, one automatically obtains the Lorentz condition

$$\partial_\alpha W^\alpha(x) = 0 \tag{16.19}$$

for $m_W \neq 0$. This is in contrast to the photon case, where the Lorentz condition must be imposed as a subsidiary condition. On account of Eq. (16.19), the Proca equation (16.18) reduces to

$$\Box W^\alpha(x) + m_W^2 W^\alpha(x) = 0. \tag{16.20}$$

A free-field Lagrangian density which leads to Eq. (16.18) is

$$\mathscr{L}(x) = -\frac{1}{2} F_{W\alpha\beta}^\dagger(x) F_W^{\alpha\beta}(x) + m_W^2 W_\alpha^\dagger(x) W^\alpha(x), \tag{16.21a}$$

where

$$F_W^{\alpha\beta}(x) \equiv \partial^\beta W^\alpha(x) - \partial^\alpha W^\beta(x). \tag{16.21b}$$

We are taking $W^\alpha(x)$ and $F_W^{\alpha\beta}(x)$ as non-Hermitian fields, since the W bosons are electrically charged particles.

To establish the connection with the particle description, we expand the W field in the usual way in a complete set of plane waves:

$$W^\alpha(x) = W^{\alpha+}(x) + W^{\alpha-}(x), \tag{16.22a}$$

where

$$W^{\alpha+}(x) = \sum_{\mathbf{k}} \sum_r \left(\frac{1}{2V\omega_\mathbf{k}} \right)^{1/2} \varepsilon_r^\alpha(\mathbf{k}) a_r(\mathbf{k}) e^{-ikx} \tag{16.22b}$$

$$W^{\alpha-}(x) = \sum_{\mathbf{k}} \sum_r \left(\frac{1}{2V\omega_\mathbf{k}} \right)^{1/2} \varepsilon_r^\alpha(\mathbf{k}) b_r^\dagger(\mathbf{k}) e^{ikx} \tag{16.22c}$$

and

$$\omega_{\mathbf{k}} \equiv \left(m_W^2 + \mathbf{k}^2\right)^{1/2}. \tag{16.23}$$

[Eqs. (16.22) are analogous to Eqs. (3.26a) and (5.16) for charged scalar mesons and photons.] The vectors $\varepsilon_r^\alpha(\mathbf{k})$, $r = 1,\ 2,\ 3$, are a complete set of real orthonormal polarization vectors, i.e.

$$\varepsilon_r^\alpha(\mathbf{k})\varepsilon_s^\alpha(\mathbf{k}) = -\delta_{rs}, \tag{16.24}$$

and the Lorentz condition (16.19) implies the conditions

$$k_\alpha \varepsilon_r(\mathbf{k}) = 0 \tag{16.25}$$

for the polarization vectors. In the frame in which $k = (\omega_{\mathbf{k}}, 0, 0, |\mathbf{k}|)$, a suitable choice of polarization vectors is

$$\left.\begin{aligned}
\varepsilon_1(\mathbf{k}) &= (0, 1, 0, 0) \\
\varepsilon_2(\mathbf{k}) &= (0, 0, 1, 0) \\
\varepsilon_3(\mathbf{k}) &= (|\mathbf{k}|, 0, 0, \omega_{\mathbf{k}})/m_W
\end{aligned}\right\}. \tag{16.26}$$

The completeness relation

$$\sum_{r=1}^{3} \varepsilon_r^\alpha(\mathbf{k})\varepsilon_r^\beta(\mathbf{k}) = -g^{\alpha\beta} + k^\alpha k^\beta / m_W^2 \tag{16.27}$$

follows directly from Eqs. (16.26).

It should be self-evident to the reader that quantization of the Lagrangian (16.21) by means of the canonical formalism allows us to interpret $a_r(\mathbf{k})$ and $b_r(\mathbf{k})$ as annihilation operators of vector mesons of mass m_W, momentum $\mathbf{k}$ and polarization vector $\varepsilon_r(\mathbf{k})$, with $a_r^\dagger(\mathbf{k})$ and $b_r^\dagger(\mathbf{k})$ the corresponding creation operators. In order that the interaction (16.7) implies conservation of electric charge, we shall want to associate $a_r(\mathbf{k})$ and $a_r^\dagger(\mathbf{k})$ with positively charged bosons (W^+), and $b_r(\mathbf{k})$ and $b_r^\dagger(\mathbf{k})$ with W^- bosons.

Lastly, we require the W boson propagator

$$\langle 0|T\{W^\alpha(x)W^{\beta\dagger}(y)\}|0\rangle \equiv iD_F^{\alpha\beta}(x - y, m_W), \tag{16.28}$$

where we have explicitly shown the dependence on the mass m_W. This propagator is derived by the same method by which the scalar meson propagator was obtained in Section 3.4. We shall omit the derivation[7] and only quote the result. With

$$iD_F^{\alpha\beta}(x, m_W) = \frac{1}{(2\pi)^4} \int d^4k\, e^{-ikx} iD_F^{\alpha\beta}(k, m_W), \tag{16.29}$$

---

[7] An explicit derivation is given on pp. 96–97 of D. Bailin's *Weak Interactions*, quoted in Section 16.1. The relation of the W propagator to the photon propagator is discussed in C. Itzykson and J. B. Zuber, *Quantum Field Theory*, McGraw-Hill, New York, 1980, Section 3-2-3.

the W propagator in momentum space is given by

$$iD_F^{\alpha\beta}(k, m_W) = \frac{i(-g^{\alpha\beta} + k^\alpha k^\beta/m_W^2)}{k^2 - m_W^2 + i\varepsilon}.$$

(α) k (β) (16.30)

16.4 The Feynman Rules for the IVB Theory

We now easily write down the Feynman rules for treating leptonic processes in perturbation theory.

The basic vertex part which arises from the interaction (16.7) is shown in Fig. 16.4. It consists of two lepton lines (representing a lepton-number-conserving current) and a W boson line. The only restrictions at the vertex are that lepton numbers and electric charge are conserved. Hence Fig. 16.4 stands for many different processes, for example those in Fig. 16.5, where in each diagram there may occur absorption or emission of a $W^\pm$ boson of the appropriate sign to conserve charge. Substituting the leptonic interaction (16.7) in the S-matrix expansion (6.23), one sees that corresponding to Feynman rule 1 of QED (see Section 7.3) we now obtain the Feynman rule:

10. For each vertex, write a factor

$$-ig_W\gamma^\alpha(1 - \gamma_5). \tag{16.31}$$

The Feynman rules for electron propagators and external lines, given in Section 7.3, only require trivial relabeling to apply to all leptons. In the Feynman rule 3 for the lepton propagator (7.48), m now stands for the mass m_l or $m_{v_1} (l = e, \mu \ldots)$ of the lepton involved. In the Feynman rule 4 for external line factors, Eqs. (7.49a) and (7.49b) become the factors appropriate to negative leptons l^- and neutrinos v_l, and Eqs.

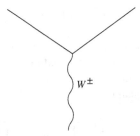

Figure 16.4 *The basic vertex part of the IVB interaction*

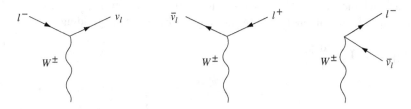

Figure 16.5 *Three particular cases of the basic vertex diagram in Figure 16.4*

(7.49c) and (7.49d) the factors appropriate to the corresponding antiparticles l^+ and $\bar{v}_l$. Hence we only need the rules for internal and external W lines, corresponding to rules 2 and 4 for photons. These are:

11. For each internal W boson line, labelled by the momentum k, write a factor

$$
iD_{F\alpha\beta}(k, m_W) = \frac{i\left(-g_{\alpha\beta} + k_\alpha k_\beta/m_W^2\right)}{k^2 - m_W^2 + i\varepsilon}. \qquad \text{(}\alpha\text{)} \qquad k \qquad \text{(}\beta\text{)} \qquad (16.30)
$$

12. For each external initial or final W boson line, write a factor

$$
\left. \begin{array}{l} \qquad k \quad \text{(}\alpha\text{)} \quad \text{(initial)} \\ \varepsilon_{r\alpha}(\mathbf{k}) \\ \qquad \text{(}\alpha\text{)} \quad k \qquad \text{(final)} \end{array} \right\}. \qquad (16.32)
$$

As for photons, if complex polarization vectors are used, we must write $\varepsilon^*_{r\alpha}(\mathbf{k})$ for a final-state W boson, instead of $\varepsilon_{r\alpha}(\mathbf{k})$.

16.5 Decay Rates

Before calculating the rate for muon decay, we first derive the general expression for the decay rate of any process in terms of the corresponding Feynman amplitude. We consider a particle P decaying into N particles $P'_1, P'_2, \ldots P'_N$:

$$
P \rightarrow P'_1 + P'_2 + \ldots + P'_N. \qquad (16.33)
$$

The relation (8.1) between the S-matrix element and the Feynman amplitude then reduces to

$$
S_{fi} = \delta_{fi} + (2\pi)^4 \delta^{(4)}\left(\sum P'_f - p\right) \frac{1}{(2VE)^{1/2}} \prod_f \left(\frac{1}{2VE'_f}\right)^{1/2} \prod_l (2m_l)^{1/2} \mathscr{M}, \qquad (16.34)
$$

where $p = (E, \mathbf{p})$ and $p'_f = \left(E'_f, \mathbf{p}'_f\right)$, $f = 1, \ldots, N$, are the four-momenta of the initial and final particles, and the index l runs over all external fermions in the process. From Eq. (16.34) one obtains for the decay rate w (i.e. the transition probability per unit time) from a given initial to a specific final quantum state

$$
w = (2\pi)^4 \delta^{(4)}\left(\sum P'_f - p\right) \frac{1}{2E} \left(\prod_f \frac{1}{2VE'_f}\right) \left(\prod_l (2m_l)\right) |\mathscr{M}|^2, \qquad (16.35)
$$

analogously to Eq. (8.6). The differential decay rate $d\Gamma$ for the process (16.33) to final states in which the particle P'_1 has momentum in the range $d^3\mathbf{p}'_1$ at $\mathbf{p}'_1$, etc., is obtained by multiplying w by the number of these states, given by Eq. (8.7), yielding

$$
d\Gamma = w \prod_f \frac{V d^3\mathbf{p}'_f}{(2\pi)^3}
$$

$$
= (2\pi)^4 \delta^{(4)}\left(\sum P'_f - p\right) \frac{1}{2E} \left(\prod_l (2m_l)\right) \left(\prod_f \frac{d^3\mathbf{p}'_f}{(2\pi)^3 2E'_f}\right) |\mathscr{M}|^2. \qquad (16.36)
$$

This equation is our general result. To take the analysis further, we must consider a specific process. The energy–momentum conserving δ-function in Eq. (16.36) is eliminated in the usual way by integrating over the appropriate final-state variables. Eq. (16.36) gives the transition rate between definite initial and final spin states of all particles. To obtain the total decay rate Γ for the process (16.33) we must sum (16.36) over all final spin states and integrate over all final-state momenta.

So far we have considered only the decay mode (16.33) of the particle P. In general, there may be several decay modes. The branching ratio B for the decay mode (16.33) with decay rate Γ is defined by

$$B = \Gamma / \sum \Gamma, \tag{16.37}$$

where $\sum \Gamma$ is the sum of the decay rates over all decay modes, i.e. $\sum \Gamma$ is the total decay rate. (It is also called the total decay width.) The lifetime of the particle P is then given by

$$\tau = \frac{1}{\sum \Gamma} = \frac{B}{\Gamma}. \tag{16.38}$$

16.6   Applications of the IVB Theory

In this section we shall apply the IVB theory to three leptonic processes: (i) muon decay, (ii) neutrino scattering and (iii) the leptonic decay of the W boson. We shall give the analysis of the muon decay in some detail, which should enable the reader to perform similar calculations, and we shall therefore treat the two other cases much more concisely.

16.6.1   Muon decay

As our first application we shall consider one of the best studied leptonic processes, the muon decay

$$\mu^-(p, r) \rightarrow e^-(p', r') + \bar{\nu}_e(q_1, r_1) + \nu_\mu(q_2, r_2), \tag{16.39}$$

where p is the four-momentum of the muon and r labels its spin, etc. In lowest order of perturbation theory, this process is represented by the Feynman graph in Fig. 16.6. The corresponding Feynman amplitude follows from the Feynman rules and is given by

$$\begin{aligned}
\mathcal{M} = &-g_W^2 [\bar{u}(\mathbf{p}')\gamma^\alpha(1 - \gamma_5)v(\mathbf{q}_1)] \\
&\times \frac{i(-g_{\alpha\beta} + k_\alpha k_\beta / m_W^2)}{k^2 - m_W^2 + i\varepsilon} [\bar{u}(\mathbf{q}_2)\gamma^\beta(1 - \gamma_5)u(\mathbf{p})],
\end{aligned} \tag{16.40}$$

where we suppressed the spin indices and

$$k = p - q_2 = p' + q_1. \tag{16.41}$$

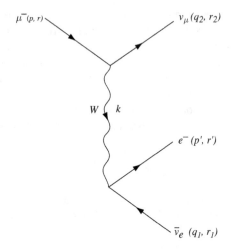

Figure 16.6 *Muon decay*

In the limit $m_W \to \infty$, the Feynman amplitude (16.40) reduces to

$$\mathcal{M} = -\frac{iG}{\sqrt{2}} [\bar{u}(\mathbf{p}')\gamma^\alpha(1 - \gamma_5)v(\mathbf{q}_1)][\bar{u}(\mathbf{q}_2)\gamma_\alpha(1 - \gamma_5)u(\mathbf{p})] \qquad (16.42)$$

where G is defined by

$$\frac{G}{\sqrt{2}} = \left(\frac{g_W}{m_W}\right)^2. \qquad (16.43)$$

The same expression would have been obtained for the Feynman amplitude of this process if we had calculated it in first-order perturbation theory from the interaction

$$\mathcal{H}_I^{(F)}(x) = \frac{G}{\sqrt{2}} J^\alpha(x) J_\alpha^\dagger(x). \qquad (16.44)$$

Eq. (16.44) represents a contact interaction of four lepton fields. Such a contact interaction was first proposed by Fermi in 1934 to describe the nuclear β-decay process (16.3) and G is known as the Fermi weak-interaction coupling constant. The Feynman diagram representing the amplitude (16.42) is shown in Fig. 16.7.

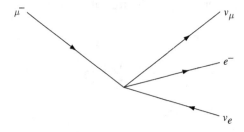

Figure 16.7   *The Feynman diagram describing muon decay in terms of the contact interaction (16.44)*

For large but finite values of m_W, the amplitude (16.40), calculated from the IVB interaction, differs from the amplitude (16.42), calculated from the contact interaction, by terms of order $(m_\mu/m_W)^2$, i.e. of order 10^{-6} for the W mass (16.1). The corresponding decay rates differ by terms of the same order. We conclude that our picture of muon decay is the same in both modes of description. This is typical of low-energy processes where, in effect, the propagator (16.30) is replaced by $ig_{\alpha\beta}/m_W^2$. Consequently, we shall use the amplitude (16.42) in calculating the muon decay rate.

The differential decay rate for the muon decay is obtained from the general expression (16.36) and is given by

$$d\Gamma = (2\pi)^4 \delta^{(4)}(p' + q_1 + q_2 - p)\frac{m_\mu m_e m_{\nu_e} m_{\nu_\mu}}{E}$$
$$\times \frac{1}{(2\pi)^9}\frac{d^3\mathbf{p}'}{E'}\frac{d^3\mathbf{q}_1}{E_1}\frac{d^3\mathbf{q}_2}{E_2}|\mathcal{M}|^2 \tag{16.45}$$

where $p \equiv (E, \mathbf{p}), p' \equiv (E', \mathbf{p}')$ and $q_i \equiv (E_i, \mathbf{q}_i), i = 1, 2$.

To obtain the total decay rate from Eq. (16.45), we must sum over all final spin states and integrate over all final momenta.

We first deal with the spins. The lifetime of the muon is, of course, independent of its spin state. Hence, we shall also average over the spin states of the initial muon, in order to express the result as a trace.[8] Using the standard techniques, developed in Section 8.2, to sum over final spin states and average over initial spin states, one obtains

$$m_\mu m_e m_{\nu_e} m_{\nu_\mu}\frac{1}{2}\sum_{\text{spins}}|\mathcal{M}|^2$$

$$= \frac{G^2}{64}\text{Tr}\left[(\not{p}' + m_e)\gamma^\alpha(1 - \gamma_5)(\not{q}_1 - m_{e_\alpha})\gamma^\beta(1 - \gamma_5)\right] \tag{16.46}$$
$$\times \text{Tr}\left[(\not{q}_2 + m_{\nu_\mu})\gamma_\alpha(1 - \gamma_5)(\not{p} + m_\mu)\gamma_\beta(1 - \gamma_5)\right]$$

$$= \frac{G^2}{64}\text{Tr}\left[\not{p}'\gamma^\alpha(1 - \gamma_5)\not{q}_1\gamma^\beta(1 - \gamma_5)\right]\text{Tr}\left[\not{q}_2\gamma_\alpha(1 - \gamma_5)\not{p}\gamma_\beta(1 - \gamma_5)\right],$$

where in the last line we have taken the limits $m_{\nu_e} \to 0, m_{\nu_\mu} \to 0$.

We evaluate the first of the traces in Eq. (16.46), i.e.

$$E^{\alpha\beta} \equiv \text{Tr}\left[\not{p}'\gamma^\alpha(1 - \gamma_5)\not{q}_1\gamma^\beta(1 - \gamma_5)\right]. \tag{16.47}$$

Using

$$[\gamma_5, \gamma^\alpha]_+ = 0, \qquad \alpha = 0, \dots, 3, \qquad (1 - \gamma_5)^2 = 2(1 - \gamma_5),$$

we obtain

$$E^{\alpha\beta} = 2p'_\mu q_{1\nu}\text{Tr}\left[\gamma^\mu\gamma^\alpha\gamma^\nu\gamma^\beta(1 - \gamma_5)\right],$$

---

[8] For zero-mass neutrinos, the emitted ν_e and ν_μ have definite helicities. By summing over the helicities of these neutrinos, leaving it to the helicity projection operators in the interaction to select the appropriate helicity states, one again ensures that the result is expressed as a trace.

and using Eqs. (A.17) and (A.21) this reduces to

$$E^{\alpha\beta} = 8p'_\mu q_1 x^{\mu\alpha\nu\beta} \qquad (16.48\text{a})$$

where

$$x^{\mu\alpha\nu\beta} \equiv g^{\mu\alpha}g^{\nu\beta} - g^{\mu\nu}g^{\alpha\beta} + g^{\mu\beta}g^{\alpha\nu} + i\varepsilon^{\mu\alpha\nu\beta}. \qquad (16.49)$$

It follows at once that the second of the traces in Eq. (16.46) is given by

$$M_{\alpha\beta} \equiv \text{Tr}\left[\not{q}_2\gamma_\alpha(1-\gamma_5)\not{p}\gamma_\beta(1-\gamma_5)\right] = 8q_2^\sigma p^\tau x_{\sigma\alpha\tau\beta}. \qquad (16.48\text{b})$$

Substituting Eqs. (16.48a) and (16.48b) into Eq. (16.46) gives

$$m_\mu m_e m_{\nu_e} m_{\nu_\mu} \frac{1}{2} \sum_{\text{spins}} |\mathcal{M}|^2 = G^2 p'_\mu q_{1\nu} x^{\mu\alpha\nu\beta} q_2^\sigma p^\tau x_{\sigma\alpha\tau\beta}. \qquad (16.50)$$

From the definition (16.49) and Eqs. (A.14c) it follows that

$$x^{\mu\alpha\nu\beta} x_{\sigma\alpha\tau\beta} = 4g_\sigma^\mu g_\tau^\nu. \qquad (16.51)$$

By means of this relation, Eq. (16.50) reduces to our final result for the spin sum

$$m_\mu m_e m_{\nu_e} m_{\nu_\mu} \frac{1}{2} \sum_{\text{spins}} |\mathcal{M}|^2 = 4G^2(pq_1)(p'q_2), \qquad (16.52)$$

in the limit $m_{\nu_e} \to 0$ and $m_{\nu_\mu} \to 0$.

Combining Eqs. (16.45) and (16.52), we obtain the unpolarized differential decay rate

$$d\Gamma = \frac{4G^2}{(2\pi)^5 E}(pq_1)(p'q_2)\delta^{(4)}(p' + q_1 + q_2 - p)\frac{d^3\mathbf{p}'}{E'}\frac{d^3\mathbf{q}_1}{E_1}\frac{d^3\mathbf{q}_2}{E_2}. \qquad (16.53)$$

We next carry out the phase space integrations, starting with integrals over the neutrino momenta, given by

$$I^{\mu\nu}(q) \equiv \int d^3\mathbf{q}_1\, d^3\mathbf{q}_2 \frac{q_1^\mu q_2^\nu}{E_1 E_2}\delta^{(4)}(q_1 + q_2 - q) \qquad (16.54)$$

where

$$q \equiv p - p'. \qquad (16.55)$$

It follows from the Lorentz covariance of the integral (16.54) that its most general form is

$$I^{\mu\nu}(q) = g^{\mu\nu}A(q^2) + q^\mu q^\nu B(q^2). \qquad (16.56)$$

From this equation it follows that

$$g_{\mu\nu}I^{\mu\nu}(q) = 4A(q^2) + q^2 B(q^2) \qquad (16.57\text{a})$$

$$q_\mu q_\nu I^{\mu\nu}(q) = q^2 A(q^2) + (q^2)^2 B(q^2). \qquad (16.57\text{b})$$

From now on we shall take the neutrino masses as zero so that $q_1^2 = q_2^2 = 0$ and, on account of the δ-function in (16.54),

$$q^2 = 2(q_1 q_2). \tag{16.58}$$

In order to find $A(q^2)$ and $B(q^2)$, we calculate the expressions on the left-hand sides of Eqs. (16.57). From Eqs. (16.54) and (16.58) we obtain

$$g_{\mu\nu}I^{\mu\nu}(q) = \frac{q^2}{2} \int \frac{d^3\mathbf{q}_1}{E_1} \frac{d^3\mathbf{q}_2}{E_2} \delta^{(4)}(q_1 + q_2 - q) \equiv \frac{1}{2} q^2 I(q^2). \tag{16.59}$$

We see from its definition that the integral $I(q^2)$ is an invariant, so that it can be evaluated in any coordinate system. We shall choose the centre-of-momentum system of the two neutrinos. In this system $\mathbf{q}_1 = -\mathbf{q}_2$, so that $\mathbf{q} = 0$, and the energy ω of either neutrino is given by

$$\omega \equiv E_1 = |\mathbf{q}_1| = E_2 = |\mathbf{q}_2|. \tag{16.60}$$

Hence,

$$I(q^2) = \int d^3\mathbf{q}_1 \frac{\delta(2\omega - q_0)}{\omega^2} = 2\pi \tag{16.61}$$

and from Eq. (16.59)

$$g_{\mu\nu}I^{\mu\nu}(q) = \pi q^2. \tag{16.62a}$$

Similarly, one finds from Eqs. (16.54), (16.58) and (16.61) that

$$q_\mu q_\nu I^{\mu\nu}(q^2) = \left(\frac{1}{2}q^2\right)^2 I = \frac{1}{2}\pi(q^2)^2. \tag{16.62b}$$

From Eqs. (16.57) and (16.62) we can find $A(q^2)$ and $B(q^2)$, and substituting these into Eq. (16.56) leads to

$$I^{\mu\nu}(q) = \frac{1}{6}\pi(g^{\mu\nu}q^2 + 2q^\mu q^\nu). \tag{16.63}$$

From Eqs. (16.63) and (16.53) we obtain the muon decay rate for emission of an electron with momentum in the range $d^3\mathbf{p}'$ at $\mathbf{p}'$:

$$d\Gamma = \frac{2\pi}{3} \frac{G^2}{(2\pi)^5 E} \frac{d^3\mathbf{p}'}{E'} [(pp')q^2 + 2(pq)(p'q)]. \tag{16.64}$$

Finally, we must integrate Eq. (16.64) over all momenta $\mathbf{p}'$ of the emitted electron. For a muon at rest, i.e. in the rest frame of the muon, we have

$$p = (m_\mu, 0), \qquad q_0 = m_\mu - E', \qquad \mathbf{q} = -\mathbf{p}', \tag{16.65}$$

and in this frame Eq. (16.64) becomes

$$\begin{aligned} d\Gamma = \frac{2\pi}{3} \frac{G^2}{(2\pi)^5 m_\mu} |\mathbf{p}'|dE' \, d\Omega' \big[& \left(m_\mu^2 + m_e^2 - 2m_\mu E'\right)m_\mu E' \\ & + 2m_\mu(m_\mu - E')(m_\mu E' - m_e^2) \big] \end{aligned} \tag{16.66}$$

where we put $d^3\mathbf{p}' = |\mathbf{p}'|E'\,dE'\,d\Omega'$. If we neglect terms of order m_e^2/m_μ^2, Eq. (16.66) reduces to

$$d\Gamma = \frac{2\pi}{3}\frac{G^2}{(2\pi)^5}m_\mu E'^2\,dE'\,d\Omega'(3m_\mu - 4E'). \tag{16.67}$$

Integrating Eq. (16.67) over all directions Ω' of the emitted electron and over its complete range of energies $0 \leq E' \leq \frac{1}{2}m_\mu$, we obtain the total decay rate

$$\Gamma = \frac{G^2 m_\mu^5}{192\pi^3}. \tag{16.68}$$

Taking $\mu^- \to e^-\bar{\nu}_e\nu_\mu$ as the only muon decay mode (the experimental branching ratio is 98.6%), we obtain the muon lifetime

$$\tau_\mu = \frac{1}{\Gamma} = \frac{192\pi^3}{G^2 m_\mu^5}. \tag{16.69}$$

Substituting the experimental values $\tau = 2.2 \times 10^{-6}$s and $m_\mu = 105.7\,\text{MeV}$ in Eq. (16.69) gives for the value of the Fermi coupling constant

$$G = 1.16 \times 10^{-5}\,\text{GeV}^{-2} \tag{16.70a}$$

Substituting this value for G into Eq. (16.43) leads to

$$\frac{g_W^2}{4\pi} = \frac{G}{4\pi\sqrt{2}}m_W^2 \approx 4 \times 10^{-3} \tag{16.71}$$

for $m_W \approx 80$ GeV. This value is comparable to that of the fine structure constant, $e^2/4\pi = 7.3 \times 10^{-3}$, so that lowest order of perturbation theory should give a good description of weak interactions.

A much more precise value of G than Eq. (16.70a) can be obtained from the most accurate current experimental value for the muon lifetime

$$\tau_\mu = (2.19703\pm0.00004) \times 10^{-6}\text{s}. \tag{16.72}$$

For this purpose one must retain the terms involving the electron mass m_e in the expression for the decay rate Γ [i.e. one must derive Γ from Eq. (16.66) instead of from Eq. (16.67)], one must include the radiative corrections of QED, and one must, of course, also allow for the 98.6% branching ratio of the $\mu^- \to e^-\bar{\nu}_e\nu_\mu$ mode. In this way one finds the value

$$G = (1.16637\pm0.00002) \times 10^{-5}\,\text{GeV}^{-2} \tag{16.70b}$$

The muon lifetime only determines the coupling constant G. A thorough test of the theory comes from comparing its predictions with detailed experiments on: (i) the energy spectrum of the emitted electrons, (ii) the energy spectrum and the angular distribution of the electrons emitted in the decay of polarized muons and (iii) the helicity of the electrons emitted from unpolarized muons. Some of these experiments are sufficiently accurate that one must again include radiative corrections in the theory. In all cases there is good agreement between theory and experiment. For example, the mean helicity of the electrons emitted in the decay of unpolarized muons is -1.008 ± 0.057, in agreement with the theoretical predictions and with the qualitative arguments we gave following Eq. (11.16).[9]

Finally, we note that evidence for the universality of the leptonic interaction (16.6)–(16.7) comes from the tauon decay

$$\tau^- \to e^- + \bar{\nu}_e + \nu_\tau. \tag{16.73}$$

The tauon life time τ_τ is at once obtained from expression (16.69) for τ_μ by replacing m_μ by m_τ and allowing for the fact that the branching ratio for the process (16.73) is $B = 0.178\pm0.001$. Using Eq. (16.38), it follows that

$$\tau_\tau = B\frac{192\pi^3}{G^2 m_\tau^5} = B\tau_\mu\left(\frac{m_\mu}{m_\tau}\right)^5, \tag{16.74}$$

and substituting for τ_μ, B and m_μ/m_τ in Eq. (16.74) leads to the prediction $\tau_\tau = 2.85 \pm 0.02 \times 10^{-13}$ s. This is in reasonable agreement with the experimental lifetime of $(2.91 \pm 0.02) \times 10^{-13}$s, given the difficulty of the experiments and the approximations made. (The quoted values of B and τ_τ are the averages of many determinations with much larger errors.)

16.6.2 Neutrino scattering

Unlike the muon decay process, one would expect neutrino scattering processes at sufficiently high energies to exhibit the effects of the intermediate vector boson. We shall illustrate this for the process

$$\nu_\mu + e^- \to \nu_e + \mu^-, \tag{16.75}$$

which is often called inverse muon decay. In lowest order of perturbation theory, this process is represented by the Feynman graph in Fig. 16.8, which also specifies the four-momenta of the particles. It is left as an exercise for the reader to show that the corresponding Feynman amplitude is given by

$$\mathcal{M} = \left\{\cdot\frac{m_W^2}{m_W^2 - k^2}\right\}(-i)\left(\frac{g_W}{m_W}\right)^2 [\bar{u}(\mathbf{p}')\gamma_\alpha(1 - \gamma_5)u(\mathbf{q})]$$
$$\times [\bar{u}(\mathbf{q}')\gamma^\alpha(1 - \gamma_5)u(\mathbf{p})], \tag{16.76}$$

when terms of order $(m_\mu m_e/m_W^2)$ are dropped. The vector k in this equation is the momentum of the intermediate W boson and is given by

$$k = p - q' = p' - q. \tag{16.77}$$

The factor in curly brackets in the Feynman amplitude (16.76) leads to the factor

$$\left\{\frac{m_W^2}{m_W^2 - k^2}\right\}^2 \tag{16.78}$$

---

[9] For details of these calculations involving polarized electrons or polarized muons see D. Bailin's *Weak Interactions*, quoted in Section 16.1.

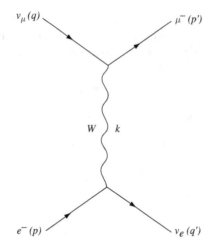

Figure 16.8   *The process $v_\mu + e^- \rightarrow v_e + \mu^-$ (inverse muon decay)*

in the differential scattering cross-section. For $m_W \rightarrow \infty$, this factor becomes unity, and the differential cross-section becomes identical with that of the contact interaction theory if we make the identification (16.43). For finite m_W, the deviations from the latter theory are of order k^2/m_W^2. With $m_W \approx 80$ GeV, such corrections are difficult to detect, even at the highest available neutrino energies, and we shall not consider the inverse muon decay process further. We only remark that the IVB theory is in good agreement with all experiments on this and similar processes, which in lowest-order perturbation theory are described by the exchange of one W boson.

16.6.3   The leptonic decay of the W boson

Lastly, we consider the leptonic decay process

$$W^+ \rightarrow l^+ + v_l \tag{16.79}$$

represented by the Feynman graph in Fig. 16.9. The derivation of the decay rate for this process is left as an exercise for the reader and gives

$$\Gamma_l = \frac{g_W^2 m_W}{6\pi}\left(1 - \frac{m_l^2}{m_W^2}\right)^2\left(1 + \frac{m_l^2}{2m_W^2}\right). \tag{16.80}$$

Figure 16.9   *The leptonic decay modes of the W^+ boson*

For $m_l^2 << m_W^2$ (which is certainly the case for the e, μ and τ leptons) this reduces to

$$\Gamma_l = \frac{g_W^2 m_W}{6\pi} \tag{16.81}$$

implying a value

$$\tau_W = \frac{6\pi B_l}{g_W^2 m_W}, \tag{16.82}$$

where B_l is the branching ratio for the leptonic decay (16.79). Apart from the three decay modes (16.79) for $l = e$, μ and τ, the W boson is also observed to decay into hadrons. The measured values for the three branching ratios B_l, for $l = e$, μ, τ, are consistent within errors, as predicted by Eq. (16.81), with the mean value $B_l = 0.1080 \pm 0.0009$. Substituting this into Eq. (16.82), and using Eqs. (16.43) and (16.70) gives a predicted lifetime of $(1.38 \pm 0.01) \times 10^{-24}$ s, in excellent agreement with the measured value of $(1.41 \pm 0.03) \times 10^{-24}$ s.

16.7 Neutrino Masses

So far, we have assumed that neutrinos are associated with Dirac fields corresponding to a free-field Lagrangian density of the usual form:

$$\mathscr{L}_0 = \sum_l \overline{\psi}_{\nu_l}(x)(i\partial\!\!\!/ - m_{\nu_l})\psi_{\nu_l}(x). \tag{16.83}$$

The values of the neutrino masses are not known, but they are less than an experimental upper limit of order 2 eV. Since the corrections to the theoretical transition rates due to nonzero neutrino masses are of order $\left[m_{\nu_e}/m_e\right]^2$, one can safely put neutrino masses equal to zero when comparing theoretical and observed transition rates. This will be our normal practice throughout this book,[10] Nonetheless, interesting phenomena can, and do, occur, which are not possible for zero-mass neutrinos. Here we shall discuss the two most important; neutrino oscillations and possible lepton-number violation arising from the possibility that the neutrino is its own antiparticle.

16.7.1 Neutrino oscillations

A new phenomenon that can occur if neutrinos have non-zero masses is neutrino mixing. This is the assumption that the neutrino states ν_e, ν_μ and ν_τ, that couple to electrons, muons and tauons respectively, do not have definite masses. Instead, they are linear combinations

$$\nu_l = \sum_i U_{li}\, \nu_i \quad l = e, \mu, \tau \quad i = 1, 2, 3 \tag{16.84}$$

---

[10] We cannot simply put the neutrino masses to zero from the beginning because of our spinor normalization convention (A.27). This gives rise to factors $m_\nu^{-1/2}$ in Feynman amplitudes, which are cancelled by the Feynman mass factors in the relation (8.1) for S-matrix elements, so that the limit can then be safely taken.

of three other neutrino states ν_1, ν_2 and ν_3, which do have definite masses m_1, m_2 and m_3. Here $U \equiv U_{li}$ is a 3×3 mixing matrix whose form must be determined from experiment.

Instead of considering mixing between all three neutrino states ν_e, ν_μ and ν_τ, it is simpler, and often a good approximation, to consider mixing between just two of them. Denoting the flavour states by ν_α and ν_β, and the two mass eigenstates by ν_1 and ν_2, the most general form of mixing (16.84) then reduces to

$$\nu_\alpha = \nu_1 \cos\theta + \nu_2 \sin\theta \qquad (16.85aa)$$

$$\nu_\beta = -\nu_1 \sin\theta + \nu_2 \cos\theta. \qquad (16.85bb)$$

If $\theta = 0$, $\nu_\alpha = \nu_1$ and $\nu_\beta = \nu_2$ and there is no mixing. However, if $\theta \neq 0$ some interesting predictions follow. If, for example, a ν_α neutrino beam is produced with momentum $\mathbf{p}$ at $t = 0$, the ν_1 and ν_2 components will have slightly different energies due to their different masses. In quantum mechanics, their associated waves will have slightly different frequencies. This gives rise to time-dependent quantum-mechanical interference effects, and the composition of the beam will change along its path. The probability that a 'neutrino-flip' $\nu_\alpha \rightarrow \nu_\beta$ has occurred when the beam has travelled a distance L from its source is given by[11]

$$P(\nu_\alpha \rightarrow \nu_\beta) = \sin^2\theta \sin^2\left[\frac{(m_2^2 - m_1^2)L}{4E}\right], \qquad (16.86)$$

where E is the beam energy and we are assuring $E \gg m_1, m_2$. Thus for $\theta \neq 0$, one will observe ν_β neutrinos, and a corresponding reduction in the intensity of ν_α neutrinos, at distances $L \gtrsim E/(m_1^2 - m_2^2)$ from the source.

Such effects are called neutrino oscillations, and there is clear experimental evidence for their existence. We will not give details here, but merely note that a full analysis, allowing for three-component mixing, yields the values

$$\begin{aligned} 7.6 \times 10^{-5} &\lesssim |m_2^2 - m_1^2| \lesssim 8.6 \times 10^{-5} \text{ eV} \\ 1.9 \times 10^{-3} &\lesssim |m_3^2 - m_2^2| \lesssim 3.0 \times 10^{-3} \text{ eV} \end{aligned} \qquad (16.87)$$

for the differences of the squared neutrino masses.

These results establish that neutrino masses are non-zero, contrary to earlier speculation. However, the mass differences (16.87) are very small, so that neutrino oscillations can only be observed in practice when neutrinos travel over very long distances.

Returning to field theory, these results imply that the free-field Lagrangian (16.83) should be replaced by

$$\mathscr{L}_0 = \sum_{j=i}^{3} \overline{\psi}_j(x)\left(i\partial\!\!\!/ - m_j\right)\psi_j(x), \qquad (16.88)$$

---

[11] For a fuller discussion of neutrino oscillations, including a derivation of Eq.(16.86), see B.R. Martin and G. Shaw, loc.cit. (2008), Section 2.3.

where the fields ψ_j refer to the mass eigenstates v_j, $j = 1, 2, 3$. When this is rewritten in terms of the fields

$$\psi_{v_l}(x) = \sum_j U_{lj} \psi_j(x), \tag{16.89}$$

corresponding to the neutrinos v_e, v_μ, v_τ, one easily shows that

$$\mathscr{L}_0 = \sum_l \bar{\psi}_{v_l}(x) i \partial\!\!\!/ \psi_{v_l}(x) - \sum_{l,l'} \bar{\psi}_{vl}(x) G_{ll'} \psi_{v_{l'}}(x), \tag{16.90a}$$

where the Hermitian matrix

$$G_{ll'} = \sum_j U_{lj} m_j U^\dagger_{jl'}. \tag{16.90b}$$

Nonetheless, since we will usually take the limit of zero-mass neutrinos in what follows, and since Eqs. (16.89) and (16.90) only differ by terms which vanish in this limit, we will usually continue to use the simpler form (16.83).

16.7.2 Dirac or Majorana neutrinos?

Neutral particles might, or might not, have distinct antiparticles. Thus there are two possible descriptions of neutral leptons, i.e. neutrinos. Neutrinos that have distinct antiparticles, as we have assumed so far, are called *Dirac neutrinos*. Neutrinos that do not have distinct antiparticles are called *Majorana neutrinos*. Due to the structure of weak interactions, these two descriptions can only be experimentally distinguished if neutrino masses are non-zero. In this section, we shall, for simplicity, discuss these possibilities in the context of a single neutrino species, neglecting neutrino mixing.

The symmetry between particles and antiparticles is most clearly exhibited by using a Majorana representation[12] [cf. Section 4.3 and Appendix A.8.]. In such a representation, the Dirac matrices are imaginary, and the Dirac wave operator (4.49a) is purely real. Hence we can distinguish two types of field, analogous to the real and complex scalar fields discussed in Sections 3.1 and 3.2. Dirac fields are complex in a Majorana representation, and the quantized field operators can be expanded in the form (4.51), where $c_r^\dagger$ and $d_r^\dagger$ are the creation operators for particles (neutrinos) and antiparticles (antineutrinos) respectively. Alternatively, we can consider Majorana fields, which are real in a Majorana representation, so that the corresponding quantized field operator is expanded in the form

$$\psi(x) = \sum_{r, \mathbf{p}} \left(\frac{m}{V E_\mathbf{p}} \right)^{1/2} [c_r(\mathbf{p}) u_r(\mathbf{p}) e^{-ipx} + c_r^\dagger(\mathbf{p}) u_r^*(\mathbf{p}) e^{ipx}], \tag{16.91}$$

where $c_r(\mathbf{p}), c_r^\dagger(\mathbf{p})$ are neutrino annihilation and creation operators respectively. In this case, there is no longer an independent field $\psi^\dagger(x)$, and there are no distinct antiparticles.

---

[12] For an alternative discussion, which uses representations of the Dirac type, satisfying $\gamma^{0^T} = \gamma^0, \gamma^{2^T} = \gamma^2$, see, for example, T-P. Cheng and L-F. Li, *Gauge Theory of Elementary Particle Physics*, Oxford University Press (1984), p412–414.

To summarize, for a Majorana neutrino of given momentum, there are two possible helicity states and no distinct antineutrino states. On the other hand, for Dirac neutrinos we have two neutrino states and two antineutrino states, with opposite values of the corresponding lepton number (16.9).

The relation between the two descriptions in the zero mass neutrino limit is made clear by considering the fields

$$\psi^L(x) \equiv P_L\psi(x), \quad \psi^R(x) = P_R\psi(x), \tag{16.92}$$

where P_R, P_L are the projection operators (16.14). By acting on the Dirac equation with these projection operators, one easily shows that ψ^L and ψ^R satisfy the coupled Dirac equations

$$i\partial\!\!\!/\psi^L(x) - m\psi^R(x) = 0, \quad i\partial\!\!\!/\psi^R(x) - m\psi^L(x) = 0. \tag{16.93}$$

These equations decouple in the zero-mass limit, and since the interaction given by (16.7) and (16.16) only involves $\psi^L(x)$, the field $\psi^R(x)$ can be eliminated from the theory. Furthermore, P_L is a helicity projection operator in this limit, and for Dirac neutrinos the argument following Eq. (16.12) shows that the emission of an electron can be accompanied by the absorption of a negative helicity neutrino or the emission of a positive helicity antineutrino, conserving lepton number. The same argument applied to a Majorana neutrino, described by the field (16.91) in a Majorana representation, shows that the emission of an electron can be accompanied by the absorption of a negative helicity neutrino or the emission of a positive helicity neutrino. Hence lepton number is still conserved, provided we modify the definition (16.9) to

$$N(e) = N(e^-) - N(e^+) + N(v_-) - N(v_+) \tag{16.94}$$

where v_- and v_+ mean negative or positive helicity electron neutrino states respectively. The same analysis applies to the electron absorption and positron emission and absorption, and in the zero-mass limit, the Majorana description in terms of two neutrino states v_- and v_+ and the Dirac description in terms of the two states v_- and $\bar{v}_+$ cannot be experimentally distinguished.

The equivalence breaks down for neutrinos of non-zero mass, since P_R and P_L are no longer helicity projection operators. However, the differences in the theoretical predictions are very difficult to detect, since they vanish in the limit of zero neutrino masses, and are consequently very small. No such effects have been detected to data, and the best prospects to observe them are in double β-decay experiments. The double β-decay process

$$(Z,A) \rightarrow (Z+2, A) + 2e^- + 2\bar{v}_e, \tag{16.95}$$

is allowed for both Dirac and Majorana neutrinos by the mechanism of Fig. 16.10(a). In contrast, the neutrinoless double β-decay process

$$(Z,A) \rightarrow (Z+2, A) + 2e^- \tag{16.96}$$

is forbidden for Dirac neutrinos with the interaction (16.7), because it violates lepton number conservation. However, it can occur for Majorana neutrinos of non-zero mass by the mechanism shown in Fig. 16.10(b). Observation of the reaction (16.96) would be strong evidence for the existence of Majorana neutrinos, although, in principle, the

Figure 16.10 (a) Double β-decay, which is allowed for both Dirac and Majorana neutrinos. (b) Neutrinoless double β-decay, which is allowed for Majorana neutrinos only

reaction could also proceed by some other unknown mechanism. At the time of writing, there is no confirmed evidence for neutrinoless double β-decay, and current experiments[13] can be interpreted as setting an upper limit of order 0.5 eV/c^2 on the mass of a possible Majorana electron neutrino. In what follows, we shall continue to treat the neutrinos as Dirac neutrinos with distinct antiparticles.

16.8 Difficulties with the IVB Theory

In spite of its successes, the IVB theory presents serious difficulties. It cannot describe such processes as

$$\nu_\mu + e^- \rightarrow \nu_\mu + e^- \tag{16.97}$$

and

$$\bar\nu_\mu + e^- \rightarrow \bar\nu_\mu + e^-, \tag{16.98}$$

which are not forbidden by any conservation laws. With the interaction (16.7), the leading contributions to the process (16.98) come from the Feynman graphs in Fig. 16.11. This is in contrast to processes such as

$$\nu_e + e^- \rightarrow \nu_e + e^- \tag{16.99}$$

and the inverse muon decay process

$$\nu_\mu + e^- \rightarrow \nu_e + \mu^-, \tag{16.100}$$

which in the lowest order involve the exchange of one W boson, as in Figs. 16.3 and 16.8. This difference is due to the fact that each term of the leptonic currents, J_α and $J_\alpha^\dagger$, Eqs. (16.6), always couples a neutral and a charged lepton. For example, the term in the interaction (16.7) which annihilates a muon–neutrino necessarily creates a μ^- or destroys a μ^+. Hence the leading contribution to the process (16.97) must involve the exchange of two W bosons, as shown in Fig. 16.11.

---

[13] For a simple account of double β-decay experiments, see Martin and Shaw, *loc. cit.*, Section 11.6.2.

Figure 16.11   *The leading contribution to the process $v_\mu + e^- \to v_\mu + e^-$, as given by the IVB theory*

We found earlier that the coupling constant g_W is small. Hence one would expect the cross-sections for two-boson-exchange processes, like Eq. (16.97), to be small compared with those for one-boson-exchange processes, like Eq. (16.89). Unfortunately, the loop integrals to which the Feynman graphs in Fig. 16.11 give rise are divergent, and the IVB theory is not renormalizable, as we shall discuss further in a moment. This means we simply do not know how to calculate with it sensibly if loop intergrals are involved. Hence IVB theory represents a phenomenological theory only; it allows one to calculate those processes which do not involve loop integrals in the lowest order of perturbation theory.

On the other hand, the measured cross-sections for processes like $v_\mu - e$ scattering, Eq. (16.97), are comparable to the cross-sections for processes like $\bar{v}_\mu - e$ scattering, Eq. (16.99). This suggests that the IVB interaction (16.7) is not complete and that the leptonic interaction contains additional terms which allow the $v_\mu - e$ scattering to occur as a one-boson-exchange process. If we wish to retain lepton number conservation, since this is well established, these extra terms will lead to Feynman diagrams like that in Fig. 16.12, involving the exchange of a neutral vector boson Z^0 between an electron–electron current and a neutrino–neutrino current. (Currents involving the emission or absorption of a neutral vector boson are called *neutral currents*, in contrast to the *charged currents*, J_α and $J_\alpha^\dagger$, which involve the emission or absorption of a charged vector boson.) The existence of the Z^0 boson and the presence of the neutral-current terms in the interaction are required by the standard electroweak theory, for this theory to be renormalizable. The prediction of processes like $v_\mu - e$ and $\bar{v}_\mu - e$ scattering, before they had been observed, was one of the great early successes of this theory.

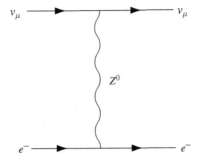

Figure 16.12   *A one-boson-exchange contribution to the process $v_\mu + e^- \to v_\mu + e^-$, not contained in the IVB theory*

Finally, we return to the question of the renormalizability of the IVB theory. We see from Eq. (16.30), i.e.

$$iD_{F\alpha\beta}(k, m_W) = \frac{i\left(-g_{\alpha\beta} + k_\alpha k_\beta / m_W^2\right)}{k^2 - m_W^2 + i\varepsilon}, \qquad (16.30)$$

that the W propagator behaves like a constant for large momenta k. Hence the loop integrals in the amplitudes of the Feynman graphs 16.11 are quadratically divergent,

$$\int^\Lambda \frac{d^4 k}{k^2} \sim \Lambda^2, \qquad (16.101)$$

where Λ is a high-energy cut-off parameter. More generally, in more complicated higher-order graphs, each additional W propagator which forms part of a loop will lead to a factor of the order Λ^2 / m_W^2, where the factor Λ^2 follows purely from dimensional arguments. Hence, as we consider a given process in higher orders of perturbation theory, new, progressively more severe divergences arise. To cancel these, we would have to introduce additional renormalization constants at each stage, ending up with infinitely many such constants. In other words it is non-renormalizable.

As we saw in Chapter 9, the situation is quite different in QED, which contains only three renormalization constants which can be absorbed into the mass and the charge of the electron, and which is renormalizable. This difference between QED and the IVB theory is reflected in the fact that in QED the degree of divergence of a primitively divergent graph depends on the external lines only and is independent of the number of vertices n [see Eq. (9.112)]. For the IVB theory, on the other hand, the degree of divergence increases with n (see Problem 16.3). The divergence difficulties of the IVB theory stem from the term $k^\alpha k^\beta / m_W^2$ in the W propagator. This term is absent from the photon propagator $-ig^{\alpha\beta}/k^2$. Hence the latter acts as a convergence factor for large k in a loop integral with respect to k. In the case of the photon propagator we were able to avoid a term in $k^\alpha k^\beta$ by exploiting gauge invariance. The renormalizability of the theory of weak interactions is similarly achieved by formulating it as a gauge theory.

Problems

16.1. Prove that for the muon decay process $\mu^- \to e^- + \bar{v}_e + v_\mu$ the maximum possible energy of the electron in the muon rest frame is

$$\left(m_\mu^2 + m_e^2\right) / 2m_\mu.$$

16.2. For the leptonic process $W^+ \to l^+ v_l$, derive the decay rate (16.80).

16.3. Show that in the IVB theory the dimensionality K of the Feynman amplitude of a Feynman graph is given by

$$K = n + 4 - \frac{3}{2} f_e - 2b_e,$$

where $f_e (b_e)$ is the number of external fermion (boson) lines of the graph, and n is the number of its vertices. (The corresponding result (9.112) for QED is independent of n.)

16.4. Show that the differential decay rate for a polarized muon is given by

$$d\Gamma = \frac{1}{3} \frac{G^2}{(2\pi)^4} d^3 \mathbf{p} m_\mu \left[3m_\mu - 4E + \cos \theta (m_\mu - 4E) \right],$$

where $(E, \mathbf{p})$ is the energy-momentum four-vector of the emitted electron in the rest-frame of the muon, θ is the angle which $\mathbf{p}$ makes with the direction of the muon spin, and the electron mass has been neglected, i.e. the above formula applies only to high-energy electrons produced in the decay. The $\cos \theta$ term in the formula demonstrates the parity violation of the process.

16.5. Show that, in the centre-of-mass frame, the differential cross-section for the inverse muon decay process

$$\nu_\mu + e^- \to \nu_e + \mu^-, \tag{A}$$

which results from the Fermi contact interaction (16.44), is given by

$$\frac{d\sigma}{d\Omega} = \frac{G^2 E^2}{\pi^2},$$

where all lepton masses have been neglected compared with the centre-of-mass energy E of the incoming neutrino.

Show that for the related process

$$\bar{\nu}_e + e^- \to \mu^- + \bar{\nu}_\mu$$

the corresponding differential cross-section is given by

$$\frac{d\sigma}{d\Omega} = \frac{G^2 E^2}{4\pi^2} (1 - \cos \theta)^2,$$

where θ is the angle between the incoming neutrino and the outgoing muon. For the second reaction, the cross-section vanishes in the forward direction $(\theta = 0)$. Deduce this directly by considering the helicities of the leptons in the process and the contact nature of the interaction. Show that the same line of argument does not rule out forward scattering for the inverse muon decay process (A).

17

A Gauge Theory of Weak Interactions

In this chapter we shall attempt to formulate a gauge theory of weak interactions in analogy to QED for electromagnetic interactions. For weak interactions, the role of the electromagnetic current is taken by the weak leptonic currents J_α and $J_\alpha^\dagger$ Eqs. (11.6), and the role of the photon is taken by the $W^\pm$ bosons. We shall find that the requirement of gauge invariance leads to the neutral leptonic current, mentioned at the end of the last chapter, and to a third, electrically neutral, vector boson. At the same time, a unification of electromagnetic and weak interactions is achieved in a natural way. The gauge invariance of the theory developed in this chapter necessitates all leptons and vector bosons to be massless. This difficulty is discussed at the end of this chapter, but it will only be resolved in the next chapter.

17.1   QED Revisited

In Section 11.1, we discussed the use of the minimal gauge principle to infer the form of the QED interaction. Since the discussion is the template for all that follows, we summarize it for convenience here.

The basic idea is to first find a set of global phase transformations which leave the free-field Lagrangian invariant, and lead to conservation of the appropriate current or currents. These transformation are then generalized to local phase (i.e. gauge) transformations, and invariance under these transformations is assumed and used to infer the form of the interaction.

Thus, in Section 12.1 we first noted that the invariance of the free-field Lagrangian

$$\mathscr{L}_0 = \overline{\psi}(x) \left(i\gamma^\mu \partial_\mu - m \right) \psi(x) \tag{17.1}$$

under the global phase transformation

$$\left.\begin{array}{l}\psi(x) \rightarrow \psi'(x) = \psi(x)e^{-i\alpha} \\ \overline{\psi}(x) \rightarrow \overline{\psi}'(x) = \overline{\psi}(x)e^{i\alpha}\end{array}\right\} \qquad (17.2)$$

led to conservation of the electromagnetic current

$$s^\mu(x) = q\overline{\psi}(x)\gamma^\mu\psi(x). \qquad (17.3)$$

That is, the current satisfies the continuity equation $\partial_\mu s^\mu(x) = 0$, leading to the conservation of electric charge

$$Q = q \int d^3\mathbf{x} \, \psi^\dagger(x)\psi(x) \qquad (17.4)$$

as shown in the discussion following Eq. (2.40).

The next step was to generalize Eq. (17.2) to a set of local phase transformations

$$\left.\begin{array}{l}\psi(x) \rightarrow \psi'(x) = \psi(x)e^{-iqf(x)} \\ \overline{\psi}(x) \rightarrow \overline{\psi}'(x) = \overline{\psi}(x)e^{iqf(x)}\end{array}\right\} \qquad (17.5a)$$

where $f(x)$ is an arbitrary function. Under these transformations, the free-field Lagrangian density $\mathscr{L}_0$ becomes

$$\mathscr{L}_0 \rightarrow \mathscr{L}'_0 = \mathscr{L}_0 + q\overline{\psi}(x)\gamma^\mu\psi(x)\partial_\mu f(x), \qquad (17.6)$$

and an interaction term must be added to restore invariance. To deduce the form of the interaction, we first introduced the covariant derivative (11.7), i.e.

$$D^\mu\psi(x) = \left[\partial_\mu + iqA_\mu(x)\right]\psi(x), \qquad (17.7)$$

where the electromagnetic field transforms under a gauge transformation in the usual way:

$$A_\mu(x) \rightarrow A'_\mu(x) = A_\mu(x) + \partial_\mu f(x). \qquad (17.5b)$$

The crucial property of the covariant derivative is that, under the combined gauge transformation (17.5a) and (17.5b), it transforms in the same way as the matter field $\psi(x)$, i.e.

$$D_\mu\psi(x) \rightarrow e^{-iqf(x)}D_\mu\psi(x). \qquad (17.8)$$

Hence by replacing the ordinary derivative $\partial_\mu\psi$ in Eq. (17.1) by the covariant derivative (17.7), we obtain the familiar gauge invariant Lagrangian of QED:

$$\begin{aligned}\mathscr{L} &= \overline{\psi}(x)\left(i\gamma^\mu D_\mu - m\right)\psi(x) \\ &= \mathscr{L}_0 - q\overline{\psi}(x)\gamma^\mu\psi(x)A_\mu(x).\end{aligned} \qquad (17.9)$$

The interaction term in Eq. (17.9) is called the 'minimal gauge interaction.' Additional gauge-invariant terms can be added to Eq. (17.9) without spoiling the symmetry, but, as shown in Section 16.3.2, they lead to a non-renormalizable theory.

17.2 Global Phase Transformations and Conserved Weak Currents

We shall now apply the programme outlined in the last section to formulate the theory of weak interactions as a gauge theory. As a first step, we must find a set of global phase transformations which leaves the free-lepton Lagrangian density invariant, leading to conservation of the weak currents $J_\alpha(x), J_\alpha^\dagger(x)$, Eqs. (16.6).

To begin with, we shall assume that all leptons are massless. At the end of the chapter we shall return to the problems to which non-vanishing lepton masses give rise. The free-lepton Lagrangian density is then given by

$$\mathscr{L}_0 = i\left[\overline{\psi}_l(x)\partial\!\!\!/\psi_l(x) + \overline{\psi}_{\nu_l}(x)\partial\!\!\!/\psi_{\nu_l}(x)\right], \tag{17.10}$$

where, as in the following, summation over all different kinds of leptons is understood: $l = e, \mu, \ldots$.

In Chapter 16 we found that the leptonic currents, and consequently the leptonic interaction, involve only the left-handed lepton fields, Eqs. (16.13)–(16.16) We shall therefore write Eq. (17.10) in terms of left- and right-handed fields. For any Dirac spinor $\psi(x)$, these fields are defined by

$$\left.\begin{array}{l}\psi^{L}(x) = P_L\psi(x) \\ \psi^{R}(x) = P_R\psi(x)\end{array}\right\} \equiv \frac{1}{2}(1\pm\gamma_5)\psi(x), \tag{17.11}$$

and Eq. (17.10) becomes

$$\mathscr{L}_0 = i\left[\overline{\psi}_l^{L}(x)\partial\!\!\!/\psi_l^{L}(x) + \overline{\psi}_{\nu_l}^{L}(x)\partial\!\!\!/\psi_{\nu_l}^{L}(x) + \overline{\psi}_l^{R}(x)\partial\!\!\!/\psi_l^{R}(x) + \overline{\psi}_{\nu_l}^{R}(x)\partial\!\!\!/\psi_{\nu_l}^{R}(x)\right]. \tag{17.12}$$

We now combine the fields ψ_l^{L} and $\psi_{\nu_l}^{L}$ into a two-component field

$$\Psi_l^{L}(x) = \begin{pmatrix}\psi_{\nu_l}^{L}(x) \\ \psi_l^{L}(x)\end{pmatrix} \tag{17.13a}$$

and, correspondingly,

$$\overline{\Psi}_l^{L}(x) = \left(\overline{\psi}_{\nu_l}^{L}(x), \overline{\psi}_l^{L}(x)\right). \tag{17.13b}$$

In terms of these fields, Eq. (17.12) becomes

$$\mathscr{L}_0 = i\left[\overline{\Psi}_l^{L}(x)\partial\!\!\!/\Psi_l^{L}(x) + \overline{\psi}_l^{R}(x)\partial\!\!\!/\psi_l^{R}(x) + \overline{\psi}_{\nu_l}^{R}(x)\partial\!\!\!/\psi_{\nu_l}^{R}(x)\right]. \tag{17.14}$$

Although Eq. (17.12) is symmetric between left- and right-handed fields, we have written Eq. (17.14) in a very unsymmetric way, i.e. we have not introduced two-component right-handed fields. We shall see that the left–right asymmetry of weak interactions can be described in terms of different transformation properties of the left- and right-handed fields. For the two-component left-handed fields, the possibility arises of two-dimensional transformations which leave bilinear forms

$\overline{\Psi}_l^L(x)(\ldots)\Psi_l^L(x)$ invariant. For this purpose, we introduce the three 2×2 Hermitian matrices

$$\tau_1 = \begin{pmatrix} 0 & 1 \\ 1 & 0 \end{pmatrix}, \quad \tau_2 = \begin{pmatrix} 0 & -i \\ i & 0 \end{pmatrix}, \quad \tau_3 = \begin{pmatrix} 1 & 0 \\ 0 & -1 \end{pmatrix}, \tag{17.15}$$

which satisfy the commutation relations

$$\left[\tau_i,\ \tau_j\right] = 2i\varepsilon_{ijk}\tau_k, \tag{17.16}$$

where ε_{ijk} is the usual completely antisymmetric tensor and, as throughout, summation over the repeated index $k\,(=1, 2, 3)$ is implied. (The τ-matrices are just the usual Pauli spin matrices.) The operator

$$U(\alpha) \equiv \exp\left(i\alpha_j\tau_j/2\right) \tag{17.17}$$

is unitary for any three real numbers $\alpha \equiv (\alpha_1, \alpha_2, \alpha_3)$, and the set of transformations

$$\left.\begin{aligned} \Psi_l^L(x) &\rightarrow \Psi_l^{L'}(x) = U(\alpha)\Psi_l^L(x) \equiv \exp\left(i\alpha_j\tau_j/2\right)\Psi_l^L(x) \\ \overline{\Psi}_l^L(x) &\rightarrow \overline{\Psi}_l^{L'}(x) = \overline{\Psi}_l^L(x)U^\dagger(\alpha) \equiv \overline{\Psi}_l^L(x)\exp\left(-i\alpha_j\tau_j/2\right) \end{aligned}\right\} \tag{17.18a}$$

leaves the term $i\overline{\Psi}_l^L(x)\not\partial\Psi_l^L(x)$ in $\mathscr{L}_0$, Eq. (17.14), invariant.

The operators $U(\alpha)$ are 2×2 unitary matrices with the special property that $\det U(\alpha) = +1$. They are therefore called SU(2) transformations.[1] The transformations (17.18a) can be regarded as two-dimensional global phase transformations, i.e. as a generalization of the one-dimensional global phase transformations (17.2).

The SU(2) transformation properties of the two-component left-handed lepton fields $\Psi_l^L(x)$ are identical with those of the two-component spinors, which describe spin-$\frac{1}{2}$ particles in the non-relativistic Pauli theory of spin, and with those of the two-component isospinors, which describe the neutron and the proton as different charge states of the nucleon. The two-component field $\Psi_l^L(x)$ is therefore called a weak isospinor. In addition to two-component spinors, the Pauli spin theory also gives rise to one-component scalars (e.g. singlet spin states), three-component vectors (e.g. triplet states), etc., and these entities are characterized by the transformations induced in them by the basic spinor transformations. For example, any quantity invariant under these transformations is a scalar. All these concepts carry over to weak isospin, leading to quantities being classified according to their transformation properties under SU(2) transformations as weak isoscalars, weak isospinors and so on.

So far, we have considered the left-handed lepton fields. We shall now define each right-handed lepton field to be a weak isoscalar, i.e. to be invariant under any SU(2) transformation:

$$\left.\begin{aligned} \psi_l^R(x) &\rightarrow \psi_l^{R'}(x) = \psi_l^R(x), \quad \psi_{\nu_l}^R(x) \rightarrow \psi_{\nu_l}^{R'}(x) = \psi_{\nu_l}^R(x) \\ \overline{\psi}_l^R(x) &\rightarrow \overline{\psi}_l^{R'}(x) = \overline{\psi}_l^R(x), \quad \overline{\psi}_{\nu_l}^R(x) \rightarrow \overline{\psi}_{\nu_l}^{R'}(x) = \overline{\psi}_{\nu_l}^R(x) \end{aligned}\right\}. \tag{17.18b}$$

----

[1] The set of all SU(2) transformations forms the SU(2) group. A group is called Abelian (non-Abelian) if its elements do (do not) commute. The SU(2) group is non-Abelian, since the τ-matrices and hence the operators (17.17) are non-commuting. The terms 'SU(2)', 'non-Abelian' and some other nomenclature which we shall introduce derive from group theory, but no knowledge of group theory is required in what follows.

It follows at once that the SU(2) transformations (17.18a) and (17.18b) of the lepton fields leave the free-lepton Lagrangian density $\mathcal{L}_0$, Eq. (17.14), invariant.

From this invariance of $\mathcal{L}_0$, the conservation of the leptonic currents $J_\alpha(x)$ and $J_\alpha^\dagger(x)$, Eqs. (16.6), follows. For infinitesimal α_j, the transformations (17.18a) reduce to

$$\left. \begin{aligned} \Psi_l^L(x) &\to \Psi_l^{L\prime}(x) = \left(1 + i\alpha_j\tau_j/2\right)\Psi_l^L(x) \\ \overline{\Psi}_l^L(x) &\to \overline{\Psi}_l^{L\prime}(x) = \overline{\Psi}_l^L(x)\left(1 - i\alpha_j\tau_j/2\right) \end{aligned} \right\}, \tag{17.19}$$

whereas Eqs. (17.18b) for the right-handed fields remain unchanged. An argument analogous to that following Eq. (2.39) leads to the three conserved currents

$$J_i^\alpha(x) = \frac{1}{2}\overline{\Psi}_l^L(x)\gamma^\alpha\tau_i\Psi_l^L(x), \qquad\qquad i = 1, 2, 3, \tag{17.20}$$

which are called weak isospin currents. The corresponding conserved quantities

$$I_i^W = \int d^3\mathbf{x} J_i^0(x) = \frac{1}{2}\int d^3\mathbf{x}\,\Psi_l^{L\dagger}(x)\tau_i\Psi_l^L(x), \qquad i = 1, 2, 3, \tag{17.21}$$

are called weak isospin charges.

The leptonic currents $J^\alpha(x)$ and $J^{\alpha\dagger}(x)$, Eqs. (16.6), in terms of which the IVB theory was formulated, can be written as linear combinations of the conserved weak isospin currents $J_1^\alpha(x)$ and $J_2^\alpha(x)$. Using Eqs. (17.11) and (17.15), one obtains

$$\left. \begin{aligned} J^\alpha(x) &= 2\left[J_1^\alpha(x) - iJ_2^\alpha(x)\right] = \overline{\psi}_l(x)\gamma^\alpha(1 - \gamma_5)\psi_{\nu_l}(x) \\ J^{\alpha\dagger}(x) &= 2\left[J_1^\alpha(x) + iJ_2^\alpha(x)\right] = \overline{\psi}_{\nu_l}(x)\gamma^\alpha(1 - \gamma_5)\psi_l(x) \end{aligned} \right\}. \tag{17.22}$$

Most remarkably, the above development necessarily led to the conservation of a *third* current, namely the weak isospin current

$$J_3^\alpha(x) = \frac{1}{2}\overline{\Psi}_l^L(x)\gamma^\alpha\tau_3\Psi_l^L(x) = -\frac{1}{2}\left[\overline{\psi}_{\nu_l}^L(x)\gamma^\alpha\psi_{\nu_l}^L(x) - \overline{\psi}_l^L(x)\gamma^\alpha\psi_l^L(x)\right]. \tag{17.23}$$

The current $J_3^\alpha(x)$ is called a *neutral* current since it couples either electrically neutral leptons or, like the electromagnetic current

$$s^\alpha(x) = -e\overline{\psi}_l(x)\gamma^\alpha\psi_l(x), \tag{17.24}$$

electrically charged leptons. This is in contrast to the *charged* currents $J^\alpha(x)$ and $J^{\alpha\dagger}(x)$, which couple electrically neutral with electrically charged leptons. Apart from a constant factor, the last term on the right-hand side of Eq. (17.23) is a part of the electromagnetic current (17.24). We have here a first indication that, in the theory we are developing, electromagnetic and weak processes will be interconnected.

The weak hypercharge current $J_Y^\alpha(x)$ is defined by

$$\begin{aligned} J_Y^\alpha(x) &= s^\alpha(x)/e - J_3^\alpha(x) \\ &= -\frac{1}{2}\overline{\Psi}_l^L(x)\gamma^\alpha\Psi_l^L(x) - \overline{\psi}_l^R(x)\gamma^\alpha\psi_l^R(x). \end{aligned} \tag{17.25}$$

The corresponding charge

$$Y = \int d^3\mathbf{x} J_Y^0(x) \tag{17.26}$$

is called the weak hypercharge.[2] We see from Eq. (17.25) that Y is related to the electric charge Q and the weak isocharge I_3^W by

$$Y = Q/e - I_3^W. \tag{17.27}$$

The conservation of the electric charge Q and of the weak isocharge I_3^W implies conservation of the weak hypercharge Y.

We next determine the weak isospin charge I_3^W and the weak hypercharge Y of the different leptons. From Eqs. (17.13a) and (17.15) we have

$$\tau_3 \Psi_l^L(x) = \begin{pmatrix} 1 & 0 \\ 0 & -1 \end{pmatrix} \begin{pmatrix} \psi_{\nu_l}^L(x) \\ \psi_l^L(x) \end{pmatrix} = \begin{pmatrix} \psi_{\nu_l}^L(x) \\ -\psi_l^L(x) \end{pmatrix}.$$

It follows that the weak isospin charge I_3^W has the value $+\frac{1}{2}$ for a left-handed ν_l neutrino, and the value $-\frac{1}{2}$ for a left-handed l^- lepton. A more proper derivation of these results is obtained by substituting the expansions (4.38) of the Dirac fields in Eq. (17.21). If $|l^-,\ L\rangle$ and $|\nu_l,\ L\rangle$ are one-particle states which respectively contain one left-handed l^- lepton and one left-handed ν_l neutrino, one finds that

$$I_3^W |l^-,\ L\rangle = -\frac{1}{2}|l^-,\ L\rangle, \quad I_3^W|\nu_l,\ L\rangle = +\frac{1}{2}|\nu_l,\ L\rangle. \tag{17.28a}$$

Since the right-handed lepton fields are isoscalars, it similarly follows that I_3^W has the value 0 for right-handed l^- leptons and for right-handed ν_l neutrinos, i.e.

$$I_3^W |l^-,R\rangle = 0, \quad I_3^W|\nu_l,R\rangle = 0. \tag{17.28b}$$

The values of the weak hypercharge Y for the different leptons follows from Eqs. (17.27) and (17.28):

$$\left. \begin{array}{c} Y|l^-,\ L\rangle = -\frac{1}{2}|l^-,\ L\rangle, \quad Y|\nu_l,\ L\rangle = -\frac{1}{2}|\nu_l,\ L\rangle \\[2mm] Y|l^-,\ R\rangle = -|l^-,\ R\rangle \\ Y|\nu_l,\ R\rangle = 0 \end{array} \right\} \tag{17.29}$$

i.e. for the left-handed states of l^- and ν_l leptons, Y has the value $-\frac{1}{2}$; for the right-handed states of l^- and ν_l leptons, it has the values -1 and 0, respectively.

Above we deduced the conservation of weak hypercharge from the conservation of Q and of I_3^W, using Eq. (17.27). The conservation of weak hypercharge also follows directly from the invariance of the free-lepton Lagrangian density (17.14) under the simultaneous transformations

$$\left. \begin{array}{l} \Psi_l^L(x) \rightarrow \Psi_l^{L'}(x) = e^{-i\beta/2}\Psi_l^L(x) \\ \psi_l^R(x) \rightarrow \psi_l^{R'}(x) = e^{-i\beta}\psi_l^R(x) \\ \psi_{\nu_l}^R(x) \rightarrow \psi_{\nu_l}^{R'}(x) = \psi_{\nu_l}^R(x) \end{array} \right\} \tag{17.30a}$$

---

[2] These names have their origin in the formal similarity which exists with corresponding quantities and relations for the strong interactions.

and the corresponding transformations of the adjoint fields, i.e.

$$\overline{\Psi}_l^{L}(x) \rightarrow \overline{\Psi}_l^{L'}(x) = \overline{\Psi}_l^{L}(x) e^{i\beta/2}, \text{ etc.,} \tag{17.30b}$$

where β is an arbitrary real number. We can write the transformation (12.30) more concisely as

$$\psi(x) \rightarrow \psi'(x) = e^{i\beta Y} \psi(x), \quad \overline{\psi}(x) \rightarrow \overline{\psi}'(x) = \overline{\psi}(x) e^{-i\beta Y}, \tag{17.31}$$

where $\psi(x)$ denotes any one of the lepton fields $\psi_{\nu_l}^{L}$, ψ_l^{L}, $\psi_{\nu_l}^{R}$ and ψ_l^{R} and Y is the weak hypercharge of the particle annihilated by the field $\psi(x)$. These transformations are similar in form to the global phase transformations (17.3) of QED and, like the latter, are called U(l) phase transformations since $U = e^{i\beta Y}$ is just a one-dimensional unitary matrix, i.e. a complex number of unit modulus. (The U (l) transformations of course also form a group.)

The line of reasoning we have used above can be inverted. We can start from the invariance of the free-lepton Lagrangian density (17.14): (i) under the global SU(2) transformations (17.18) leading to conservation of the weak isospin charges I_i^{W}, and (ii) under the global U(l) transformations (17.30), leading to conservation of the weak hypercharge Y. The conservation of the electric charge then follows from Eq. (17.27).

17.3   The Gauge-Invariant Electroweak Interaction

We next generalize the above SU(2) and U(1) transformations from global to local phase transformations. The development will be very similar to that for QED in Section 17.1. In order to retain invariance under local phase transformations, we shall have to introduce gauge fields, and this will automatically generate the interactions.

We shall start with the SU(2) transformations and replace the global transformations (17.18) by the local phase transformations

$$\left. \begin{aligned} \Psi_l^{L}(x) &\rightarrow \Psi_l^{L'}(x) = \exp\left[ig\tau_j \omega_j(x)/2\right] \Psi_l^{L}(x) \\ \overline{\Psi}_l^{L}(x) &\rightarrow \overline{\Psi}_l^{L'}(x) = \overline{\Psi}_l^{L}(x) \exp\left[-ig\tau_j \omega_j(x)/2\right] \\ \psi_l^{R}(x) &\rightarrow \psi_l^{R'}(x) = \psi_l^{R}(x), \quad \psi_{\nu_l}^{R}(x) \rightarrow \psi_{\nu_l}^{R'}(x) = \psi_{\nu_l}^{R}(x) \\ \overline{\psi}_l^{R}(x) &\rightarrow \overline{\psi}_l^{R'}(x) = \overline{\psi}_l^{R}(x), \quad \overline{\psi}_{\nu_l}^{R}(x) \rightarrow \overline{\psi}_{\nu_l}^{R'}(x) = \overline{\psi}_{\nu_l}^{R}(x). \end{aligned} \right\} \tag{17.32a}$$

Here $\omega_j(x), j = 1, 2, 3$, are three arbitrary real differentiable functions of x, and g is a real constant which will later be identified with a coupling constant.

Applying the transformations (17.32a) to the free-lepton Lagrangian density (17.14), the differential operator $\partial\!\!\!/$ in the spinor term will also act on the functions $\omega_j(x)$ in the exponent. Hence Eq. (17.14) is not invariant under this transformation and for infinitesimal $\omega_j(x)$ transforms according to

$$\mathcal{L}_0 \rightarrow \mathcal{L}'_0 = \mathcal{L}_0 + \delta\mathcal{L}_0 \equiv \mathcal{L}_0 - \frac{1}{2} g \overline{\Psi}_l^{L}(x) \tau_j \partial\!\!\!/ \omega_j(x) \Psi_l^{L}(x). \tag{17.33}$$

We shall obtain an invariant Lagrangian density if, analogously to the replacement (17.7) in QED, we replace the ordinary derivatives $\partial^\mu \Psi_l^L(x)$ in Eq. (17.14) by the covariant derivatives,

$$\partial^\mu \Psi_l^L(x) \rightarrow D^\mu \Psi_l^L(x) = \left[\partial^\mu + ig\tau_j W_j^\mu(x)/2\right]\Psi_l^L(x), \tag{17.34}$$

so that $\tilde{\mathscr{L}}_0$ goes over into

$$\tilde{\mathscr{L}}_0 = i\left[\overline{\Psi}_l^L(x)\slashed{D}\,\Psi_l^L(x) + \overline{\psi}_l^R(x)\slashed{\partial}\psi_l^R(x) + \overline{\psi}_{\nu_l}^R(x)\slashed{\partial}\psi_{\nu_l}^R(x)\right]. \tag{17.35}$$

In Eq. (17.34), we had to introduce three real gauge fields $W_j^\mu(x)$, compared to the one gauge field $A^\mu(x)$ of QED, as there are now three conserved charges I_j^W and as the gauge transformation (17.32a) now contains three arbitrary functions $\omega_j(x)$.

For the modified Lagrangian density $\tilde{\mathscr{L}}_0$ to be invariant, the transformations (17.32a) of the lepton fields must be coupled to transformations of the gauge fields,

$$W_j^\mu(x) \rightarrow W_j^{\mu'}(x) = W_j^\mu(x) + \delta W_j^\mu(x),$$

such that the covariant derivatives $D^\mu \Psi_l^L(x)$ transform in the same way as the fields $\Psi_l^L(x)$ themselves, i.e.

$$D^\mu \Psi_l^L(x) \rightarrow \exp\left[ig\tau_j\omega_j(x)/2\right]D^\mu \Psi_l^L(x). \tag{17.36}$$

For finite functions $\omega_j(x)$, the resulting transformation law for the gauge fields $W_j^\mu(x)$ is quite complicated. However, it suffices to consider the transformations for infinitesimal functions $\omega_j(x)$, when the required infinitesimal transformations are given by

$$\begin{aligned}W_i^\mu(x) \rightarrow W_i^{\mu'}(x) &= W_i^\mu(x) + \delta W_i^\mu(x) \\ &\equiv W_i^\mu(x) - \partial^\mu\omega_i(x) - g\varepsilon_{ijk}\omega_j(x)W_k^\mu(x) \quad \left[\text{small } \omega_j(x)\right].\end{aligned} \tag{17.32b}$$

This result is closely analogous to the corresponding transformation law (11.26b) in QCD, and its derivation follows closely the derivation of Eq. (11.26b) given in Appendix 11.4

Before discussing the implications of the transformation laws (17.32), we shall consider the global U(l) transformations (17.31). The corresponding local phase transformations are

$$\left.\begin{aligned}\psi(x) &\rightarrow \psi'(x) = \exp\left[ig'Yf(x)\right]\psi(x) \\ \overline{\psi}(x) &\rightarrow \overline{\psi}'(x) = \overline{\psi}(x)\exp\left[-ig'Yf(x)\right]\end{aligned}\right\} \tag{17.37a}$$

where g' is a real number which will be determined later, $f(x)$ is an arbitrary real differentiable function, and $Y = -\frac{1}{2},\ -1,\ 0$ is the weak hypercharge associated with the fields $\Psi_l^L(x)$, $\psi_l^R(x)$ and $\psi_{\nu_l}^R(x)$ respectively. The analogy with QED is even closer in this case. One obtains a Lagrangian density invariant under the local phase transformations (17.37a) if in $\mathscr{L}_0$, Eq. (17.14), one replaces the ordinary derivatives by covariant derivatives,

$$\partial^\mu\psi(x) \rightarrow D^\mu\psi(x) = \left[\partial^\mu + ig'YB^\mu(x)\right]\psi(x), \tag{17.38}$$

where ψ is any one of the four lepton fields ψ_l^L, $\psi_{\nu_l}^L$, ψ_l^R and $\psi_{\nu_l}^R$, and the real gauge field $B^\mu(x)$, which has been introduced, transforms like

$$B^\mu(x) \rightarrow B^{\mu'}(x) = B^\mu(x) - \partial^\mu f(x). \tag{17.37b}$$

We have seen that making the replacement (17.34) in Eq. (17.14) yields a Lagrangian density which is invariant under the SU(2) gauge transformations (17.32a) and (17.32b), while the replacement (17.38) yields a Lagrangian density invariant under the U(l) gauge transformations (17.37a) and (17.37b). If we make both replacements (17.34) and (17.38) simultaneously in Eq. (17.14), we obtain the leptonic Lagrangian density

$$\mathscr{L}^{L} = i[\overline{\Psi}_{l}^{L}(x)\slashed{D}\Psi_{l}^{L}(x) + \overline{\psi}_{l}^{R}(x)\slashed{D}\psi_{l}^{R}(x) + \overline{\psi}_{\nu_{l}}^{R}(x)\slashed{D}\psi_{\nu_{l}}^{R}(x)], \tag{17.39}$$

where

$$D^{\mu}\Psi_{l}^{L}(x) = \left[\partial^{\mu} + ig\tau_{j}W_{j}^{\mu}(x)/2 - ig'B^{\mu}(x)/2\right]\Psi_{l}^{L}(x) \tag{17.40a}$$

$$D^{\mu}\psi_{l}^{R}(x) = \left[\partial^{\mu} - ig'B^{\mu}(x)\right]\psi_{l}^{R}(x) \tag{17.40b}$$

$$D^{\mu}\psi_{\nu_{l}}^{R}(x) = \partial^{\mu}\psi_{\nu_{l}}^{R}(x). \tag{17.40c}$$

We now define the fields $W_{i}^{\mu}(x)$ to be invariant under U(1) gauge transformations, and $B^{\mu}(x)$ to be invariant under SU(2) gauge transformations. It then follows that the Lagrangian density (17.39) is invariant under both SU(2) and U(l) gauge transformations. Such a Lagrangian density is said to be SU(2) × U(1) gauge invariant.

We can write the Lagrangian density (17.39) in the form

$$\mathscr{L}^{L} = \mathscr{L}_{0} + \mathscr{L}_{1}, \tag{17.41}$$

where $\mathscr{L}_0$ is the free-lepton Lagrangian density (17.14) and

$$\mathscr{L}_{1} = -gJ_{i}^{\mu}(x)W_{i\mu}(x) - g'J_{Y}^{\mu}(x)B_{\mu}(x) \tag{17.42}$$

represents the interaction of the weak isospin currents and the weak hypercharge current, Eqs. (17.20) and (17.25), with the gauge fields $W_{i\mu}(x)$ and $B_{\mu}(x)$.

In order to interpret the interaction $\mathscr{L}_1$ we rewrite Eq. (17.42). Using Eqs. (17.22), we write the weak isospin currents $J_{1}^{\mu}(x)$ and $J_{2}^{\mu}(x)$ in terms of the charged leptonic currents $J^{\mu}(x)$ and $J^{\mu\dagger}(x)$ and we introduce the non-Hermitian gauge field

$$W_{\mu}(x) = \frac{1}{\sqrt{2}}\left[W_{1\mu}(x) - iW_{2\mu}(x)\right] \tag{17.43}$$

and its adjoint $W_{\mu}^{\dagger}(x)$, in place of $W_{1\mu}(x)$ and $W_{2\mu}(x)$. In this way, one obtains for the first two terms of $\mathscr{L}_1$

$$-g\sum_{i=1}^{2}J_{i}^{\mu}(x)W_{i\mu}(x) = \frac{-g}{2\sqrt{2}}\left[J^{\mu\dagger}(x)W_{\mu}(x) + J^{\mu}(x)W_{\mu}^{\dagger}(x)\right]. \tag{17.44}$$

In the remaining two terms of $\mathscr{L}_1$, we write $W_{3\mu}(x)$ and $B_{\mu}(x)$ as linear combinations of two different Hermitian fields $A_{\mu}(x)$ and $Z_{\mu}(x)$, defined by

$$\left.\begin{array}{l}W_{3\mu}(x) = \cos\theta_{W}Z_{\mu}(x) + \sin\theta_{W}A_{\mu}(x) \\ B_{\mu}(x) = -\sin\theta_{W}Z_{\mu}(x) + \cos\theta_{W}A_{\mu}(x)\end{array}\right\}. \tag{17.45}$$

The angle θ_W, which specifies the mixture of Z_μ and A_μ fields in $W_{3\mu}$ and B_μ, is known as the weak mixing angle (or the Weinberg angle), and we shall consider its determination later. From Eqs. (17.45) and

$$J_Y^\mu(x) = s^\mu(x)/e - J_3^\mu(x), \tag{17.25}$$

we obtain

$$
\begin{aligned}
-gJ_3^\mu(x)&W_{3\mu}(x) - g'J_Y^\mu(x)B_\mu(x) \\
&= -\frac{g'}{e}s^\mu(x)\left[-\sin\theta_W Z_\mu(x) + \cos\theta_W A_\mu(x)\right] \\
&\quad - J_3^\mu(x)\{g\left[\cos\theta_W Z_\mu(x) + \sin\theta_W A_\mu(x)\right] \\
&\quad - g'\left[-\sin\theta_W Z_\mu(x) + \cos\theta_W A_\mu(x)\right]\}.
\end{aligned}
\tag{17.46}
$$

We now demand that the gauge field $A_\mu(x)$, defined by Eqs. (17.45), is the electromagnetic field and is coupled to electric charges in the usual way, i.e. through the usual term $-s^\mu(x)A_\mu(x)$ in the interaction Lagrangian density. This means that in Eq. (17.46) the coefficient of $J_3^\mu(x)A_\mu(x)$ must vanish, and that of $s^\mu(x)A_\mu(x)$ must be (-1), i.e. we require

$$g\sin\theta_W = g'\cos\theta_W = e. \tag{17.47}$$

Substituting Eqs. (17.44) and (17.46) in Eq. (17.42), and eliminating g' by means of Eq. (17.47), we obtain as the final expression for the interaction Lagrangian density

$$
\begin{aligned}
\mathscr{L}_I = &-s^\mu(x)A_\mu(x) - \frac{g}{2\sqrt{2}}\left[J^{\mu\dagger}(x)W_\mu(x) + J^\mu(x)W_\mu^\dagger(x)\right] \\
&- \frac{g}{\cos\theta_W}\left[J_3^\mu(x) - \sin^2\theta_W s^\mu(x)/e\right]Z_\mu(x).
\end{aligned}
\tag{17.48}
$$

The $SU(2) \times U(1)$ gauge-invariant interaction (17.48), first introduced by Glashow in 1961, is eminently satisfactory as a Lagrangian density describing the electromagnetic and weak interactions of leptons. The first term, obtained by imposing the conditions (17.47) on the coupling constants, is the familiar interaction of QED. The second term is just the IVB interaction Lagrangian density corresponding to Eq. (16.7) provided we set

$$g_W = \frac{g}{2\sqrt{2}}. \tag{17.49}$$

Thus the quanta of the gauge field $W(x)$ are just the $W^\pm$ vector bosons. The third term in Eq. (17.48) represents a neutral current

$$
\begin{aligned}
J_3^\mu(x) &- \sin^2\theta_W s^\mu(x)/e \\
&= \frac{1}{4}\overline{\psi}_{\nu_l}(x)\gamma^\mu(1-\gamma_5)\psi_{\nu_l}(x) - \frac{1}{4}\overline{\psi}_l(x)\gamma^\mu\left[(1-4\sin^2\theta_W)-\gamma_5\right]\psi_l(x)
\end{aligned}
\tag{17.50}
$$

coupled to a real vector field $Z_\mu(x)$. If we interpret the quanta of this field as the electrically neutral vector bosons Z^0 of Section 16.7 then this term will lead to neutral current processes of the type shown in Fig. 16.12 Many neutral current processes have now been observed, as will be discussed in Chapter 19, and good agreement between theory and experiment is obtained by taking

$$\sin^2\theta_W = 0.23122 \pm 0.00015, \tag{17.51}$$

i.e. the last term in Eq. (17.50) is almost a pure axial current. This agreement between theory and experiment is strong support for the unified theory of electromagnetic and weak interactions. On the other hand, taking $\theta_W=0$ in Eqs. (17.45) leads to a SU(2) gauge-invariant theory of weak interactions alone, i.e. weak and electromagnetic interactions are decoupled. Such a theory is ruled out by experiment.

17.4   Properties of the Gauge Bosons

The Lagrangian density (17.41), which we have considered, describes the free leptons and their interactions with the gauge fields. The complete Lagrangian density must also contain terms which describe these gauge bosons when no leptons are present. These terms must also be SU(2) × U(1) gauge-invariant. As for the leptons, we shall for the moment assume the gauge bosons to have zero masses, and we shall return to the question of non-vanishing masses in the next section.

For the $B^\mu(x)$ field, it is easy to construct suitable terms. The U(l) gauge transformation law (17.37b) of this field has the same form as that of the electromagnetic field $A^\mu(x)$, Eq. (17.5b). Hence in analogy to the electromagnetic case, a U(l) gauge-invariant Lagrangian density for the $B^\mu(x)$ field is given by

$$-\frac{1}{4}B_{\mu\nu}(x)B^{\mu\nu}(x),\tag{17.52}$$

where

$$B^{\mu\nu}(x) \equiv \partial^\nu B^\mu(x) - \partial^\mu B^\nu(x)\tag{17.53}$$

is the analogue of the electromagnetic field tensor $F^{\mu\nu}(x)$, Eq. (5.5). The SU(2) gauge invariance of the expression (17.52) follows from that of the $B^\mu(x)$ field.

The situation is more complicated for the $W_i^\mu(x)$ fields. The expression analogous to Eq. (17.52) is

$$-\frac{1}{4}F_{i\mu\nu}(x)F_i^{\mu\nu}(x),\tag{17.54}$$

where

$$F_i^{\mu\nu}(x) \equiv \partial^\nu W_i^\mu(x) - \partial^\mu W_i^\nu(x).\tag{17.55}$$

However, this expression is not invariant under the transformation (17.32b), on account of the term

$$-g\varepsilon_{ijk}\omega_j(x)W_k^\mu(x)$$

in Eq. (17.32b). To restore invariance, additional interaction terms must again be introduced. This is achieved by replacing $F_i^{\mu\nu}(x)$ by

$$G_i^{\mu\nu}(x) \equiv F_i^{\mu\nu}(x) + g\varepsilon_{ijk}W_j^\mu(x)W_k^\nu(x)\tag{17.56}$$

in Eq. (17.54), i.e. the expression

$$\mathscr{L}_G = -\frac{1}{4}G_{i\mu\nu}(x)G_i^{\mu\nu}(x),\tag{17.57}$$

is SU(2) gauge-invariant. Eqs. (17.56) and (17.57) are closely analogous to the corresponding Eqs. (11.33) and (11.34) in QCD, and the demonstration of the gauge invariance of Eq. (17.57) follows closely that given for Eq. (11.34) in Appendix 11.4.2. The invariance of $\mathscr{L}_G$ under U(1) transformations follows trivially from that of the $W_i^\mu(x)$ fields.

Combining expressions (17.52) and (17.57), and substituting Eq. (17.56), we obtain the complete SU(2) × U(1) gauge-invariant Lagrangian density for the gauge bosons

$$\mathscr{L}^B = -\frac{1}{4}B_{\mu\nu}(x)B^{\mu\nu}(x) - \frac{1}{4}G_{i\mu\nu}(x)G_i^{\mu\nu}(x) \tag{17.58a}$$

$$= -\frac{1}{4}B_{\mu\nu}(x)B^{\mu\nu}(x) - \frac{1}{4}F_{i\mu\nu}(x)F_i^{\mu\nu}(x)$$

$$+ g\varepsilon_{ijk}W_{i\mu}(x)W_{j\nu}(x)\partial^\mu W_k^\nu(x)$$

$$- \frac{1}{4}g^2\varepsilon_{ijk}\varepsilon_{ilm}W_j^\mu(x)W_k^\nu(x)W_{l\mu}(x)W_{m\nu}(x). \tag{17.58b}$$

The first two terms in Eq. (17.58b) represent the Lagrangian density $\mathscr{L}_0^B$ of the free (i.e. non-interacting) gauge fields. In terms of the fields $A^\mu(x)$, $Z^\mu(x)$ and $W^\mu(x)$, $\mathscr{L}_0^B$ becomes

$$\mathscr{L}_0^B = -\frac{1}{4}F_{\mu\nu}(x)F^{\mu\nu}(x) - \frac{1}{2}F^\dagger_{W\mu\nu}(x)F_W^{\mu\nu}(x) - \frac{1}{4}Z_{\mu\nu}(x)Z^{\mu\nu}(x), \tag{17.59}$$

where $F^{\mu\nu}(x)$ is the electromagnetic field tensor (5.5), $F_W^{\mu\nu}(x)$ is the corresponding tensor (16.21b) for the $W^\mu(x)$ field, and

$$Z^{\mu\nu}(x) \equiv \partial^\nu Z^\mu(x) - \partial^\mu Z^\nu(x) \tag{17.60}$$

is similarly associated with the $Z^\mu(x)$ field. Eq. (17.59) is thus the free-field Lagrangian density for mass zero, spin one γ, $W^\pm$ and Z^0 bosons.

In contrast, the third and fourth terms in Eq. (17.58b) represent interactions of the gauge bosons amongst themselves. In perturbation theory, these terms generate three- or four-line vertices. Some examples of such vertices are shown in Fig. 17.1. These boson self-interactions are one of the most striking features of the theory. They arise because the $W_i^\mu(x)$ fields, which transmit the interactions between the weak isospin currents, themselves are weak isospin vectors, i.e. they carry weak isospin charge. This is in contrast to QED, where the electromagnetic interactions are transmitted by the photons, but the latter

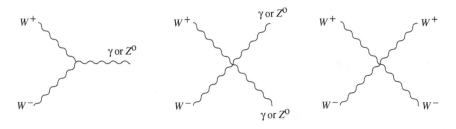

Figure 17.1   *Some examples of three-line and four-line vertices generated by the boson self-interaction terms in Eq. (17.58)*

do not carry electric charge. Consequently, there are no photon self-coupling terms in QED.

17.5   Lepton and Gauge Boson Masses

So far we have assumed that all leptons and gauge bosons are massless. Except for the photon, this assumption is certainly unrealistic.

To describe massive $W^\pm$ and Z^0 bosons, we can add the mass terms

$$m_W^2 W_\mu^\dagger(x)W^\mu(x) + \frac{1}{2}\, m_Z^2 Z_\mu(x)Z^\mu(x) \tag{17.61}$$

to the Lagrangian density (17.59). For the $W^\pm$ particles, one obtains in this way the Lagrangian density (16.21a) of the IVB theory of weak interactions. Adding such mass terms results in a Lagrangian density which is not invariant under the transformations (17.32) and (17.37), i.e. it violates both SU(2) and U(1) gauge invariance. It also reintroduces all the renormalization problems associated with massive vector bosons, as discussed in Section 16.7.

We could similarly introduce non-zero lepton masses. For example, for the electron one could add the mass term,

$$-m_e \overline{\psi}_e(x)\psi_e(x) \tag{17.62a}$$

to the Lagrangian density. Unfortunately, this term is again not SU(2) × U(1) gauge invariant. This follows since the expression (17.62a) can be written

$$-m_e \overline{\psi}_e(x)[P_R + P_L]\psi_e(x) = -m_e\left[\overline{\psi}_e^L(x)\psi_e^R(x) + \overline{\psi}_e^R(x)\psi_e^L(x)\right], \tag{17.62b}$$

and ψ_e^R and $\overline{\psi}_e^R$ are isoscalars while ψ_e^L and $\overline{\psi}_e^L$ are isospinors.

We thus arrive at the conclusion that, if we wish to preserve SU(2) × U(1) gauge invariance, we must set the masses of the leptons and of the $W^\pm$ and Z^0 bosons equal to zero. We could at this point simply add the required lepton and boson mass terms, as outlined above, violating SU(2) × U(1) gauge invariance. The resulting model is known as the Glashow model. In lowest-order perturbation theory calculations, this model yields results which are in good agreement with present experiments for appropriately chosen values of $\sin\theta_W$, m_W, and m_Z. However, the Glashow model is not renormalizable. We shall therefore insist on retaining the gauge invariance of the Lagrangian density. In the next chapter, we shall learn how to introduce non-zero masses by the mechanism of spontaneous symmetry breaking. In this way we shall obtain a renormalizable theory which is in agreement with experiment and which, for given $\sin\theta_W$, predicts the gauge boson masses m_W and m_Z.

18

Spontaneous Symmetry Breaking

In the previous chapter, a gauge-invariant and renormalizable unified theory of weak and electromagnetic interactions was obtained. However, all leptons and gauge bosons had to have zero mass. In reality, only photons are massless, but the leptons and the $W^{\pm}$ and Z^0 bosons have non-zero masses. We have seen that the *ad hoc* addition of mass terms to the Lagrangian density spoils the gauge invariance and the renormalizability of the theory. In order to obtain a renormalizable theory, it is essential to introduce the masses by a mechanism which retains the gauge invariance of the Lagrangian density. In this chapter we shall develop a quite remarkable such mechanism, that of spontaneous symmetry breaking.

As in the last chapter, we shall proceed from global to local phase invariance, i.e. to gauge invariance. In Section 18.1, we shall introduce the idea of spontaneous symmetry breaking and shall consider the simplest field-theoretic example of it: the Goldstone model, which is a field theory invariant under global U(1) phase transformations. The Goldstone model necessarily leads to zero-mass bosons (other than photons) which are not observed in nature. This undesirable feature is absent when applying spontaneous symmetry breaking to a theory which is invariant under *local* phase transformations, i.e. to a gauge theory. In Section 18.2, this will be demonstrated for a field theory invariant under U(l) gauge transformations (the Higgs model). It will then be easy (in Section 18.3) to generalize these results for the SU(2) × U(1) gauge-invariant electroweak theory of the previous chapter. In this way, we shall obtain a Lagrangian density which is gauge-invariant, renormalizable and contains mass terms for leptons and for the $W^{\pm}$ and Z^0 bosons, while the photon remains massless. The resulting theory is known as the standard electroweak theory. It was first formulated in this way, independently, by Weinberg in 1967 and by Salam in 1968. Its interpretation and applications will form the topic of the last chapter of this book.

18.1   The Goldstone Model

In order to explain the idea of spontaneous symmetry breaking, we consider a system whose Lagrangian L possesses a particular symmetry, i.e. it is invariant under the corresponding symmetry transformations.[1] (For example, L might be spherically symmetric, i.e. invariant under spatial rotations.) In classifying the energy levels of this system, essentially two situations can occur. Firstly, if a given energy level is non-degenerate, the corresponding energy eigenstate is unique and invariant under the symmetry transformations of L. Secondly, the energy level may be degenerate and the corresponding eigenstates are not invariant, but transform linearly amongst themselves under the symmetry transformations of L. In particular, consider the lowest energy level of the system. If it is non-degenerate, the state of lowest energy of the system (its ground state) is unique and possesses the symmetries of L. In the second case, of degeneracy, there is no unique eigenstate to represent the ground state. If we arbitrarily select one of the degenerate states as the ground state, then the ground state no longer shares the symmetries of L. This way of obtaining an asymmetric ground state is known as *spontaneous symmetry breaking*. The asymmetry is not due to adding a non-invariant asymmetric term to L, but to the arbitrary choice of one of the degenerate states.

Ferromagnetism represents a familiar example of spontaneous symmetry breaking. In a ferromagnetic material, the forces which couple the electronic spins and hence the Hamiltonian of the system are rotationally invariant. However, in the ground state, the spins are aligned in some definite direction resulting in a non-zero magnetization **M**. The orientation of **M** is arbitrary and we are clearly dealing with a case of degeneracy. **M** could equally well point in any other direction and all properties of the system, other than the direction of **M**, would remain unchanged. An important feature of this asymmetric ground state is that excited states obtained from it by small perturbations also display this asymmetry.

In field theory, the state of lowest energy is the vacuum. Spontaneous symmetry breaking is only relevant to field theory if the vacuum state is non-unique. This very bold and startling idea was first suggested by Nambu and his co-workers. It implies that some quantity in the vacuum is non-vanishing, is not invariant under the symmetry transformations of the system, and can therefore be used to characterize a particular vacuum state as *the* ground state. In the following we shall assume this quantity to be the vacuum expectation value of a quantized field. If we require the vacuum states to be invariant under Lorentz transformations and under translations, then this field must be a scalar field, $\phi(x)$, and its vacuum expectation value must be constant:

$$\langle 0|\phi(x)|0\rangle = c \neq 0. \tag{18.1}$$

In contrast, the vacuum expectation value of any spinor field $\psi(x)$ or any vector field $V^\mu(x)$ must vanish:

$$\langle 0|\psi(x)|0\rangle = 0, \qquad \langle 0|V^\mu(x)|0\rangle = 0. \tag{18.2}$$

---

[1] Instead of the symmetries of L, we could talk of those of the Hamiltonian or of the equations of motion or, in the case of a field theory, of the Lagrangian density.

The simplest example of a field theory exhibiting spontaneous symmetry breaking is the Goldstone model. Its Lagrangian density is

$$\mathcal{L}(x) = [\partial^\mu \phi^*(x)][\partial_\mu \phi(x)] - \mu^2 |\phi(x)|^2 - \lambda |\phi(x)|^4, \tag{18.3}$$

with

$$\phi(x) = \frac{1}{\sqrt{2}}[\phi_1(x) + i\phi_2(x)] \tag{18.4}$$

a complex scalar field, and μ^2 and λ arbitrary real parameters. To begin with, we shall consider a classical field theory, i.e. $\phi(x)$ is a classical and not a quantized field, and μ is not to be interpreted as a particle mass.

The Lagrangian density (18.3) is invariant under the global U(1) phase transformations

$$\phi(x) \rightarrow \phi'(x) = \phi(x)e^{i\alpha}, \quad \phi^*(x) \rightarrow \phi^{*\prime}(x) = \phi^*(x)e^{-i\alpha}. \tag{18.5}$$

We shall see that this symmetry is spontaneously broken in this model.

The Hamiltonian density of this theory follows from Eq. (18.3) and the general relations (2.22) and (2.25), and is

$$\mathcal{H}(x) = [\partial^0 \phi^*(x)][\partial_0 \phi(x)] + [\nabla \phi^*(x)] \cdot [\nabla \phi(x)] + \mathcal{V}(\phi) \tag{18.6}$$

where

$$\mathcal{V}(\phi) = \mu^2 |\phi(x)|^2 + \lambda |\phi(x)|^4 \tag{18.7}$$

is the potential energy density of the field. For the energy of the field to be bounded from below, we require $\lambda > 0$. The first two terms in Eq. (18.6) are positive definite and vanish for constant $\phi(x)$. It follows that the minimum value of $\mathcal{H}(x)$, and hence of the total energy of the field, corresponds to that constant value of $\phi(x)$ which minimizes $\mathcal{V}(\phi)$. Two different situations occur, depending on the sign of μ^2:

(i) $\mu^2 > 0$. In this case, the two terms in $\mathcal{V}(\phi)$ are also positive definite. In Fig. 18.1(a) we sketch the corresponding potential energy surface $\mathcal{V}(\phi)$ as a function of $\phi_1(x)$ and $\phi_2(x)$. $\mathcal{V}(\phi)$ has an absolute minimum for the unique value $\phi(x) = 0$, i.e. spontaneous symmetry breaking cannot occur. This is the situation with which we are familiar. Omitting the quartic term $\lambda |\phi(x)|^4$, the above expressions for $\mathcal{L}(x)$ and $\mathcal{H}(x)$ are those of the free complex Klein–Gordon field. Classically, they give rise to normal modes of oscillation about the stable equilibrium position $\phi(x) = 0$. On quantization, they give rise to charged spin-0 particles of mass μ (see Section 3.2). The ground state, i.e. the vacuum state, is unique and $\langle 0|\phi(x)|0 \rangle = 0$. We can think of Eqs. (18.3), (18.6) and (18.7) as expansions in powers of $\phi(x)$ and $\phi^*(x)$ about the stable equilibrium configuration $\phi(x) = 0$ and, in our approach, treat $\lambda |\phi(x)|^4$ by perturbation theory. In the quantized theory, this term represents a self-interaction of the particles.

(ii) $\mu^2 < 0$. The potential energy surface for this case is shown in Fig. 18.1(b). $\mathcal{V}(\phi)$ possesses a local maximum at $\phi(x) = 0$ and a whole circle of absolute minima at

$$\phi(x) = \phi_0 = \left(\frac{-\mu^2}{2\lambda}\right)^{1/2} e^{i\theta}, \quad 0 \leq \theta < 2\pi, \tag{18.8}$$

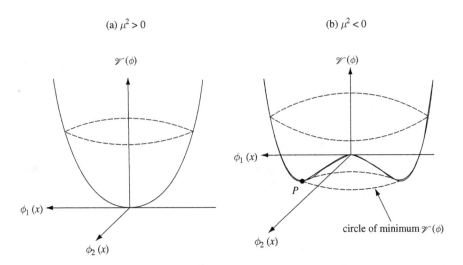

Figure 18.1　*The potential energy density $\mathscr{V}(\phi) = \mu^2|\phi(x)|^2 + \lambda|\phi(x)|^4$, Eq. (18.7), for $\lambda > 0$. In case (b), our choice $(\phi_1, \phi_2) = (v, 0)$ for the vacuum ground state corresponds to the point labeled P on the circle of minimum $\mathscr{V}(\phi)$*

where the phase angle θ defines a direction in the complex ϕ-plane. We see that the state of lowest energy, the vacuum state, is not unique in this case. This arbitrariness in the direction θ is analogous to that in the direction of the magnetization $\mathbf{M}$ of a ferromagnet. Analogously to the latter case, spontaneous symmetry breaking will occur if we choose one particular direction θ to represent the vacuum ground state. Because of the invariance of the Lagrangian density (18.3) under the global phase transformations (18.5), the value of θ chosen is not significant and we shall take $\theta = 0$, so that

$$\phi_0 = \left(\frac{-\mu^2}{2\lambda}\right)^{1/2} = \frac{1}{\sqrt{2}}v \quad (>0) \tag{18.9}$$

is purely real.

We now introduce two real fields $\sigma(x)$ and $\eta(x)$ through the equation

$$\phi(x) = \frac{1}{\sqrt{2}}[v + \sigma(x) + i\eta(x)]. \tag{18.10}$$

$\sigma(x)$ and $\eta(x)$ measure the deviations of the field $\phi(x)$ from the equilibrium ground state configuration $\phi(x) = \phi_0$. In terms of these fields, the Lagrangian density (18.3) becomes

$$\begin{aligned}
\mathscr{L}(x) = {} & \frac{1}{2}[\partial^\mu \sigma(x)][\partial_\mu \sigma(x)] - \frac{1}{2}(2\lambda v^2)\sigma^2(x) \\
& + \frac{1}{2}[\partial^\mu \eta(x)][\partial_\mu \eta(x)] \\
& - \lambda v \sigma(x)[\sigma^2(x) + \eta^2(x)] - \frac{1}{4}\lambda[\sigma^2(x) + \eta^2(x)]^2,
\end{aligned} \tag{18.11}$$

where we have omitted a constant term which is of no consequence.

Eqs. (18.3) and (18.11) are the same Lagrangian density expressed in terms of different variables. Thus they are entirely equivalent and must lead to the same physical results. This equivalence only holds for exact solutions of the theory. We shall be using perturbation theory, and for approximate solutions the picture is very different. For $\mu^2 < 0$, we cannot proceed as we did for $\mu^2 > 0$, i.e. by treating the quartic term $\lambda|\phi(x)|^4$ in Eqs. (18.3), (18.6) and (18.7) as a perturbation of the other terms which are bilinear in $\phi(x)$ and $\phi^*(x)$. For $\mu^2 < 0, \phi(x) = 0$ is an unstable equilibrium configuration, and one cannot carry out perturbation calculations about an unstable solution. That this procedure leads to nonsense for $\mu^2 < 0$ shows up very clearly in the quantized theory where the unperturbed system corresponds to particles of imaginary mass, and no finite order of perturbation theory can put this right.

In contrast, Eq. (18.11) suggests a different quantization procedure. We shall now treat the terms which are quadratic in $\sigma(x)$ and $\eta(x)$, i.e. the first three terms in Eq. (18.11), as the free Lagrangian density

$$\mathscr{L}_0(x) = \frac{1}{2}[\partial^\mu \sigma(x)][\partial_\mu \sigma(x)] - \frac{1}{2}(2\lambda v^2)\sigma^2(x)$$
$$+ \frac{1}{2}[\partial^\mu \eta(x)][\partial_\mu \eta(x)] \tag{18.12}$$

and the remaining terms, which are cubic and quartic in $\sigma(x)$ and $\eta(x)$, as interactions. The fields $\sigma(x)$ and $\eta(x)$ measure the deviations from the stable equilibrium configuration $\phi(x) = \phi_0$, and one would expect to be able to treat the interaction terms by perturbation theory about this stable solution.

The Lagrangian density (18.12) contains no terms which couple $\sigma(x)$ and $\eta(x)$, i.e. the fields $\sigma(x)$ and $\eta(x)$ are normal coordinates of Eq. (18.12). Comparing Eqs. (18.12) and (3.4), we see that $\sigma(x)$ and $\eta(x)$ are real Klein–Gordon fields. On quantization, both fields lead to neutral spin-0 particles: the σ boson with the (real positive) mass $\sqrt{(2\lambda v^2)}$, and the η boson which has zero mass since there is no term in $\eta^2(x)$ in Eq. (18.12). Since, by definition, there are no particles present in the vacuum, it follows from Eqs. (18.9) and (18.10) that

$$\langle 0|\phi(x)|0\rangle = \phi_0. \tag{18.13}$$

This is the condition for spontaneous symmetry breaking in the quantized theory, analogous to Eqs. (18.8) and (18.9) in the classical theory.

The origin of the above mass spectrum can be understood from Fig. 18.1(b) by considering small displacements $\sigma(x)$ and $\eta(x)$ from the equilibrium configuration $\phi(x) = \phi_0$. $\sigma(x)$ represents a displacement in the radial plane $\phi_2(x) = 0$ in which the potential energy density $\mathscr{V}(\phi)$ increases quadratically with $\sigma(x)$. On the other hand, $\eta(x)$ represents a displacement along the valley of minimum potential energy where $\mathscr{V}(\phi)$ is constant, so that the corresponding quantum excitations – the η bosons – are massless. Thus the zero mass of the η bosons is a consequence of the degeneracy of the vacuum. Such zero-mass bosons frequently occur in theories with spontaneous symmetry breaking and they are known as *Goldstone bosons*.

No Goldstone bosons are observed in nature. It is therefore of particular interest that gauge theories with spontaneous symmetry breaking do not generate Goldstone bosons. In the next section we shall demonstrate this for a simple model, before turning to the realistic $SU(2) \times U(1)$ gauge theory of the previous chapter.

18.2   The Higgs Model

The Goldstone model is easily generalized to be invariant under $U(1)$ gauge transformations. As in the last chapter, we introduce a gauge field $A_\mu(x)$, replace the ordinary derivatives in the Goldstone Lagrangian density (18.3) by the covariant derivatives

$$D_\mu\phi(x) = [\partial_\mu + iqA_\mu(x)]\phi(x)$$

and add the Lagrangian density of the free gauge field

$$-\frac{1}{4}F_{\mu\nu}(x)F^{\mu\nu}(x), \qquad (18.14)$$

where, as usual,

$$F_{\mu\nu}(x) = \partial_\nu A_\mu(x) - \partial_\mu A_\nu(x).$$

In this way we obtain the Lagrangian density

$$\mathscr{L}(x) = [D^\mu\phi(x)]^*[D_\mu\phi(x)] - \mu^2|\phi(x)|^2 - \lambda|\phi(x)|^4 - \frac{1}{4}F_{\mu\nu}(x)F^{\mu\nu}(x), \qquad (18.15)$$

which defines the Higgs model. Eq. (18.15) is invariant under the $U(1)$ gauge transformations

$$\left. \begin{array}{l} \phi(x) \rightarrow \phi'(x) = \phi(x)e^{-iqf(x)} \\ \phi^*(x) \rightarrow \phi^{*\prime}(x) = \phi^*(x)e^{iqf(x)} \\ A_\mu(x) \rightarrow A'_\mu(x) = A_\mu(x) + \partial_\mu f(x) \end{array} \right\}. \qquad (18.16)$$

The further analysis parallels that for the Goldstone model. We again start from a classical theory. Taking $\lambda > 0$, two situations arise. For $\mu^2 > 0$, the state of lowest energy corresponds to both $\phi(x)$ and $A_\mu(x)$ vanishing, so that spontaneous symmetry breaking cannot occur.

For $\mu^2 > 0$, the vacuum state is not unique, leading to spontaneous symmetry breaking. To ensure Lorentz invariance, the vector field $A_\mu(x)$ must vanish for the vacuum, but we again obtain a circle of minimum $\mathscr{H}(x)$ corresponding to $\phi(x)$ taking on the values ϕ_0, given by Eq. (18.8). As for the Goldstone model, we choose the real value (18.9) for ϕ_0 and define the real fields $\sigma(x)$ and $\eta(x)$ by Eq. (18.10). In terms of these fields, the Lagrangian density (18.15) becomes

$$\begin{aligned} \mathscr{L}(x) = {} & \frac{1}{2}[\partial^\mu\sigma(x)][\partial_\mu\sigma(x)] - \frac{1}{2}(2\lambda v^2)\sigma^2(x) \\ & - \frac{1}{4}F_{\mu\nu}(x)F^{\mu\nu}(x) + \frac{1}{2}(qv)^2 A_\mu(x)A^\mu(x) \\ & + \frac{1}{2}[\partial^\mu\eta(x)][\partial_\mu\eta(x)] \\ & + qvA^\mu(x)\partial_\mu\eta(x) + \text{`interaction terms'}, \end{aligned} \qquad (18.17)$$

where the 'interaction terms', which we have not given explicitly, are cubic and quartic in the fields, and where an insignificant constant term has been discarded.

The direct interpretation of Eq. (18.17) leads to difficulties. The first line of this equation describes a real Klein–Gordon field which on quantization gives uncharged spin-0 bosons with mass $\sqrt{(2\lambda v^2)}$. However, the product term $A^\mu(x)\partial_\mu\eta(x)$ shows that $A^\mu(x)$ and $\eta(x)$ are not independent normal coordinates, and one cannot conclude that the second and third lines of Eq. (18.17) describe massive vector bosons and massless scalar bosons respectively.[2] This difficulty also shows up if we count degrees of freedom for the Lagrangian densities (18.15) and (18.17). Eq. (18.15) has four degrees of freedom: two from the complex scalar field $\phi(x)$, and two from the real massless vector field $A^\mu(x)$ (i.e. for massless photons there are only two independent polarization states, the third is eliminated by gauge invariance). In Eq. (18.17), the real scalar fields $\sigma(x)$ and $\eta(x)$ each represent one degree and the real massive vector field $A^\mu(x)$ contributes three degrees (corresponding to three independent polarization states), i.e. the transformed Lagrangian density (18.17) appears to have five degrees of freedom. Of course, a change of variables cannot alter the number of degrees of freedom of a system. We must conclude that the Lagrangian density (18.17) contains an unphysical field which does not represent real particles and which can be eliminated.

The scalar field $\eta(x)$ can be eliminated from Eq. (18.17). For any complex field $\phi(x)$, a gauge transformation of the form (18.16) can be found which transforms $\phi(x)$ into a real field of the form

$$\phi(x) = \frac{1}{\sqrt{2}}[v + \sigma(x)]. \tag{18.18}$$

The gauge in which the transformed field has this form is called the *unitary gauge*. (We shall continue to label the transformed field $\phi(x)$ and not $\phi'(x)$, etc.) Substituting Eq. (18.18) into Eq. (18.15) gives

$$\mathscr{L}(x) = \mathscr{L}_0(x) + \mathscr{L}_1(x), \tag{18.19a}$$

where we have separated the quadratic terms

$$\mathscr{L}_0(x) = \frac{1}{2}[\partial^\mu\sigma(x)][\partial_\mu\sigma(x)] - \frac{1}{2}(2\lambda v^2)\sigma^2(x)$$
$$- \frac{1}{4}F_{\mu\nu}(x)F^{\mu\nu}(x) + \frac{1}{2}(qv)^2 A_\mu(x)A^\mu(x) \tag{18.19b}$$

from the higher-order interaction terms

$$\mathscr{L}_1(x) = -\lambda v\sigma^3(x) - \frac{1}{4}\lambda\sigma^4(x)$$
$$+ \frac{1}{2}q^2 A_\mu(x)A^\mu(x)[2v\sigma(x) + \sigma^2(x)]. \tag{18.19c}$$

$\mathscr{L}_0(x)$ contains no terms which couple $\sigma(x)$ and $A_\mu(x)$. Hence, treating $\mathscr{L}_1(x)$ in perturbation theory, we can interpret $\mathscr{L}_0(x)$ as the free-field Lagrangian density of a real

---

[2] We do not include this term as a part of the interaction, to be treated in perturbation theory, since it is of the same (second) order in the fields as the first five terms in Eq. (18.17).

Klein–Gordon field $\sigma(x)$ and a real massive vector field $A_\mu(x)$. On quantizing $\mathscr{L}_0(x)$, $\sigma(x)$ gives rise to neutral scalar bosons of mass $\sqrt{(2\lambda v^2)}$, and $A_\mu(x)$ to neutral vector bosons of mass $|qv|$.

This is a remarkable result! Having started from the Lagrangian density (18.15) for a *complex* scalar field and a *massless* real vector field, we have ended up with the Lagrangian density (18.19) for a *real* scalar field and a *massive* real vector field. The number of degrees of freedom is four in both cases. Of the two degrees of freedom of the complex field $\phi(x)$, one has been taken up by the vector field $A_\mu(x)$ which has become massive in the process; the other shows up as the real field $\sigma(x)$. This phenomenon by which a vector boson acquires mass without destroying the gauge invariance of the Lagrangian density is known as the *Higgs mechanism*, and the massive spin-0 boson associated with the field $\sigma(x)$ is called a Higgs boson or a Higgs scalar. The Higgs mechanism does not generate Goldstone bosons, in contrast to the spontaneous symmetry breaking of the global phase invariance of the Goldstone model. In essence, the field $\eta(x)$ in Eq. (18.17), which in the Goldstone model was associated with the massless Goldstone bosons, has been eliminated by gauge invariance, and the degree of freedom of $\eta(x)$ has been transferred to the vector field $A_\mu(x)$.

The Higgs mechanism also works for non-Abelian gauge theories. In the next section we shall apply it to the $SU(2) \times U(1)$ gauge theory of Chapter 17 and this will lead directly to the standard electroweak theory. First, however, we comment briefly on the renormalizability of such theories.

Unlike the IVB theory or the Glashow model of weak interactions, gauge theories have the great merit of being renormalizable, allowing meaningful calculations to be made in higher orders of perturbation theory. The proof of renormalizability has been given by 't Hooft, Veltman and others. Unfortunately, it is very complicated. We shall content ourselves with briefly indicating the underlying ideas, using the Higgs model for illustration.[3]

The second line of Eq. (18.19b) is identical with the Lagrangian density of a massive neutral vector boson field and leads to the propagator

$$\mathrm{i}D_{\mathrm{F}}^{\alpha\beta}(k, m) = \frac{\mathrm{i}\left(-g^{\alpha\beta} + k^\alpha k^\beta/m^2\right)}{k^2 - m^2 + \mathrm{i}\varepsilon}, \tag{18.20}$$

where $m = |qv|$. We met this propagator in the IVB theory [see Eq. (16.30)] and it was suggested in Section 16.7 that the $k^\alpha k^\beta/m^2$ in this propagator makes the IVB theory non-renormalizable. However, we saw in Section 9.8 that this kind of dimensional argument (i.e. counting powers of momenta in loop integrals) predicts the maximum possible degree of divergence. For gauge theories with spontaneously broken symmetry, the actual divergences are less severe and the theories are renormalizable. This could be due to the propagator being coupled to exactly conserved currents or to the exact cancellation of divergences arising from different Feynman graphs of the same order.

---

[3] The interested reader will find proper treatments in E. S. Abers and B. W. Lee, *Gauge Theories*, Physics Reports, 9C, No. 1 (1973), and J. C. Taylor, *Gauge Theories and Weak Interactions*, Cambridge University Press, Cambridge, 1976.

The renormalizability of the Higgs model is difficult to prove from the Lagrangian density (18.19) which employs the unitary gauge. Instead, 't Hooft proceeds in a way reminiscent of QED. There we replaced the gauge-invariant Lagrangian density

$$\mathscr{L} = -\frac{1}{4}F_{\mu\nu}(x)F^{\mu\nu}(x) \tag{18.21}$$

by

$$\mathscr{L} = -\frac{1}{4}F_{\mu\nu}(x)F^{\mu\nu}(x) - \frac{1}{2}[\partial_\mu A^\mu(x)]^2, \tag{18.22}$$

which is equivalent to Eq. (18.21) provided we work in a Lorentz gauge, i.e.

$$\partial_\mu A^\mu(x) = 0.^4 \tag{18.23}$$

In the present case, 't Hooft imposes the gauge condition

$$\partial_\mu A^\mu(x) - m\eta(x) = 0 \tag{18.24}$$

on the fields, where $m = |qv|$ and $\eta(x)$ is defined by Eq. (18.10). 't Hooft shows that, if condition (18.24) holds, one may add the term

$$-\frac{1}{2}[\partial_\mu A^\mu(x) - m\eta(x)]^2 \tag{18.25}$$

to the Lagrangian density (18.17), i.e. the modified Lagrangian density and Eq. (18.17) leads to the same predictions for observable quantities.[5] Adding the 'gauge-fixing' term (18.25) to Eq. (18.17) gives the modified Lagrangian density

$$\mathscr{L}(x) = \frac{1}{2}[\partial^\mu\sigma(x)][\partial_\mu\sigma(x)] - \frac{1}{2}(2\lambda v^2)\sigma^2(x)$$

$$- \frac{1}{4}F_{\mu\nu}(x)F^{\mu\nu}(x) + \frac{1}{2}m^2 A_\mu(x)A^\mu(x) - \frac{1}{2}[\partial_\mu A^\mu(x)]^2$$

$$+ \frac{1}{2}[\partial^\mu\eta(x)][\partial_\mu\eta(x)] - \frac{1}{2}m^2\eta^2(x)$$

$$+ \text{'interaction terms'} \tag{18.26}$$

where $m = |qv|$, as before, and we have again omitted an irrelevant four-divergence $m\partial_\mu[A^\mu(x)\eta(x)]$.

Eq. (18.26) no longer contains the troublesome bilinear term $A^\mu(x)\partial_\mu\eta(x)$ which was present in Eq. (18.17). Hence we can, in perturbation theory, treat $\sigma(x)$, $\eta(x)$ and $A_\mu(x)$ as three independent free fields which may be quantized in the usual way. We see from the first and third lines of Eq. (18.26) that $\sigma(x)$ and $\eta(x)$ are real Klein–Gordon fields which on quantization give the usual equations of motion and propagators for such fields. The

---

[4] In Section 5.1 we replaced Eq. (18.21) by the Lagrangian density (5.10) and not by Eq. (18.22). However, Eqs. (5.10) and (18.22) are equivalent since they differ by a four-divergence only.
[5] Eq. (18.24) represents a particular choice of gauge, 't Hooft, more generally, defined a whole class of gauges by conditions similar to Eq. (18.24). These are known as 't Hooft gauges.

second line of Eq. (18.26) differs from the Lagrangian density of a vector field in the unitary gauge by the term $-\frac{1}{2}[\partial_\mu A^\mu(x)]^2$.From Eq. (18.26) one obtains

$$(\Box + m^2)A^\mu(x) = 0 \qquad (18.27)$$

as the equation of motion of $A^\mu(x)$. Eq. (18.27) is like the Klein-Gordon equation for a scalar field. When quantized, it leads to the propagator[6]

$$iD_F^{\alpha\beta}(k,m) = \frac{-ig^{\alpha\beta}}{k^2 - m^2 + i\varepsilon}. \qquad (18.28)$$

The troublesome term $k^\alpha k^\beta/m^2$ which occurred in the vector propagator (18.20) is absent from Eq. (18.28). For large k^2, the propagator (18.28) behaves as $1/k^2$, just like the photon propagator, and acts as a convergence factor. Dimensional arguments, like those used in Section 9.8, suggest that, like QED, the Higgs model is renormalizable. This is confirmed by the detailed analysis.

Working in a 't Hooft gauge reintroduces the $\eta(x)$ field which had been eliminated in the unitary gauge. There exist no real particles corresponding to the quantized $\eta(x)$ field although the Feynman propagator for this field can be interpreted in terms of the exchange of virtual scalar bosons. The properties of these 'ghost particles', as they are called, are analogous to those of the longitudinal and scalar photons in QED, which also do not exist as real free particles, but contribute as virtual intermediate quanta to the photon propagator. The detailed properties of the $\eta(x)$ field are quite complicated and, of course, gauge dependent. However, all observable quantities are gauge invariant. This situation is again quite analogous to that in QED.

18.3   The Standard Electroweak Theory

In the last chapter we developed a unified model of electromagnetic and weak interactions of massless leptons and massless gauge bosons ($W^\pm, Z^0$ bosons and photons). The Lagrangian density of this model is

$$\mathcal{L} = \mathcal{L}^L + \mathcal{L}^B, \qquad (18.29)$$

where $\mathcal{L}^L$ is the leptonic Lagrangian density (17.39) and $\mathcal{L}^B$ is the gauge-boson Lagrangian density (17.28). The Lagrangian density (18.29) is exactly invariant under the SU(2) × U(1) gauge transformations (17.32a), (17.32b) and (17.37a), (17.37b). We now apply the Higgs mechanism to this model to generate non-vanishing masses for the $W^\pm$ and Z^0 bosons, and we shall see how this also enables one to introduce lepton masses. In this way we shall finally arrive at the standard electroweak theory of Weinberg and Salam.

The necessary formalism is an immediate extension of that of the Higgs model. To break the gauge invariance spontaneously, we must again introduce a *Higgs field*, i.e. a scalar

---

[6] An explicit derivation of Eq. (18.28) is given in C. Itzykson and J. B. Zuber, *Quantum Field Theory*, McGraw-Hill, New York, 1980, Section 3-2-3.

field with non-vanishing vacuum expectation value which is not invariant under the gauge transformations.

Since we now want to break the SU(2) symmetry, we must introduce, not a single such field, but a field with several components and non-zero isospin. The simplest possibility is a weak isospin doublet

$$\Phi(x) = \begin{pmatrix} \phi_a(x) \\ \phi_b(x) \end{pmatrix} \tag{18.30}$$

where $\phi_a(x)$ and $\phi_b(x)$ are scalar fields under Lorentz transformations.

The transformation laws of $\Phi(x)$ under SU(2) × U(1) gauge transformations are, of course, the same as those of the isospin doublet $\Psi_l^L(x)$. The latter were given in Eqs. (17.32a) and (17.37a). Analogously, $\Phi(x)$ transforms under SU(2) transformations according to

$$\left.\begin{aligned} \Phi(x) &\to \Phi'(x) = \exp\left[ig\tau_j\omega_j(x)/2\right]\Phi(x) \\ \Phi^\dagger(x) &\to \Phi^{\dagger\prime}(x) = \Phi^\dagger(x)\exp\left[-ig\tau_j\omega_j(x)/2\right] \end{aligned}\right\}, \tag{18.31}$$

and under U(1) weak hypercharge transformations according to

$$\left.\begin{aligned} \Phi(x) &\to \Phi'(x) = \exp[ig'Yf(x)]\Phi(x) \\ \Phi^\dagger(x) &\to \Phi^{\dagger\prime}(x) = \Phi^\dagger(x)\exp[-ig'Yf(x)] \end{aligned}\right\}, \tag{18.32}$$

where Y is the weak hypercharge of the field $\Phi(x)$. We shall determine its value shortly. [The corresponding global phase transformations are obtained from Eqs.(18.31) and (18.32) through the replacements $g\omega_j(x) \to \alpha_j, g'f(x) \to \beta$, where α_j and β are real constants; compare Eqs. (17.l8a) and (17.31).]

We now want to generalize the Lagrangian density (18.29) to include the Higgs field $\Phi(x)$ and its interactions with the gauge-boson fields, and to continue to be SU(2) × U(1) gauge invariant. The two terms in Eq. (18.29) already possess this invariance property. A generalization which obviously shares this property is

$$\mathscr{L} = \mathscr{L}^L + \mathscr{L}^B + \mathscr{L}^H, \tag{18.33}$$

where

$$\mathscr{L}^H(x) = [D^\mu\Phi(x)]^\dagger\left[D_\mu\Phi(x)\right] - \mu^2\Phi^\dagger(x)\Phi(x) - \lambda\left[\Phi^\dagger(x)\Phi(x)\right]^2. \tag{18.34}$$

Here the covariant derivative $D^\mu\Phi(x)$ is defined by

$$D^\mu\Phi(x) = \left[\partial^\mu + ig\tau_jW_j^\mu(x)/2 + ig'YB^\mu(x)\right]\Phi(x), \tag{18.35}$$

in analogy with Eq. (17.40a) for $\Psi_l^L(x)$ which has hypercharge $-\frac{1}{2}$.

The expression $\mathscr{L}^B + \mathscr{L}^H$ in Eq. (18.33) is a direct generalization of the Higgs model Lagrangian density (18.15), and the further analysis closely follows that for the Higgs model. For $\lambda > 0$ and $\mu^2 < 0$, the classical energy density is a minimum for a constant Higgs field

$$\Phi(x) = \Phi_0 = \begin{pmatrix} \phi_a^0 \\ \phi_b^0 \end{pmatrix}, \tag{18.36}$$

with

$$\Phi_0^\dagger\Phi_0 = |\phi_a^0|^2 + |\phi_b^0|^2 = \frac{-\mu^2}{2\lambda}, \tag{18.37}$$

and all other fields vanishing. Choosing for the ground state a particular value Φ_0, compatible with Eq. (18.37), again leads to spontaneous symmetry breaking. Without loss of generality, we can choose

$$\Phi_0 = \begin{pmatrix} \phi_a^0 \\ \phi_b^0 \end{pmatrix} = \begin{pmatrix} 0 \\ v/\sqrt{2} \end{pmatrix}, \tag{18.38}$$

where

$$v = (-\mu^2/\lambda)^{1/2} \quad (>0), \tag{18.39}$$

since any other choice of Φ_0 is related to the value (18.38) by a global phase transformation.

The Higgs field of the vacuum ground state, Eq. (18.38), is, in general, not invariant under SU(2) × U(l) gauge transformations. However, it must be invariant under U(l) electromagnetic gauge transformations, in order to ensure zero mass for the photon and conservation of the electric charge. If we assign the weak hypercharge $Y = \frac{1}{2}$ to the Higgs field, then it follows from Eq. (17.27) that the lower component $\phi_b(x)$ of the Higgs field is electrically neutral, so that spontaneous symmetry breaking occurs only in the electrically neutral component of the vacuum field (18.38), and charge conservation holds exactly. Alternatively, we see from Eqs. (18.16) and (17.27) that an electromagnetic gauge transformation of the Higgs field is given by

$$\Phi(x) \rightarrow \Phi'(x) = \exp\left[-i\left(Y + I_3^W\right)ef(x)\right]\Phi(x). \tag{18.40}$$

Applied to the vacuum field (18.38), this transformation gives

$$\Phi_0 \rightarrow \Phi'_0 = \Phi_0, \tag{18.41}$$

i.e. the vacuum field is invariant under electromagnetic gauge transformations.

An arbitrary Higgs field $\Phi(x)$ can again be parameterized in terms of its deviations from the vacuum field Φ_0 in the form

$$\Phi(x) = 2^{-1/2}\begin{pmatrix} \eta_1(x) + i\eta_2(x) \\ v + \sigma(x) + i\eta_3(x) \end{pmatrix}. \tag{18.42}$$

By means of this equation, we can express the Lagrangian density $\mathscr{L}^H$, Eq. (18.34), in terms of the four real fields $\sigma(x)$ and $\eta_i(x)$, $i = 1, 2, 3$. The interpretation and quantization of these fields lead to the same difficulties which we met for the Higgs model. The way these difficulties are resolved and the further analysis are closely analogous to our treatment of the Higgs model. The interpretation becomes particularly simple if we employ a special gauge, the unitary gauge. We shall give the analysis in the next chapter (Section 19.1). However, the similarity to the Higgs model is so close that we can anticipate the results. We shall find that the fields $\eta_i(x)$, $i = 1, 2, 3$, are unphysical fields. In the unitary gauge they are transformed away, and the $W^\pm$ and Z^0 bosons are seen to acquire mass. The photon remains massless since the electromagnetic gauge symmetry has not been spontaneously broken. This is also reflected in the fact that only three unphysical fields $\eta_i(x)$ occur. In contrast to the fields $\eta_i(x)$ disappearing, the field $\sigma(x)$ survives in the unitary gauge and, on quantization, gives rise to massive, electrically neutral, spin-0 particles (Higgs scalars).

To obtain non-vanishing lepton masses, we must augment the Lagrangian density (18.33), by adding a suitable term $\mathscr{L}^{\text{LH}}$, to

$$\mathscr{L} = \mathscr{L}^{\text{L}} + \mathscr{L}^{\text{B}} + \mathscr{L}^{\text{H}} + \mathscr{L}^{\text{LH}}. \tag{18.43}$$

We shall couple the lepton and Higgs fields through Yukawa interactions,[7] described by the Lagrangian density

$$\begin{aligned}\mathscr{L}^{\text{LH}}(x) = & - g_l \left[\overline{\Psi}_l^{\text{L}}(x) \psi_l^{\text{R}}(x) \Phi(x) + \Phi^\dagger(x) \overline{\psi}_l^{\text{R}}(x) \Psi_l^{\text{L}}(x) \right] \\ & - g_{\nu_l} \left[\overline{\Psi}_l^{\text{L}}(x) \psi_{\nu_l}^{\text{R}}(x) \widetilde{\Phi}(x) + \widetilde{\Phi}^\dagger(x) \overline{\psi}_{\nu_l}^{\text{R}}(x) \Psi_l^{\text{L}}(x) \right]. \end{aligned} \tag{18.44}$$

Here g_l, and g_{ν_l} are dimensionless coupling constants, summations over $l = e, \mu, \ldots$ are implied, as usual, and $\widetilde{\Phi}(x)$ is defined by

$$\widetilde{\Phi}(x) = - i \left[\Phi^\dagger(x) \tau_2 \right]^{\text{T}} = \begin{pmatrix} \phi_b^*(x) \\ - \phi_a^*(x) \end{pmatrix}. \tag{18.45}$$

In Eq. (18.45), τ_2 is the Pauli matrix (17.15) and T denotes the transpose.

We shall now show that the Lagrangian density $\mathscr{L}^{\text{LH}}$ is invariant under SU(2) × U(1) gauge transformations. The invariance of the first line of Eq. (18.44) follows from the transformation laws (17.32a) and (17.37a) of the lepton fields and the transformation laws (18.31) and (18.32) of the Higgs field. For the second line of Eq. (18.44), we require the transformation properties of $\widetilde{\Phi}(x)$. Under U(1) transformations, it follows from Eqs. (18.32) and (18.45) that

$$\widetilde{\Phi}(x) \longrightarrow \widetilde{\Phi}'(x) = \exp[- i g'f(x)/2]\widetilde{\Phi}(x). \tag{18.46}$$

We shall show below that, under SU(2) transformations, $\widetilde{\Phi}(x)$ transforms in exactly the same way as $\Phi(x)$. From these transformation properties of $\widetilde{\Phi}$ and of the lepton fields, the SU(2) × U(1) gauge invariance of the second line of Eq. (18.44) follows.

In order to derive the SU(2) transformation properties of $\widetilde{\Phi}(x)$, it suffices to consider infinitesimal transformations. For these, we obtain from Eq. (18.31) that

$$\Phi(x) \rightarrow \Phi(x) + \delta \Phi(x) = [1 + i \frac{1}{2} g \tau_j \omega_j(x) + \cdots] \Phi(x)$$

i.e.

$$\delta \Phi(x) = i \frac{1}{2} g \tau_j \omega_j(x) \Phi(x). \tag{18.47}$$

---

[7] A term of the form $\overline{\psi}(x)\phi(x)\psi(x)$, where $\psi(x)$ and $\phi(x)$ are spinor and scalar fields, respectively, is called a Yukawa interaction or a Yukawa coupling.

Hence

$$\delta\mathbf{\Phi}^{\dagger}(x) = -\,\mathrm{i}\frac{1}{2}g\omega_{j}(x)\mathbf{\Phi}^{\dagger}(x)\tau_{j}$$

and from Eq. (18.45)

$$\delta\widetilde{\mathbf{\Phi}}(x) = -\,\mathrm{i}[\delta\mathbf{\Phi}^{\dagger}(x)\tau_{2}]^{\mathrm{T}} = -\,\mathrm{i}[-\,\mathrm{i}\frac{1}{2}g\omega_{j}(x)\mathbf{\Phi}^{\dagger}(x)\tau_{j}\tau_{2}]^{\mathrm{T}}. \tag{18.48}$$

For the Pauli matrices (17.15) $\tau_{j}\tau_{2} = -\tau_{2}\tau_{j}^{\mathrm{T}}$. Substituting this relation in Eq. (18.48), the resulting expression simplifies to

$$\delta\widetilde{\mathbf{\Phi}}(x) = \mathrm{i}\frac{1}{2}g\omega_{j}(x)\tau_{j}[-\,\mathrm{i}\mathbf{\Phi}^{\dagger}(x)\tau_{2}]^{\mathrm{T}} = \mathrm{i}\frac{1}{2}g\omega_{j}(x)\tau_{j}\widetilde{\mathbf{\Phi}}(x). \tag{18.49}$$

Comparing Eqs. (18.49) and (18.47), we see that $\widetilde{\mathbf{\Phi}}(x)$ and $\mathbf{\Phi}(x)$ transform in the same way under SU(2) transformations.

In the next chapter, we shall transform the Lagrangian density of the electroweak theory into the unitary gauge, and we shall see that the term (18.44) not only generates interactions between leptons and Higgs bosons, but also leads to non-zero lepton masses. In their original formulation of the theory, Weinberg and Salam took $g_{\nu_{l}} = 0$, leading to zero neutrino masses. The second line of Eq. (18.44) represents the simplest way of introducing non-vanishing neutrino masses, but does not take account of the subtle effects know as neutrino oscillations, which were discussed in Section 16.7.1. To do this [of Problem 19.5], we must replace the second line of Eq. (18.44) by

$$-\,G_{ll'}\overline{\mathbf{\Psi}}_{l'}^{\mathrm{L}}(x)\psi_{\nu l}^{\mathrm{R}}(x)\widetilde{\mathbf{\Phi}}(x) - G_{ll'}^{*}\widetilde{\mathbf{\Phi}}^{\dagger}(x)\overline{\psi}_{\nu l}^{\mathrm{R}}(x)\mathbf{\Psi}_{l'}^{\mathrm{L}}(x), \tag{18.44a}$$

where G is the Hermitian matrix (16.101b).

Nonetheless, we shall for simplicity use the simpler form (18.44) as our 'standard model', since we shall usually take the limit of zero neutrino masses in application, when the second line of Eq. (18.44) and the expression (18.44a) both vanish.

In this book, we are restricting ourselves to the electroweak interactions of leptons only. However, the theory is easily extended to include quarks and hence hadrons, enabling one to treat semi-leptonic processes. Indeed, this extension, as well as the gauge invariance of the Lagrangian density (18.44), is necessary in order to be able to prove the renormaliz-ability of the theory. The proof again relies on the use of Hooft gauges, mentioned in Section 18.2 on the Higgs model, rather than of the unitary gauge which we shall employ in the next chapter.[8]

We have now completed the derivation of the basic equations of the standard electro-weak theory for purely leptonic processes and we shall conclude the chapter by summarizing these equations.

---

[8] For the extension of the theory to quarks, see D. Bailin, *Weak Interactions*, 2nd edn, A. Hilger, Bristol, 1982, Section 6.5; for a full discussion of renormalizability, see the articles by Abers and Lee or the book by J. C. Taylor, cited in Section 13.2.

The Lagrangian density (18.43) is from Eqs. (17.39), (17.58), (18.34) and (18.44) given by

$$\mathscr{L}(x) = \mathscr{L}^{\mathrm{L}}(x) + \mathscr{L}^{\mathrm{B}}(x) + \mathscr{L}^{\mathrm{H}}(x) + \mathscr{L}^{\mathrm{LH}}(x)$$

$$= \mathrm{i}[\overline{\Psi}_l^{\mathrm{L}}(x)\not{D}\,\Psi_l^{\mathrm{L}}(x) + \overline{\psi}_l^{\mathrm{R}}(x)\not{D}\psi_l^{\mathrm{R}}(x) + \overline{\psi}_{\nu_l}^{\mathrm{R}}(x)\not{D}\psi_{\nu_l}^{\mathrm{R}}(x)]$$

$$+ \left[-\frac{1}{4}B_{\mu\nu}(x)B^{\mu\nu}(x) - \frac{1}{4}G_{i\mu\nu}(x)G_i^{\mu\nu}(x) \right]$$

$$+ \left\{ [D^{\mu}\mathbf{\Phi}(x)]^{\dagger}[D_{\mu}\mathbf{\Phi}(x)] - \mu^2\mathbf{\Phi}^{\dagger}(x)\mathbf{\Phi}(x) - \lambda[\mathbf{\Phi}^{\dagger}(x)\mathbf{\Phi}(x)]^2 \right\}$$

$$+ \left\{ -g_1[\overline{\Psi}_l^{\mathrm{L}}(x)\psi_l^{\mathrm{R}}(x)\mathbf{\Phi}(x) + \mathbf{\Phi}^{\dagger}(x)\overline{\psi}_l^{\mathrm{R}}(x)\Psi_l^{\mathrm{L}}(x)] \right.$$

$$\left. - g_{\nu_l}[\overline{\Psi}_l^{\mathrm{L}}(x)\overline{\psi}_{\nu_l}^{\mathrm{R}}(x)\widetilde{\mathbf{\Phi}}(x) + \widetilde{\mathbf{\Phi}}^{\dagger}(x)\overline{\psi}_{\nu_l}^{\mathrm{R}}(x)\Psi_l^{\mathrm{L}}(x)] \right\}. \tag{18.50}$$

Here $B^{\mu\nu}(x)$ and $G_i^{\mu\nu}(x)$ are defined in Eqs. (17.53), (17.56) and (17.55) as

$$B^{\mu\nu}(x) \equiv \partial^{\nu}B^{\mu}(x) - \partial^{\mu}B^{\nu}(x) \tag{18.51}$$

$$G_i^{\mu\nu}(x) \equiv F_i^{\mu\nu}(x) + g\varepsilon_{ijk}W_j^{\mu}(x)W_k^{\nu}(x), \tag{18.52}$$

with

$$F_i^{\mu\nu}(x) \equiv \partial^{\nu}W_i^{\mu}(x) - \partial^{\mu}W_i^{\nu}(x). \tag{18.53}$$

The covariant derivatives in Eq. (18.50) are defined in Eqs. (17.40) and (18.35) as

$$D^{\mu}\Psi_l^{\mathrm{L}}(x) = \left[\partial^{\mu} + \mathrm{i}g\tau_j W_j^{\mu}(x)/2 - \mathrm{i}g'B^{\mu}(x)/2 \right]\Psi_l^{\mathrm{L}}(x) \tag{18.54}$$

$$D^{\mu}\psi_l^{\mathrm{R}}(x) = [\partial^{\mu} - \mathrm{i}g'B^{\mu}(x)]\psi_l^{\mathrm{R}}(x) \tag{18.55}$$

$$D^{\mu}\psi_{\nu_l}^{\mathrm{R}}(x) = \partial^{\mu}\psi_{\nu_l}^{\mathrm{R}}(x) \tag{18.56}$$

$$D^{\mu}\mathbf{\Phi}(x) = \left[\partial^{\mu} + \mathrm{i}g\tau_j W_j^{\mu}(x)/2 + \mathrm{i}g'B^{\mu}(x)/2 \right]\mathbf{\Phi}(x). \tag{18.57}$$

The Lagrangian density (18.50) is invariant under SU(2) × U(l) gauge transformations. However, the terms $\mathscr{L}^{\mathrm{H}} + \mathscr{L}^{\mathrm{B}}$ are just the direct generalization of the Higgs model Lagrangian density (18.15), and for $\lambda > 0$ and $\mu^2 < 0$ the SU(2) × U(1) gauge symmetry is spontaneously broken. In the quantized theory, the vacuum expectation value of the Higgs field

$$\langle 0|\mathbf{\Phi}(x)|0\rangle = \mathbf{\Phi}_0 = \begin{pmatrix} 0 \\ v/\sqrt{2} \end{pmatrix}, \tag{18.58}$$

where

$$v = (-\mu^2/\lambda)^{1/2} \quad (> 0), \tag{18.59}$$

is, in general, not invariant under SU(2) × U(l) gauge transformations. However, it is invariant under electromagnetic gauge transformations (which are contained in the full set of SU(2) × U(l) transformations), so that the photon remains massless and electric charge is conserved.

19

The Standard Electroweak Theory

In the last chapter, we derived the basic equations of the electroweak theory. In this chapter, we shall consider its physical interpretation and applications. In Section 19.1 we shall transform the theory into the unitary gauge, which facilitates the interpretation. In particular, it will become apparent that the Lagrangian density, expressed in this gauge, describes photons, charged and neutral leptons, W and Z^0 bosons, and Higgs bosons. We shall see that, with exception of the photon, all these particles have acquired mass, and we shall obtain expressions for these masses in terms of the basic parameters of the theory. In Section 19.2 we shall derive the Feynman rules for the electroweak theory, which are an extension of those for QED.

In the following two sections we shall consider some applications. The electroweak theory is renormalizable and so allows calculations to be performed to all orders of perturbation theory. This opens up a rich field of phenomena for accurate calculations and asks for high-precision experiments to test the theory, particularly the rediative corrections. The latter require elaborate calculations for which a 't Hooft gauge would be more appropriate than the unitary gauge. We shall restrict ourselves to lowest-order calculations.[1] In Section 19.3 we shall consider neutrino–electron scattering, which led historically to a determination of the weak mixing angle θ_W from experiment and hence to predictions of the masses of the W and Z^0 bosons. In Section 19.4 we shall discuss several electron–positron annihilation processes at increasingly higher energies. These experiments represent some of the cleanest tests of the electro–weak theory and demonstrate how electroweak unification manifests itself at high energies.

---

[1] Further applications, including ones to semi-leptonic processes, will be found in the following books: D. Bailin, *Weak Interactions*, 2nd edn, A. Hilger, Bristol, 1982; S. M. Bilenky, *Introduction to the Physics of Weak Interactions*, Pergamon, Oxford, 1982; E. Leader and E. Predazzi, *An Introduction to Gauge Theories and the New Physics*, Cambridge University Press, Cambridge, 1982.

Quantum Field Theory, Second Edition   Franz Mandl and Graham Shaw
© 2010 John Wiley & Sons, Ltd

In the last section of the chapter we shall discuss the most important unresolved puzzle of the theory: the mystery of the Higgs particle, which has not been observed so far, yet its existence is essential for the renormalizability of the theory. We shall see that the nature of the interactions of the Higgs particle are such that, after all, it is not surprising that it has not yet been detected, and we shall discuss which processes are most likely to lead to its experimental discovery in the near future.

19.1   The Lagrangian Density in the Unitary Gauge

In the last chapter we wrote the Higgs field $\Phi(x)$ in an arbitrary gauge as

$$\Phi(x) = 2^{-1/2}\begin{pmatrix} \eta_1(x) + i\eta_2(x) \\ v + \sigma(x) + i\eta_3(x) \end{pmatrix} \tag{18.42}$$

This isospinor can always be transformed into the form

$$\Phi(x) = 2^{-1/2}\begin{pmatrix} 0 \\ v + \sigma(x) \end{pmatrix}, \tag{19.1}$$

which no longer contains the fields $\eta_i(x)$. Eqs. (18.42) and (19.1) are analogous to Eqs. (18.10) and (18.18) for the Higgs model, and the gauge in which the Higgs field has the form (19.1) is again called the unitary gauge. The gauge transformation which transforms Eq. (18.42) into Eq. (19.1) consists of an SU(2) transformation (18.31), which converts the isospinor (18.42) into a 'down' isospinor, followed by a U(1) transformation (18.32) which makes this 'down' isospinor a real quantity. Under this SU(2) × U(1) gauge transformation, all other fields transform according to the corresponding Eqs. (17.32a), (17.32b) and (17.37a), (17.37b), and we shall assume that all fields are already expressed in the unitary gauge.

To transform the Lagrangian density (18.50) into the unitary gauge, we substitute Eq. (19.1) into it. We shall also use Eq. (17.43), its complex conjugate and Eqs. (17.45) to replace the fields $W_j^\mu(x)$ and $B^\mu(x)$, by $W^\mu(x)$, $W^{\dagger\mu}(x)$, $Z^\mu(x)$ and $A^\mu(x)$. At this stage we are still dealing with classical fields, so that $W^{\dagger\mu}(x)$ stands for the complex conjugate field. In using this notation, we are anticipating that these fields will presently be quantized, and $W^{\dagger\mu}(x)$ will denote the Hermitian adjoint field operator, as usual. The motivation for this transformation is, of course, that the quantized fields $W^\mu(x)$, $W^{\dagger\mu}(x)$, $Z^\mu(x)$ and $A^\mu(x)$ describe $W^\pm$ bosons, Z^0 bosons and photons. The transformation of the Lagrangian density is straightforward, but tedious, and we shall only quote the results.

On transforming the terms $\mathscr{L}^B + \mathscr{L}^H$ in Eq. (18.50) in this way one obtains[2]

---

[2] In the following equations, all fields have the same argument x, and we shall therefore omit it.

$$\mathcal{L}^{\mathrm{B}} + \mathcal{L}^{\mathrm{H}} = -\frac{1}{4} F_{\mu\nu} F^{\mu\nu}$$

$$-\frac{1}{2} F^{\dagger}_{W\mu\nu} F^{\mu\nu}_{W} + m^2_{\mathrm{W}} W^{\dagger}_{\mu} W^{\mu}$$

$$-\frac{1}{4} Z_{\mu\nu} Z^{\mu\nu} + \frac{1}{2} m^2_{\mathrm{Z}} Z_{\mu} Z^{\mu} \tag{19.2}$$

$$+\frac{1}{2} (\partial^{\mu}\sigma)(\partial_{\mu}\sigma) - \frac{1}{2} m^2_{\mathrm{H}} \sigma^2$$

$$+ \mathcal{L}^{\mathrm{BB}}_{\mathrm{I}} + \mathcal{L}^{\mathrm{HH}}_{\mathrm{I}} + \mathcal{L}^{\mathrm{HB}}_{\mathrm{I}},$$

where we have dropped a constant term, and $\mathcal{L}^{\mathrm{BB}}_{\mathrm{I}}$, etc., are given by

$$\begin{aligned}
\mathcal{L}^{\mathrm{BB}}_{\mathrm{I}} = & \; i g \cos\theta_{\mathrm{W}} [(W^{\dagger}_{\alpha} W_{\beta} - W^{\dagger}_{\beta} W_{\alpha}) \partial^{\alpha} Z^{\beta} \\
& + (\partial_{\alpha} W_{\beta} - \partial_{\beta} W_{\alpha}) W^{\dagger\beta} Z^{\alpha} - (\partial_{\alpha} W^{\dagger}_{\beta} - \partial_{\beta} W^{\dagger}_{\alpha}) W^{\beta} Z^{\alpha}] \\
& + i e [(W^{\dagger}_{\alpha} W_{\beta} - W^{\dagger}_{\beta} W_{\alpha}) \partial^{\alpha} A^{\beta} \\
& + (\partial_{\alpha} W_{\beta} - \partial_{\beta} W_{\alpha}) W^{\dagger\beta} A^{\alpha} - (\partial_{\alpha} W^{\dagger}_{\beta} - \partial_{\beta} W^{\dagger}_{\alpha}) W^{\beta} A^{\alpha}] \\
& + g^2 \cos^2\theta_{\mathrm{W}} [W_{\alpha} W^{\dagger}_{\beta} Z^{\alpha} Z^{\beta} - W_{\beta} W^{\dagger\beta} Z_{\alpha} Z^{\alpha}] \\
& + e^2 [W_{\alpha} W^{\dagger}_{\beta} A^{\alpha} A^{\beta} - W_{\beta} W^{\dagger\beta} A_{\alpha} A^{\alpha}] \\
& + e g \cos\theta_{\mathrm{W}} [W_{\alpha} W^{\dagger}_{\beta} (Z^{\alpha} A^{\beta} + A^{\alpha} Z^{\beta}) - 2 W_{\beta} W^{\dagger\beta} A_{\alpha} Z^{\alpha}] \\
& + \frac{1}{2} g^2 W^{\dagger}_{\alpha} W_{\beta} [W^{\dagger\alpha} W^{\beta} - W^{\alpha} W^{\dagger\beta}]
\end{aligned} \tag{19.3a}$$

$$\mathcal{L}^{\mathrm{HH}}_{\mathrm{I}} = -\frac{1}{4} \lambda \sigma^4 - \lambda v \sigma^3 \tag{19.3b}$$

$$\mathcal{L}^{\mathrm{HB}}_{\mathrm{I}} = \frac{1}{2} v g^2 W^{\dagger}_{\alpha} W^{\alpha} \sigma + \frac{1}{4} g^2 W^{\dagger}_{\alpha} W^{\alpha} \sigma^2 \\
+ \frac{v g^2}{4\cos^2\theta_{\mathrm{W}}} Z_{\alpha} Z^{\alpha} \sigma + \frac{g^2}{8\cos^2\theta_{\mathrm{W}}} Z_{\alpha} Z^{\alpha} \sigma^2. \tag{19.3c}$$

The parameters m_{W}, m_{Z} and m_{H}, which have been introduced in Eq. (19.2), are defined by

$$m_{\mathrm{W}} = \frac{1}{2} v g, \quad m_{\mathrm{Z}} = m_{\mathrm{W}} / \cos\theta_{\mathrm{W}}, \quad m_{\mathrm{H}} = \sqrt{(-2\mu^2)}, \tag{19.4}$$

and v and the weak mixing angle θ_{W} are, from Eqs. (18.59) and (17.47), given by

$$v = (-\mu^2/\lambda)^{1/2} \quad (>0) \tag{19.5}$$

$$g \sin\theta_{\mathrm{W}} = g' \cos\theta_{\mathrm{W}} = e. \tag{19.6}$$

We now consider the remaining two terms $\mathcal{L}^{\mathrm{L}} + \mathcal{L}^{\mathrm{LH}}$ in Eq. (18.50). We again transform from the fields W^{μ}_{j} and B^{μ} to W^{μ}, $W^{\dagger\mu}$, Z^{μ} and A^{μ}. In addition, we use Eq. (17.11) to replace left- and right-handed lepton fields ψ^{L} and ψ^{R} by the complete fields ψ. In this way one finds that

$$\mathcal{L}^{\mathrm{L}} + \mathcal{L}^{\mathrm{LH}} = \overline{\psi}_l (i\partial\!\!\!/ - m_l) \psi_l + \overline{\psi}_{\nu_l} (i\partial\!\!\!/ - m_{\nu_l}) \psi_{\nu_l} + \mathcal{L}^{\mathrm{LB}}_{\mathrm{I}} + \mathcal{L}^{\mathrm{HL}}_{\mathrm{I}} \tag{19.7}$$

where $\mathscr{L}_I^{LB}$ and $\mathscr{L}_I^{HL}$ are given by

$$
\begin{aligned}
\mathscr{L}_I^{LB} = {} & e\bar{\psi}_l\gamma^\alpha\psi_l A^\alpha \\
& - \frac{g}{2\sqrt{2}}[\bar{\psi}_{\nu_l}\gamma^\alpha(1-\gamma_5)\psi_l W_\alpha + \bar{\psi}_l\gamma^\alpha(1-\gamma_5)\psi_{\nu_l}W_\alpha^\dagger] \\
& - \frac{g}{4\cos\theta_W}\bar{\psi}_{\nu_l}\gamma^\alpha(1-\gamma_5)\psi_{\nu_l}Z_\alpha \\
& + \frac{g}{4\cos\theta_W}\bar{\psi}_l\gamma^\alpha(1-4\sin^2\theta_W-\gamma_5)\psi_l Z_\alpha
\end{aligned}
\tag{19.3d}
$$

$$
\mathscr{L}_I^{HL} = -\frac{1}{v}m_l\bar{\psi}_l\psi_l\sigma - \frac{1}{v}m_{\nu_l}\bar{\psi}_{\nu_l}\psi_{\nu_l}\sigma.
\tag{19.3e}
$$

The parameters m_l and m_{ν_l} which have been introduced in Eq. (19.7) are defined by

$$
m_l = vg_l/\sqrt{2}, \quad m_{\nu_l} = vg_{\nu_l}/\sqrt{2}.
\tag{19.8}
$$

Finally, combining Eqs. (19.2) and (19.7), one obtains the complete Lagrangian density of the standard electroweak theory in the unitary gauge

$$
\mathscr{L} = \mathscr{L}_0 + \mathscr{L}_I,
\tag{19.9}
$$

where

$$
\begin{aligned}
\mathscr{L}_0 = {} & \bar{\psi}_l(i\slashed{\partial}-m_l)\psi_l + \bar{\psi}_{\nu_l}(i\slashed{\partial}-m_{\nu_l})\psi_{\nu_l} \\
& - \frac{1}{4}F_{\mu\nu}F^{\mu\nu} \\
& - \frac{1}{2}F_{W\mu\nu}^\dagger F_W^{\mu\nu} + m_W^2 W_\mu^\dagger W^\mu \\
& - \frac{1}{4}Z_{\mu\nu}Z^{\mu\nu} + \frac{1}{2}m_Z^2 Z_\mu Z^\mu \\
& + \frac{1}{2}(\partial^\mu\sigma)(\partial_\mu\sigma) - \frac{1}{2}m_H^2\sigma^2
\end{aligned}
\tag{19.10}
$$

and

$$
\mathscr{L}_I = \mathscr{L}_I^{LB} + \mathscr{L}_I^{BB} + \mathscr{L}_I^{HH} + \mathscr{L}_I^{HB} + \mathscr{L}_I^{HL},
\tag{19.11}
$$

with the individual terms $\mathscr{L}_I^{LB},\ldots$ given by Eqs. (19.3a)–(19.3e).

Eq. (19.10) clearly admits interpretation as the Lagrangian density of free fields, which can be quantized in the usual way. The two terms in the first line of this equation are just the Lagrangian densities of charged leptons with mass m_l and of neutrinos with mass m_{ν_l}. The second, third and fourth lines respectively describe photons, charged vector bosons ($W^\pm$) of mass m_W and neutral vector bosons (Z^0) of mass m_Z. The last line of Eq. (19.10) is just the Lagrangian density of a neutral Klein–Gordon field, and the quantized $\sigma(x)$ field describes neutral spin-0 bosons (Higgs scalars) of mass m_H. The mass terms of the $W^\pm$ and Z^0 bosons arise from the spontaneous symmetry breaking of the SU(2) × U(1) gauge invariance of the Lagrangian density (18.50). On the other hand, the photon remains massless [i.e. there is no term in $A_\mu A^\mu$ in Eq. (19.10)], since the electromagnetic gauge invariance is not broken spontaneously. The lepton mass terms in the first line of Eq. (19.10) have their origin in the Yukawa coupling terms $\mathscr{L}^{LH}$ in Eq. (18.50).

Eq. (19.11) is the interaction Lagrangian density of the standard electroweak theory, with the individual terms (19.3a)–(19.3e) describing the interactions between pairs of particle types. For example, $\mathscr{L}_1^{LB}$ is the interaction of leptons with gauge bosons, etc. These terms will be discussed in the next section, where the Feynman rules for treating them in perturbation theory will be derived.

Eqs. (19.4) and (19.8) relate the boson and lepton masses to the basic parameters

$$g,\ g',\ -\mu^2,\ \lambda,\ g_l,\ g_{v_l} \tag{19.12}$$

of the theory. Rather remarkably, these relations allow the masses of the $W^\pm$ and Z^0 bosons to be determined in terms of three experimentally well-known quantities: the fine structure constant

$$\alpha = e^2/4\pi = 1/137.04, \tag{19.13a}$$

the Fermi coupling constant

$$G = 1.166 \times 10^{-5}\text{GeV}^{-2} \tag{19.13b}$$

[see Eq. (16.70b)], and the weak mixing angle θ_W, which has the measured value:

$$\sin^2\theta_W = 0.23122 \pm 0.00015. \tag{19.13c}$$

From Eqs. (16.43), (17.49) and (19.4), the parameter v can be expressed in terms of G as

$$v = (G\sqrt{2})^{-1/2}, \tag{19.14}$$

and combining Eqs. (19.4), (19.6) and (19.14) gives

$$m_W = \left(\frac{\alpha\pi}{G\sqrt{2}}\right)^{1/2}\frac{1}{\sin\theta_W}, \qquad m_Z = \left(\frac{\alpha\pi}{G\sqrt{2}}\right)^{1/2}\frac{2}{\sin 2\theta_W}. \tag{19.15}$$

However, on substituting Eqs (19.13) into Eqs. (19.15), one obtains the values

$$m_W = 77.5 \text{ GeV} \qquad m_Z = 88.4 \text{ GeV},$$

which are slightly different from the measured values (16.1). This is because in deriving Eq. (19.15) we have neglected radiative corrections, and both (19.15) and the values obtained from it will be modified by terms of order α when these are taken into account.[3] The calculation of such radiative corrections requires a discussion of renormalization which goes beyond the scope of this book, and we shall merely note that the predictions for the renormalized (i.e. physical) masses are in agreement with experiment when radiative corrections are included.

We see from Eq. (19.6) that the fine structure constant and the weak mixing angle also determine the coupling constants g and g'. Similarly, the Higgs–lepton coupling constants g_l and g_{v_l} are known from Eqs. (19.8) and (19.14), provided the mass of the corresponding lepton is known.

---

[3] This is because Eq. (19.15) is obtained from relations between the masses and coupling constants which appear in the Lagrangian (19.10) rather than between the renormalized (i.e. physical) masses and couplings; and because we have used Eq. (16.43), which was derived from the lowest-order diagram Fig. (16.6).

This leaves only the parameter λ in the set (19.12) to be determined since, from Eq. (19.5), $(-\mu^2) = \lambda v^2$. λ occurs as coupling constant in the Higgs self-coupling terms $\mathscr{L}_1^{HH}$, Eq. (19.3b). It also determines the mass of the Higgs boson, since from Eqs. (19.4) and (19.5)

$$m_H = \sqrt{(-2\mu^2)} = \sqrt{(2\lambda v^2)}. \tag{19.16}$$

Since the Higgs boson has never been detected, its mass or, equivalently λ, cannot be measured directly. However, for any assumed value, we can calculate the contribution to any given electroweak process from higher-order diagrams in which Higgs bosons are emitted and re-absorbed. These diagrams, together with other higher-order diagrams, give small corrections to the leading order predictions of the theory. These corrections have been measured in many processes, and the requirement that the measured and calculated corrections agree leads to the value

$$m_H = 84^{+34}_{-26}\text{GeV}$$

at 68% confidence level, with a corresponding upper limit on the Higgs mass of 154 GeV at 95% confidence level. On the other hand, the Higgs boson would have been detected experimentally if its mass were below 113.5 GeV [See Section 19.5]. Together these results restrict the mass of the Higgs boson to the range

$$113.5 \ \text{GeV} < m_H \leq 160 \text{ GeV}. \tag{19.17}$$

19.2 Feynman Rules

We shall now derive the Feynman rules for treating the standard electroweak theory in perturbation theory, restricting ourselves to the rules for obtaining the Feynman amplitudes of only the lowest-order non-vanishing graphs for a process. We shall work in the unitary gauge in which the Lagrangian density is given by Eqs. (19.9)–(19.11). As discussed in the context of QED [see Eqs. (6.9) and (6.10)], the Lagrangian densities $\mathscr{L}_0$ and $\mathscr{L}_1$ are to be interpreted as normal products in the quantized theory. In the following, normal products will always be implied. The Lagrangian density (19.9)–(19.11) is invariant under electromagnetic gauge transformations, and we shall choose to work in the Feynman gauge, which has been used throughout this book. In this gauge, the Lagrangian density of the free electromagnetic field is given by Eq. (18.22), i.e. the free-field Lagrangian density (19.10) is replaced by

$$\mathscr{L}_0 - \frac{1}{2}[\partial_\mu A^\mu(x)]^2. \tag{19.18}$$

In the following, we shall denote this augmented Lagrangian density by $\mathscr{L}_0$.

In Section 6.2, we found that in the interaction picture, the interacting fields satisfy the same equations of motion and the same commutation relations as the free fields, provided the interaction Lagrangian density $\mathscr{L}_1(x)$ does not involve derivatives of the field operators. This result enabled us to obtain the S-matrix expansion (6.23) and hence to derive the Feynman rules for the amplitudes in Chapter 7. For the electroweak theory, the interaction

Lagrangian density $\mathcal{L}_I(x)$ involves derivatives of the fields. However, irrespective of whether $\mathcal{L}_I(x)$ involves derivatives or not, a generally valid form of S-matrix expansion is

$$S = \sum_{n=0}^{\infty} \frac{i^n}{n!} \int \ldots \int d^4x_1 \ldots d^4x_n T\{\mathcal{L}_I(x_1)\ldots\mathcal{L}(x_n)\}, \tag{19.19}$$

provided one uses the free-field commutation relations in evaluating the matrix elements of S.[4] For QED we have

$$\mathcal{L}_I(x) = -\mathcal{H}_I(x), \tag{19.20}$$

and Eq. (19.19) reduces to our earlier expansion (6.23). From these results it follows that the same procedures which were used in Chapter 7 for QED can at once be applied to Eq. (19.19) to obtain the Feynman rules for the electroweak theory.

With our choice of electromagnetic gauge, the Feynman rules of QED are taken over directly.[5] The arguments used in extending these rules to the electroweak theory are largely analogous to those for QED. We shall discuss the new features which occur. Otherwise, we shall only quote the results in Appendix B, leaving it to the reader to fill in the details.

We first consider the internal lines (propagators) and external lines (initial- and final-state particles) of the additional particles which now occur. For the neutral leptons

Table 19.1 *The 18 basic interactions of the standard electroweak theory*

Reference		Type of interaction	Term no.[*]
$\mathcal{L}_I^{BB}$ Eq. (19.3a)	lines 1 and 2	$W^\dagger W Z$	1
	lines 3 and 4	$W^\dagger W A$	2
	line 5	$W^\dagger W Z^2$	3
	line 6	$W^\dagger W A^2$	4
	line 7	$W^\dagger W A Z$	5
	line 8	$(W^\dagger W)^2$	6
$\mathcal{L}_I^{HH}$ Eq. (19.3b)	term 1	σ^4	7
	term 2	σ^3	8
$\mathcal{L}_I^{HB}$ Eq. (19.3c)	term 1	$W^\dagger W \sigma$	9
	term 2	$W^\dagger W \sigma^2$	10
	term 3	$Z^2 \sigma$	11
	term 4	$Z^2 \sigma^2$	12
$\mathcal{L}_I^{LB}$ Eq. (19.3d)	line 1	$\bar{l} l A$	13
	line 2	$\bar{\nu}_l l W + \text{h.c.}$	14
	line 3	$\bar{\nu}_l \nu_l Z$	15
	line 4	$\bar{l} l Z$	16
$\mathcal{L}_I^{HL}$ Eq. (19.3e)	term 1	$\bar{l} l \sigma$	17
	term 2	$\bar{\nu}_l \nu_l \sigma$	18

[*]The corresponding Feynman diagrams and vertex factors are given in Appendix B and numbered (B.1)–(B.18) in the same order.

---

[4] For the derivation of this S-matrix expansion, see N. N. Bogoliubov and D. V. Shirkov, *Introduction to the Theory of Quantized Fields*, 3rd edn, Wiley, New York, 1979, Section 2.1.
[5] In Appendix B at the end of this book we summarize the Feynman rules for QED, as well as the additional rules which will now be derived for the electroweak theory.

(i.e. neutrinos and antineutrinos) the propagators and external line factors have already been included in Feynman rules 3 and 4 in Appendix B. The propagators and external line factors for $W^{\pm}$ bosons were given when we considered the IVB theory in Section 16.4 (Feynman rules 11 and 12). The extension to Z^0 bosons is trivial, and Feynman rules 11 and 12 in Appendix B state these results for both $W^{\pm}$ and Z^0 bosons. Lastly, we must deal with the Higgs particle. This is a massive neutral spin-0 particle, such as was considered in Section 3.1. The expansion of the field $\sigma(x)$ in terms of creation and annihilation operators is given by Eqs. (3.7). Since the Higgs boson has spin 0, Eqs. (3.7) contain no polarization vectors and there are no external line factors for Higgs scalars. The Feynman propagator for Higgs scalars follows from Eq. (3.58) and is given by the Feynman rule:

13. For each internal Higgs line, labelled by the momentum k, write a factor

$$i\Delta_F(k,\ m_H) = \frac{i}{k^2 - m_H^2 + i\varepsilon}, \qquad \overset{k}{\bullet\text{-}\text{-}\text{-}\text{-}\text{-}\text{-}\blacktriangleright\text{-}\text{-}\text{-}\text{-}\text{-}\bullet} \qquad (19.21)$$

where m_H is the mass of the Higgs scalar. (We shall represent Higgs scalars by dashed lines.)

It remains to discuss the basic vertices to which the interaction (19.11) gives rise. There are now 18 types of vertices, stemming from the 18 terms in Eqs. (19.3a)–(19.3e). Let us reassure the alarmed reader that the Feynman rules for these vertices are easy to write down, and we shall shortly do so.

To begin with, we catalogue the terms of Eqs. (19.3a)–(19.3e) in Table 19.1. The first column in this table shows where in these equations each term occurs. In the last column we number the terms for easier cross-reference in Appendix B, where the corresponding vertex factors are numbered (B.1) to (B.18) in the same order. In the middle column of the table we show the fields involved in each term. (For conciseness we put l for ψ_l and $\bar{l}$ for $\bar{\psi}_l$, etc.)

It must be remembered that each term corresponds to many possibilities, consistent with conservation of electric charge and of lepton numbers, just as the basic QED vertex stands for any one of the eight basic processes shown in Fig. 7.1. All the interactions are local interactions at one space–time point, and the type of vertices to which the 18 terms give rise are obvious from the middle column of Table 19.1 and are shown in Feynman diagrams (B.1)–(B.18). For example, terms 1 and 2 describe the interaction of three gauge bosons [Figs. (B.1) and (B.2)], terms 3–6 the interactions of four gauge bosons [Figs. (B.3)–(B.6)], and so on. The lepton–boson interaction terms 13–16 we met previously in Eq. (17.48). Term 13 is the usual QED interaction, term 14 represents the interaction of charged leptonic currents with $W^{\pm}$ bosons, and terms 15 and 16 the interaction of neutral leptonic currents with the Z^0 boson.

The remaining interactions involve Higgs particles. Terms 7 and 8 correspond to self-interactions of three or four Higgs particles, terms 9–12 to the interactions of Higgs particles with massive vector bosons. Having no charge, the Higgs particle does not couple directly to the photon. Finally, terms 17 and 18 originate from $\mathscr{L}^{LH}$, Eq. (18.44), and represent the interactions of the Higgs particle with charged leptons and with neutrinos. The latter interactions only occur for non-zero neutrino masses.

The derivation of the vertex factors of these 18 interaction terms is quite similar to the corresponding calculation for QED in Section 7.2.1. We shall therefore only remind the reader of the latter calculation and discuss the new features which arise for the electroweak interactions. The reader should then easily be able to derive the 18 vertex factors which are given in Feynman rule 14 in Appendix B.

In Section 7.2.1 we obtained the QED vertex factor $ie\gamma^\alpha$ by calculating the matrix element of the first-order term

$$S^{(1)} = i \int d^4x \mathscr{L}_1(x) \tag{19.22}$$

in the S-matrix expansion (19.19) between free-particle states. This leads to the first-order Feynman amplitude $\mathscr{M}$ and, omitting the external line factors, to the corresponding vertex factor. For QED

$$\mathscr{L}_1(x) = e(\bar{\psi}\gamma^\alpha\psi A_\alpha)_x. \tag{19.23}$$

In Section 7.2.1 we considered the basic process $e^- \rightarrow e^- + \gamma$ and obtained the first-order Feynman amplitude

$$\mathscr{M} = ie\bar{u}(\mathbf{p}')\gamma^\alpha u(\mathbf{p})\varepsilon_\alpha(\mathbf{p}' - p) \tag{7.32}$$

and hence the vertex factor $ie\gamma^\alpha$. This vertex factor can be read off directly from Eq. (19.23) by omitting the field operators and multiplying by a factor i [which is just the factor i in Eq. (19.22)]. The same vertex factor is, of course, obtained from any of the other basic processes shown in Fig. 7.1.

Next, we discuss the new features which arise when deriving the vertex factors for the electroweak theory.

First, some of the interactions in Eqs. (19.3a)–(19.3e) contain a particular field operator, not linearly, but to a higher power, resulting in extra combinatorial factors in the Feynman amplitude. Consider, for example, the third term in Eq. (19.3c), i.e.

$$\frac{vg^2}{4\cos^2\theta_W} g^{\alpha\beta} Z_\alpha(x) Z_\beta(x)\sigma(x). \tag{19.24}$$

We shall calculate the vertex factor for this term by considering the first-order process

$$Z^0(k_1, r_1) + Z^0(k_2, r_2) \rightarrow H(k_3) \tag{19.25}$$

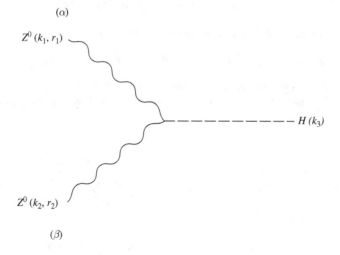

(α)

$Z^0(k_1, r_1)$

$H(k_3)$

$Z^0(k_2, r_2)$

(β)

Figure 19.1   The first-order process $Z^0(k_1, r_1) + Z^0(k_2, r_2) \rightarrow H(k_3)$

where H stands for the Higgs particle, k_i ($i = 1, 2, 3$) are the four-momenta of the three particles, and r_1 and r_2 label the polarization states of the two vector bosons. The Feynman diagram for this process is shown in Fig. 19.1. Either of the two operators $Z(x)$ in Eq. (19.24) can annihilate the $Z^0(k_1, r_1)$ boson, and the other operator $Z(x)$ will then annihilate the $Z^0(k_2, r_2)$ boson. This leads to a combinatorial factor 2! and to the vertex factor

$$\frac{i \upsilon g^2}{2 \cos^2 \theta_W} g^{\alpha \beta}, \tag{19.26}$$

which is quoted in Eq. (B.11). In Figs. 19.1 and (B.11) we have attached the tensor indices of the vertex factor (19.26) to the external Z^0 boson lines to which they belong. The same vertex factor would, of course, have been obtained from any of the related processes, e.g. $Z^0 + Z^0 + H \rightarrow$ vacuum. The Feynman diagram of the basic vertex part, i.e. the Feynman diagram in which only the characteristic features of the vertex are retained, is shown in Fig. (B.11) in Appendix B.

The corresponding combinatorial factors for all other interaction terms are derived in the same way. For graphs containing radiative corrections, additional factors occur. Consider, for example, the modification of a Higgs boson line, shown in Fig. 19.2. In this case, the above type of argument leads to a combinatorial factor 3! for each vertex, i.e. $(3!)^2$ in all. It is easy to see that the correct weight factor is $(3!)^2/2$. Choosing any one of the three operators $\sigma(x_1)$ and any one of the three operators $\sigma(x_2)$ in $[\sigma(x_1)]^3[\sigma(x_2)]^3$ to take care of the two external lines, there are just two ways of contracting the remaining operators $[\sigma(x_1)]^2[\sigma(x_2)]^2$ into two propagators. Since we shall not be considering the calculations of radiative corrections in the electroweak theory, we shall not derive the rules for these additional combinatorial factors.

A second point, requiring care, concerns the ordering of the tensor indices associated with the vector fields. We illustrate this for the $WW^\dagger Z^2$ interaction term which occurs in line 5 of Eq. (19.3a), i.e. the term

$$g^2 \cos^2 \theta_W [W_\alpha W_\beta^\dagger Z^\alpha Z^\beta - W_\beta W^{\dagger \beta} Z_\alpha Z^\alpha]. \tag{19.27}$$

To find the vertex part, we consider the first-order process

$$Z^0(k_1, r_1) + Z^0(k_2, r_2) \rightarrow W^+(k_1', r_1') + W^-(k_2', r_2'). \tag{19.28}$$

The corresponding Feynman diagram is shown in Fig. 19.3. We note, first of all, that the final $W^+(k_1', r_1')$ and $W^-(k_2', r_2')$ bosons must be created by the operators $W^\dagger(x)$ and

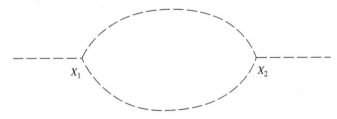

Figure 19.2   *A modification of the Higgs boson line*

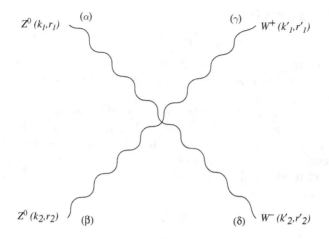

Figure 19.3 *The first-order process* $Z^0(k_1, r_1) + Z^0(k_2, r_2) \rightarrow W^+(k'_1, r'_1) + W^-(k'_2, r'_2)$. *The order of the tensor indices* $(\alpha) \dots (\delta)$ *corresponds to the vertex factor (19.30)*

$W(x)$, respectively [see Eqs. (16.22)]. On the other hand, the initial $Z^0(k_1, r_1)$ boson can be annihilated by either operator $Z(x)$ in Eq. (19.27), with $Z^0(k_2, r_2)$ being annihilated by the other operator $Z(x)$. This leads to the Feynman amplitude

$$\mathcal{M} = ig^2 \cos^2\theta_W \{\varepsilon_\alpha(2')\varepsilon_\beta(1')[\varepsilon^\alpha(1)\varepsilon^\beta(2) + \varepsilon^\alpha(2)\varepsilon^\beta(1)]$$
$$- \varepsilon_\beta(2')\varepsilon^\beta(1')[\varepsilon_\alpha(1)\varepsilon^\alpha(2) + \varepsilon_\alpha(2)\varepsilon^\alpha(1)]\}, \qquad (19.29)$$

where we abbreviated $\varepsilon_{r'_1\beta}(\mathbf{k}'_1)$, to $\varepsilon_\beta(1')$ etc. Relabelling tensor indices so that all terms have the common factor $\varepsilon_\alpha(1)\varepsilon_\beta(2)\varepsilon_\gamma(1')\varepsilon_\delta(2')$, Eq. (19.29) reduces to

$$\mathcal{M} = \varepsilon_\alpha(1)\varepsilon_\beta(2)\varepsilon_\gamma(1')\varepsilon_\delta(2')ig^2 \cos^2\theta_W[g^{\alpha\delta}g^{\beta\gamma} + g^{\alpha\gamma}g^{\beta\delta} - 2g^{\alpha\beta}g^{\gamma\delta}].$$

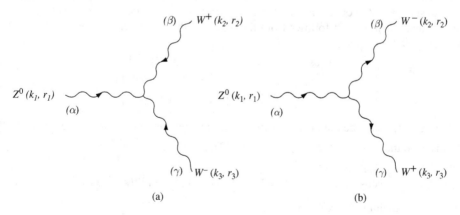

Figure 19.4 *The related first-order processes: (a)* $Z^0(k_1, r_1) + W^+(k_2, r_2) + W^-(k_3, r_3) \rightarrow$ *vacuum, (b)* $Z^0(k_1, r_1) \rightarrow W^-(k_2, r_2) + W^+(k_3, r_3)$

The corresponding vertex factor is

$$ig^2 \cos^2\theta_{\mathrm{W}} [g^{\alpha\delta}g^{\beta\gamma} + g^{\alpha\gamma}g^{\beta\delta} - 2g^{\alpha\beta}g^{\gamma\delta}], \qquad (19.30)$$

which is the result given in Eq. (B.3). In Figs. 19.3 and (B.3) we have attached tensor indices to the external lines in the correct order to correspond to the vertex factor (19.30).

The third new feature of the electroweak vertices stems from the derivative couplings which occur in the $W^\dagger WZ$ and $W^\dagger WA$ terms of $\mathscr{L}_I^{\mathrm{BB}}$, Eq. (19.3a). One sees from the plane wave expansions of the fields, e.g. Eqs. (16.22), that these derivative couplings introduce momentum factors into the vertex functions. To obtain the exact result for the term

$$ig\cos\theta_{\mathrm{W}} [(W_\alpha^\dagger W_\beta - W_\beta^\dagger W_\alpha)\,\partial^\alpha Z^\beta$$
$$+ (\partial_\alpha W_\beta - \partial_\beta W_\alpha)W^{\dagger\beta}Z^\alpha - (\partial_\alpha W_\beta^\dagger - \partial_\beta W_\alpha^\dagger)W^\beta Z^\alpha] \qquad (19.31)$$

in Eq. (19.3a), we consider the process

$$Z^0(k_1, r_1) + W^+(k_2, r_2) + W^-(k_3, r_3) \to \text{vacuum}. \qquad (19.32)$$

The Feynman graph for this process is shown in Fig. 19.4(a). We have chosen the process (19.32) in which all three bosons are annihilated at the vertex on account of its greater symmetry compared with related processes, such as $Z^0 \to W^+ + W^-$. Below we shall show how to obtain the vertex factors for these related processes from that for the reaction (19.32). In Fig. 19.4(a), we have marked the boson lines with arrows which show the direction of the flow of energy–momentum (and, for the $W^\pm$ bosons, of charge) towards the vertex where the particles are annihilated.

The operators $W(x)$ and $W^\dagger(x)$ are linear in the absorption operators of W^+ and W^- bosons, respectively. It follows from Eqs. (16.22) that, in annihilating a W^+ boson with four-momentum k and polarization index r, the operator $\partial_\alpha W_\beta(x)$ leads to a factor

$$(-ik_\alpha)\varepsilon_{r\beta}(\mathbf{k}).$$

The same factor occurs if we replace the operator $\partial_\alpha W_\beta(x)$ by $\partial_\alpha W_\beta^\dagger(x)$ or $\partial_\alpha Z_\beta(x)$ and the W^+ boson by a W^- or Z^0 boson, as can be seen from the plane wave expansions of the $W^\dagger(x)$ and $Z(x)$ fields. It follows from these results that the Feynman amplitude for the process (19.32) is given by

$$\mathscr{M} = -g\cos\theta_{\mathrm{W}} \big\{ [\varepsilon_\alpha(3)\varepsilon_\beta(2) - \varepsilon_\beta(3)\varepsilon_\alpha(2)](-ik_1^\alpha)\varepsilon^\beta(1) \qquad (19.33)$$
$$+ [(-ik_{2\alpha})\varepsilon_\beta(2) - (-ik_{2\beta})\varepsilon_\alpha(2)]\varepsilon^\beta(3)\varepsilon^\alpha(1)$$
$$- [(-ik_{3\alpha})\varepsilon_\beta(3) - (-ik_{3\beta})\varepsilon_\alpha(3)]\varepsilon^\beta(2)\varepsilon^\alpha(1)] \big\}$$

where $\varepsilon_{r_1}^\alpha(\mathbf{k}_1)$ has been abbreviated to $\varepsilon^\alpha(1)$, etc. Relabelling tensor indices so that all terms have the common factor $\varepsilon_\alpha(1)\varepsilon_\beta(2)\varepsilon_\gamma(3)$, Eq. (19.33) reduces to

$$M = \varepsilon_\alpha(1)\varepsilon_\beta(2)\varepsilon_\gamma(3)$$
$$\times ig\cos\theta_{\mathrm{W}} [g^{\alpha\beta}(k_1 - k_2)^\gamma + g^{\beta\gamma}(k_2 - k_3)^\alpha + g^{\gamma\alpha}(k_3 - k_1)^\beta]. \qquad (19.34)$$

The corresponding vertex factor is

$$ig\cos\theta_{\mathrm{W}} [g^{\alpha\beta}(k_1 - k_2)^\gamma + g^{\beta\gamma}(k_2 - k_3)^\alpha + g^{\gamma\alpha}(k_3 - k_1)^\beta], \qquad (19.35)$$

Figure 19.5   *The leading contributions to $(\nu_\mu e)$ scattering*

Figure 19.6   *The leading contributions to $(\nu_e e)$ scattering*

which is the result quoted in Eq. (B.1). Figs. 19.4(a) and (B.1) show the tensor indices of the boson lines appropriate to the vertex factor (19.35). Since this vertex factor depends on the four-momenta of the particles, we require an arrow on each boson line to show the direction of the energy–momentum flow.

In Fig. 19.4(a), all particles are annihilated at the vertex, but the other cases are easily derived from Eq. (19.35). If the Z^0 boson is created at the vertex, one replaces k_1 by $-k_1$ in Eq. (19.35). If the $W^+ W^-$ pair is created at the vertex, one replaces k_i by $-k_i (i = 2, 3)$, and one interchanges W^+ and W^- since the flow of four-momentum and charge of the $W^\pm$ bosons now is away from the vertex instead of towards it. Thus the vertex function for the Feynman diagram 19.4(b) is given by

$$ig\cos\theta_W [g^{\alpha\beta}(k_1 + k_2)^\gamma + g^{\beta\gamma}(k_3 - k_2)^\alpha - g^{\gamma\alpha}(k_1 + k_3)^\beta]. \tag{19.36}$$

Lastly, term 6 in Table 19.1, i.e. the term in $(W^\dagger W)^2$ which occurs in L_I^{BB}, Eq. (19.3a), requires a comment. The vertex factor (B.6) for this term is obtained by the same technique by which we derived the vertex factor (19.30) for the vertex in Fig. 19.3. However, unlike all other cases involving $W^\pm$ bosons, the vertex factor (B.6) is altered if one exchanges the index α with β, or γ with δ. Consequently, this vertex factor depends on the direction of flow of the $W^\pm$ charges, towards or away from the vertex. In Fig. (B.6), these directions are indicated by the arrows on the boson lines and correspond to the vertex factor given with Fig. (B.6).

The above results enable us to write down by inspection of Eqs. (19.3a)–(19.3e) the vertex factors of all terms in Table 19.1.

This completes our discussion of the Feynman rules of the standard electroweak theory. These rules enable one straightforwardly to calculate processes in lowest non-vanishing order of perturbation theory. The calculation of higher-order corrections involves very many graphs and, in general, becomes extremely complicated. Except for sometimes quoting results, we shall not consider such calculations further and shall limit ourselves to lowest-order calculations.

19.3   Elastic Neutrino–Electron Scattering

As our first application of the electroweak theory, we shall consider the four elastic neutrino–electron scattering processes

$$\nu_\mu + e^- \rightarrow \nu_\mu + e^-, \quad \bar{\nu}_\mu + e^- \rightarrow \bar{\nu}_\mu + e^-, \qquad (19.37a)$$
$$\nu_e + e^- \rightarrow \nu_e + e^-, \quad \bar{\nu}_e + e^- \rightarrow \bar{\nu}_e + e^-. \qquad (19.37b)$$

More briefly, we shall refer to the first of these reactions as $(\nu_\mu e)$ scattering, etc. If lepton numbers are conserved – as in the electroweak theory – the leading one-boson-exchange contributions to the $(\nu_\mu e)$ and $(\nu_e e)$ processes arise from the Feynman graphs in Figs. 19.5 and 19.6 respectively, with similar graphs in which neutrinos are replaced by antineutrinos for the other two processes. The significance of these four processes is due to the fact that all of them involve neutral currents [diagrams (a) in Figs. 19.5 and 19.6] and can therefore be employed to determine the weak mixing angle θ_W. The $(\nu_e e)$ and $(\bar{\nu}_e e)$ processes have additional contributions from the exchange of one W boson, Fig. 19.6(c).

We first consider the $(\nu_\mu e)$ process. From the Feynman graphs 19.5 and the Feynman rules in Appendix B, its Feynman amplitude is given by

$$\mathscr{M}(\nu_\mu e) = \mathscr{M}_Z(\nu_\mu e) + \mathscr{M}_H(\nu_\mu e), \qquad (19.38a)$$

where

$$\mathscr{M}_Z(\nu_\mu e) = \frac{-g^2}{8\cos^2\theta_w} [\bar{u}'_{\nu_\mu}\gamma^\alpha(1-\gamma_5)u_{\nu_\mu}]iD_{F\alpha\beta}(k, m_Z)[\bar{u}'_e\gamma^\beta(g_V - g_A\gamma_5)u_e] \qquad (19.38b)$$

and

$$\mathscr{M}_H(\nu_\mu e) = \frac{-1}{v^2}m_{\nu_\mu}m_e(\bar{u}'_{\nu_\mu}u_{\nu_\mu})i\Delta_F(k, m_H)(\bar{u}'_e u_e). \qquad (19.38c)$$

Here u and $\bar{u}'$ are the spinors of the initial and final leptons respectively, with the kind of lepton (electron or ν_μ) labelled, but momenta and spin quantum numbers suppressed, and we have introduced the abbreviations

$$g_V \equiv 2\sin^2\theta_w - \frac{1}{2}, \quad g_A \equiv -\frac{1}{2}. \qquad (19.39)$$

The momentum k of the intermediate boson is given by

$$k = q - q' = p' - p, \qquad (19.40)$$

where q and q' (p and p') are the momenta of the initial and final neutrino (electron) respectively.

With $m_Z \approx 91$ GeV and our assumption (19.17) for m_H, we have

$$k^2 \ll m_Z^2, \quad k^2 \ll m_H^2, \qquad (19.41a)$$

even for quite high energies. In this limit, the Feynman amplitudes (19.38b) and (19.38c) become

$$\mathscr{M}_Z(\nu_\mu e) = \frac{-iG}{\sqrt{2}} [\bar{u}'_{\nu_\mu}\gamma^\alpha(1-\gamma_5)u_{\nu_\mu}][\bar{u}'_e\gamma_\alpha(g_V - g_A\gamma_5)u_e] \qquad (19.42a)$$

$$\mathscr{M}_H(\nu_\mu e) = iG\sqrt{2}\,\frac{1}{m_H^2}\,m_{\nu_\mu}m_e(\bar{u}'_{\nu_\mu}\,u_{\nu_\mu})(\bar{u}'_e u_e), \tag{19.42b}$$

where we used Eqs. (19.4) and (19.14). We see from Eqs. (19.42a) and (19.42b) that $\mathscr{M}_H(\nu_\mu e)$ is of order $m_{\nu_\mu}m_e/m_H^2$ relative to $\mathscr{M}_Z(\nu_\mu e)$ and so can be neglected. We shall also assume that the lepton masses are negligible; more precisely, that

$$s \equiv (p+q)^2 \gg m_e^2. \tag{19.41b}$$

Here s is the square of the centre-of-mass momentum and is given by

$$s = 4E_{\text{CoM}}^2 = 2m_e E_{\text{Lab}} + m_e^2, \tag{19.43}$$

where E_{CoM} and E_{Lab} are the neutrino energy in the centre-of-mass system and in the laboratory system in which the target electron is at rest, respectively.

If the conditions (19.41a) and (19.41b) hold, a straightforward calculation leads to the total elastic ($\nu_\mu e$) cross-section

$$\sigma_T(\nu_\mu e) = \frac{G^2 s}{3\pi}\,(g_V^2 + g_V g_A + g_A^2). \tag{19.44a}$$

In the same way, one obtains the total elastic ($\bar{\nu}_\mu e$) cross-section

$$\sigma_T(\bar{\nu}_\mu e) = \frac{G^2 s}{3\pi}\,(g_V^2 - g_V g_A + g_A^2). \tag{19.44b}$$

We next obtain the Feynman amplitude for the ($\nu_e e$) process

$$\mathscr{M}(\nu_e e) = \mathscr{M}_Z(\nu_e e) + \mathscr{M}_H(\nu_e e) + \mathscr{M}_W(\nu_e e) \tag{19.45}$$

where the three terms stem from the three Feynman graphs in Fig. 19.6. The first two of these graphs lead to the same amplitudes [Eqs. (19.42a) and (19.42b)] as before, with m_{ν_μ} replaced by m_{ν_e}, so that the Higgs contribution $\mathscr{M}_H(\nu_e e)$ is again negligible. Assuming

$$k^2 \ll m_W^2 \tag{19.41c}$$

and that lepton masses may again be neglected leads to the amplitude

$$\mathscr{M}_W(\nu_e e) = \frac{-iG}{\sqrt{2}}\,[\bar{u}'_e \gamma^\alpha (1 - \gamma_5) u_{\nu_e}][\bar{u}'_{\nu_e} \gamma_\alpha (1 - \gamma_5) u_e]. \tag{19.46}$$

[The relative signs of the amplitudes (19.46) and (19.42a) follow from the Feynman rule 8 by arranging the lepton operators, which lead to $\mathscr{M}_Z(\nu_e e)$ and $\mathscr{M}_W(\nu_e e)$, in the same order.] Using the Fierz identity[6]

$$(\bar{u}_1 \gamma^\alpha (1 - \gamma_5) u_2)(\bar{u}_3 \gamma_\alpha (1 - \gamma_5) u_4) = (\bar{u}_1 \gamma^\alpha (1 - \gamma_5) u_4)(\bar{u}_3 \gamma_\alpha (1 - \gamma_5) u_2), \tag{19.47}$$

---

[6] For the derivation of this and other Fierz identities, see C. Itzykson and J. B. Zuber, *Quantum Field Theory*, McGraw-Hill, 1980, pp. 160–162.

where $u_1, \ldots, u_4$ are arbitrary spinors, we can write the amplitude (19.46) as

$$\mathcal{M}_W(\nu_e e) = \frac{-iG}{\sqrt{2}} \, [\bar{u}_{\nu_e}' \gamma^\alpha (1 - \gamma_5) u_{\nu_e}][\bar{u}_e' \gamma_\alpha (1 - \gamma_5) u_e]. \qquad (19.48)$$

Combining these results in Eq. (19.45), it follows that the Feynman amplitude $M(\nu_e e)$ is obtained from Eq. (19.42a) by replacing ν_μ by ν_e, together with the replacements

$$g_V \to g_V + 1, \qquad g_A \to g_A + 1. \qquad (19.49)$$

Making the replacements (19.49) in Eq. (19.44a) leads to the total cross-section for elastic $(\nu_e e)$ scattering. A similar argument shows that the same replacements made in Eq. (19.44b) give the total cross-section for elastic $(\bar{\nu}_e e)$ scattering.

Experimentally, the above processes are investigated by scattering neutrinos from atomic electrons and detecting the recoil electrons. The $(\nu_\mu e)$ and $(\bar{\nu}_\mu e)$ reactions have been studied by several groups using ν_μ and $\bar{\nu}_\mu$ beams originating from pion decay. The $(\bar{\nu}_e e)$ process has been studied using $\bar{\nu}_e$ beams produced from neutron decays in nuclear reactors. Historically such experiments, together with neutrino-scattering experiments on nucleons, were used to fix the value of $\sin^2\theta_W$, and by 1983 the best value was[7]

$$\sin^2\theta_W = 0.227 \pm 0.014. \qquad (19.50)$$

On substituting the value into Eq. (19.5) the predictions

$$m_W = 78.3 \pm 2.4 \text{ GeV} \qquad m_Z = 89.0 \pm 2.0 \text{ GeV} \qquad (19.51)$$

were obtained for the weak boson masses, prior to their detection in 1983 with masses[8] of approximately

$$m_W = 81 \pm 3 \text{ GeV} \quad m_Z = 94 \pm 4 \text{ GeV}, \qquad (19.52)$$

in good agreement with Eq. (19.51). Today, the enormously improved precision of the data means that radiative corrections must be taken into account to understand the relation between the weak mixing angle and the boson masses, as we have seen in Section 19.1; and the precise value (19.13c) for $\sin^2\theta_W$ is obtained by fitting the theory to a large number of precise measurements, of which the $W^\pm$ and Z^0 boson masses are but two.

Figure 19.7   The leading contributions to $e^+ + e^- \to l^+ + l^-$

[7] See the review by W. Marciano, *Proceedings of the 1983 International Symposium on Lepton and Photon Interactions at High Energies*, Cornell University, 1983.

[8] These values are a rough combination of the values reported in the papers by B. Sadoulet (UA1) and A. Clark (UA2) in the Proceedings of the International Symposuim on Lepton and Photon Interactions at High Energies, Cornell University, 1983.

19.4   Electron–Positron Annihilation

In Section 8.4 we discussed electron–positron annihilation in the context of QED. We now reconsider this process in the wider framework of the electroweak theory, starting with the reaction

$$e^+ + e^- \rightarrow l^+ + l^- \quad (l = \mu, \tau). \tag{19.53}$$

In lowest order, this process is described by the Feynman graphs in Fig. 19.7. The corresponding Feynman amplitude is given by

$$\mathcal{M} = \mathcal{M}_\gamma + \mathcal{M}_Z + \mathcal{M}_H, \tag{19.54}$$

where

$$\mathcal{M}_\gamma = ie^2 (\bar{u}_l \gamma^\alpha v_l) \frac{1}{k^2 + i\varepsilon} (\bar{v}_e \gamma_\alpha u_e) \tag{19.55a}$$

$$\mathcal{M}_Z = \frac{ig^2}{4\cos^2\theta_W} [\bar{u}_l \gamma^\alpha (g_V - g_A \gamma_5) v_l] \frac{1}{k^2 - m_Z^2 + i\varepsilon} [\bar{v}_e \gamma_\alpha (g_V - g_A \gamma_5) u_e] \tag{19.55b}$$

$$\mathcal{M}_H = \frac{-i}{v^2} m_e m_l (\bar{u}_l v_l) \frac{1}{k^2 - m_H^2 + i\varepsilon} (\bar{v}_e u_e) \tag{19.55c}$$

and $\bar{v}_e$, u_e, and v_l, $\bar{u}_l$ are the spinors of the initial e^+e^- pair and the final l^+l^- pair. In writing Eq. (19.55b), we have omitted the term $k_\alpha k_\beta / m_Z^2$ from the boson propagator $D_{F\alpha\beta}(k, m_Z)$, since this term gives contributions of order $m_e m_l / m_Z^2$.[9]

We see from Eqs. (19.55b), (19.55c) and (19.4) that $\mathcal{M}_H / \mathcal{M}_Z$ is of order

$$\frac{m_e m_l}{m_Z^2} \frac{k^2 - m_Z^2}{k^2 - m_H^2}. \tag{19.57}$$

Hence $\mathcal{M}_H$ may be neglected, unless k^2 is very close to m_H^2, and the Feynman amplitude (19.54) reduces to

$$\mathcal{M} = \mathcal{M}_\gamma + \mathcal{M}_Z. \tag{19.58}$$

A long, but straightforward, calculation gives the cross-section for the process $e^+e^- \rightarrow l^+l^-$. Assuming that the electron's energy E in the CoM system is sufficiently high so that

$$s \equiv k^2 = 4E^2 \gg m_l^2 \tag{19.59}$$

(i.e. all lepton masses may be neglected), one obtains the CoM differential cross-section

$$\sigma(\theta) = F(s)(1 + \cos^2\theta) + G(s)\cos\theta. \tag{19.60a}$$

---

[9] To see this, one writes the intermediate boson momenta in $k_\alpha k_\beta / m_Z^2$ as

$$k = p_1 + p_2 = p'_1 + p'_2, \tag{19.56}$$

where $p_1(p_2)$ and $p'_1(p'_2)$ are the momenta of the positive (negative) initial and final leptons respectively, and uses the Dirac equation to replace $\bar{v}_e \not{p}_1$ by $(-\bar{v}_e m_e)$, etc.

Here θ is the angle (in the CoM system) between the initial positron direction and the final l^+ direction (compare Fig. 8.1), and $F(s)$ and $G(s)$ are given by

$$F(s) = \frac{\alpha^2}{4s}\left[1 + \frac{g_v^2}{\pi\sqrt{2}}\frac{m_Z^2}{s - m_Z^2}\left(\frac{sG}{\alpha}\right) + \frac{(g_v^2 + g_A^2)^2}{8\pi^2}\left(\frac{m_Z^2}{s - m_Z^2}\right)^2\left(\frac{sG}{\alpha}\right)^2\right] \tag{19.60b}$$

$$G(s) = \frac{\alpha^2}{4s}\left[\frac{\sqrt{2}g_A^2}{\pi}\frac{m_Z^2}{s - m_Z^2}\left(\frac{sG}{\alpha}\right) + \frac{g_A^2 g_v^2}{\pi^2}\left(\frac{m_Z^2}{s - m_Z^2}\right)^2\left(\frac{sG}{\alpha}\right)^2\right]. \tag{19.60c}$$

In these equations the terms proportional to α^2, αG and G^2 stem respectively from $|\mathcal{M}_\gamma|^2$, the $\mathcal{M}_\gamma - \mathcal{M}_Z$ interference term and $|\mathcal{M}_Z|^2$ in the expression $|\mathcal{M}_\gamma + \mathcal{M}_Z|^2$ from which the cross-section is derived.

We distinguish three energy regions. First, we have the *low-energy region*, in which

$$s \equiv k^2 \ll m_Z^2 \approx 800 \quad \text{GeV}^2$$

and $sG/\alpha \ll 1$. We may then neglect the terms in Eqs. (19.60b) and (19.60c) arising from $\mathcal{M}_Z$ (i.e. the effects of weak interactions) and Eqs. (19.60) reduce to the cross-section (8.46) of pure QED.

As s increases, $|\mathcal{M}_\gamma|$ falls off as $1/s$ while $|\mathcal{M}_Z|$ increases as $|s - m_Z^2|^{-1}$, i.e. the expansion parameter

$$\frac{m_Z^2}{|s - m_Z^2|}\left(\frac{sG}{\alpha}\right) \tag{19.61}$$

in Eqs. (19.60b) and (19.60c) increases. Hence we reach an *intermediate energy region* where the $\mathcal{M}_\gamma - \mathcal{M}_Z$ interference term is no longer negligible but $|\mathcal{M}_Z|^2$ is; in Eqs. (19.60b) and (19.60c) we retain the terms in αG and drop those in G^2. In particular, $G(s)$ is different from zero so that the cross-section (19.60a) contains a term in $\cos\theta$ and displays a forward–backward asymmetry

$$A(\theta) \equiv \frac{\sigma(\theta) - \sigma(\pi - \theta)}{\sigma(\theta) + \sigma(\pi - \theta)} = \frac{G(s)}{F(s)}\frac{\cos\theta}{1 + \cos^2\theta}, \tag{19.62}$$

where $0 \leq \theta \leq \pi/2$. From Eqs. (19.60b) and (19.60c) this becomes, on neglecting terms proportional to G^2,

$$A(\theta) = \frac{\sqrt{2}}{\pi}g_A^2\frac{m_Z^2}{s - m_Z^2}\left(\frac{sG}{\alpha}\right)\frac{\cos\theta}{1 + \cos^2\theta}. \tag{19.63}$$

For $s < m_Z^2$, $A(\theta)$ is negative (i.e. there is excess backward scattering) and its magnitude increases rapidly with energy. To gain an idea of the size of the asymmetry, we calculate $A(\theta = 60°)$. With $g_A = -\frac{1}{2}$ and $m_Z \approx 91$ GeV, $A(\theta = 60°)$ has the approximate values -0.03, -0.07 and -0.14 for the total CoM energies $\sqrt{s} = 2E = 20$, 30 and 40 GeV, respectively.

The asymmetry $A(\theta)$ has been measured in this energy range for both $\mu^+\mu^-$ and $\tau^+\tau^-$ production. In comparing theory with experiment, it is essential to take into account electromagnetic radiative corrections associated with soft photons, which are typically of the order of 10% in this region and which can themselves generate positive asymmetries

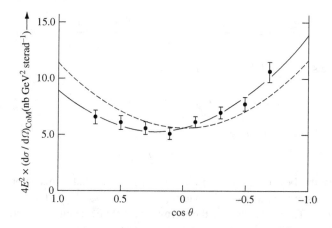

Figure 19.8 *The differential cross-section* $(d\sigma/d\Omega)_{\text{CoM}}$ *for the process* $e^+e^- \to \mu^+\mu^-$, *at the total CoM energy* $2E = 34$ *GeV. [After R. Brandelik et al., Phys. Lett.* **110B** *(1982), 173.]* $\dotmark$: *experimental data;* — *electroweak theory;* ---- *QED*

of the order of 2%. As noted in Section 8.9, the precise value of the radiative corrections depends on the experimental set-up. For this reason, it is usual to subtract the radiative corrections from the experimental data before comparing with theory, so that the results of different experiments can be compared more easily.[10]

Fig. 19.8 shows a typical result: the 'radiatively corrected' angular distribution for $\mu^+\mu^-$ production at $\sqrt{s} = 34$ GeV, obtained by the TASSO Collaboration.[11] The agreement of the experimental points with the electroweak theory (the continuous curve) is seen to be excellent, and the deviation from the symmetric dashed curve, which corresponds to the lowest-order QED distribution, is apparent. Similar results have been obtained in several other experiments and at different energies for both $\mu^+\mu^-$ and $\tau^+\tau^-$ production.

Going to higher energies, we reach the Z^0 *resonance region:* $s \approx m_Z^2$, i.e. the total CoM energy $\sqrt{s}$ is in the vicinity of the mass m_Z of the Z^0 boson. In this region, sG/α is of order unity and it follows that, for $s \approx m_Z^2$, the cross-section for the process $e^+ + e^- \to l^+ + l^-$ is dominated by the terms proportional to G^2 in Eqs. (19.60b) and (19.60c), i.e. by the terms originating from $|\mathcal{M}_Z|^2$. Retaining only these terms, we obtain from Eqs. (19.60a)–(19.60c) the total cross-section

$$\sigma_{\text{T}}(e^+e^- \to l^+l^-) = \frac{1}{6\pi}(g_{\text{v}}^2 + g_{\text{A}}^2)^2 \frac{G^2 m_Z^6}{(s - m_Z^2)^2 + \varepsilon^2} \tag{19.64}$$

for $s \approx m_Z^2$, where we have temporarily restored the infinitesimal parameter ε which occurs in the denominator of the amplitude $\mathcal{M}_Z$, Eq. (19.55b). Eq. (19.64) can be written

---

[10] For a detailed discussion, giving typical results, see F.A. Berends and R. Gastmans, Radiative Corrections in e^+e^- Collisions, in: *Electromagnetic Interactions of Hadrons*, vol. 2 (edited by A. Donnachie and G. Shaw), Plenum Press, New York, 1978.
[11] R. Brandelik *et al., Phys. Lett.* **110B** (1982), 173.

$$\sigma_T(e^+e^- \to l^+l^-) = \frac{12\pi\Gamma(Z^0 \to e^+e^-)\Gamma(Z^0 \to l^+l^-)}{\left(s - m_Z^2\right)^2 + \varepsilon^2} \quad (s \approx m_Z^2), \quad (19.65)$$

where we have substituted the expression

$$\Gamma(Z^0 \to l^+l^-) = \frac{1}{\pi 6\sqrt{2}} Gm_Z^3 (g_v^2 + g_A^2). \quad (19.66)$$

This is the width for the decay process

$$Z^0 \to l^+ + l^- \quad (l = e, \mu, \tau), \quad (19.67)$$

represented by the Feynman graph in Fig. 19.9, assuming that terms of order m_l^2/m_Z^2 are negligible. In this approximation, the width (19.66) does not depend on the type of lepton pair, $l = e$, μ, τ. The derivation of the decay rate (19.66), using the method of Section 12.5, is straightforward and is left as an exercise for the reader.

Eq. (19.65) corresponds to an infinitesimally narrow peak in the cross-section at the resonance energy $\sqrt{s} = m_Z$. Of course, the peak in the experimental cross-section is not infinitesimally narrow. From the uncertainty principle, it must have a width of order $\tau^{-1} = \Gamma_t$, where τ and Γ_t are the lifetime and total decay width of the Z^0 boson. More formally, near $s = m_Z^2$, higher-order corrections to the denominator in Eq. (19.65) cannot be neglected, since in lowest order this denominator vanishes as $\varepsilon \to 0$. The relevant corrections arise from the propagator modifications, shown in Fig. 19.10, which involve

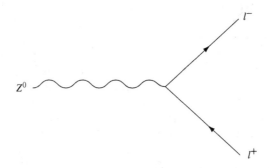

Figure 19.9 *The decay process* $Z^0 \to l^+ + l^-$

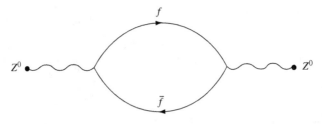

Figure 19.10 *Modifications of the Z^0 propagator due to intermediate states of fermion-antifermion pairs $f\bar{f}$*

intermediate fermion–antifermion states, like l^+l^-, $\nu_l\bar\nu_l$, to which the Z^0 boson can decay. The calculation of these terms is similar to that for the vacuum polarization graph, Fig. 9.8, given in Section 10.4. We shall not discuss it here, but merely quote the final result, which is that Eq. (19.65) becomes modified to

$$\sigma_T(e^+e^- \to l^+l^-) = \frac{12\pi\Gamma(Z^0 \to e^+e^-)\Gamma(Z^0 \to l^+l^-)}{\left(s - m_Z^2\right)^2 + m_Z^2\Gamma_t^2} \quad (s \approx m_Z^2), \qquad (19.68)$$

where we have again set $\varepsilon = 0$.

Eq. (19.68) is a special case of the one-level Breit–Wigner formula[12]

$$\sigma_T(i \to X) = 4\pi\frac{m^2}{p^2}\frac{2J + 1}{(2s_1 + 1)(2s_2 + 1)}\frac{\Gamma_i\Gamma_X}{(s - m^2)^2 + m^2\Gamma_t^2} \quad (s \approx m^2)$$

for the contribution of an unstable particle or resonance (of spin J, mass m and total decay with Γ_t) to the total CoM cross-section of a reaction $i \to X$, in the vicinity of the resonance energy $\sqrt{s} = m$. Γ_i and Γ_X are the partial widths of this resonance for decay to the incident channel (i) and the exit channel (X) respectively, s_1 and s_2 are the spins of the colliding particles in the incident channel, and p is the CoM three-momentum of either colliding particle. For highly relativistic energies, the total CoM energy $\sqrt{s}$ is very large compared with the masses of the colliding particles and $p = \sqrt{s}/2 = m/2$. In this approximation and with $s_1 = s_2 = \frac{1}{2}$ for the case of e^+e^- collisions which we are considering, the Breit–Wigner formula becomes

$$\sigma_T(i \to X) = \frac{4\pi(2J + 1)\Gamma_i\Gamma_X}{(s - m^2)^2 + m^2\Gamma_t^2} \quad (s \approx m^2). \qquad (19.69)$$

We can apply Eq. (19.69) to any reaction $e^+e^- \to X$ which can proceed via an intermediate Z^0 boson: $e^+e^- \to Z^0 \to X$. At energies in the vicinity of the resonance energy $\sqrt{s} = m_Z$, where the Feynman graph 19.11 will dominate, the total cross-section is, from Eq. (19.69), given by

$$\sigma_T(e^+e^- \to X) = \frac{12\pi\Gamma(Z^0 \to e^+e^-)\Gamma(Z^0 \to X)}{\left(s - m_Z^2\right)^2 + m_Z^2\Gamma_t^2} \quad (s \approx m_Z^2). \qquad (19.70)$$

Figure 19.11 *The dominant contribution to the reaction $e^+ + e^- \to X$, in the vicinity of the Z^0 resonance peak*

[12] See H. M. Pilkuhn, *Relativistic Particle Physics*, Springer, New York, 1979, p. 168.

We see from Eq. (19.70) that the total cross-section for the reaction $e^+e^- \to X$ exhibits a peak at $s = m_Z^2$, with half-width $m_Z\Gamma_t$ and a height which is proportional to the product of the partial decay widths

$$\Gamma(Z^0 \to e^+e^-)\Gamma(Z^0 \to X).$$

Hence an electron–positron colliding beam facility, operating at a total CoM energy in the region of $\sqrt{s} = m_Z \approx 91$ GeV, is ideally suited for studying the properties of the Z^0 boson. Two such facilities were constructed for this purpose: the Stanford Linear Collider (SLC) and the LEP electron-positron storage rings at CERN. In the latter machine in particular, the beam intensities were such that of order one thousand Z^0 production events per hour could be observed at the Z^0 peak, enabling its properties to be measured with considerable precision. Fig. 19.12 shows a typical result: the cross-section for $\mu^+\mu^-$ production obtained by the OPAL collaboration.[13] Similar results have been obtained in several experiments for e^+e^-, $\mu^+\mu^-$ and $\tau^+\tau^-$ production, and for hadron production. These cross-sections are analysed by comparing them to the Breit–Wigner formulae (19.70), where the mass and decay widths of the Z^0 are treated as free parameters to be fitted to experiment.[14] The resulting values, obtained by averaging over several different experiments, are:

$$m_Z = 91.188 \pm 0.002 \text{ GeV}; \tag{19.71a}$$

$$\Gamma_t = 2.495 \pm 0.002 \text{ GeV}; \tag{19.71b}$$

$$\Gamma(Z^0 \to \text{ hadrons}) = 1.744 \pm 0.002 \text{ GeV}; \tag{19.71c}$$

while the leptonic decay rates are all consistent with the value

$$\Gamma(Z^0 \to l^+l^-) = 0.084 \pm 0.001 \text{ GeV} \tag{19.71d}$$

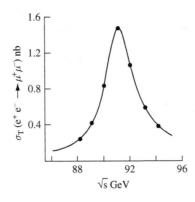

Figure 19.12   *The measured total cross-section $\sigma_T(e^+e^- \to \mu^+\mu^-)$ as a function of the total CoM energy $\sqrt{s} = 2E$. [After W. Bartel et al. Zeit. Phys. **C52** (1991) 175]. The curve shows the prediction of the standard electroweak theory*

---

[13] *Zeit. f. Phys.* **C52** (1991) 175.
[14] In this analysis, the radiative corrections associated with the emission of soft photons must again be included.

independent of the lepton type $l = e, \mu, \tau$. This again confirms the universality of leptonic interactions, and is in good agreement with the lowest order theoretical prediction

$$\Gamma\left(Z^0 \to l^+ l^-\right) = 0.0834 \text{ GeV}$$

obtained by substituting Eqs. (19.13b), (19.13c) and (19.39) into (19.66).

Decays to hadrons and to charged lepton pairs account for just over 80% of all Z^0 decays, as can be seen by comparing the observed total decay width (19.71b) with the partial decay widths (19.71c, d). The remaining decays are ascribed to the reactions

$$Z^0 \to v_l + \overline{v}_l,$$

represented by the Feynman diagram of Fig. 19.13, since these are the only other decays allowed in leading order electroweak interactions. Neglecting higher-order decays, we thus obtain

$$\Gamma_t = \Gamma\left(Z^0 \to \text{ hadrons}\right) + 3\Gamma\left(Z^0 \to l^+ l^-\right) + N_v \Gamma\left(Z^0 \to v_l \overline{v}_l\right),$$

where we now allow for an arbitrary number $N_v \geq 3$ of neutrinos in the sequence $v_e, v_\mu, v_\tau, \ldots$ for reasons to be discussed shortly. The total neutrino contribution is therefore

$$\begin{aligned} N_v \Gamma\left(Z^0 \to v_l \overline{v}_l\right) &= \Gamma_t - \Gamma\left(Z^0 \to \text{ hadrons}\right) - 3\Gamma\left(Z^0 \to l^+ l^-\right) \\ &= 0.499 \pm 0.004 \text{ GeV}, \end{aligned} \qquad (19.72a)$$

where we have substituted the experimental values (19.71b, c, d). The decay rate to neutrinos cannot be measured directly, but can be calculated from the diagram Fig. 19.13 and is found to be [cf. problem 19.1]

$$\Gamma\left(Z^0 \to v_l \overline{v}_l\right) = 0.166 \text{ GeV}. \qquad (19.72b)$$

This is compatible with (19.72a) if and only if the number of neutrino types $N_v = 3$.

This is a significant result. Three so-called 'generations' of fermions (e^-, v_e), (μ^-, v_μ) and (τ^-, v_τ) are known to exist and further generations with the same properties are easily incorporated within the standard electroweak theory. It was for a long time a matter of speculation whether such extra generations exist, since neutrinos are difficult to detect and additional charged leptons would have escaped detection if they were very heavy. In particular, any such charged leptons must be too heavy to contribute to Z^0 decays or they would have been detected. However, if additional neutrinos existed in the sequence $v_e, v_\mu, v_\tau, \ldots$, with masses much less than m_Z, then the Z^0 would decay to them with a rate given by (19.72b) according to the standard electroweak theory. Thus, in

Figure 19.13   *The decay process* $Z^0 \to v_l \overline{v}_l$

Figure 19.14   *The reaction $e^+ + e^- \rightarrow W^+ + W^-$*

the framework of this theory, the experimental result (19.72a) restricts the number of lepton generations to the three already known, if we assume that neutrino masses are not very large.

So far, we have considered those reactions $e^+ e^- \rightarrow X$ which, in lowest order of perturbation theory, depend only on the lepton–boson interactions (19.3d) and (19.3e), represented by the Feynman graphs (B.13)–(B.18) in Appendix B. The boson self-coupling terms (19.3a), Figs. (B.1)–(B.6), are a most striking and characteristic feature of a non-Abelian gauge theory, and electron–positron annihilation processes allow us to explore the three-boson self-coupling terms [lines 1–4 of Eq. (19.3a) and Figs. (B.1) and (B.2)].

The most direct way of doing this is to study the reaction

$$e^+ + e^- \rightarrow W^+ + W^- \tag{19.73}$$

In lowest order, the main contribution to this process comes from the Feynman diagrams in Fig. 19.14. (There is an additional diagram in which the photon in Fig. 19.14(b) is replaced by a Higgs boson. This contribution can be neglected because of the very weak Higgs–lepton coupling; see Section 19.5.) The cross-section predicted by the electroweak theory for the reaction (14.73) has been calculated.[15] In the immediate vicinity of the threshold, the cross-section is dominated by the contribution from diagram 19.14(a). However, at somewhat higher energies, the amplitudes for the Feynman graphs 19.14(a)–(c) become of comparable importance, and the cross-section becomes extremely sensitive to the large interference terms between these amplitudes. The measured cross-sections are in excellent agreement with the theoretical predictions, and confirm the detailed form of the $\gamma W^+ W^-$ and $Z^0 W^+ W^-$ vertices predicted by the standard electroweak theory.

19.5   The Higgs Boson

The electroweak theory has been extremely successful in accounting for a wide range of high-precision data. However one of its most important predictions – the existence of the Higgs boson – is yet to be confirmed. It's mass, as we have seen, is not predicted by the theory. However, for any given mass, its interactions are completely specified. They are summarized by the Feynman graphs and vertex factors of Figs. (B.9)–(B.12) and (B.17).

---

[15] O. P. Sushkov *et. al.*, *Sov. J. Nucl. Phys.* **20** (1975), 537 and W. Alles *et. al.*, *Nucl. Phys.* **B119** (1977) 125.

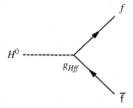

Figure 19.15 *The basic vertex part of the Higgs–fermion interaction, where $f = l, \nu_l$ or q*

The Higgs boson does not couple directly to photons, and the couplings to leptons are specified by the vertex factor $- im\ /\upsilon$, where m is the relevant lepton mass. The coupling of Higgs bosons to quarks $q = u, d, s, \ldots$ is of the same form, so that the basic vertex part shown in Fig. 19.15 and the associated vertex factor

$$\frac{-i}{\upsilon} m_f \quad (f = l, \nu_l, q) \tag{19.74}$$

apply generally to leptons and quarks of mass m_f. Using Eqs. (19.4) and (19.6), we can write this vertex factor

$$\frac{-i}{\upsilon} m_f = \frac{-ie}{2 \sin \theta_W} \frac{m_f}{m_W}, \tag{19.75}$$

i.e. the Higgs–fermion coupling is of order m_f / m_W compared with the QED coupling, so that the coupling of the Higgs boson to light fermions $m_f \ll m_W$ is heavily suppressed.

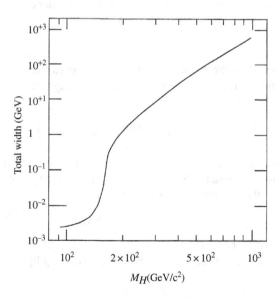

Figure 19.16 *Width of the Higgs boson as a function of its mass. With kind permission from Springer Science & Business Media. Kunst et al, Zeitschrift f. physik C74, 1997, 479*

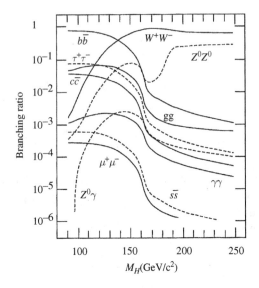

Figure 19.17 *Branching ratios of the Higgs boson as a function of its mass. (Z. Kunst, S. Moretti and W.J. Stirling, Zeit. Phys.* **C74** *(1997) 479.)*

19.5.1   Higgs boson decays

The predicted decay properties of the Higgs boson as a function of its mass are shown in Figures 19.16 and 19.17. Here we shall just consider their main features.

For $m_H < 2m_W$, the only decays that can occur in lowest order electroweak interactions are the decays to fermion pairs:

$$H^0 \rightarrow f + \bar{f} \qquad f = l,\; \nu_e \text{ or } q. \tag{19.76}$$

These reactions proceed via the mechanism of Fig. 19.15 and a straightforward calculation, left as an exercise for the reader, leads to the predicted decay width,

$$\Gamma\left(H \rightarrow f\bar{f}\right) = \frac{N_f}{\pi 4\sqrt{2}} G m_H m_f^2 \left(1 - \frac{4m_f^2}{m_H^2}\right)^{3/2}. \tag{19.77}$$

Here $N_f = 1$ for leptons, but for quarks $N_f = 3$ to allow for the fact that each quark flavor u, d, s, ... exists in three color states. The decay rate (19.77) is proportional to the square of the fermion mass, so that the dominant decay mode is

$$H \rightarrow b + \bar{b}, \qquad m_b \approx 4.5 \text{ GeV}. \tag{19.78}$$

Here the b-quarks would be observed as hadron jets in the laboratory. Neglecting other small contributions, the total decay rate is given approximately by

$$\Gamma_t \approx \frac{3G m_H m_b^2}{4\sqrt{2}\pi} \approx 10^{-5} m_H, \tag{19.79}$$

Figure 19.18   *Decays of the Higgs boson to (a) W^+W^- and (b) Z^0Z^0 pairs*

where we have ignored terms of order m_f^2/m_H^2. In other words, the predicted Higgs boson decay width is quite small in this mass region, and it is much longer-lived than the $W^\pm$ and Z^0 bosons. However, as the Higgs mass increases beyond the W^+W^- threshold, the decays

$$H^0 \rightarrow W^+ + W^- \quad (m_H > 2m_W) \tag{19.80a}$$

and

$$H^0 \rightarrow Z^0 + Z^0 \quad (m_H > 2m_{Z^0}) \tag{19.80b}$$

are allowed, as shown in Fig. 19.18. Furthermore, because the Higgs boson couples much more strongly to heavy particles, these decay modes completely dominate the decay modes (19.76).

This leads to a rapid increase in the total decay width as the Higgs mass increases above the W^+W^- threshold energy, as shown in Fig. 19.17.

So far, we have only considered decays which occur in leading-order electroweak interactions. Two important classes of second-order decay modes are

$$H^0 \rightarrow W^\pm + f + \bar{f}' \quad H^0 \rightarrow Z^0 + f + \bar{f}, \tag{19.81}$$

where f and f' are any combination of fermions which satisfies the appropriate conservation laws. These decays proceed by the mechanisms of Fig. 19.19, involving a real and a virtual intermediate vector boson. For this reason they are often classified as W^+W^- and Z^0Z^0 decays, which is the reason why the branching ratios for these decays shown in Fig. 19.17 are non-zero below the W^+W^- and Z^0Z^0 thresholds.

Figure 19.19   *Decays of the Higgs boson to (a) $W^\pm f \bar{f}'$ and (b) $Z^0 f \bar{f}'$, where f, f' are any quarks and leptons allowed by the conservation laws*

Figure 19.20   *The dominant mechanisms for the decay $H^0 \rightarrow \gamma + \gamma$*

In addition, there are many relatively rare decay modes predicted in higher order, as shown in Fig. 19.17. For example, the decay mode

$$H^0 \rightarrow \gamma + \gamma \tag{19.82}$$

can proceed by the mechanisms of Fig. 19.20 with a branching ratio of order 10^{-3}. Nonetheless, it may well be important in detecting the Higgs boson if its mass is between 110 GeV and 140 GeV, as we shall see below.

19.5.2   Higgs boson searches

The particles which are readily available in the laboratory as beams or targets are photons, electrons, muons and some hadrons composed of light quarks, like protons. The Higgs boson does not couple directly to photons, and the couplings to electrons, muons and light quarks are very small. Hence, although it is possible to produce Higgs bosons in reactions initiated by these particles, the production rates are usually very small. For example, the contribution to the process $e^+e^- \rightarrow f\bar{f}$ coming from the Feynman diagram 19.7c will be extremely small, even for e^+e^- collisions at the total centre of mass energy $\sqrt{s} = m_H$, where the probability of the H resonance being formed is largest. The fact that the Higgs particles have not been observed in this way can be attributed to the weakness of the interactions of Higgs particles with light fermions ($m_f \ll m_W$) and so does not lead to a bound on the mass of the Higgs bosons.

This problem can be overcome by first producing heavy particles, like the $W^\pm$ and Z^0 bosons and the t quark, to which the Higgs boson couples strongly. The best results to date come from the large electron–positron (LEP II) accelerator at CERN. This machine had a maximum energy of 208 GeV at the time of its closure in November 2002. This energy is high enough to produce Higgs bosons with masses up to almost 120 GeV in the reaction

$$e^+ + e^- \rightarrow H^0 + Z^0, \tag{19.83}$$

which is expected to occur by the dominant mechanism of Fig. 19.21. Attempts were made to detect Higgs bosons by their decay to $b\bar{b}$ pairs, where the quarks would be observed as jets. The failure to observe such events implied a lower limit

$$m_H > 113.5 \ GeV \tag{19.84}$$

on the mass of the Higgs boson, assuming that it exists.

Figure 19.21   *The dominant mechanism for Higgs boson production in e^+e^- annihilation*

Higgs boson searches have also been carried out at the Tevatron, which is a proton–antiproton collider with beam energies of 1 TeV, situated at the FNAL laboratory near Chicago. At the Tevatron, the searches concentrate on the associated production reaction $p + \bar{p} \to V + H + X$, where X is any hadron state consistent with the appropriate conservation laws and where the vector boson $V \equiv W^{\pm}, Z^0$ decays into charged leptons and/or neutrinos. At present, the sensitivity of these experiments is rather limited, but they may well lead to significant results very soon. Higgs boson searches are also just beginning at the LHC, which is a proton–proton collider with beam energies of 7 TeV situated at CERN. At the LHC, the main reaction of interest will be

$$p + \bar{p} \to H^0 + X, \tag{19.85}$$

where X is any hadron state consistent with the appropriate conservation laws. The dominant mechanism for this reaction at the LHC is 'gluon fusion' at all Higgs boson masses. This is shown in Fig. 19.22, where the ggH^0 vertex is dominated by a loop involving top quarks because of the strong coupling of the H^0 to the very heavy $t\,\bar{t}$ state, as shown in Fig. 19.23(a). Other production processes are shown in Figures 19.23(b), (c) and (d). These are also of interest because of the different experimental signatures they provide for identifying the Higgs boson. In Figures 19.23(a) – (d), for simplicity the spectator particles are not shown.

Referring to Eq. (19.17), we see that the most likely mass range for a standard model Higgs boson is $113.5 < M_H < \sim 160$ GeV. Referring to Fig 19.17, we see that for much of this mass range, the obvious mode to study would seems to be the dominant $H^0 \to b\bar{b}$ channel, with the quarks fragmenting to jets, as used in the LEP experiment. Unfortunately, it is very difficult to distinguish these jets from those produced more copiously by other means. Because of this it is likely that other decay modes will be more useful, even though they have much smaller branching ratios. For example, the decay

Figure 19.22   *The dominant 'gluon fusion' mechanism for the reaction $p + \bar{p} \to H^0 + X$*

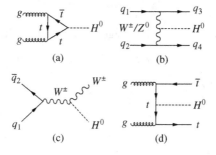

Figure 19.23   *Production mechanisms for a standard model Higgs boson at the LHC*

$H^0 \to \gamma\gamma$ mentioned above, with a branching ratio of order 10^{-3}, would have a signature of two isolated electromagnetic clusters in the detector, with much lower backgrounds than the $H^0 \to b\bar{b}$ channel.

On the other hand, if the Higgs boson is somewhat heavier, it will decay predominantly into the W^+W^- channel. Either way, the whole of the mass range (19.17), and well beyond, will be thoroughly explored over the next few years at the LHC.

Problems

19.1. For the process $Z^0 \to l^+l^-$, derive the decay rate (19.66), assuming that terms of order m_l^2/m_Z^2 are negligible. Show that the decay rate for the process $Z^0 \to \nu_l\bar{\nu}_l$ is given by

$$\Gamma(Z^0 \to \nu_l\bar{\nu}_l) = \frac{1}{\pi 12\sqrt{2}} Gm_Z^3,$$

which, for $m_Z = 91$ GeV, has the value 0.17 GeV.

19.2. The Higgs boson decays to quark–antiquark and lepton–antilepton pairs. In lowest order of perturbation theory these decays, $H \to f\bar{f}$ ($f = l, \nu_1, q$), are described by the Feynman diagram in Fig. 19.15 Derive the corresponding decay widths

$$\Gamma(H \to f\bar{f}) = \frac{1}{\pi 4\sqrt{2}} Gm_H m_f^2 \left(1 - 4\frac{m_f^2}{m_H^2}\right)^{3/2}.$$

19.3. Derive the total cross-sections (19.44a) and (19.44b) for $\nu_\mu e^-$ and $\bar{\nu}_\mu e^-$ scattering.

19.4. Starting from the Feynman amplitude, Eqs. (19.58) and (19.55), derive the differential cross-section for the process $e^+e^- \to l^+l^-$ ($l \neq e$), making the reasonable simplifying approximation $\sin^2\theta_W = 0.25$, i.e. $g_V = 0$ and $g_A = -\frac{1}{2}$. Check your result against Eqs. (19.60a)–(19.60c).

19.5. Show that, in the unitary gauge, the generalized Higgs–neutrino coupling term (18.44a) reduces to

$$\sum_j \frac{-1}{\sqrt{2}} \lambda_j \bar{\psi}_j(x)\psi_j(x)[v + \sigma(x)],$$

where

$$\psi_j(x) = \sum_l U_{jl} \psi_{\nu_l}(x)$$

and U is the unitary matrix which diagonalized the Hermitian coupling matrix G:

$$(UGU^\dagger)_{ij} = \lambda_i\delta_{ij}.$$

Hence show that Eq. (18.44a) leads to eigenstate neutrinos ν_j associated with the fields $\psi_j(x)$, with masses

$$m_j = \lambda_j v/\sqrt{2},$$

and that the Higgs–neutrino interactions are given by vertices of the form of Fig. 19.15, where now $f = v_j$ are eigenstate neutrinos, and by vertex factors

$$\frac{-i}{v} m_f \qquad (f = v_j).$$

For $G_{ij} = \delta_{ij}$, the eigenstate neutrinos v_j become indentical with the leptonic neutrinos, v_e, v_μ, v_τ, and we regain the results of Chapter 19.

19.6. Write down the Feynman amplitude for the decay process

$$Z^0(p,\, r) \rightarrow H(p') + e^+(q_1,\, r_1) + e^-(q_2,\, r_2),$$

and show that, taking $\sin^2\theta_W = 0.25$ as a reasonable first approximation and neglecting electron masses, it reduces to

$$\mathcal{M} = \frac{-iC}{k^2 - m_Z^2} \bar{u}(\mathbf{q}_2)\gamma^\alpha\gamma^5 v(\mathbf{q}_1)\varepsilon_{r\alpha}(\mathbf{p}),$$

where $k = p - p'$ and

$$C = vg^3/(8 \cos^3\theta_W).$$

Hence show that, neglecting electron masses,

$$(2m_e)^2 \frac{1}{3}\sum_r \sum_{r_1 r_2} |\mathcal{M}|^2 = \frac{C^2}{(k^2 - m_Z^2)^2} \frac{4}{3}\left[(q_1 q_2) + \frac{2}{m_Z^2}(pq_1)(pq_2)\right].$$

From this expression, the differential decay rate $d\Gamma$ to final states in which the momentum of the Higgs particle lies in the range $d^3\mathbf{p}'$ at $\mathbf{p}'$ can be obtained by integrating over the lepton momenta q_1, q_2 in a manner closely analogous to that used to integrate over the neutrino momenta in muon decay, Eqs. (16.53)–(16.63). Exploit this to show that

$$d\Gamma = \frac{C^2}{36(2\pi)^4} \frac{d^3\mathbf{p}'}{EE'} \frac{1}{(k^2 - m_Z^2)^2}\left[2k^2 + \frac{1}{m_Z^2}(pk)^2\right].$$

Appendix A
The Dirac Equation

In this appendix we shall derive the main results relating to the Dirac equation.[1]

A.1   The Dirac Equation

The Dirac equation can be written

$$i\hbar \frac{\partial \psi(x)}{\partial t} = \left[c\boldsymbol{\alpha}.(-i\hbar \boldsymbol{\nabla}) + \beta mc^2 \right] \psi(x) \tag{A.1}$$

where $\boldsymbol{\alpha} = (\alpha_1, \alpha_2, \alpha_3)$ and β are 4×4 Hermitian matrices satisfying

$$[\alpha_i, \alpha_j]_+ = 2\delta_{ij}, \quad [\alpha_i, \beta]_+ = 0, \quad \beta^2 = 1, \quad i,j = 1,2,3. \tag{A.2}$$

With

$$\gamma^0 = \beta, \quad \gamma^i = \beta \alpha_i \tag{A.3}$$

the Dirac equation becomes

$$i\hbar \gamma^\mu \frac{\partial \psi(x)}{\partial x^\mu} - mc\psi(x) = 0 \tag{A.4}$$

---

[1] The reader is assumed familiar with the elementary ideas of the Dirac equation as given in, for example, L. I. Schiff, *Quantum Mechanics*, 3rd edn, McGraw-Hill, New York, 1968, pp. 472–488. A more complete treatment, suitable for complementary background reading to this appendix, will be found in H. A. Bethe and R. W. Jackiw, *Intermediate Quantum Mechanics*, 3rd edn, Addison Wesley, 1997 p349–390.

Quantum Field Theory, Second Edition   Franz Mandl and Graham Shaw
© 2010 John Wiley & Sons, Ltd

with the 4×4 matrices γ^{μ}, $\mu = 0, \ldots, 3$, satisfying the anticommutation relations

$$[\gamma^{\mu}, \gamma^{\nu}]_{+} = 2g^{\mu\nu} \tag{A.5}$$

and the Hermiticity conditions

$$\gamma^{\mu\dagger} = \gamma^{0}\gamma^{\mu}\gamma^{0}. \tag{A.6}$$

Except for Section A.8 at the end of the appendix, the following properties are consequences of Eqs. (A.4)–(A.6) only and do not depend on choosing a particular representation for the γ-matrices.

A fifth anticommuting-matrix is defined by

$$\gamma^{5} \equiv i\gamma^{0}\gamma^{1}\gamma^{2}\gamma^{3}, \tag{A.7}$$

and γ^{5} has the properties

$$[\gamma^{\mu}, \gamma^{5}]_{+} = 0, \quad (\gamma^{5})^{2} = 1, \quad \gamma^{5\dagger} = \gamma^{5}. \tag{A.8}$$

Note that Greek indices will *always* stand for the values $0, \ldots, 3$ only, and not for 5.

The 4×4 spin matrices

$$\sigma^{\mu\nu} = \frac{i}{2}[\gamma^{\mu}, \gamma^{\nu}] \tag{A.9}$$

satisfy

$$\sigma^{\mu\nu\dagger} = \gamma^{0}\sigma^{\mu\nu}\gamma^{0} \tag{A.10}$$

and, with $\boldsymbol{\sigma} = (\sigma^{23}, \sigma^{31}, \sigma^{12})$ and $i,j,k = 1, 2, 3$ in cyclic order, we can write

$$\sigma^{ij} = -\gamma^{0}\gamma^{5}\gamma^{k}. \tag{A.11}$$

So far the γ-matrices, and matrices derived from them, have been defined with upper indices: $\gamma^{\mu}, \sigma^{\mu\nu}$, etc. We now define corresponding matrices with lower indices by the relation

$$\gamma_{\mu} = g_{\mu\nu}\gamma^{\nu}, \tag{A.12}$$

etc.[2]

We also define the matrix γ_{5} through

$$\gamma_{5} \equiv \frac{i}{4!}\varepsilon_{\lambda\mu\nu\pi}\gamma^{\lambda}\gamma^{\mu}\gamma^{\nu}\gamma^{\pi} = \gamma^{5}, \tag{A.13}$$

where the completely antisymmetric alternating symbol $\varepsilon_{\lambda\mu\nu\pi}$ is equal to $+1$ for (λ,μ,ν,π) an even permutation of $(0, 1, 2, 3)$, is equal to -1 for an odd permutation, and vanishes if two or more indices are the same.

---

[2] However, these matrices do not transform as tensors. We shall see below that it is bilinear forms of spinors containing these matrices which have the transformation properties of tensors. See Eqs. (A.53).

A.2 Contraction Identities

The manipulation of expressions involving γ-matrices is often greatly facilitated by the use of the following algebraic identities, which follow easily from the anticommutation relations (A.5):

$$\left.\begin{array}{l} \gamma_\lambda \gamma^\lambda = 4, \quad \gamma_\lambda \gamma^\alpha \gamma^\lambda = -2\gamma^\alpha \\[4pt] \gamma_\lambda \gamma^\alpha \gamma^\beta \gamma^\lambda = 4g^{\alpha\beta}, \quad \gamma_\lambda \gamma^\alpha \gamma^\beta \gamma^\gamma \gamma^\lambda = -2\gamma^\gamma \gamma^\beta \gamma^\alpha \\[4pt] \gamma_\lambda \gamma^\alpha \gamma^\beta \gamma^\gamma \gamma^\delta \gamma^\lambda = 2\left(\gamma^\delta \gamma^\alpha \gamma^\beta \gamma^\gamma + \gamma^\gamma \gamma^\beta \gamma^\alpha \gamma^\delta\right) \end{array}\right\} . \tag{A.14a}$$

If $A, B, \ldots$ denote four-vectors and 'A slash' is defined by $\slashed{A} \equiv \gamma^\alpha A_\alpha$, etc., we obtain the following contraction identifies from Eqs. (A.14a):

$$\left.\begin{array}{l} \gamma_\lambda \slashed{A} \gamma^\lambda = -2\slashed{A} \\[4pt] \gamma_\lambda \slashed{A} \slashed{B} \gamma^\lambda = 4AB, \quad \gamma_\lambda \slashed{A} \slashed{B} \slashed{C} \gamma^\lambda = -2\slashed{C}\slashed{B}\slashed{A} \\[4pt] \gamma_\lambda \slashed{A} \slashed{B} \slashed{C} \slashed{D} \gamma^\lambda = 2(\slashed{D}\slashed{A}\slashed{B}\slashed{C} + \slashed{C}\slashed{B}\slashed{A}\slashed{D}) \end{array}\right\} . \tag{A.14b}$$

The completely antisymmetric alternating symbol $\varepsilon^{\alpha\beta\gamma\delta}$, introduced in Eq. (A.13), satisfies the following contraction identities:

$$\left.\begin{array}{l} \varepsilon^{\alpha\beta\mu\nu}\varepsilon_{\alpha\beta\sigma\tau} = -2(g^\mu_\sigma g^\nu_\tau - g^\mu_\tau g^\nu_\sigma) \\[4pt] \varepsilon^{\alpha\beta\gamma\nu}\varepsilon_{\alpha\beta\gamma\tau} = -6g^\nu_\tau \\[4pt] \varepsilon^{\alpha\beta\gamma\delta}\varepsilon_{\alpha\beta\gamma\delta} = -24 \end{array}\right\} . \tag{A14c}$$

A.3 Traces

We next list some rules and relations which are extremely useful in evaluating the trace of a product of γ-matrices. Some comments on the derivation of these results are given at the end of the list.

(i) For any two $n \times n$ matrices U and V

$$\mathrm{Tr}(UV) = \mathrm{Tr}(VU). \tag{A.15}$$

(ii) If $\left(\gamma^\alpha \gamma^\beta \cdots \gamma^\mu \gamma^\nu\right)$ contains an odd number of γ-matrices, then

$$\mathrm{Tr}\left(\gamma^\alpha \gamma^\beta \cdots \gamma^\mu \gamma^\nu\right) = 0. \tag{A.16}$$

(iii) For a product of an even number of γ-matrices:

$$\left.\begin{array}{l} \mathrm{Tr}\left(\gamma^\alpha \gamma^\beta\right) = 4g^{\alpha\beta}, \quad \mathrm{Tr}\,\sigma^{\alpha\beta} = 0 \\[4pt] \mathrm{Tr}\left(\gamma^\alpha \gamma^\beta \gamma^\gamma \gamma^\delta\right) = 4\left(g^{\alpha\beta}g^{\gamma\delta} - g^{\alpha\gamma}g^{\beta\delta} + g^{\alpha\delta}g^{\beta\gamma}\right) \end{array}\right\} , \tag{A.17}$$

and from Eqs. (A.l7):

$$\mathrm{Tr}(\slashed{A}\slashed{B}) = 4(AB) \tag{A.18a}$$

$$\mathrm{Tr}(\slashed{A}\slashed{B}\slashed{C}\slashed{D}) = 4\{(AB)(CD) - (AC)(BD) + (AD)(BC)\}, \tag{A.18b}$$

and generally, if $A_1, A_2, \ldots, A_{2n}$ are four-vectors, then

$$\text{Tr}(\not{A}_1\not{A}_2 \ldots \not{A}_{2n}) = \{(A_1A_2)\text{Tr}(\not{A}_3 \ldots \not{A}_{2n}) - (A_1A_3)\text{Tr}(\not{A}_2\not{A}_4 \ldots \not{A}_{2n})$$
$$+ \ldots + (A_1A_{2n})\text{Tr}(\not{A}_2\not{A}_3 \ldots \not{A}_{2n-1})\}. \tag{A.18c}$$

In many specific cases one can evaluate traces more simply than by direct repeated use of Eq. (A.18c). The contraction relations, Eqs. (A.14), are particularly useful in this connection, as is

$$\not{A}\not{B} = AB - i\sigma^{\alpha\beta}A_\alpha B_\beta = 2AB - \not{B}\not{A}, \tag{A.19a}$$

and the particular cases of this equation:

$$\not{A}\not{A} = A^2; \quad \not{A}\not{B} = -\not{B}\not{A}, \quad \text{if } AB = 0. \tag{A.19b}$$

(iv) For any product of γ-matrices

$$\text{Tr}\left(\gamma^\alpha\gamma^\beta \ldots \gamma^\mu\gamma^\nu\right) = \text{Tr}\left(\gamma^\nu\gamma^\mu \ldots \gamma^\beta\gamma^\alpha\right) \tag{A.20a}$$

whence

$$\text{Tr}(\not{A}_1\not{A}_2 \ldots \not{A}_{2n}) = \text{Tr}(\not{A}_{2n} \ldots \not{A}_2\not{A}_1). \tag{A.20b}$$

(v) The above results can be extended to products involving the γ^5 matrix, the most important relations being

$$\text{Tr}\,\gamma^5 = \text{Tr}\left(\gamma^5\gamma^\alpha\right) = \text{Tr}\left(\gamma^5\gamma^\alpha\gamma^\beta\right) = \text{Tr}\left(\gamma^5\gamma^\alpha\gamma^\beta\gamma^\gamma\right) = 0$$
$$\text{Tr}\left(\gamma^5\gamma^\alpha\gamma^\beta\gamma^\gamma\gamma^\delta\right) = -4i\varepsilon^{\alpha\beta\gamma\delta} \tag{A.21}$$

Other results involving γ^5 are easily obtained from Eqs. (A.15)–(A.21), using Eqs. (A.7) and (A.8) which define γ^5 and state its main properties.

The following comments should suffice to enable the reader to derive Eqs. (A.16)–(A.21).

We obtain Eq. (A.16) by using $\left(\gamma^5\right)^2 = 1$ and Eq. (A.15), whence

$$\text{Tr}(\gamma^\alpha \ldots \gamma^\nu) = \text{Tr}\left[\left(\gamma^5\right)^2\gamma^\alpha \ldots \gamma^\nu\right] = \text{Tr}\left(\gamma^5\gamma^\alpha \ldots \gamma^\nu\gamma^5\right).$$

In the last trace, we use $\left[\gamma^\mu, \gamma^5\right]_+ = 0$ to commute the left-hand γ^5 matrix through to the right-hand side, with the result $-\text{Tr}(\gamma^\alpha \ldots \gamma^\nu)$ for an odd number of factors in $(\gamma^\alpha \ldots \gamma^\nu)$. Hence Eq. (A.16) follows.

Eqs. (A.17) are derived by repeated use of the anticommutation relations (A.5) of the γ-matrices and of the cyclic property (A.15) of traces.

To derive Eq. (A.20a), we introduce the transposed matrices $\gamma^{\alpha T}$. It follows from Eqs. (A.5) and (A.6) that the matrices $(-\gamma^{\alpha T})$ also satisfy the anticommutation relations (A.5) and the Hermiticity conditions (A.6). Hence by Pauli's fundamental theorem (A.61) there exists a unitary matrix C such that

$$C\gamma^\alpha C^{-1} = -\gamma^{\alpha T}. \tag{A.22}$$

Eq. (A.20a) then follows from

$$\text{Tr}\left(\gamma^\alpha\gamma^\beta\ldots\gamma^\nu\right) = \text{Tr}\left[\left(-C^{-1}\gamma^{\alpha T}C\right)\left(-C^{-1}\gamma^{\beta T}C\right)\ldots\left(-C^{-1}\gamma^{\nu T}C\right)\right]$$
$$= \text{Tr}\left(\gamma^{\alpha T}\gamma^{\beta T}\cdots\gamma^{\nu T}\right) = \text{Tr}\left[\left(\gamma^\nu\cdots\gamma^\beta\gamma^\alpha\right)^T\right].$$

Lastly, Eqs. (A.21) follow from $\gamma^5 = i\gamma^0\gamma^1\gamma^2\gamma^3$, $\left[\gamma^\mu,\gamma^5\right]_+ = 0$, the cyclic property (A. 15) of traces, and Eqs. (A. 16) and (A.17). E.g.

$$\text{Tr }\gamma^5 = \text{Tr}\left[\left(\gamma^5\gamma^0\right)\gamma^0\right] = \text{Tr}\left[\left(-\gamma^0\gamma^5\right)\gamma^0\right] = -\text{Tr }\gamma^5 = 0.$$

A.4   Plane Wave Solutions

The Dirac equation (A.4) possesses plane wave solutions

$$\psi(x) = \text{const.}\begin{Bmatrix} u_r(\mathbf{p}) \\ v_r(\mathbf{p}) \end{Bmatrix}e^{\pm ipx/\hbar} \tag{A.23}$$

where $p = (E_\mathbf{p}/c, \mathbf{p})$ and $E_\mathbf{p} = +(m^2c^4 + c^2\mathbf{p}^2)^{1/2}$. In the single-particle theory, $u_r(\mathbf{p})$ corresponds to a particle of momentum $\mathbf{p}$ and positive energy $E_\mathbf{p}$, $v_r(\mathbf{p})$ to momentum $-\mathbf{p}$ and negative energy $-E_\mathbf{p}$, The index $r = 1, 2$ labels two independent solutions for each four-momentum p, which we shall choose to be orthogonal.

The constant four-spinors $u_r(\mathbf{p})$ and $v_r(\mathbf{p})$, and their adjoints

$$\bar{u}_r(\mathbf{p}) = u_r^\dagger(\mathbf{p})\gamma^0, \quad \bar{v}_r(\mathbf{p}) = v_r^\dagger(\mathbf{p})\gamma^0, \tag{A.24}$$

satisfy the equations

$$(\not{p} - mc)u_r(\mathbf{p}) = 0 \quad (\not{p} + mc)v_r(\mathbf{p}) = 0 \tag{A.25}$$

$$\bar{u}_r(\mathbf{p})(\not{p} - mc) = 0 \quad \bar{v}_r(\mathbf{p})(\not{p} + mc) = 0. \tag{A.26}$$

With the normalization of these spinors defined by

$$u_r^\dagger(\mathbf{p})u_r(\mathbf{p}) = v_r^\dagger(\mathbf{p})v_r(\mathbf{p}) = \frac{E_\mathbf{p}}{mc^2}, \tag{A.27}$$

one derives the following *orthonormality relations* from Eqs. (A.25) and (A.26):

$$\left.\begin{aligned} u_r^\dagger(\mathbf{p})u_s(\mathbf{p}) = v_r^\dagger(\mathbf{p})v_s(\mathbf{p}) = \frac{E_\mathbf{p}}{mc^2}\delta_{rs} \\ u_r^\dagger(\mathbf{p})v_s(-\mathbf{p}) = 0 \end{aligned}\right\} \tag{A.28}$$

and

$$\left.\begin{aligned} \bar{u}_r(\mathbf{p})u_s(\mathbf{p}) = -\bar{v}_r(\mathbf{p})v_s(\mathbf{p}) = \delta_{rs} \\ \bar{u}_r(\mathbf{p})v_s(\mathbf{p}) = \bar{v}_r(\mathbf{p})u_s(\mathbf{p}) = 0 \end{aligned}\right\}. \tag{A.29}$$

The spinors $u_r(\mathbf{p})$ and $v_r(\mathbf{p})$, $r = 1, 2$, satisfy the *completeness relation*

$$\sum_{r=1}^{2}\left[u_{r\alpha}(\mathbf{p})\bar{u}_{r\beta}(\mathbf{p}) - v_{r\alpha}(\mathbf{p})\bar{v}_{r\beta}(\mathbf{p})\right] = \delta_{\alpha\beta}. \tag{A.30}$$

This relation can be established by showing that it holds for the four basis states $u_s(\mathbf{p})$ and $v_s(\mathbf{p})$, $s = 1, 2$.

A.5   Energy Projection Operators

The energy projection operators are defined by

$$\Lambda^{\pm}(\mathbf{p}) = \frac{\pm \not{p} + mc}{2mc}. \tag{A.31}$$

They have the properties, which follow from Eqs. (A.25) and (A.26), of projecting out the positive/negative energy solutions from a linear combination of the four plane wave states $u_r(\mathbf{p})$ and $v_r(\mathbf{p})$, i.e.

$$\left. \begin{array}{ll} \Lambda^+(\mathbf{p})u_r(\mathbf{p}) = u_r(\mathbf{p}), & \Lambda^-(\mathbf{p})v_r(\mathbf{p}) = v_r(\mathbf{p}) \\ \bar{u}_r(\mathbf{p})\Lambda^+(\mathbf{p}) = \bar{u}_r(\mathbf{p}), & \bar{v}_r(\mathbf{p})\Lambda^-(\mathbf{p}) = \bar{v}_r(\mathbf{p}) \end{array} \right\}, \tag{A.32}$$

and

$$\Lambda^+(\mathbf{p})v_r(\mathbf{p}) = \Lambda^-(\mathbf{p})u_r(\mathbf{p}) = 0, \quad \bar{v}_r(\mathbf{p})\Lambda^+(\mathbf{p}) = \bar{u}_r(\mathbf{p})\Lambda^-(\mathbf{p}) = 0. \tag{A.33}$$

From Eqs. (A.31) one verifies directly the property characteristic of projection operators

$$\left[\Lambda^{\pm}(\mathbf{p})\right]^2 = \Lambda^{\pm}(\mathbf{p}) \tag{A.34a}$$

(since $\not{p}\not{p} = p^2 = m^2c^2$), as well as

$$\Lambda^{\pm}(\mathbf{p})\Lambda^{\mp}(\mathbf{p}) = 0, \quad \Lambda^+(\mathbf{p}) + \Lambda^-(\mathbf{p}) = 1. \tag{A.34b}$$

Using the completeness relation (A.30), one easily shows that the projection operators $\Lambda^{\pm}(\mathbf{p})$ can be written

$$\Lambda^+_{\alpha\beta}(\mathbf{p}) = \sum_{r=1}^{2} u_{r\alpha}(\mathbf{p})\bar{u}_{r\beta}(\mathbf{p}), \quad \Lambda^-_{\alpha\beta}(\mathbf{p}) = -\sum_{r=1}^{2} v_{r\alpha}(\mathbf{p})\bar{v}_{r\beta}(\mathbf{p}). \tag{A.35}$$

A.6   Helicity and Spin Projection Operators

In Section 4.2 we chose the plane wave solutions $u_r(\mathbf{p})$ and $v_r(\mathbf{p})$ of the Dirac equation as eigenstates of the 4×4 spin matrix

$$\sigma_{\mathbf{p}} = \frac{\boldsymbol{\sigma}.\mathbf{p}}{|\mathbf{p}|}, \tag{A.36}$$

satisfying the equations

$$\sigma_{\mathbf{p}}u_r(\mathbf{p}) = (-1)^{r+1}u_r(\mathbf{p}), \quad \sigma_{\mathbf{p}}v_r(\mathbf{p}) = (-1)^r v_r(\mathbf{p}), \quad r = 1, 2. \tag{4.35}$$

Correspondingly, we now define the operators

$$\Pi^{\pm}(\mathbf{p}) = \frac{1}{2}\left(1 \pm \sigma_{\mathbf{p}}\right) \tag{A.37}$$

which are easily seen to have the properties

$$\left[\Pi^\pm(\mathbf{p})\right]^2 = \Pi^\pm(\mathbf{p}), \quad \Pi^\pm(\mathbf{p})\Pi^\mp(\mathbf{p}) = 0, \quad \Pi^+(\mathbf{p}) + \Pi^-(\mathbf{p}) = 1, \tag{A.38}$$

and

$$\left[\Lambda^+(\mathbf{p}), \Pi^\pm(\mathbf{p})\right] = \left[\Lambda^-(\mathbf{p}), \Pi^\pm(\mathbf{p})\right] = 0. \tag{A.39}$$

It follows from Eqs. (4.35) and (A.37) that the spinors $u_r(\mathbf{p})$ and $v_r(\mathbf{p})$ satisfy the equations

$$\left.\begin{array}{ll} \Pi^+(\mathbf{p})u_r(\mathbf{p}) = \delta_{1r}u_r(\mathbf{p}), & \Pi^+(\mathbf{p})v_r(\mathbf{p}) = \delta_{2r}v_r(\mathbf{p}) \\ \Pi^-(\mathbf{p})u_r(\mathbf{p}) = \delta_{2r}u_r(\mathbf{p}), & \Pi^-(\mathbf{p})v_r(\mathbf{p}) = \delta_{1r}v_r(\mathbf{p}) \end{array}\right\} \quad r = 1, 2. \tag{A.40}$$

We see from Eqs. (A.38) that the operators $\Pi^\pm(\mathbf{p})$ are projection operators of mutually orthogonal states. From the first of Eqs. (4.35), it follows that $u_1(\mathbf{p})[u_2(\mathbf{p})]$ represents a positive energy electron with spin parallel [antiparallel] to its direction of motion $\mathbf{p}$, i.e. it is a positive [negative] helicity state, as defined in Section 4.3, following Eq. (4.48). Hence the operators $\Pi^\pm(\mathbf{p})$ are called helicity projection operators.

The corresponding interpretation for the spinors $v_r(\mathbf{p})$ is also possible in the single-particle theory, either in terms of negative energy states of electrons or in terms of hole theory. Although our discussion of positrons in the quantized field theory in Chapter 4 does not depend on hole theory, a reader familiar with the latter may like to see the connection. Consider, for example, the spinor $v_1(\mathbf{p})$. In the language of negative energy states, $v_1(\mathbf{p})$ represents a negative-energy electron with momentum $-\mathbf{p}$ and, according to Eqs. (4.35), spin parallel to $-\mathbf{p}$, so that $v_1(\mathbf{p})$ is a positive helicity state of the negative-energy electron. Translated into hole theory language, the absence of this negative-energy electron represents a positron with momentum $+\mathbf{p}$ and spin parallel to $+\mathbf{p}$, i.e. it continues to be a positive helicity state of the positron.

For a zero-mass Dirac particle, we can express the helicity projection operators in terms of the γ^5 matrix. For $m = 0$, we have $p_0 = |\mathbf{p}|$, and Eqs. (A.25) become

$$\gamma^0|\mathbf{p}|w_r(\mathbf{p}) = -\gamma^k p_k w_r(\mathbf{p}) = \gamma^k p^k w_r(\mathbf{p}), \tag{A.41}$$

where $w_r(\mathbf{p})$ stands for either $u_r(\mathbf{p})$ or $v_r(\mathbf{p})$. Premultiplying the last equation by $\gamma^5\gamma^0$ and using Eq. (A.11), we obtain

$$\gamma^5 w_r(\mathbf{p}) = \sigma_\mathbf{p} w_r(\mathbf{p}) \tag{A.42}$$

i.e. the helicity projection operators (A.37) become, for $m = 0$,

$$\Pi^\pm(\mathbf{p}) = \frac{1}{2}\left(1\pm\gamma^5\right). \tag{A.43}$$

For particles of non-zero mass m, this result holds to $O(m/p_0)$ in the high-energy limit.

So far we have considered helicity projection operators only. For a Dirac particle, the spin component in an arbitrary direction is a good quantum number in the rest frame of the particle only. We shall see that in the rest frame one can define spin projection operators for an arbitrary axis of quantization in a covariant way, and by carrying out a Lorentz transformation to an arbitrary coordinate frame one can then define spin projection operators in this frame.

With the axis of quantization in the rest frame of the particle specified by the unit vector **n**, we define the unit vector n^μ to be given in the rest frame by

$$n^\mu = (0, \mathbf{n}). \tag{A.44}$$

From the invariance of scalar products, it follows that in any other frame also

$$n^2 = -1, \quad np = 0, \tag{A.45}$$

where p is the four-momentum of the particle in this frame. The required spin projection operators are then given by

$$\Pi^\pm(n) = \frac{1}{2}\left(1 \pm \gamma^5 \not{n}\right). \tag{A.46}$$

One verifies easily that these operators satisfy equations like Eqs. (A.38), characteristic of projection operators, and that they commute with the energy projection operators $\Lambda^\pm(\mathbf{p})$ for all vectors **p** satisfying Eq. (A.45). It is left to the reader to verify that, in the rest frame of the particle, the operators (A.46) have the desired properties. A matrix representation suitable for this purpose is given below, Eqs. (A.63)–(A.66).

A.7   Relativistic Properties

The Dirac equations for $\psi(x)$ and its adjoint $\overline{\psi}(x)$,

$$i\hbar\gamma^\mu \frac{\partial \psi(x)}{\partial x^\mu} - mc\psi(x) = 0, \quad i\hbar \frac{\partial \overline{\psi}(x)}{\partial x^\mu}\gamma^\mu + mc\overline{\psi}(x) = 0, \tag{A.47}$$

are Lorentz covariant, provided the spinors $\psi(x)$ and $\overline{\psi}(x)$ transform appropriately. We shall consider homogeneous orthochronous Lorentz transformations

$$x^\mu \to x'^\mu = \Lambda^\mu_{\ \nu} x^\nu, \tag{A.48}$$

i.e. $\Lambda^0_{\ 0} > 0$ and $\det \Lambda^\mu_{\ \nu} = \pm 1$, so that the sense of time is not reversed, but the transformation may or may not involve spatial inversion. It can be shown[3] that corresponding to each such transformation one can construct a non-singular 4×4 matrix $S = S(\Lambda)$ with the properties

$$\gamma^\nu = \Lambda^\nu_{\ \mu} S \gamma^\mu S^{-1} \tag{A.49}$$

and

$$S^{-1} = \gamma^0 S^\dagger \gamma^0. \tag{A.50}$$

If the transformation properties of the Dirac spinor $\psi(x)$ are defined by

$$\psi(x) \to \psi'(x') = S\psi(x), \tag{A.51}$$

then the covariance of the Dirac equations (A.47) is easily established.

---

[3] See, for example, Bethe and Jackiw, quoted at the beginning of this appendix, pp. 360–365.

From Eqs. (A.50) and (A.51) one derives the corresponding transformation property of the adjoint spinor $\overline{\psi}(x)$ as

$$\overline{\psi}(x) \to \overline{\psi}'(x') = \overline{\psi}(x)S^{-1}. \tag{A.52}$$

From Eqs. (A.49), (A.51) and (A.52), one obtains the five basic bilinear covariants of the Dirac theory. Under a Lorentz transformation

$$\left.\begin{array}{l} \overline{\psi}\psi \\ \overline{\psi}\gamma^\mu\psi \\ \overline{\psi}\sigma^{\mu\nu}\psi \\ \overline{\psi}\gamma^5\gamma^\mu\psi \\ \overline{\psi}\gamma^5\psi \end{array}\right\} \text{transforms as a} \left\{\begin{array}{l} \text{scalar} \\ \text{vector} \\ \text{antisymmetric second} - \text{rank tensor} \\ \text{pseudo-vector} \\ \text{pseudo-scalar} \end{array}\right\}. \tag{A.53}$$

Finally, we obtain the explicit form of the transformation (A. 51) for an infinitesimal Lorentz rotation (2.46), i.e.

$$x_\mu \to x'_\mu = \Lambda_{\mu\nu}x^\nu = (g_{\mu\nu} + \varepsilon_{\mu\nu})x^\nu \tag{A.54}$$

where $\varepsilon_{\mu\nu} = -\varepsilon_{\nu\mu}$. Eq. (2.47) now becomes

$$\psi_\alpha(x) \to \psi'_\alpha(x') = S_{\alpha\beta}\psi_\beta(x) = \left(\delta_{\alpha\beta} + \frac{1}{2}\varepsilon_{\mu\nu}S^{\mu\nu}_{\alpha\beta}\right)\psi_\beta(x)$$

or in matrix form

$$\psi(x) \to \psi'(x') = S\psi(x) = \left(1 + \frac{1}{2}\varepsilon_{\mu\nu}S^{\mu\nu}\right)\psi(x) \tag{A.55}$$

where $S^{\mu\nu}$ is antisymmetric.
Using the orthogonality relation (2.6) for $\Lambda_{\mu\nu}$, we can rewrite Eq. (A.49) as

$$S\gamma^\lambda S^{-1} = \gamma^\nu\Lambda_\nu^{\ \lambda}. \tag{A.56}$$

Using Eq. (A.55) we can write the left-hand side of Eq. (A.56), to first order in $\varepsilon_{\mu\nu}$, as

$$\gamma^\lambda + \frac{1}{2}\varepsilon_{\mu\nu}[S^{\mu\nu},\gamma^\lambda]. \tag{A.57}$$

From (A.54) the right-hand side of Eq. (A.56) can be written

$$\gamma^\nu(g_\nu^{\ \lambda} + \varepsilon_\nu^{\ \lambda}) = \gamma^\lambda + \gamma^\nu g^{\lambda\mu}\varepsilon_{\nu\mu}$$
$$= \gamma^\lambda + \frac{1}{2}\varepsilon_{\mu\nu}(\gamma^\mu g^{\lambda\mu} - \gamma^\nu g^{\lambda\mu}). \tag{A.58}$$

Equating the last two expressions, we obtain

$$[S^{\mu\nu},\gamma^\lambda] = \gamma^\mu g^{\lambda\nu} - \gamma^\nu g^{\lambda\mu},$$

and one verifies directly that this equation has the solution

$$S^{\mu\nu} = \frac{1}{2}\gamma^\mu\gamma^\nu. \tag{A.59}$$

Hence, from Eqs. (A.55) and (A.59), and the definition (A.9) of $\sigma^{\mu\nu}$, one finds that under Lorentz rotations a Dirac spinor transforms as

$$\psi(x) \rightarrow \psi'(x') = \psi(x) - \frac{i}{4}\varepsilon_{\mu\nu}\sigma^{\mu\nu}\psi(x). \tag{A.60}$$

A.8   Particular Representations of the γ-Matrices

So far we have developed the Dirac theory in a representation-free way, relying only on the anticommutation relations (A.5) and the Hermiticity conditions (A.6) of the γ-matrices. There are many ways of writing γ^{μ}, $\mu = 0, \ldots, 3$, as 4×4 matrices such that Eqs. (A.5) and (A.6) hold. If γ^{μ}, $\mu = 0, \ldots, 3$, and $\tilde{\gamma}^{\mu}$, $\mu = 0, \ldots, 3$, are two such sets of matrices, i.e. each set satisfies Eqs. (A.5) and (A.6), then Pauli's fundamental theorem[4] states that

$$\tilde{\gamma}^{\mu} = U\gamma^{\mu}U^{\dagger}, \tag{A.61}$$

where U is a unitary matrix.

We shall state two particular representations which are useful in practice.

(i) *Dirac-Pauli representation.* This representation has a simple non-relativistic limit. In terms of the Pauli 2×2 spin matrices

$$\sigma_1 = \begin{pmatrix} 0 & 1 \\ 1 & 0 \end{pmatrix} \quad \sigma_2 = \begin{pmatrix} 0 & -i \\ i & 0 \end{pmatrix} \quad \sigma_3 = \begin{pmatrix} 1 & 0 \\ 0 & -1 \end{pmatrix} \tag{A.62}$$

the Dirac matrices can in this representation be written as

$$\left. \begin{array}{ll} \gamma^0 = \beta = \begin{pmatrix} 1 & 0 \\ 0 & -1 \end{pmatrix}, & \alpha_k = \begin{pmatrix} 0 & \sigma_k \\ \sigma_k & 0 \end{pmatrix}, \quad k = 1, 2, 3 \\[4mm] \gamma^k = \beta\alpha_k = \begin{pmatrix} 0 & \sigma_k \\ -\sigma_k & 0 \end{pmatrix}, & k = 1, 2, 3 \end{array} \right\}, \tag{A.63}$$

whence

$$\sigma^{ij} = \begin{pmatrix} \sigma_k & 0 \\ 0 & \sigma_k \end{pmatrix}, \quad i, j, k = 1, 2, 3 \text{ in cyclic order}, \tag{A.64}$$

$$\sigma^{0k} = i\alpha_k = i\begin{pmatrix} 0 & \sigma_k \\ \sigma_k & 0 \end{pmatrix}, \quad k = 1, 2, 3, \tag{A.65}$$

and

$$\gamma^5 = \begin{pmatrix} 0 & 1 \\ 1 & 0 \end{pmatrix}. \tag{A.66}$$

A complete set of plane wave states is now easily constructed. With the two component non-relativistic spinors defined by

$$\chi_1 \equiv \chi_2' \equiv \begin{pmatrix} 1 \\ 0 \end{pmatrix}, \quad \chi_2 \equiv \chi_1' \equiv \begin{pmatrix} 0 \\ 1 \end{pmatrix}, \tag{A.67}$$

---

[4] For its derivation, see Bethe and Jackiw, quoted at the beginning of this appendix, pp. 358–359.

the positive and negative energy solutions of the Dirac equation for a particle at rest can be written

$$u_r(\mathbf{0}) = \begin{pmatrix} \chi_r \\ 0 \end{pmatrix}, \quad v_r(\mathbf{0}) = \begin{pmatrix} 0 \\ \chi'_r \end{pmatrix}, \quad r = 1, 2. \tag{A.68}$$

Since

$$(mc \pm \not{p})(mc \mp \not{p}) = (mc)^2 - p^2 = 0,$$

it follows that

$$u_r(\mathbf{p}) = \frac{(mc + \not{p})}{\sqrt{(2mE_\mathbf{p} + 2m^2c^2)}} u_r(\mathbf{0}), \quad r = 1, 2, \tag{A.69}$$

and

$$v_r(\mathbf{p}) = \frac{(mc - \not{p})}{\sqrt{(2mE_\mathbf{p} + 2m^2c^2)}} v_r(\mathbf{0}), \quad r = 1, 2, \tag{A.70}$$

are solutions of the Dirac equation with energy-momentum vectors $\pm p = (\pm E_\mathbf{p}/c, \pm \mathbf{p})$ respectively. The denominators in Eqs. (A.69) and (A.70) ensure the normalization (A.27). From the representation (A.63) we can write Eq. (A.69) as

$$u_r(\mathbf{p}) = A \begin{pmatrix} \chi_r \\ B\mathbf{p} \cdot \boldsymbol{\sigma} \chi_r \end{pmatrix}, \quad r = 1, 2, \tag{A.71}$$

where

$$A \equiv \left(\frac{E_\mathbf{p} + mc^2}{2mc^2} \right)^{1/2}, \quad B \equiv \frac{c}{E_\mathbf{p} + mc^2}. \tag{A.72}$$

Using Eqs. (A.62) and (A.67) we finally obtain from (A.71)

$$u_1(\mathbf{p}) = A \begin{pmatrix} 1 \\ 0 \\ Bp^3 \\ B(p^1 + ip^2) \end{pmatrix}, \quad u_2(\mathbf{p}) = A \begin{pmatrix} 0 \\ 1 \\ B(p^1 - ip^2) \\ -Bp^3 \end{pmatrix}. \tag{A.73}$$

In the same way, the negative energy spinors (A.70) can be written as

$$v_r(\mathbf{p}) = A \begin{pmatrix} B\mathbf{p} \cdot \boldsymbol{\sigma} \chi'_r \\ \chi'_r \end{pmatrix}, \quad r = 1, 2, \tag{A.74}$$

and

$$v_1(\mathbf{p}) = A \begin{pmatrix} B(p^1 - ip^2) \\ -Bp^3 \\ 0 \\ 1 \end{pmatrix}, \quad v_2(\mathbf{p}) = A \begin{pmatrix} Bp^3 \\ B(p^1 + ip^2) \\ 1 \\ 0 \end{pmatrix}. \tag{A.75}$$

We see that the spinors (A.73) and (A.75) are, in general, not helicity eigenstates, but eigenstates of the z-component of spin in the rest frame of the particle.

The behavior of these solutions for non-relativistic velocities v (i.e. $v/c \approx |\mathbf{p}|/mc \ll 1$) is easily seen. For the positive energy solutions u_r, the upper two components are very large compared to the lower two components, while for the negative energy solutions it is the lower two components which dominate.

(ii) *Majorana representation.* We showed in Section 4.3 that the symmetry of the quantized Dirac field between particles and antiparticles becomes particularly obvious if one works in a Majorana representation, distinguished by the property that the four γ-matrices are pure imaginary, i.e. using a subscript M to denote a Majorana representation we require

$$\gamma_M^{\mu*} = -\gamma_M^\mu, \quad \mu = 0, \ldots, 3. \tag{A.76}$$

A particular Majorana representation γ_M^μ is obtained from the Dirac-Pauli representation, Eqs. (A.63)–(A.66), by the unitary transformation [see Eq. (A.61)]

$$\gamma_M^\mu = U\gamma^\mu U^\dagger, \tag{A.77}$$

with

$$U = U^\dagger = U^{-1} = \frac{1}{2}\gamma^0(1+\gamma^2). \tag{A.78}$$

Explicitly, the matrices in this Majorana representation are given by

$$\left.\begin{array}{ll} \gamma_M^0 = \gamma^0\gamma^2 = \begin{pmatrix} 0 & \sigma_2 \\ \sigma_2 & 0 \end{pmatrix}, & \gamma_M^1 = \gamma^2\gamma^1 = i\sigma^{12} = \begin{pmatrix} i\sigma_3 & 0 \\ 0 & i\sigma_3 \end{pmatrix} \\[2mm] \gamma_M^2 = -\gamma^2 = \begin{pmatrix} 0 & -\sigma_2 \\ \sigma_2 & 0 \end{pmatrix}, & \gamma_M^3 = \gamma^2\gamma^3 = -i\sigma^{23} = \begin{pmatrix} -i\sigma_1 & 0 \\ 0 & -i\sigma_1 \end{pmatrix} \\[2mm] \gamma_M^5 = -i\gamma^0\gamma^1\gamma^3 = \gamma^0\sigma^{31} = \begin{pmatrix} \sigma_2 & 0 \\ 0 & -\sigma_2 \end{pmatrix} & \end{array}\right\}. \tag{A.79}$$

We see that in this representation all five γ-matrices are pure imaginary, since the Pauli matrices σ_1 and σ_3 are real and σ_2 is purely imaginary.

Problems

1.1. From Eq. (A.49) prove that

$$S^{-1}\gamma^5 S = \gamma^5 \det\Lambda$$

and hence that $\overline{\psi}(x)\gamma^5\psi(x)$ transforms as a pseudo-scalar under Lorentz transformations. Establish the transformation properties of the other four covariants in Eq. (A.53) in a similar way.

1.2. For any two positive energy solutions $u_r(\mathbf{p})$ and $u_s(\mathbf{p}')$ of the Dirac equation prove that

$$2m\overline{u}_s(\mathbf{p}')\gamma^\mu u_r(\mathbf{p}) = \overline{u}_s(\mathbf{p}')[(p'+p)^\mu + i\sigma^{\mu\nu}(p'-p)_\nu]u_r(\mathbf{p}). \tag{A.80}$$

Eq. (A.80) is known as Gordon's identity. [*Hint*: Consider the identity

$$\overline{u}_s(\mathbf{p}')[\not{a}(\not{p}-m) + (\not{p}'-m)\not{a}]u_r(\mathbf{p}) = 0$$

for an arbitrary four-vector a_μ.]

Appendix B

Feynman Rules and Formulae
for Perturbation Theory

In this appendix we collect together the principal results of covariant perturbation theory, i.e. the expressions for cross-sections, lifetimes and Feynman amplitudes, and the Feynman rules for writing down these amplitudes directly from the Feynman diagrams. This summary is intended for readers who have assimilated these methods and wish to apply them. Explanations have been kept to a minimum, but the frequent cross-references should help readers in difficulty.

In this Appendix, neutrinos are assumed to be Dirac particles with non-zero masses, but mixing is neglected.[1] The limit $m_{\nu_l} \to 0$ is easily obtained for observable quantities like transition rates, etc.

(i) *Feynman amplitude.* The Feynman amplitude $\mathcal{M}$ for the transition $|i\rangle \to |f\rangle$ is defined in terms of the corresponding S-matrix element S_{fi} by

$$S_{fi} = \delta_{fi} + (2\pi)^4 \delta^{(4)} \left(\sum p'_f - \sum p_i \right)$$
$$\times \prod_i \left(\frac{1}{2VE_i} \right)^{1/2} \prod_f \left(\frac{1}{2VE'_f} \right)^{1/2} \prod_l (2m_l)^{1/2} \mathcal{M} \qquad (8.1)$$

where $p_i = (E_i, \mathbf{p}_i)$ and $p'_f = (E'_f, \mathbf{p}'_f)$ are the four-momenta of the initial and final particles and l runs over all charged and neutral external leptons in the process. m_l is the mass of lepton l.

---

[1] Mixing and Majorana neutrinos are discussed in Sections 16.7 and 18.3, and references therein.

Quantum Field Theory, Second Edition Franz Mandl and Graham Shaw
© 2010 John Wiley & Sons, Ltd

(ii) *Cross-section.* The differential cross-section for the collision of two particles ($i = 1, 2$) moving collinearly with relative velocity v_{rel} and resulting in N final particles ($f = 1, 2, \ldots, N$) is given by

$$d\sigma = (2\pi)^4 \delta^{(4)} \left(\sum p'_f - \sum p_i \right) \frac{1}{4E_1 E_2 v_{rel}} \left(\prod_l (2m_l) \right) \left(\prod_f \frac{d^3 \mathbf{p}'_f}{(2\pi)^3 2E'_f} \right) |\mathcal{M}|^2.$$

$$(8.8)$$

In Eq. (8.8) all initial and final particles are in definite spin/polarization states. Eq. (8.8) holds in any Lorentz frame and

$$v_{rel} = \left[(p_1 p_2)^2 - m_1^2 m_2^2 \right]^{1/2} \Big/ (E_1 E_2);$$

$$(8.9)$$

in particular:

$$v_{rel} = |\mathbf{p}_1| \frac{E_1 + E_2}{E_1 E_2} \quad \text{(CoM system)}$$

$$(8.10a)$$

$$v_{rel} = |\mathbf{p}_1| / E_1 \quad \text{(laboratory system)}.$$

$$(8.10b)$$

(iii) *Lifetime.* The differential decay rate $d\Gamma$ for the decay of a particle P with four-momentum $p = (E, \mathbf{p})$ into N particles with four-momenta $p'_f = (E'_f, \mathbf{p}'_f)$ is given by

$$d\Gamma = (2\pi)^4 \delta^{(4)} \left(\Sigma p'_f - p \right) \frac{1}{2E} \left(\prod_l (2m_l) \right) \left(\prod_f \frac{d^3 \mathbf{p}'_f}{(2\pi)^3 2E'_f} \right) |\mathcal{M}|^2.$$

$$(16.36)$$

Eq. (16.36) refers to definite initial and final spin/polarization states for all particles. The lifetime τ of the particle P is given by

$$\tau = B/\Gamma,$$

$$(16.38)$$

where Γ is the total decay rate for the above decay mode, and B is the branching ratio for this mode, Eq. (16.37).

(iv) *The Feynman rules for QED.* The Feynman amplitude for a given graph in QED is obtained from the Feynman rules of Sections 7.3 and 8.7. Some of the Feynman rules for charged leptons $l^{\pm}$ apply with trivial changes to neutral leptons ν_l and $\bar{\nu}_l$. To avoid lengthy repetition later, we write these rules at once for both charged and neutral leptons.

1. For each QED vertex, write a factor

2. For each internal photon line, labelled by the momentum k, write a factor

$$iD_{F\alpha\beta}(k) = i \frac{-g_{\alpha\beta}}{k^2 + i\varepsilon}. \qquad (\alpha) \qquad k \qquad (\beta)$$

$$(7.47)$$

3. For each internal lepton line, labelled by the momentum p, write a factor

$$iS_F(p) = i \frac{1}{\not{p} - m + i\varepsilon} . \qquad \underset{p}{\bullet \!-\!\!\!-\!\!\!-\!\!\!\longrightarrow\!\!\!-\!\!\!-\!\!\!- \bullet} \qquad (7.48)$$

Here m stands for the mass m_l, or m_{ν_l} of the particular lepton considered.
4. For each external line, write one of the following factors:
   (a) for each initial lepton l^- or $\nu_l : u_r(\mathbf{p})$ $p \longrightarrow\!\!\!-\!\!\!- \bullet$ (7.49a)

   (b) for each final lepton l^- or $\nu_l : \bar{u}_r(\mathbf{p})$ $\bullet \!-\!\!\!-\!\!\!\longrightarrow\!\!\!-\!\!\!- p$ (7.49b)

   (c) for each initial lepton l^+ or $\bar{\nu}_l : \bar{v}_r(\mathbf{p})$ $p \longrightarrow\!\!\!\longleftarrow\!\!\!- \bullet$ (7.49c)

   (d) for each final lepton l^+ or $\bar{\nu}_l : v_r(\mathbf{p})$ $\bullet \!-\!\!\!-\!\!\!\longleftarrow\!\!\!- p$ (7.49d)

   (e) for each initial photon: $\varepsilon_{r\alpha}(\mathbf{k})$ $\underset{k}{\wwww} \overset{(\alpha)}{\bullet}$ (7.49e)

   (f) for each final photon:[2] $\varepsilon_{ra}(\mathbf{k})$ $\overset{(\alpha)}{\bullet} \wwww_k$ (7.49f)

   In Eqs. (7.49) $\mathbf{p}$ and $\mathbf{k}$ denote the three-momenta of the external particles, and $r\,(=1,2)$ their spin and polarization states.
5. The spinor factors (γ-matrices, S_F-functions, four-spinors) for each fermion line are ordered so that, reading from right to left, they occur in the same sequence as following the fermion line in the direction of its arrows.
6. For each closed fermion loop, take the trace and multiply by a factor (-1).
7. The four-momenta associated with the lines meeting at each vertex satisfy energy–momentum conservation. For each four-momentum q which is not fixed by energy–momentum conservation carry out the integration $(2\pi)^{-4} \int d^4q$. One such integration with respect to an internal momentum variable q occurs for each closed loop.
8. Multiply the expression by a phase factor δ_p which is equal to $+1\,(-1)$ if an even (odd) number of interchanges of neighbouring fermion operators is required to write the fermion operators in the correct normal order.
   To allow for the interaction with an *external static electromagnetic field* $A_{e\alpha}(x)$:
   (a) In Eq. (8.1), relating $\mathscr{M}$ to S_{fi} make the replacement

$$(2\pi)^4 \delta^{(4)} \left(\sum P_f' - \sum P_i \right) \to (2\pi)\delta \left(\sum E_f' - \sum E_i \right) \qquad (8.89)$$

   (b) Add the following Feynman rule:
9. For each interaction of a charged particle with an external static field $A_e(\mathbf{x})$, write a factor

$$A_{e\alpha}(\mathbf{q}) = \int d^3x\, e^{-i\mathbf{q}\cdot\mathbf{x}} A_{e\alpha}(\mathbf{x}) \qquad \overset{(\alpha)}{\bullet} \wwww \overset{\mathbf{q}}{\longleftarrow} \times \qquad (8.90)$$

Here $\mathbf{q}$ is the momentum transferred from the field source (x) to the particle.

---

[2] For linear polarization states, $\varepsilon_{ra}(\mathbf{k})$ is real. In general it is complex and we must then replace $\varepsilon_{ra}(\mathbf{k})$ by $\varepsilon_{ra}^*(\mathbf{k})$ for a final-state photon.

(v) *Additional Feynman rules for the standard electroweak theory.* The following rules allow calculations to be carried out in lowest non-vanishing order of perturbation theory only.[3]

11. For each internal massive vector boson line, labelled by the momentum k, write a factor

$$iD_{F\alpha\beta}(k,m) = \frac{i\left(-g_{\alpha\beta} + k_\alpha k_\beta/m^2\right)}{k^2 - m^2 + i\varepsilon} \qquad (\alpha) \quad\overset{k}{\sim\!\!\sim\!\!\sim\!\!\sim}\quad (\beta) \qquad (16.30)$$

where $m = m_W$ for a $W^\pm$ boson, and $m = m_Z$ for a Z^0 boson.

12. For each external line, representing an initial or final $W^\pm$ or Z^0 vector boson, write a factor

$$\varepsilon_{r\alpha}(\mathbf{k}) \quad\begin{array}{c} \overset{k}{\sim\!\!\sim\!\!\sim\!\!\sim}\;(\alpha) \quad \text{(initial)} \\ (\alpha)\;\overset{k}{\sim\!\!\sim\!\!\sim\!\!\sim} \quad \text{(final)} \end{array}\Bigg\}. \qquad (16.32)$$

If complex polarization vectors are used, we must instead write $\varepsilon^*_{r\alpha}(\mathbf{k})$ for a final-state vector boson.

13. For each internal Higgs line, labelled by the momentum k, write a factor

$$i\Delta_F(k,m_H) = \frac{i}{k^2 - m_H^2 + i\varepsilon} \qquad \bullet\; \text{--}\,\text{--}\,\overset{k}{\blacktriangleright}\,\text{--}\,\text{--}\;\bullet \qquad (19.21)$$

where m_H is the mass of the Higgs scalar.

14. The standard electroweak theory gives rise to 18 basic vertex parts, resulting from the terms in Eqs. (19.3a)–(19.3e), which were listed in Table 19.1. Below we give the Feynman diagrams of these basic vertex parts and the corresponding vertex factors, numbered (B.1)–(B.18) in the same order as the terms in Table 19.1.[4]

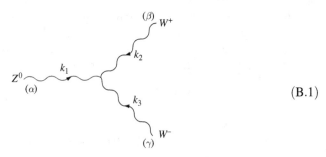

(B.1)

$$ig\cos\theta_W[g^{\alpha\beta}(k_1 - k_2)^\gamma + g^{\beta\gamma}(k_2 - k_3)^\alpha + g^{\gamma\alpha}(k_3 - k_1)^\beta]$$

---

[3] Feynman rule 10 of Section 16.4 dealt with the vertex factor of the IVB theory. This rule will now be incorporated in rule 14 for the vertex factors of the electroweak theory.
[4] Feynman rule 14 lists all the interactions of the electroweak theory. In order to have a complete, self-contained list of Feynman rules for QED in (iv) above, we also included the QED vertex factor in Feynman rule 1.

$$ie\left[g^{\alpha\beta}(k_1-k_2)^\gamma + g^{\beta\gamma}(k_2-k_3)^\alpha + g^{\gamma\alpha}(k_3-k_1)^\beta\right]$$

(B.2)

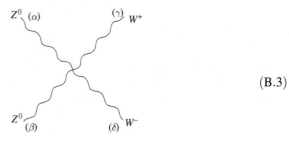

$$ig^2\cos^2\theta_W[g^{\alpha\delta}\,g^{\beta\gamma} + g^{\alpha\gamma}g^{\beta\delta} - 2g^{\alpha\beta}g^{\gamma\delta}]$$

(B.3)

$$ie^2\left[g^{\alpha\delta}g^{\beta\gamma} + g^{\alpha\gamma}g^{\beta\delta} - 2g^{\alpha\beta}g^{\gamma\delta}\right]$$

(B.4)

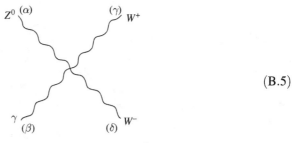

$$ieg\cos\theta_W[g^{\alpha\delta}g^{\beta\gamma} + g^{\alpha\gamma}g^{\beta\delta} - 2g^{\alpha\beta}g^{\gamma\delta}]$$

(B.5)

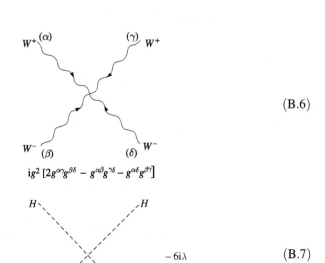

$$ig^2 \left[2g^{\alpha\gamma}g^{\beta\delta} - g^{\alpha\beta}g^{\gamma\delta} - g^{\alpha\delta}g^{\beta\gamma} \right]$$

(B.6)

$-6i\lambda$ (B.7)

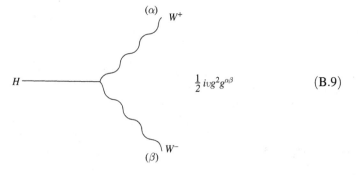

$-6i\lambda v$ (B.8)

$\frac{1}{2} i v g^2 g^{\alpha\beta}$ (B.9)

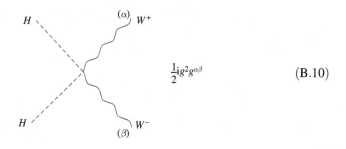

$$\frac{1}{2}ig^2g^{\alpha\beta} \tag{B.10}$$

$$\frac{i v g^2}{2\cos^2\theta_{\mathrm{w}}}g^{\alpha\beta} \tag{B.11}$$

$$\frac{ig^2}{2\cos^2\theta_{\mathrm{w}}}g^{\alpha\beta} \tag{B.12}$$

$$ie\gamma^{\alpha} \tag{B.13}$$

$$\frac{-ig}{2\sqrt{2}}\,\gamma^{\alpha}(1-\gamma_5) \tag{B.14}$$

$$\frac{-ig}{4 \cos \theta_{\rm w}} \gamma^\alpha (1-\gamma_5) \tag{B.15}$$

$$\frac{-ig\gamma^\alpha}{4 \cos \theta_{\rm w}} (1 - 4 \sin^2 \theta_{\rm w} - \gamma_5) \tag{B.16}$$

$$= \frac{-ig\gamma^\alpha}{2 \cos \theta_{\rm w}} (g_{\rm V} - g_{\rm A}\gamma_5) \tag{B.16a}$$

where we defined

$$g_{\rm V} = 2 \sin^2 \theta_{\rm W} - \frac{1}{2}, \quad g_{\rm A} = -\frac{1}{2} \tag{B.16b}$$

$$\frac{-i}{v} m_l \tag{B.17}$$

$$\frac{-i}{v} m_{v_l} \tag{B.18}$$

The coupling constants satisfy the relations

$$g \sin \theta_{\rm W} = g' \cos \theta_{\rm W} = e \tag{19.6}$$

and the particle masses are, in lowest order, given by

$$m_W = \frac{1}{2} vg, \quad m_Z = m_W / \cos \theta_{\rm W}, \quad m_H = \sqrt{(-2\mu^2)} \tag{19.4}$$

$$m_l = vg_l / \sqrt{2}, \quad m_{v_l} = vg_{v_l} / \sqrt{2}, \tag{19.8}$$

where

$$v = \left(\frac{-\mu^2}{\lambda}\right)^{1/2} = (G\sqrt{2})^{-1/2} \quad (>0),$$

(19.5) and (19.14)

and hence

$$\frac{G}{\sqrt{2}} = \frac{g^2}{8m_W^2}.$$

(vi) *Feynman rules for QCD.* The Feynman rules for QCD are summarized in Section 14.4 in a closely analogous way to that given in (iv) and (v) for QED and the unified electroweak theory, respectively.

Index